Remote Sensing and Global Climate Change

NATO ASI Series

Advanced Science Institutes Series

A series presenting the results of activities sponsored by the NATO Science Committee, which aims at the dissemination of advanced scientific and technological knowledge, with a view to strengthening links between scientific communities.

The Series is published by an international board of publishers in conjunction with the NATO Scientific Affairs Division

A Life Sciences **B Physics**	Plenum Publishing Corporation London and New York
C Mathematical and Physical Sciences **D Behavioural and Social Sciences** **E Applied Sciences**	Kluwer Academic Publishers Dordrecht, Boston and London
F Computer and Systems Sciences **G Ecological Sciences** **H Cell Biology** **I Global Environmental Change**	Springer-Verlag Berlin Heidelberg New York London Paris Tokyo Hong Kong Barcelona Budapest

NATO-PCO DATABASE

The electronic index to the NATO ASI Series provides full bibliographical references (with keywords and/or abstracts) to more than 30000 contributions from international scientists published in all sections of the NATO ASI Series. Access to the NATO-PCO DATABASE compiled by the NATO Publication Coordination Office is possible in two ways:

- via online FILE 128 (NATO-PCO DATABASE) hosted by ESRIN, Via Galileo Galilei, I-00044 Frascati, Italy.
- via CD-ROM "NATO Science & Technology Disk" with user-friendly retrieval software in English, French and German (© WTV GmbH and DATAWARE Technologies Inc. 1992).

The CD-ROM can be ordered through any member of the Board of Publishers or through NATO-PCO, Overijse, Belgium.

Series I: Global Environmental Change, Vol. 24

Remote Sensing and Global Climate Change

Edited by

Robin A. Vaughan

Arthur P. Cracknell

Department of Applied Physics and
Electronic & Mechanical Engineering
University of Dundee
Dundee DD1 4HN
Scotland, UK

Springer-Verlag
Berlin Heidelberg New York London Paris Tokyo
Hong Kong Barcelona Budapest
Published in cooperation with NATO Scientific Affairs Division

Proceedings of the NATO Advanced Study Institute on Remote Sensing and Global Climate Change, held at Dundee, Scotland, July 19 - August 8, 1992

Additional material to this bood can be downloaded from http://extra.springer.com.

ISBN-13:978-3-642-79289-2 e-ISBN-13:978-3-642-79287-8
DOI: 10.1007/978-3-642-79287-8

Library of Congress Cataloging-in-Publication Data. Arthur P. Cracknell. p. cm. – (NATO ASI series. Series I, Global environmental change; v. 24) "Published in cooperation with NATO Scientific Affairs Division. Proceedings of the NATO Advanced Study Institute on Remote Sensing and Global Climate Change, held at Dundee, Scotland, July 19-August 8, 1992" – T. p. verso. Includes bibliographical references and index.ISBN-13:978-3-642-79289-21.Climatic changes–Remote sensing–Congresses.2. Climatology-Technique–Congresses. I. Vaughan, R. A. (Robin Antony) II. Cracknell, Arthur P. III. North American Treaty Association. Scientific Affairs Division. IV. NATO Advanced Study Institute on Remote Sensing and Global Climate Change (1992: Dundee, Scotland) V. Series. QC981.8.C5R44 1994 551.6–dc20 94-36780

Softcover reprint of the hardcover 1st edition 1994

Typesetting: Camera ready by authors
SPIN 10044810 31/3140 - 5 4 3 2 1 0 - Printed on acid-free paper

Directors' Preface

Throughout the history of the world, the atmosphere, the oceans and the land have interacted to produce global environmental change to which life has contributed and adapted. Legitimate concern is now felt about the possible impact that mankind may be having on the environment and the consequences of human activities on our changing climate. The enhanced greenhouse effect, and global warming, which may affect local temperatures, sea level, and ecological infrastructures, and the reduction of atmospheric ozone, with the consequent danger to life, have concentrated the minds of people and politicians alike. Global environmental monitoring and climate change studies are now considered legitimate areas of research and are receiving financial and moral support from many quarters. Huge international programmes have been established involving the collaboration of scientists from many disciplines and many countries. Climate is a global phenomenon, and changes in climate affect everyone in the world. Remote sensing, with its synoptic, continuous and versatile monitoring capability, is an obvious source of data for the environmental scientist, but an appreciation of its limitations and inaccuracies, and the problems involved in extracting geophysical parameters from the data and correlating them with other data sets, is essential if meaningful predictions are to be made. Management systems will need to be devised to handle types of data which differ not only in format but in accuracy, subjectivity, resolution, classification and stratification. Differences in culture, political regimes, language, population and development will affect the uniformity of data from different parts of the world and provide logistic problems in assimilation into a global data base or global model.

Remote sensing is providing, and will continue to provide, valuable ammunition to the world's scientists in their massive task of understanding our climate and the factors which affect it. The potential of Earth observation is now recognised by politicians from most countries to the extent that they are now willing to invest in very expensive and extensive programmes. The breaking down of political barriers in Europe can only hasten the collaboration which is essential if this exercise is to succeed.

It was against this background that the present Summer School was devised. The Dundee Summer Schools in Remote Sensing, of which this is the seventh, have attracted support and interest from most parts of the world.

This was no exception. Nearly 80 participants and about 15 lecturers from 28 countries gathered in Dundee in Scotland (a country well known for its own particular climate!). Their scientific backgrounds were as diverse as their nationalities, but they were all united by the desire to learn about remote sensing and how it might help them to understand climate change.

Lectures covered the principles of remote sensing, the management of data, data requirements in climatology, the principles of modelling, the input of data into models, and the applications of remote sensing to the atmosphere, ice and snow, seas and land.

Seminars were given on a number of case studies and on the use of CD ROM for storing environmental data. A visit was made to the AVHHR data receiving station for which Dundee University is well known, and a lively discussion on *Politics and Global Climate Change* also took place.

A particular feature was not only the presence of participants from distant countries such as China, India, Thailand, Malaysia, Australia, Brazil, Canada, etc. but the number from geographically close, but previously almost inaccessible, countries in Central and Eastern Europe. One Committee member, who also gave a seminar, and one lecturer both came from St Petersburg. Participants came from Russia, Estonia, Poland, Czechoslovakia, Bulgaria and Hungary. The opportunity to discuss at first hand their different problems and ideas, many of which were only vaguely known in the West until recently, was a memorable experience for most participants.

As with several of our previous summer schools, this one was supported as an Advanced Study Institute by the Scientific Affairs Division of NATO for whose support the Directors are particularly grateful. Other support, however, came from the European Commission, the British Council in a number of countries, and the European Space Agency. This additional source of funding enabled us to increase the international participation and to provide a wider lecture programme. We are also grateful to those agencies and institutions which provided support to individuals which enabled them to make the often long journey to Scotland, including the US National Science Foundation and the UK Natural Environmental Research Council.

At a very late stage in the preparation of the material for this volume, a very interesting book was published which must be mentioned. This is "Atlas of satellite observations related to global change" edited by R J Gurney, J L Foster and C L Parkinson (Cambridge: University Press) 1993. It is a very informative and beautifully produced volume and would provide valuable reading material that is complementary to these Summer School Proceedings.

Readers, reviewers and librarians may be concerned that this volume is being published rather a long time after the Summer School itself was held. This is through no fault of the Publishers, who have, in fact, acted very speedily once they received the material from us. The main reason for the delay is that a substantial amount of updating and editorial work has been carried out on several chapters by both authors and editors. The material is therefore more up to date than might appear at first sight. Moreover, while new results from particular climate models or satellite systems continually become available, the underlying principles of the use of remote sensing in climate studies, the substance of this volume, is less ephemeral.

We would like to express our sincere gratitude to those colleagues who gave of their valuable time to come to Dundee to teach, and for their cooperation in providing the material which makes up this book, to the participants who attended, as without them the exercise would have been pointless, to our colleagues on the organising committee for helping to make the experience an enjoyable one, to the technical and secretarial staff for making things work and in particular to Pat Cunningham for shouldering most of the administrative burden. The success of such a meeting is a team effort - and we have a superb team.

Lastly the editors would like to thank Xiuhong Sun and his wife Xiaoping Shi for the many hours they have devoted to converting the material for this book, which arrived in a variety of digital formats, to LaTeX. Anyone who has undertaken such a task will appreciate that this is not trivial! We hope the reader will find the contents interesting and useful.

R A Vaughan
A P Cracknell
Dundee.

1
2
3
4
5
6
7
8
9
10
11
12
13
14
15
16
17
18
19
20
21
22
23
24
25
26
27
28
29
30
31
32
33
34
35
36
37
38
39
40
41
42
43
44
45
46
47
48
49
50
51
52
53
54
55
56
57
58
58
59
60
61
62
63
64
65
66
67
68
69
70
71
72
73
74
75
76
77
78
79
80
81

1. S V Victorov
2. K R Varatsos
3. P Cunningham
4. R A Vaughan
5. A P Cracknell
6. K McGuffie
7. K Y Kondratyev
8. L V de Santis
9. M Kampel
10. Deshpande V.R.
11. R Engelen
12. M A Haque
13. Kumaran T.V.
14. A Grossinho
15. A A Samah
16. P Ashcroft
17. J R De A Franca
18. G B Franca
19. B Lesht
20. R Fraile
21. J Merayo
22. A N Fox
23. A Van Lammeren
24. A Yurtaikin
25. T Mcnally
26. E Benson
27. T Davis
28. P Singhasaneh
29. F Colao
30. I Csiszar
31. S K Soo
32. M I H Mohamed
33. K Furmahczyk
34. M V Weele
35. K Milner
36. M Saraber
37. U Lourens
38. S Beaven
39. D Low
40. B Lapete
41. J P Marrero
42. P Hampl
43. H Erdun
44. C G Selles
45. J Wang
46. J C Fortea
47. A Plidini
48. T Palmer
49. V Kostopoulos
50. M Varinou
51. K Tsolakidis
52. J Bauman
53. S Mansor
54. W Jinghua
55. H Sui
56. S Zientek
57. M Bizanzadez
58. P Fernandez
59. I Ivanova
60. N Chisholm
61. L Schuller
62. N Chisholm
63. T de Jong
64. I Woodhouse
65. B Shammery
66. R Benz
67. J A P Almeida
68. K Abdulnasser
69. J M F Morais
70. P Post
71. M O Domingue
72. R Biagio
73. D M Casey
74. D Gray
75. G A Serritella
76. F Caiffa
77. T Heinemann
78. A Khiuddin
79. D Dong
80. B P Jallow
81. A Van Riet

Organisation

Organising Committee

Dr. R A Vaughan (Dundee), Director
Professor A P Cracknell (Dundee), Co-Director
Professor C Varotsos (Athens)
Dr. C O Justice (NASA, USA)
Professor A Henderson-Sellers (Macqarie, Australia)
Dr. J T Houghton (Meteological Office, Bracknell)
Dr. I Gardiner (Dundee), Social Programme
Mr G R Whyte (Dundee), Technical Arrangements

Technical Assistance

Mrs P Cunningham
Mr I Durajczyk
Miss A Paladini

Editorial Assistance

Mrs Xiaoping Shi

Contents

Climate Change - The Background

Arthur P Cracknell

1 Introduction

I started the preparation of this set of lectures by looking up the two words "weather" and "climate" in the Concise Oxford Dictionary. I found:

WEATHER: Atmospheric conditions prevailing at a place and time, combination, produced by heat or cold, clearness or cloudiness, dryness or moisture, wind or calm, high or low pressure and electrical state, of local air and sky.

CLIMATE: (Region with certain) conditions of temperature, dryness, wind, light, etc.

The important difference is that weather refers to a certain place and a certain time. It may be bright and sunny here just now now but 50 or 100 km away it might be wet and windy there at this time. It may be bright, warm and sunny here today but it may be cold, wet and windy here tomorrow. However, we would use the word climate to represent a *temporal average* of the weather conditions and we would use it to refer to a reasonably large geographical area, such as the east of Scotland. If we were to talk of the climate of a small area, e.g. the city of Dundee, which differed significantly from that of the larger region in which it is situated, e.g. because it is surrounded by a ring of low hills, we would probably speak of a micro-climate.

Over the last ten or twelve years we have, in Dundee, run a series of postgraduate summer schools in remote sensing and we have in different schools concentrated on different aspects of the applications of remote sensing that seemed to us, at the time, to be important. It seemed very topical, at this time, to address the question of the possible use of remote sensing in relation to the study of global climate change, whether that change is natural or whether it is induced by the activities of mankind. Basically, therefore, the topic of the course on which this book is based is to study the question of how we can use remote sensing to study some of the typical parameters that are fed into the various climatological models.

I am not an expert on climate. We had asked one or two famous experts to come and give the first few lectures in the course but, after some delay, they declined and

NATO ASI Series, Vol. I 24
Remote Sensing and Global Climate Change
Edited by Robin A. Vaughan and Arthur P. Cracknell

eventually it became too late to ask anyone else to give such lectures at short notice. Therefore, I felt that I ought to give them myself. I should say that I have relied heavily on a number of sources, which will be quoted as appropriate, and that the most useful ones to me were the books by Tickell (1977), Gribbin (1978, 1988), Leggett (1990) and Parry (1990), the printed version of the 1991 Bakerian Lecture by Houghton (1991) and the report and supplementary report of the Intergovernmental Panel on Climate Change (IPCC) (Houghton *et al.* 1990, 1992). I should also like to mention the book on global climate by Kondratyev (1992) which unfortunately is not yet available in English.

The reason why climate change is topical just now is that it is part of the current environmental debate. Throughout the world, people are becoming very conscious of mankind's effect on the environment; especially one recalls the recent UNCED in Rio de Janeiro. However, the first point that needs to be made is that although at the present time we are conscious of, and largely concerned with, possible anthropogenic changes in the climate, there are very large natural changes in the climate anyway. At one time Scotland was covered by glaciers for example. The question is how significant are these changes and what can we do, or should we do, to attempt to control the anthropogenic changes or, indeed, to control the natural changes as well. The second point that needs to be made is that climate is a temporal average taken over a long time scale, whereas reliable detailed observations of the weather (from conventional records and from satellite data) have only been made over a very short and recent period of time. Archaeological or geological evidence that we have for longer time periods does exist but is of a much less detailed nature. Consequently it is difficult to determine the extent to which mankind's activities over the last century or two are actually already affecting the climate significantly. In view of this, then, the third point is that given a shortage of detailed experimental evidence about the climate itself and a lack of unequivocal direct evidence regarding the extent to which mankind's activities are already affecting the climate, it is extremely important to try to construct models in which we make use of the environmental parameters (such as the increase in CO_2 concentration) that can be quantified and use them to calculate predicted effects in the climate. Constructing such models is possible, but the really serious difficulties begin to arise when one tries to use these models. This is because the atmosphere/earth system is very complicated and one is trying to extrapolate its behaviour over a very long time period, when we cannot even predict the weather more than a few days in advance with any degree of reliability! It is, perhaps, not surprising that different groups of people working with different models often produce different predictions of climate change; nevertheless, as we shall see, there is a large enough measure of agreement that a number of clear conclusions, even now, have been obtained.

2 The scale in time and space

Perhaps one of the best sources for a general introduction to the subject of climate change is the book by Sir Crispin Tickell which I have already mentioned in my list of sources. I quoted the original 1977 edition but there is a revised edition dated 1986 which is what I have actually been following. It is, indeed, the book which I would most strongly recommend to anyone as background preparatory reading on climate change. It has several advantages. First, it is short, only "xvi + 76 pp" in the notation of the book reviews; however it is published at a price (£14.75) which corresponds to about £0.16 per page and at this price one sometimes feels that the publishers are almost inviting readers to break the copyright laws and photocopy the book rather than purchase a copy legally! Secondly, the other nice feature of the book is that Sir Crispin Tickell is not a scientist. He was a diplomat in the British Foreign Service who made his foray into science in 1976. He took 12 months off from his diplomatic work after helping to mastermind Britain's entry into the EC and after three years as the UK Foreign Office's top man in charge of NATO and arms reduction talks. He began his year out in Boston, by studying astronomy at Harvard and meteorology at MIT; he ended it by giving a series of lectures at both institutions on the links between politics and climate change, later turning them into the book already mentioned. Tickell recalls, as quoted by Pearce (1992) "I found I was virtually the first in the field, I read the entire literature on climate change in half a term, and there was nothing on the policy implications." Following that year, Tickell continued his diplomatic career and has been instrumental in helping to get climate change on to the political agenda at both national and international level. From his position Tickell was able to study the problem of climate change from a scientific point of view but with an overall perspective and not, like many professional scientists, ending up being unable to see the wood for the trees.

In my lectures to first-year environmental science students at Dundee University I try to stress the extremely fragile nature of life. We occupy a very thin spherical shell on the surface of a small and insignificant planet! In an aircraft on a commercial flight at, say, an altitude of 10,000 m (10 km) the aircraft cabin has to be pressurised and heated; at the outside pressure and temperature the crew and passengers would die very quickly. Similarly, if we go very far down below the surface of the sea we need to be protected from the surrounding environment. Taking this 10 km thickness as an order of magnitude, it corresponds to 10/6400, i.e. 1/640 or about 0.16%, of the radius of the Earth. In relative terms this layer is thinner than the skin of an apple and certainly thinner than the skin of an orange; it is more like the layer of cling film or tissue paper in which an orange might be wrapped in a supermarket. Another way of looking at the question of scale is given by Tickell; if the Sun is represented by an orange, the Earth would be represented at the same scale by a speck of dust at a distance of 30ft (10 m) away and the nearest star could be represented by another orange a thousand miles (1,600 km) away. It happens that in our very thin layer at the surface of the Earth all the necessary conditions for the development and maintenance of life have been satisfied. But quite small changes in the conditions could lead to the partial or total destruction of life. A change of only a few degrees in average temperature leads either to ice ages or to melting of the polar ice and either of these would destroy much of our present civilisation. Smaller changes lead to productive agricultural land being turned into a desert.

Perhaps the time scale is even more difficult to envisage. Tickell scales it all down. We currently think of the age of the universe as being about 15,000 million years, with the Earth being formed about 4,600 million years ago. If we remove all the zeros and represent 4,600 million years by 46 years (almost an average life span) then the dinosaurs died just over 6 months ago, *homo sapiens* emerged about a week ago, the zero of our western counting system, at the birth of Christ, was less than a quarter of an hour ago and the industrial revolution has lasted just over a minute. In terms of the climatic situation, on the same time scale, there were major ice ages on the Earth about $9\frac{1}{2}$, $7\frac{3}{4}$, $6\frac{1}{4}$, $4\frac{1}{2}$ and 3 years ago, the most recent series of glaciations began less than a week ago and the last glaciers retreated about an hour ago. The temperature variations are illustrated in Figure 1. As Tickell puts it ".... we live in a tiny, damp, curved space at a pleasantly warm moment." Notice from Figure 1 that the temperature variations are quite small; however, a rise or fall of only a few tenths of a degree may correspond to quite a large change in the weather conditions. The difference in average temperature of about 1 °C between A.D. 1200 and A.D. 1600 corresponds to some quite significant differences in lifestyle, vineyards in southern England and fairs on the frozen river Thames in London, respectively. The range of temperature corresponding to what we would regard as extreme conditions is also quite small. At the height of the last ice age, when most of Britain was covered by ice, the mean global temperature was only about 6 °C below the present value.

3 Weather forecast models

What we have seen over the last decade or two has been a transition from the traditional methods of weather forecasting to forecasts based on models run on computers (see for instance Houghton 1991). The objective of a weather forecast model is to be able to predict the weather at a particular place and time. The idea is that one writes down the mathematical equations that describe the various physical processes that occur and tries to solve those equations for some future time. In an atmospheric model the behaviour of the atmosphere is represented by the values of appropriate parameters specified on a three-dimensional grid of points. The UK Meteorological Office uses both a global model operating with a grid with spacing of 90km (i.e. of the order of 1 ° of latitude or longitude) in the horizontal and also a limited area model with a grid spacing of about 40 km; in each model there are about 20 levels in the vertical. The parameters involved are indicated in figure 2. The physics involved includes the following:

horizontal momentum equations

the hydrostatic equation

the equation of continuity (i.e. matter is neither created nor destroyed)

the equation of state

the thermodynamic equations

along with parametric descriptions of the following processes:

evaporation

condensation

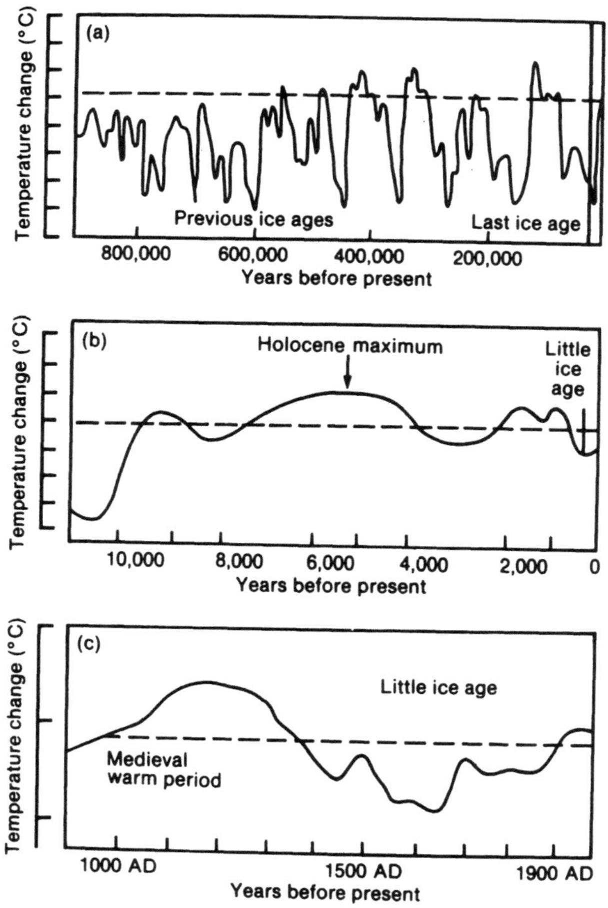

Figure 1: Schematic diagrams of global temperature variations since the Pleistocene on three time-scales: (a) the last million years, (b) the last ten thousand years and (c) the last thousand years. The dotted line nominally represents conditions near the beginning of the twentieth century. (From Houghton 1991).

cloud formation and dispersal

radiation and convection in the atmosphere

the exchange of momentum, heat and water vapour at the Earth's surface.

It is also necessary to take into account effects or processes that occur on a scale which is smaller than the grid spacing that is used in one's model. We suppose that the results of meteorological observations allow the values of the atmospheric parameters to be specified at some starting time, $t = 0$. The forecasting problem then is to solve the various equations describing the physical principles and processes that we have just

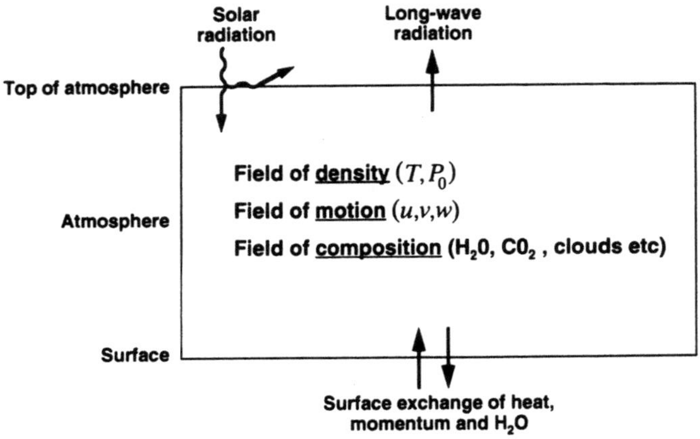

Figure 2: Schematic illustrating the parameters and physical processes involved in atmospheric models beginning of the twentieth century. (From Houghton 1991).

mentioned for a subsequent time t. That is, one seeks to determine new values of these atmospheric parameters at a subsequent time, in practice up to a few days ahead. Because there are several of these equations, and especially because they are non-linear equations, the problem has to be tackled by numerical methods using a computer. Hence the terms "computer model" or "numerical model".

The operational use of numerical models at the UK Meteorological Office began in 1965. Although their usefulness at that stage was limited, there have been rapid developments since then and within ten years the models were able to provide better forecasts of the basic motion field than could be achieved by an unaided human forecaster. These improvements have come about for two reasons. First, the power of the computing facilities available has increased enormously, see Figure 3. Secondly, the amount of data available to describe the present atmospheric conditions has been greatly increased in recent years. In addition to conventional surface measurements and radiosonde measurements, there are now also satellite observations including satellite soundings and satellite-derived wind speeds obtained from cloud tracking from geostationary satellite images, see Figure 4.

If at present the weather forecast models are successful in providing forecasts a few days ahead, then one can ask what improvements we can expect to see in the future. The European Centre for Medium-Range Weather Forecasting (ECMWF) produces forecasts for four weeks ahead. The use of models along the lines we have outlined is basically a deterministic approach. We have the present conditions, we have a set of mathematical equations describing the various atmospheric processes and we rely on the mathematics to predict the future conditions. At present this can be done successfully a few days ahead; the specification of what one means by a "few" days is not very easy but, at present, a figure of five days might be a reasonable one. It is not a simple cut-off situation, where forecasts up to a certain time ahead are accurate and those beyond that time are not. Rather one should say that accuracy, as specified by rms values in the errors of forecasts of particular parameters, is improving all the time. This

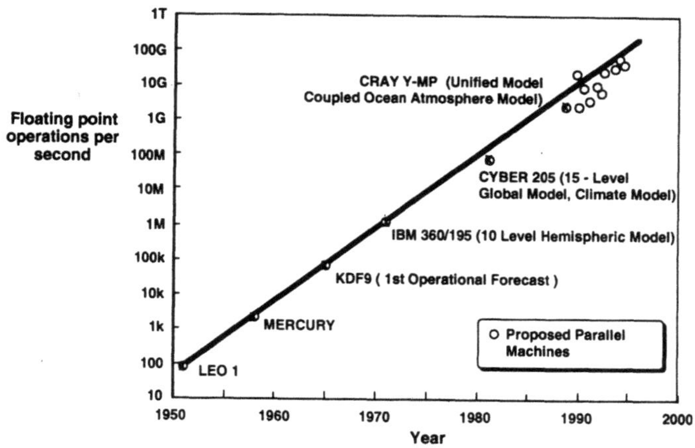

Figure 3: Computers used by the UK Meteorological Office for numerical weather prediction (NWP) research and since 1965 for operational forecasting. (From Houghton 1991).

is illustrated by Figure 5; this shows that three-day forecasts of surface pressure now are, on average, as good as the two-day forecasts were in 1980. The question arises as to whether this improvement can go on indefinitely. It is held to be likely that the ultimate limit of forecasts from deterministic models is about 2–3 weeks ahead. In making this statement is one simply being pessimistic, or realistic, about possible future developments of computers or is there some more fundamental reason for this limitation? This brings us to the idea of chaos.

4 Chaos

The term chaos is now used with a rather specific meaning; this originated in the meteorological context but the idea of chaos is now used in a much wider context. An early paper on the subject was by Lorenz (1963), who was a meteorologist. He considered the variation in the behaviour of a simple three-parameter model, the simplest numerical model he could formulate to describe convection in the atmosphere, with variations in the input conditions from which the model integrations began. He developed a description of what has become known as the Lorenz attractor, which describes the evolution of the values of these three parameters from different starting conditions. It shows the regions of phase space in which the system's behaviour is most likely to lie and also shows how solutions from very close starting conditions can diverge into very different regions. For the Lorenz attractor there are some evolving trajectories that remain close together for a long time. Some other trajectories may diverge initially but finally end in the same overall region of the phase space. Yet other trajectories may start close together but then diverge into very different regions of phase space. Looked at from another point of view this means that a very small change in the initial conditions of a system may, or may not, lead to enormous differences in the state of the system at some

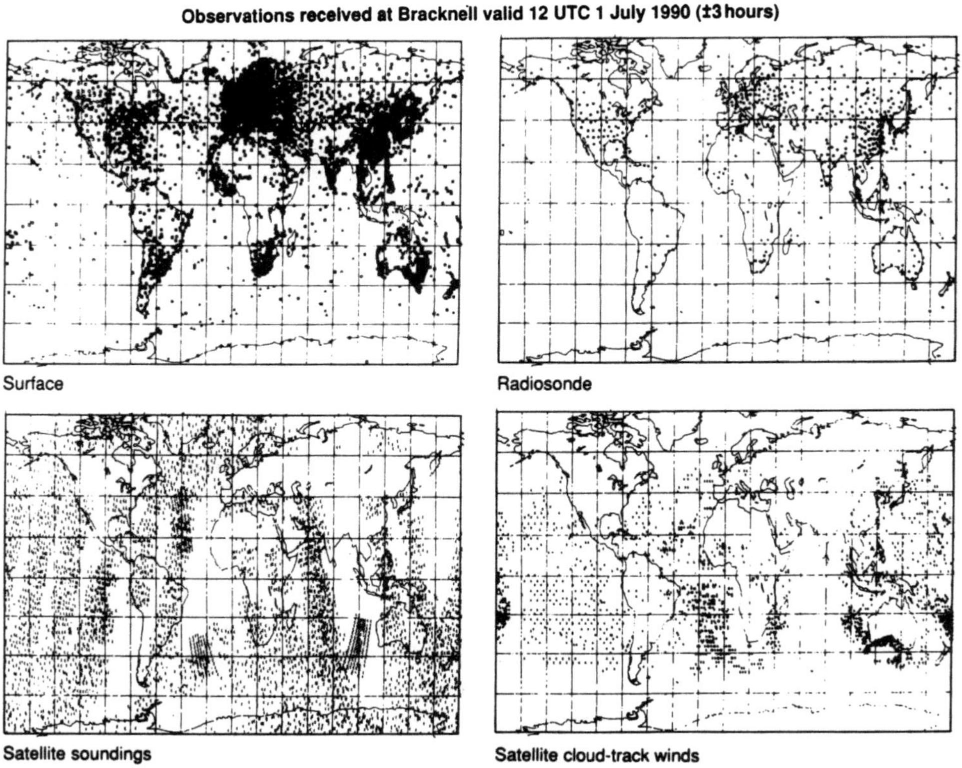

Figure 4: Some of the sources of data for input into the UK Meteorological Office weather forecasting numerical model on a typical day. (From Houghton 1991).

subsequent time, t. It is the critical dependence of a system's behaviour on the precise starting conditions which is known as chaotic behaviour. Given the grid spacings that we mentioned above as being used in meteorological models, it is apparent that many events occur which are very small and localised so that their immediate effects on the values of the parameters at the grid points of the model are negligible. Yet over a period of time their effects may build up to be very large. This is typified by the title of a lecture given by Lorenz in 1979: "Predictability: does the flap of a butterfly's wings in Brazil set off a tornado in Texas?" If such small events can have such large effects, although they frequently do not, and if we regard climate as a long-term average of weather, then there is no future in trying to extend the deterministic approach along these lines to make climatic forecasts. Thus climate models should not just be regarded as weather forecast models extended to work over a longer time span.

Although weather-forecast and climate-forecast models have much in common, particularly in terms of their physical basis, there are considerable differences in basic philosophy and detailed implementation.

Although chaos theory owes much of its early development to the work of meteorol-

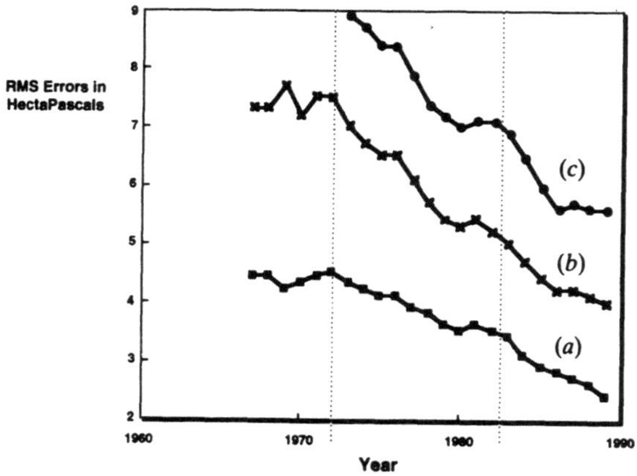

Figure 5: Errors (r.m.s. differences of forecasts of surface pressure compared with analyses) of UK Meteorological Office forecasting models since 1966 for (a) 24 h, (b) 48 h and (c) 72 h forecasts. Dotted lines show where major changes of model or computer have occurred. (From Houghton 1991).

ogists attempting to model the atmosphere, the theory has now found applications in a wide variety of situations. It is, perhaps, worthwhile to mention a few of these; a good account will be found in the series of articles from the *New Scientist* collected together and edited by Hall (1991). I shall paraphrase the editor's introduction and one or two other items from that book.

We have already mentioned the question "Does the flap of a butterfly's wings in Brazil set off a tornado in Texas?" Consider another example which is found early in Hall's book and which, for addicts of Winnie the Pooh, is a variation of the game of Pooh Sticks. If you watch from a bridge as a leaf floats down a stream, you may see it trapped by a small whirlpool, circulate a few times and escape, only to be trapped again further down the stream. Trying to guess what will happen to a leaf as it comes into view from under the bridge is an idle pursuit in more senses than one; the tiniest shift in the leaf's position can completely change its future course. Small changes lead to bigger changes later. This behaviour is the signature of chaos. If we move away from these physical situations we can think of many other everyday examples of small events that can drastically change the course of history: the assassin's bullet that triggers a revolution, a chance meeting with a stranger at a party, the impulsive decision to catch a plane doomed to crash. Romantic novels are full of such stories. We expect life to be complicated and uncertain, scattered with random events that make the future difficult to predict. For this reason, many people find the mechanistic predictability of traditional science unattractive and difficult to relate to their own lives. To a considerable extent the theory of chaos has changed that. This seems to link our everyday experiences to the laws of nature by revealing that, in spite of underlying deterministic equations or relations, there can nevertheless be an overall unpredictability in practice.

Scientists have spent many centuries searching for simple rules, or laws, that govern the Universe. For example, Newton succeeded in explaining how the stars appeared to move across the sky with his simple laws of motion and theory of gravitation. At the beginning of the 19th century Laplace believed firmly in a Newtonian universe that worked on clockwork principles. He proposed that if you knew the position and velocities of all the particles in the Universe, you could predict its future for all time. Such a deterministic view received its first blow with the development of quantum mechanics in the 1920s to describe the world of the very small. Quantum mechanics is used to describe the behaviour of fundamental particles; it is a statistical theory based on probability. According to quantum mechanics it is impossible to measure the position and momentum (or the position and velocity) of a particle at the same time. This is an example of the uncertainty principle.

In spite of the statistical aspects of quantum mechanics, physicists have used quantum mechanics to construct a reasonably robust theoretical framework for describing the fundamental properties of matter and the forces at work in the Universe. Thus one may hope to explain how the Universe has evolved and indeed even how it came into existence. The reductionist view is that once one has formulated the theory completely it should be possible to explain the more complicated natural phenomena - how atoms and molecules behave (chemistry) and how they organize into self-replicating entities (biology). In principle it is just a matter of time and effort, or of computing power. Since human beings are subject to the same laws of nature as the galaxies, so eventually we may be able to make predictions about human concerns, such as the fluctuations of the stock market or the spread of an epidemic. In theory, life can be supposed to be predictable. We have, however, seen already in the case of weather forecasting that such an approach is not very useful; it is unlikely to be useful in these other, more general, areas either. What has been realised recently is that in addition to weather systems there is a whole range of other systems to which chaos theory can usefully be applied.

Chaos theory presents a universe that is deterministic, obeying the fundamental physical laws, but with a predisposition for disorder, complexity and unpredictability. It reveals that many systems which are constantly changing are extremely sensitive to their initial state: position, velocity, and so on. As the system evolves in time, minute changes amplify rapidly through feedback. This means that systems starting off with only slightly differing conditions rapidly diverge in character at a later stage. Such behaviour imposes strict limitations on predicting a future state, since the prediction depends on how accurately you can measure the initial conditions. It is only through the widespread availability of powerful computers that it has been possible to follow the development of chaotic systems and demonstrate their behaviour. The results are often presented in computer graphics; starting with quite simple equations, one can produce breathtaking patterns of increasing complexity. Within the patterns, there are shapes that repeat themselves on smaller and smaller scales - a phenomenon which is called *self-similarity.* Mandelbrot coined the word "fractal" for such shapes (see Mandelbrot 1982). This provides an immediate link with nature, for trees and mountains are examples of fractals.

Chaos theory provides an interpretive tool for describing many of the complexities of the world, from brain rhythms to gold futures. The book by Hall (1991) contains a series of contributions on chaos theory and its implications. The various chapters explain what chaos is, the mathematics behind chaos, how chaos can be found in virtually every discipline from astronomy to population dynamics, and how chaos theory can be applied practically to areas such as chemistry, turbulence in fluids, engineering and the financial and commodities markets. They also look at the role of fractal geometry in chaos theory and how computers are now changing the way science is being done.

5 The causes of climate change

Many of the more recently published discussions of climate change concentrate almost exclusively on the changes that occur, or may be predicted to occur, as a result of human activities. For example in the article by Houghton (1991) which has already been cited he says "the particular climate changes which I consider in this paper are those which may occur over the next century as a result of human activities". These are the main changes of interest in the current active environmental debate. However, I shall follow the approach of Tickell (1977) because his account was written before this debate developed and he gives a balanced account of natural and anthropogenic causes of change.

The variables which are commonly used in studying the climate are concerned mainly with the atmosphere. However, we cannot look at the atmosphere alone. This is because processes in the atmosphere are strongly coupled to the oceans, to the land surface and to the parts of the Earth which are covered with ice, i.e. the cryosphere. There is also strong coupling to the biosphere i.e. to the vegetation and other living systems on the land and in the sea. It is convenient to consider three categories of events that may affect the climate,

(i) events that occur outside the Earth

(ii) natural events on the surface of, or within, the Earth

and

(iii) human activities.

5.1 Events that occur outside the Earth

The source of virtually all the energy that drives the atmosphere is the Sun. Thus any changes in the intensity of the solar radiation that is incident on the atmosphere from above will clearly have a significant effect on the weather and on the climate. There is a variety of possible causes of changes in the intensity of this radiation. The amount of energy emitted by the Sun itself may vary, the transmission of the radiation through space from the Sun to the Earth may change since space is not absolutely empty, and the separation and the relative orientation of the Sun and the Earth may change. Let us consider these possibilities in turn.

First there is the variation of the Sun's radiation itself. There are regular periodic variations in the Sun's behaviour; these include (a) pulsation with a period of about two hours and forty minutes, (b) rotation on its axis every twenty seven days and (c) sunspot activity with an eleven year cycle. Tickell, perhaps seeing things as a historian, has a nice comment on sunspots. "The connection between sunspots and the weather on Earth is part of universal folklore and cannot easily be dismissed, although the means

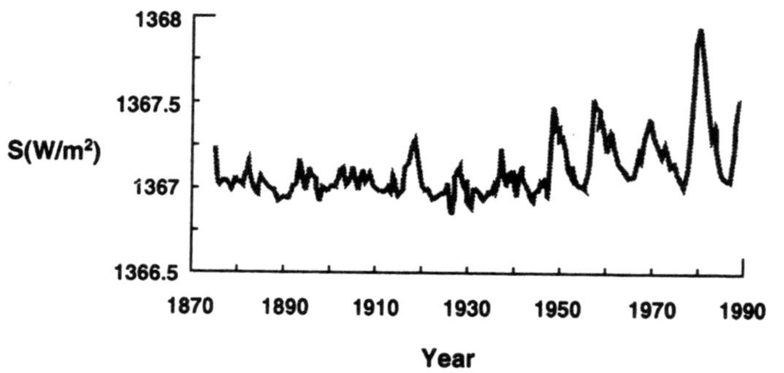

Figure 6: Changes in solar irradiance 1870-1990 as estimated by Foukal and Lean (1990).

by which one could affect the other has not been convincingly demonstrated". He goes on to relate that between 1645 and 1715, the coldest part of the Little Ice Age, was a period of very low sunspot activity, that the estimated intensity of the solar radiation was about 1% lower than its present value and that this is consistent with an average temperature of about 1 °C or so less than at present, see Figure 1. It is only within the last decade that accurate measurements of the radiation output of the Sun have been made by spacecraft which operate outside the Earth's atmosphere, i.e. by remote sensing. Evidence from several carefully calibrated spacecraft instruments indicates that the total solar energy has varied by about 0.1% during the last solar cycle (Wilson and Hudson 1988). Foukal and Lean (1990) studied the spacecraft measurements and by correlating them with sunspot data were able to construct the total solar irradiance from 1874 to 1988, see Figure 6. In addition to the three short-term variations which we have mentioned, there are also long-term variations in the intensity of the Sun's radiation. These are, inevitably, not well understood because of lack of data. For example, the gravitational field experienced by the Sun due to the planets, including the Earth, varies because of the varying positions of the planets as they rotate in their orbits around the Sun. This gives rise to tidal effects on the surface of the Sun, leading possibly to some effects on the Sun's radiation.

Secondly, the radiation travelling from the Sun to the Earth may encounter dust clouds and thus undergo some attenuation. Or showers of micro-meteors may enter the Earth's atmosphere causing dust and ice particles in the stratosphere; these then act as a screen and prevent some of the solar radiation from reaching the Earth.

Thirdly, there are the variations in the relative positions and orientations of the Sun and the Earth; this in turn involves three factors. First, the eccentricity of the elliptical orbit of the Earth around the Sun varies with a period of about 100,000 years. At present the intensity of the solar radiation received at the Earth varies by about 7% during the year, but when the eccentricity of the Earth's orbit is greatest the intensity of the solar radiation received at the Earth varies by up to 30% during the year. A second source of variation arises from the oscillation of the tilt of the Earth's axis of rotation relative to the plane of its orbit; this oscillation has a period of about 40,000 years. The greater the tilt, the greater will be the difference between summer and winter (the last maximum was 10,000 years ago). Thirdly, there is the precession of the equinoxes, i.e. the variation in the time of the year when the night and day are of equal length. This arises from the wobble of the Earth around its axis and has a period of 21,000 years. These three variations clearly have an important effect on the intensity of the solar radiation that reaches the Earth, see Figure 7. It was first pointed out by Croll (1867), whose ideas were later developed by Milankovitch (1920), that the major glacial-interglacial cycles over the past few hundred thousand years might be linked with these regular variations in the distribution of solar radiation reaching the Earth. There is a quite impressive correlation between low intensity of solar radiation and large volumes of ice on the Earth's surface and this provides strong evidence to support the Milankovitch theory, see Figure 7 (b) and (c).

It is important to have accurate measurements of the intensity of solar radiation (the solar "constant") over a period of years but, like all other measurements of parameters related to climatology, it has to be remembered that in climatological terms any period over which we are able to acquire measurements is infinitesimally short.

5.2 Natural events on the Earth

5.2.1 Plate tectonics

It is as well to recall once more that we are, at the moment, not just concerned with short-term climate changes attributable to human activities but, rather, to the whole history of the Earth. Thus, although to us the motions of the various tectonic plates (continental drift) on the surface of the Earth may seem small and of little immediate relevance to climate change, they are - over long periods of time - very important. The transfer of energy between the surface of the Earth and the atmosphere depends on whether that surface is ocean, ice or land and, if it is land, on the nature and elevation of the land surface and land cover. Thus the changes in the distribution of the land surfaces and the ocean areas of the Earth arising from the movements of the tectonic plates will cause changes in the distribution of the input of energy into the atmosphere. Another aspect of crustal movements is associated with variations in the gravitational effect of the Sun and the Moon on the surface of the Earth. Apart from causing the familiar ocean tides, these effects give rise to stresses on the land surface as well so that there is a greater chance of volcanic eruption when these stresses are at their maximum (the question of volcanic eruptions will be discussed in section 5.2.3). Another factor that, in principle, needs to be considered in long-term studies is the effect of the variation, and indeed the reversal, of the Earth's magnetic field on incoming radiation.

5.2.2 Polar ice caps.

The arctic and antarctic ice caps play an important role in the global weather system, profoundly affecting ocean currents and winds and the transfer of heat between them. The presence of land or landlocked sea at the poles seems to be indispensable for the growth of the glaciers which make up the present ice caps. Perennial ice now covers

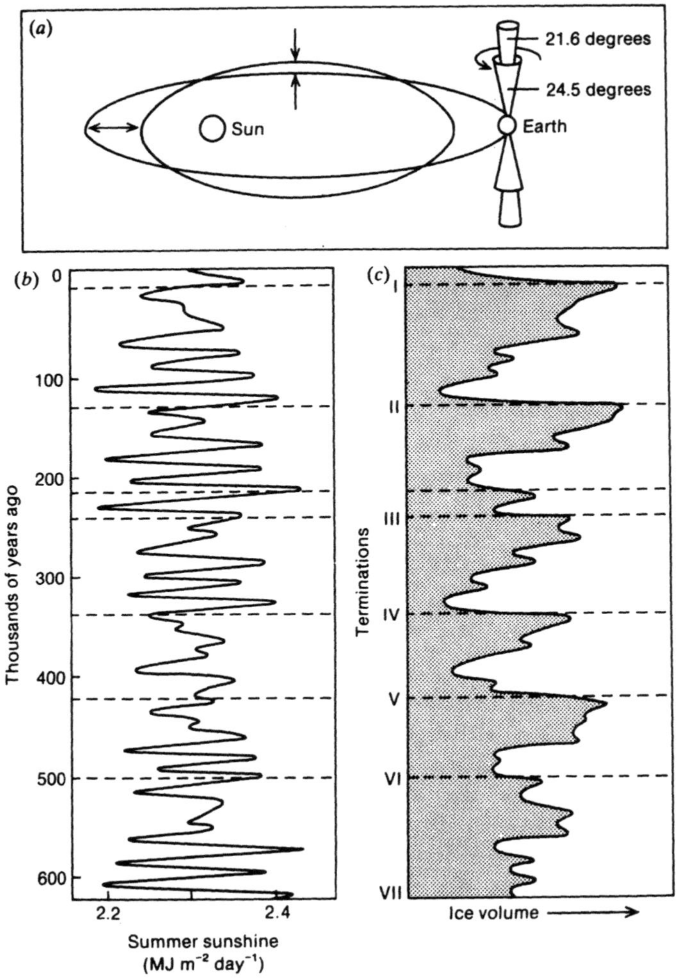

Figure 7: Variations in the Earth's orbit (a) in its eccentricity, the orientation of its spin axis (between 21.6° and 24.5°) and the longitude of perihelion cause changes in the incidence of solar radiation at the poles (b) which appear as cycles in the climate record and (c) a mechanism for the triggering of climate change suggested by Milankovitch (1920). (After Broecker and Denton 1990).

11% of the land surface and 7% of the oceans. The drift of the Antarctic continent to its present isolated position and the subsequent partial enclosure of the Arctic Ocean from most of the warm currents moving north from the equator helped the slow accretion of ice sheets. It is the existence of these ice sheets in the polar regions and their expansion in the various ice ages and their contraction in the periods in between which has been the dominant feature of the climate over the last million years or so. The Greenland and East Antarctic ice sheets, being formed over land above sea level, are relatively stable and form anchors of the systems which, from time to time, have expanded to cover enormous areas. Once an ice sheet has become established, then positive feedback occurs. The reflectivity of snow and ice is very high and so the bulk of the solar radiation incident on the ice sheet is reflected and passes through the atmosphere back into space. This leads to colder weather and the expansion of the ice sheet. The expansion of the ice sheet can eventually be halted or reversed when lack of evaporation from the ice and snow over a large area reduces the amount of precipitation of fresh snow. The expansion of the ice sheet may also be halted or reversed as a result of wind-blown dust settling on the ice; this reduces the reflectivity (or albedo) of the surface and leads to warmer weather. The transfer of heat around the Earth by ocean currents also plays an important role. That the ice ages and the warmer periods between them correspond to oscillations of this system seems to be agreed, but the detailed description of the mechanisms and their incorporation into a usable mathematical or numerical model is not simple. We shall turn later, in chapter 6, to the question of attempts to carry out detailed modelling.

5.2.3 Volcanic eruptions

In discussing extra-terrestrial causes of changes in the intensity of solar radiation that reaches the surface of the Earth we mentioned dust clouds and meteor showers. Volcanic eruptions produce similar effects to those of showers of micro-meteors, though the origin of the particles is terrestrial rather than extra-terrestrial. A volcano blasts large quantities of matter into the sky during an eruption. Some of this reaches the stratosphere as dust particles around which ice can form. Dust is produced in the atmosphere by many causes other than volcanoes, e.g. by strong winds over desert areas. But the quantities arising from major volcanic eruptions are much larger than the quantities arising from other causes. The fate of volcanic dust depends on the latitude of the eruption. A belt of dust particles forms a screen around the Earth at the latitude of the erupting volcano and then widens out in a north-south direction as well. If the erupting volcano is near to the equator the screen spreads out over the whole Earth towards both poles; if the volcano is more than about 20 °N or 20 °S from the equator the stratospheric screen tends to be confirmed to one hemisphere. This screen reduces the intensity of solar radiation reaching the surface of the Earth and also reduces the intensity of the terrestrial radiation escaping to outer space; however, the former effect is dominant. Thus the general effect of a volcanic eruption is to lead to a cooling of the Earth. A reduction of 2% in the direct solar radiation and a 0.5 °C reduction in average world temperatures resulting from the 1963 eruption of Mount Agung in Bali is quoted by Tickell (1977). Other major eruptions have produced similar effects. Volcanic eruptions appear to affect rainfall as well as the temperature. If the flap of a butterfly's wings in Brazil can lead to a tornado in Texas then it would not be surprising if an intense volcanic activity were to lead to an ice age. Evidence quoted by Tickell (1977) from an ice core taken from the Byrd glacier in Antarctica shows a steep increase in deposits of micro-particles during the coldest years of the last ice age between 18,000 and 22,000 years ago, see Figure 8. Measurement made on air from small bubbles that were trapped in the ice during that ice age show a steady natural decline in atmospheric CO_2 down to around 200 parts per million (ppm). This may be compared with the present level of around 340 ppm which is partly generated by human activities (see section 5.3.1). Of course, a correlation between high particle concentration in the ice, corresponding to

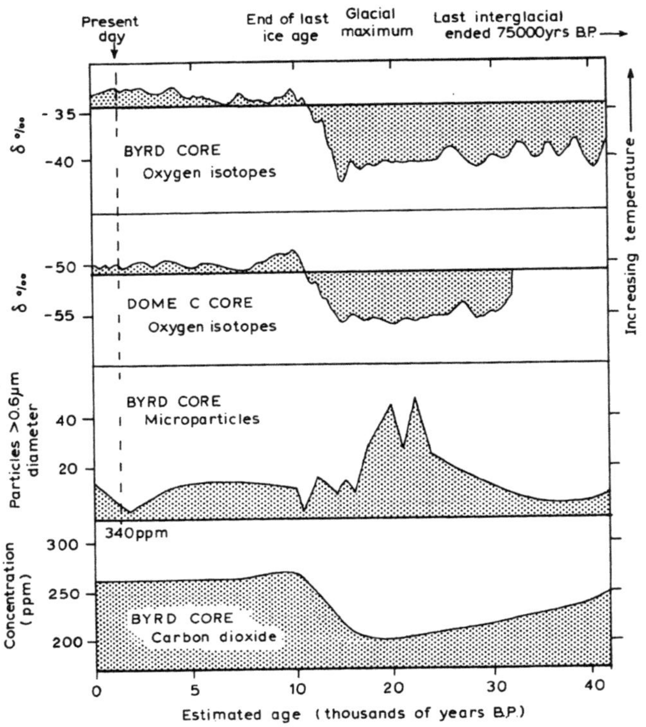

Figure 8: Findings from Antarctic ice cores, showing the relation between concentrations of oxygen isotopes, microparticles and CO_2 in the ice and temperatures over the last 40,000 years. (Adapted from Tickell 1977).

high particle concentration in the atmosphere and low concentration of CO_2 does not establish either the particle concentration or the CO_2 concentration as the cause of the low temperatures.

5.2.4 Ocean circulation

Ocean circulation is another major factor affecting the weather and therefore the climate. More than half the solar radiation reaching the surface of the Earth is absorbed by the sea, largely in the top 100 m. This layer of water acts as a giant reservoir of heat; some heat is transferred directly back into the atmosphere by evaporation, some is moved and mixed downwards and some remains in the surface layer and travels around the world with the ocean surface circulation. Such heat transfer by ocean circulation can have very important effects on weather and climate. The main ocean currents are indicated in Figure 9. The solid arrows indicate warm currents and the broken arrows indicate corresponding cold currents. One well-known example of a warm curent is the Gulf Stream which leads to Britain having a much milder climate than many other places at the same latitude. The Gulf Stream is relatively steady but over the millennia its position has varied, see Figure 10. Other situations, however, are much more variable. One well-known example is the El-Niño phenomenon. In normal circumstances a strong (cold) Pacific current runs from the south northwards along and away from the coast of Peru, see Figure 9. This creates an upwelling of deep cold water rich in nutrient salts, which nourishes the fish on which much of the Peruvian economy depends and determines the local weather. From time to time, roughly once in two to ten years, this current moves down and offshore; this is the phenomenon known as El-Niño (the Child) because of its appearance near Christmas time. Warm water moves in from the north with catastrophic results for men and fish alike. Changes of sea surface temperatures in the eastern tropical Pacific of up to 7 °C from the normal climatological average can occur. Associated with these El-Niño events are anomalies in the circulation and rainfall in all tropical regions and also to a lesser extent at mid-latitudes. A particularly intense El-Niño event occurred in 1982-83 associated with which were extreme events (droughts and floods) somewhere in almost all the continents (Figure 11). Some successes have been obtained with modelling to forecast rainfall in the Sahel, to forecast rainfall in northern Brazil, to simulate winter rainfall in Australia and to predict the consequences of the El-Niño events and, indeed, the occurrence of El-Niño itself (for further details see section 10 of Houghton (1991). For a few months the world's weather is affected by something which appeared at first to be a fairly local event (albeit somewhat larger than the flap of a butterfly's wings); after a while the effect disappears and the weather reverts to its more normal patterns.

5.3 Human activities

5.3.1 Carbon dioxide, CO_2

The human activity of which we are most aware, in common culture, in terms of its effect on the global environment is the production of CO_2 by the burning of fossil fuels. The concentration of CO_2 in the atmosphere has risen from about 280 ppm at the time of the Industrial Revolution to over 350 ppm at the present time; from Figure 8 we see that in earlier times the concentration of CO_2 was by no means constant and was probably as low as 200 ppm during the last ice age. At present the concentration of CO_2 in the atmosphere is rising fast because of the rapid increase in the consumption of fossil fuels and the destruction of forests leading to a reduction in the removal of CO_2 from the atmosphere by photosynthesis, see Figure 12.

This increase in the concentration of CO_2 in the atmosphere arises from a net imbalance in the carbon cycle, see Figure 13. This diagram shows the main reservoirs (in terms of 10^9 t of carbon) and the main fluxes (in terms of 10^9 t of carbon per year). Expressed in terms of the net quantity of carbon released into the atmosphere, based on the fluxes shown in Figure 13, this corresponds to a net release of ca 3×10^9 t of carbon into the atmosphere per annum. The seasonal variations that can be seen in Figure 12(b) are due primarily to the withdrawal and production of CO_2 by the terrestial biota.

The main effect of increasing the concentration of CO_2 in the atmosphere is to change the balance between incoming radiation from the Sun and outgoing radiation from the Earth. The CO_2 acts like the glass or transparent plastic in a greenhouse. The bulk of the incoming solar radiation is in the visible wavelength range. After

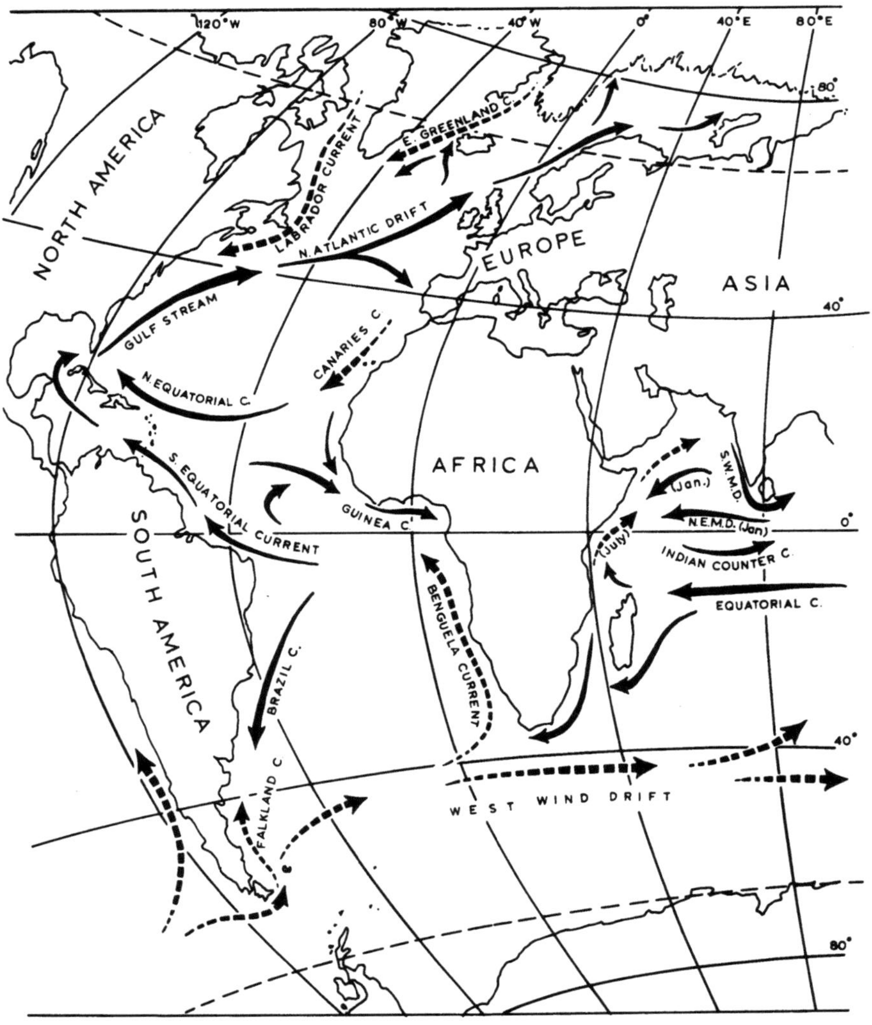

Figure 9: The main ocean currents; warm currents are indicated by continuous arrows and cold currents by broken arrows (NEMD = North East Monsoon Drift, SWMD = South West Monsoon Drift). (After Barraclough 1989).

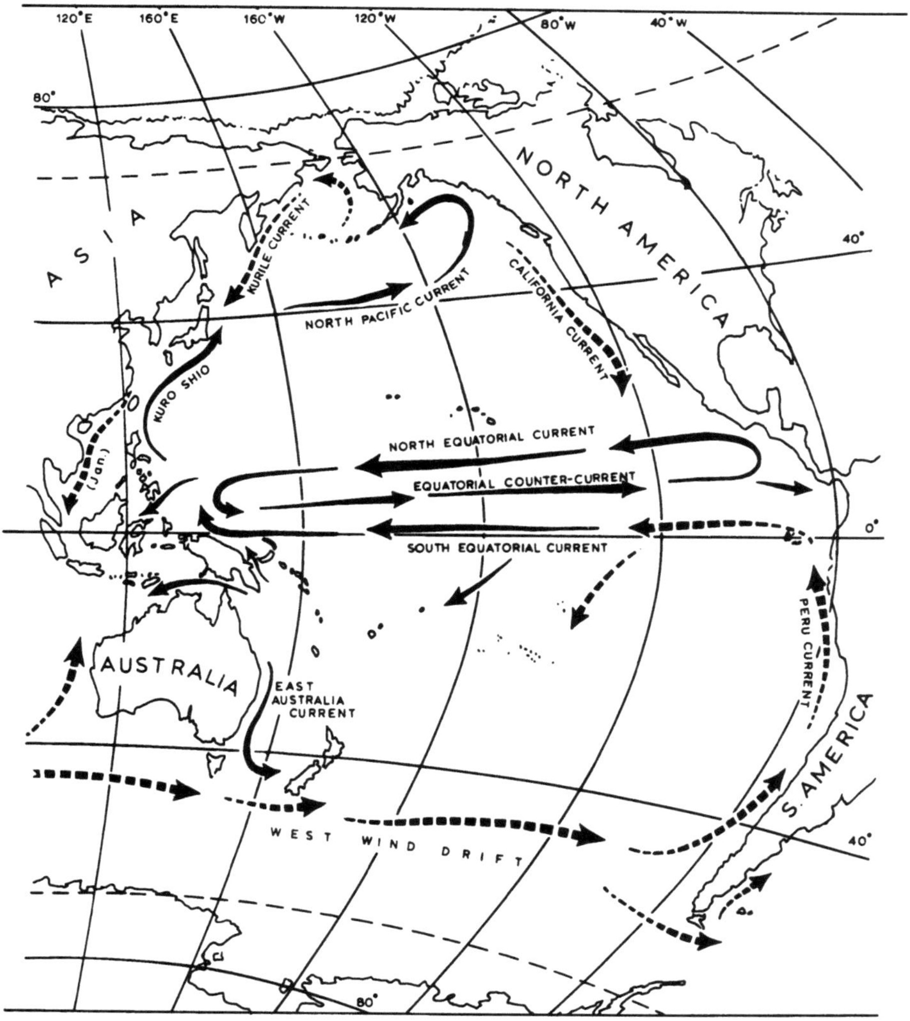

120°E
160°E
160°W
120°W
80°W
40°W
80°
40°
0°
40°
80°
ASIA
NORTH AMERICA
KURILE CURRENT
NORTH PACIFIC CURRENT
CALIFORNIA CURRENT
KURO SHIO
(Jan.)
NORTH EQUATORIAL CURRENT
EQUATORIAL COUNTER-CURRENT
SOUTH EQUATORIAL CURRENT
AUSTRALIA
EAST AUSTRALIA CURRENT
PERU CURRENT
S. AMERICA
WEST WIND DRIFT

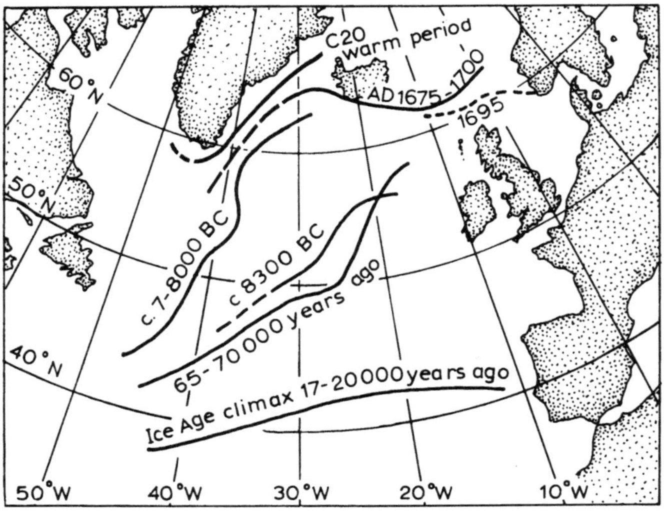

Figure 10: Historical variations of the boundary in the ocean surface between the water of the Gulf Stream and of the polar ocean current. (After Tickell 1977).

interaction with the Earth's surface much of the energy is converted to longer wavelength (infrared) radiation which is blocked by the CO_2 (or by the glass or transparent plastic in a greenhouse). The greater the concentration of CO_2, the greater the amount of infrared radiation that is trapped near to the Earth's surface. The CO_2 thus warms the lower atmosphere and cools the upper atmosphere. This in turn increases moisture, which traps more infrared radiation. The effect of rising temperature is to reduce the areas covered by snow and ice, thereby diminishing the amount of heat reflected back into space and increasing the absorption of solar radiation. CO_2 is not the only gas contributing to the greenhouse effect in the atmosphere but it is the most important one.

5.3.2 Ozone

Probably the second most well known effect of human activity on the atmosphere is the depletion of the ozone layer. This is a long story and I do not wish to go into it in great detail here; good accounts are given by Gribbin (1988), Roan (1989) and Jones and Wigley (1989). Basically ozone (O_3) is being created by the action of sunlight on the ordinary oxygen molecules O_2 and it is being destroyed, principally by oxides of nitrogen and by the decomposition products of chlorofluorocarbons (CFCs - principally CCl_3F and CCl_2F_2). The importance of the ozone is that in the stratosphere it ab-

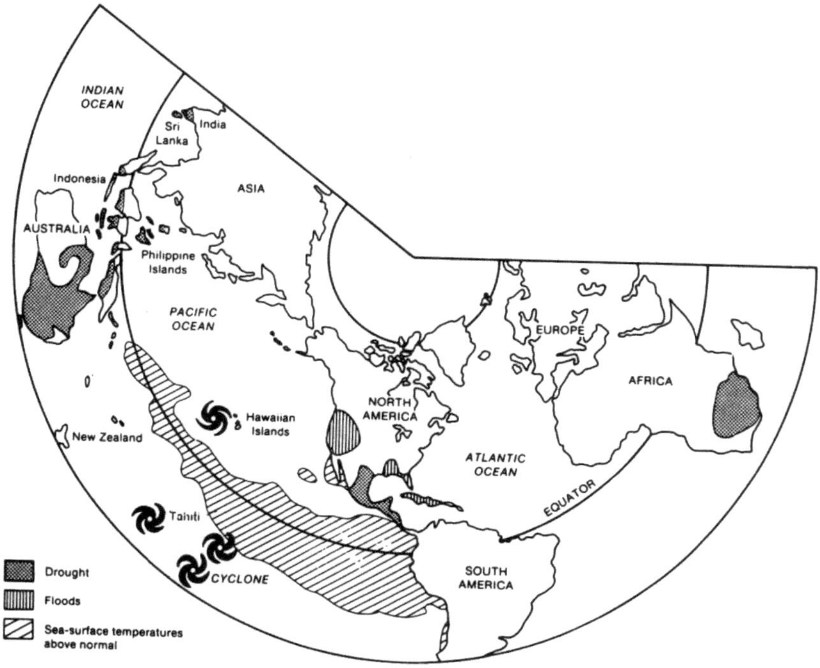

Figure 11: Regions where droughts and floods occurred associated with the 1982-83 El Niño. (After Canby 1984).

sorbs ultraviolet (UV) from the incoming solar radiation. Without the ozone layer the UV intensity at ground level would be so high as to render life impossible for many species, including *homo sapiens*. Even a modest reduction in the concentration of the layer would lead to significant increases in skin cancers and to various other medical problems as well. The other effect of O_3 in the stratosphere is that, by absorbing ultraviolet radiation, it increases the temperature of the stratosphere. Thus a reduction in O_3 concentration would lead to a lowering of temperature in the stratosphere with consequential effects on weather and climate.

Measurements of concentrations of O_3, as well as of the chemical species that destroy the O_3, in the stratosphere have to be made by remote instruments, either ground-based or satellite-flown systems. In the context of this book it is perhaps interesting to recall the story of the discovery of the notorious "ozone hole" over the Antarctic. The first thing to say about the ozone hole is that it is not a complete hole, corresponding to a zero concentration of O_3, but it does correspond to a massive reduction in the O_3 concentration. The second point to make is that the hole is not permanent. O_3 concentrations vary enormously throughout the year. What is meant by the ozone hole over Antarctica is a rather massive reduction in the O_3 concentration over a large area of Antarctica at the end of the Antarctic winter. The hole was first discovered by scientists of the British Antarctic Survey working at the Halley Bay base in Antarctica in 1982; its existence was checked in 1983 and 1984 and eventually published in 1985

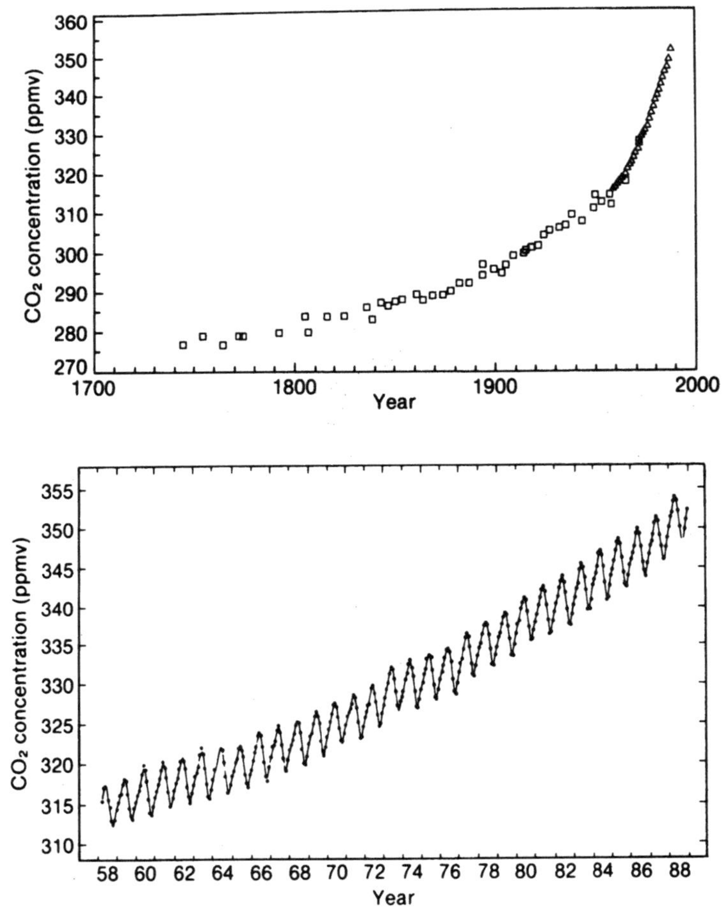

Figure 12: (a) Atmospheric CO_2 increase in the past 250 years, as indicated by measurements on air trapped in ice from Siple Station, Antarctica (squares: Neftel *et al.* 1985; Friedli *et al.* 1986) and by direct atmospheric measurements at Mauna Loa, Hawaii (triangles; Keeling *et al.* 1989). (b) Monthly average CO_2 concentration in parts per million of dry air, observed continuously at Mauna Loa, Hawaii (Keeling *et al.* 1989).

(Farman *et al.* 1985). In October 1984 Farman *et al.* had found a depletion of the ozone layer over Halley Bay of 30%, see Figure 14. By October 1987 the depletion had reached 50%. The Halley Bay work was done by ground-based measurements using a Dobson spectrophotometer which is a standard instrument widely used for measuring the integrated O_3 concentration between the ground and the top of the atmosphere. However, the Nimbus-7 satellite which was launched in 1978 carried two instruments, the Total Ozone Mapping Spectrometer (TOMS) and the Solar Backscatter Ultraviolet (SBUV) experiment, and the data from these instruments should easily have detected such a large depletion in the O_3 concentration in Antarctica. It was discovered, after the Halley Bay work was published in 1985, that at the time the programs for processing

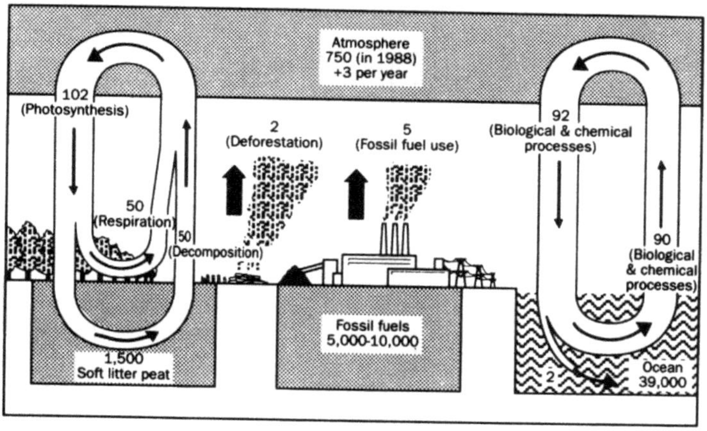

Figure 13: The carbon cycle with quantities of carbon in 10^9 t y^{-1} (fluxes). (From Leggett 1990).

the Nimbus-7 data were developed O_3 concentrations lower than about 200 Dobson units had not been observed in "ground truth" data. (The Dobson unit is a total amount of O_3 in the atmospheric column that, when converted to 0 °C and sea-level atmospheric pressure, would be 10^{-5} m thick). The data received from Nimbus-7 at the Goddard Space Flight Center were processed automatically and any measurements of less than 180 Dobson units were regarded as anomalous and reset to 180 Dobson units. Fortunately the raw Nimbus-7 data had been kept and after the publication of the British Antarctic Survey's results the TOMS data were re-examined. Those results were confirmed and indeed from the TOMS data it was seen that the depletion was not localised at Halley Bay but extended over the whole of Antarctica.

5.3.3 Development of land areas

As a result of media coverage, it is the increasing concentration of CO_2 and the depletion of O_3 in the stratosphere that spring most immediately to mind when we think about possible effects of human activities on weather and the climate. A few decades ago, at the time of extensive above-ground nuclear testing programmes, we were concerned about the effect of those nuclear explosions on the weather and, possibly, the climate. But there are many other factors which should be considered and which, in the long term, may be very important.

Human beings and domestic animals have been affecting the surface of the Earth for a long time. Tickell quotes something like 20% of the total area of the continents as having been drastically changed as a result of human activities. The cutting down of trees for settlement and agriculture, the slash-and-burn method of cultivation in primitive societies, the overgrazing of land by such animals as goats and cattle, and the overuse, impoverishment and erosion of top soils have in the past affected the heat and water balance in various areas. The amount of solar radiation absorbed by the grass, crop, or in some cases desert land is less than that absorbed by the forests which

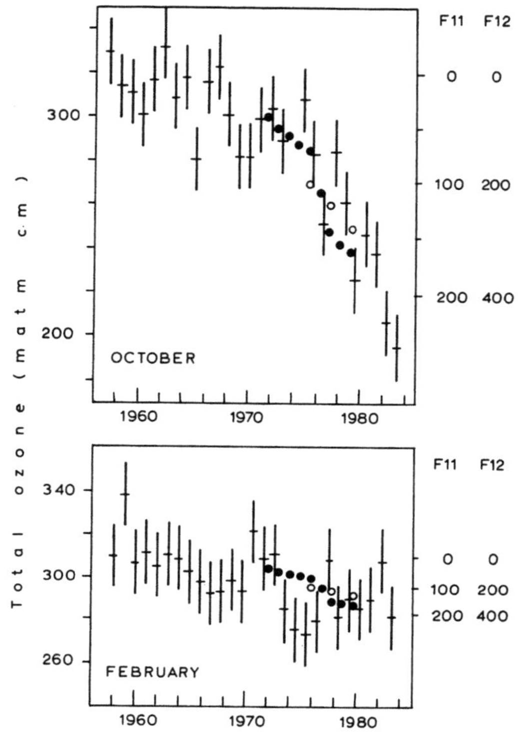

Figure 14: Monthly means of total O_3 at Halley Bay and Southern Hemisphere measurements of F-11 (• pptv (parts per thousand by volume) $CFCl_3$) and F-12 (o pptv CF_2Cl_2). (From Farman *et al.* 1985).

were there before. More energy is reflected back into space, less moisture is evaporated and so less rain is produced and of the rain which does fall more is run off the land. The inevitable increase of dust blown up from the surface can produce effects similar to those of volcanic eruptions or sandstorms.

From the point of view of food production, or general agricultural production, the most important effects have been a decline in rainfall in some areas. Tickell, again taking the western historian's approach, cites the progressive aridity in historical times of the swath of land from the Mediterranean to northern India. This area was once covered by dense forest and later became the site of successive civilizations; its increasing aridity seems to have been caused mostly by human destruction of the natural environment. Most of the grain-growing parts of the area which once supported large populations, including those of the Roman Empire, are now largely scrub or desert. Estimates are that about 7% of the Earth's surface is man-made desert and that the process of progressive desertification is continuing. This inevitably contributes to some extent to changing weather and climate.

In more recent times other human activities have become important. For instance there are now situations in which human intervention occurs to modify the weather rather directly, for instance in the case of rainmaking by cloud seeding. The short-term local effects may be desired and beneficial. The effects elsewhere, or in the longer term, may be less obvious and may not necessarily be beneficial. Another human influence is the interference with the hydrology of the Earth. The creation of artificial lakes for irrigation or hydro-electric purposes will affect the local heat balance and the evaporation rate. Changing the flow of rivers for irrigation purposes and the consequent change in land use and land cover will also affect the local heat balance and evaporation rate as well. The development of large cities in recent years is also important. Large cities are well known to form heat islands with temperatures significantly higher than their surroundings. This arises from two factors, (i) the replacement of vegetated surfaces by more highly reflecting road and roof surfaces and (ii) the heat generated by human activities within the cities. Locally the cities' effects on the heat balance will affect the weather although, in global terms, their contribution to the heat balance is relatively small (but still a good deal more significant than the flap of a butterfly's wings).

5.3.4 Other greenhouse gases

We have already mentioned the substantial increase in CO_2 concentration in the atmosphere, as a result of human activities, and the greenhouse effect associated with the atmospheric CO_2. In terms of concentration, CO_2 is by far the most important of the greenhouse gases, see Table 1. However, CO_2 is not the only important product of the combustion of fossil fuels. Other gases, principally SO_2 and oxides of nitrogen (denoted generally by NO_x) are produced by the combustion of fossil fuels. Some fuels are cleaner than others in terms of the pollution they produce. Thus gas and oil can be treated so as to remove sulphur before they are burned and therefore to prevent SO_2 being produced when they are burned; coal, being a solid, cannot be treated in this way before it is burned and so expensive scrubbing facilities have to be used to remove the SO_2 from the gases produced by the combustion process. The production of oxides of nitrogen is more difficult to prevent because the nitrogen comes from the air used in the combustion process and not from the fuel. When considering the combustion of fossil fuels to produce greenhouse gases, particularly NO_x, which lead to the destruction of O_3 in the stratosphere, we should perhaps give particular attention to high-flying aircraft since they are injecting these gases *directly* into the stratosphere whereas other combustion processes occur at ground level and produce gases which subsequently travel up to the stratosphere. It is interesting to note the story of the role of high-flying aircraft among the bogey men of ecologists. At the time of the development of the supersonic airliner Concorde it was widely argued that these, and similar aircraft, would lead to a dramatic reduction in the concentration of stratospheric O_3. Subsequently it was argued that the US Shuttle programme would be responsible for producing HCl in the stratosphere which would reduce the stratospheric O_3 concentration. The story, as recounted along with the story of Concorde in chapter 2 of the book by Gribbin (1988), is well worth reading. Basically the supersonic airliner threat came to be regarded as unreal and the Shuttle problem came to be regarded as relatively insignificant in comparison with the CFC problem. The figure given by Houghton *et al.* (1992) for the contribution of high-flying aircraft to the atmospheric concentrations of oxides of nitrogen is relatively small, see Table 2.

Parameter	CO_2	CH_4	CFC-11	CFC-12	N_2O
Pre-industrial atmospheric concentration (1750-1800)	280 ppmv[2]	0.8 ppmv	0	0	288 ppbv[2]
Current atmospheric concentration (1990)	353 ppmv	1.72 ppmv	280 pptv[2]	484 pptv	310 ppbv
Current rate of annual atmospheric accumulation	1.8 ppmv (0.5%)	0.015 ppmv (0.9%)	9.5 pptv (4%)	17 pptv (4%)	0.8 ppbv (0.25%)
Atmospheric Lifetime[3] (years)	(50-200)	10	65	130	150

1. Ozone has not been included in the table because of lack of precise data.

2. ppmv = parts per million by volume; ppbv = parts per billion by volume; pptv = parts per trillion by volume.

3. For each gas in the table, except CO_2, the "lifetime" is defined here as the ratio of the atmospheric content to the total rate of removal. This time scale also characterises the rate of adjustment of the atmospheric concentrations if the emission rates are changed abruptly. CO_2 is a special case since it has no real sinks, but is merely circulated between various reservoirs (atmosphere, ocean, biota). The "lifetime" of CO_2 given in the table is a rough indication of the time it would take for the CO_2 concentration to adjust to changes in the emissions. (From Houghton *et al.* 1990).

Table 1: Summary of Key Greenhouse Gases Influenced by Human Activities

Natural	
• Soils	5-20
• Lightning	2-20
• Transport from Stratosphere	~ 1
Anthropogenic	
• Fossil Fuel Combustion	24
• Biomass Burning	2.5-13
• Tropospheric Aircraft	0.6

Table 2: Estimated Sources of Nitrogen Oxides (Tg N per year) (From Houghton *et al.* 1992)

The effect of NO_x as far as O_3 is concerned actually appears to depend on the height in the atmosphere. At low altitudes the oxides of nitrogen favour the production of O_3 whereas at higher altitudes they lead to the destruction of O_3. Apart from their role as greenhouse gases, SO_2 and NO_x are also undesirable products of combustion because through various reactions with water they produce mineral acids; these lead to acid rain which has catastrophic effects on the vegetation in some areas. As far as SO_2 is concerned, it should be pointed out that not all the gaseous sulphur compound emissions into the atmosphere arise from anthropogenic causes. SO_2 is liberated by volcanoes, H_2S and DMS (dimethyl sulphide) are liberated by plants and DMS is liberated by plankton in the ocean. The relative magnitudes of the emissions from these various sources are indicated in Table 3. It has been suggested (Charlson *et al.* 1987) that, apart from producing acid rain, the DMS emissions by plankton lead to sulphate particles which constitute hygroscopic nuclei for cloud drops and significantly alter the cloud patterns and, thereby, the radiation balance (for an early account see Slingo 1988).

5.3.5 Human activities and the greenhouse effect

We have alluded to the greenhouse effect on several occasions in this section; the time has come to be a little more specific. First, from the point of view of life on Earth, the greenhouse effect is good, not bad. The average temperature of the surface of the Earth is 15 °C and in the absence of any greenhouse gases in the atmosphere it would be reduced by 33 °C, i.e. to -18 °C. -18 °C may not sound too cold and many of us will have experienced that temperature briefly. But as an average temperature it would make the Earth a rather inhospitable place and any semblance of modern agriculture would only be possible in a very few locations. In past ice ages the drop in average temperature has been less than 10 °C (not 33 °C). However, what is worrying people at the moment is the possibility of a substantial global warning as a result of changes in the greenhouse effect induced by human activities. We have touched on

Anthropogenic emissions (mainly SO_2)	70-80
Biomass burning (SO_2)	0.8-2.5
Oceans (DMS)	10-50
Soils and plants (DMS and H_2S)	0.2-4
Volcanic emissions (mainly SO_2)	7-10

Table 3: Estimated Sources of Short-lived Sulphur Gases (Tg S year) (From Houghton *et al.* 1992)

many of these already in this section. These are the production of CO_2 and N_2O (one of the oxides of nitrogen we denoted by NO_x) from combustion processes and there is the release of CFCs, principally CFC-11 and CFC-12. In addition to this there is methane, CH_4, which is produced by a whole variety of sources, some of which are natural and some of which are not, see Table 4. The main greenhouse gases (except O_3), their concentrations, their rate of increase in concentration and their lifetime in the atmosphere are given in Table 1. The increases in the concentrations of these gases since the Industrial Revolution are shown in Figure 15. Note that O_3 is not included in this table because of lack of precise data. A discussion of the problem of obtaining global data over a long timescale for O_3 concentrations will be found in Sections 1.6 and 1.7 of Houghton *et al.* (1990). SO_2 is also not included in Table 1 because it does not survive very long as a gas. It does, however, react with water to produce sulphuric acid aerosols; if present in sufficient quantities in the stratosphere, where the half-life is about 1 year, these aerosols can significantly affect the radiation budget and thus alter the greenhouse effect. The whole question of aerosol arising from SO_2 from fossil fuel combustion, from SO_2 and dust from volcanic eruptions and from dust from other sources is important but very difficult to quantify. What is more important than the actual concentrations, or emissions, of greenhouse gases is their actual contribution to global warning. We shall see later, in chapter 6, that for climate modelling one simply represents all the greenhouse gases by an amount of CO_2 that gives rise to an equivalent warming effect to that of all the gases.

We have mentioned briefly the question of above-ground nuclear explosions in the era when such tests were common. The effects of those tests were probably not very different, as far as weather and climate are concerned, from those of volcanic eruptions. The effects depend on the size and altitude of the explosion and the amount of dust and other particles released into the atmosphere. A major nuclear war would be much more disastrous. It has been suggested that the use of less than 1% of available nuclear weapons on 100 or 200 cities would surround the Earth with a thick cloud of dust and soot and plunge the world into such darkness that it would create a prolonged Arctic or nuclear winter. The survival of our species and other higher organisms from that

	Annual Release ($Tg\ CH_4$)	Range ($Tg\ CH_4$)
Source		
Natureal Wetlands (bogs, swampes, tundra, etc)	115	100-200
Rice Paddies	110	25-170
Enteric Fermentation (animals)	80	65-100
Gas Drilling, Venting, transmission	45	25-50
Biomass Burning	40	20-80
Termites	40	10-100
Landfills	40	20-70
Coal Mining	35	19-50
Oceans	10	5-20
Freshwaters	5	1-25
CH_4 Hydrate Destabilization	5	0-100
Sink		
Removal by soils	30	15-45
Reaction with OH in the atmosphere	500	400-600
Atmospheric Increase	44	40-48

Table 4: Estimated Sources and Sinks of Methane. (From Houghton *et al.* 1990)

could not be assured. Even when the dust eventually settled and the ozone layer was restored, the devastation of the Earth's surface might increase its reflectivity to solar radiation leading to a continuation of glacial conditions over large areas. The long-term consequences of a nuclear winter cannot be assessed; but they could well be even more devastating for life than the direct effects of the increased radioactivity.

6 Politics and Climate Change

This chapter, this book and the summer school on which this book is based are all concerned with the scientific aspects of climate change. It should, however, be appreciated that this is only part of a wider activity within the World Climate Programme (WCP). The WCP was established in 1979 following the first World Climate Conference which had been convened in Geneva by the World Meteorological Organization (WMO) and the United Nations Environment Programme (UNEP). A Second World Climate Conference was held towards the end of 1990 to provide an end-of-the-decade review of the World Climate Programme. We have already mentioned in earlier sections the change which has occurred in recent years in that global change has now become a major public and political issue. We have also already mentioned the Intergovernmental Panel on Climate Change (IPCC), which was established in 1988 by WMO and

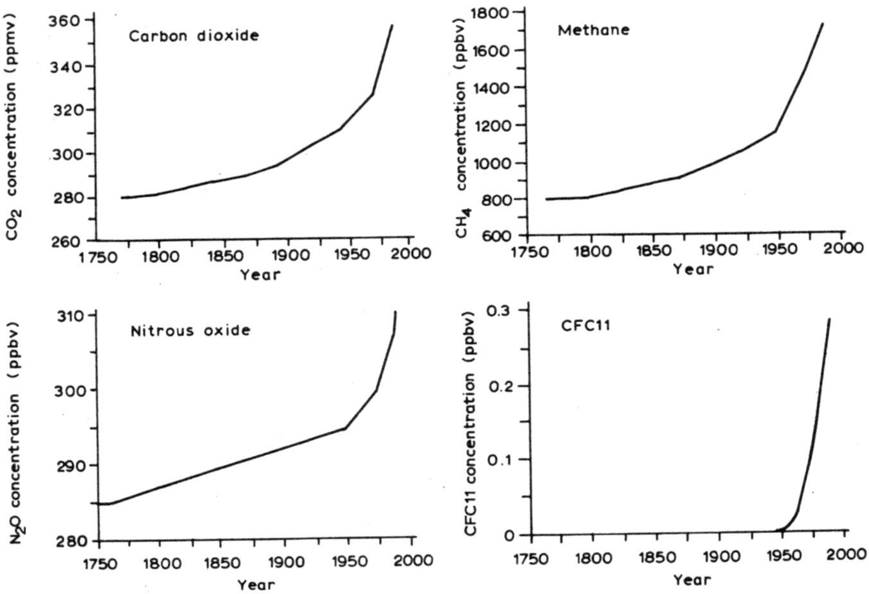

Figure 15: The increases in concentrations of (a) CO_2, (b) CH_4, (c) N_2O and (d) $CFCl_3$ since the industrial revolution; CFCs were not present before the 1930s.(From Houghton *et al.* 1990).

UNEP and we have made extensive reference to that panel's first report (Houghton *et al.* 1990). Another important conference which is worth mentioning, although it was not of the same international status as the Second World Climate Conference, was held in Washington, DC, in 1989 and its proceedings, too, have been published (Ginsberg and Angelo 1990) and, of course, there was the UNCED Conference in Rio de Janeiro in 1992. There is also a chapter in this book on Politics and Climate Change which is based on discussions among participants at the summer school on which this book is based (see chapter 28).

The Second World Climate Conference was thus scheduled for October-November 1990 so as to follow the publication of the first IPCC Report; the Proceedings of this Conference have now been published (Jäger and Ferguson 1991). It is instructive to look at the balance of papers presented and published in those Proceedings and to summarise briefly the final Conference Statement. Following the texts of the various opening addresses, there are rather more than 200 pages giving overall reports of various aspects of the World Climate Programme, summarising the IPCC's first Report and considering some specific aspects of climate research. There are then nearly as many pages again devoted to papers on climate change and socio-economic activities; the topics considered include water, agriculture and food, oceans and fisheries, energy, land use and urban planning, human disease and population, the environment and development, forestry, risk management and general economic considerations. There are also reports from various task groups and consultation groups on these and various other socio-economic matters. Thus, it can be seen that rather less than half of the

Conference Proceedings was actually concerned with the strictly scientific aspects of climate change and rather more than half was concerned with the political, economic and social aspect of climate change.

The Conference Statement runs to seven pages and is well worth reading, but it is too long to reproduce here (see Jäger and Ferguson 1991, p 497). I would, however, like to quote the summary of the Conference Statement verbatim:

"1. Climate issues reach far beyond atmospheric and oceanic sciences, affecting every aspect of life on this planet. The issues are increasingly pivotal in determining future environmental and economic well-being. Variations of climate have profound effects on natural and managed systems, the economies of nations and the well-being of people everywhere. A clear scientific consensus has emerged on estimates of the range of global warming which can be expected during the 21st century. If the increase of greenhouse gas concentrations is not limited, the predicted climate change would place stresses on natural and social systems unprecedented in the past 10,000 years.

2. At the First World Climate Conference in 1979, nations were urged "to foresee and to prevent potential manmade changes in climate that might be adverse to the well-being of humanity". The Second World Climate Conference concludes that, notwithstanding scientific and economic uncertainties, nations should now take steps towards reducing sources and increasing sinks of greenhouse gases through national and regional actions, and negotiation of a global convention on climate change and related legal instruments. The long-term goal should be to halt the build-up of greenhouse gases at a level that minimizes risks to society and natural ecosystems. The remaining uncertainties must not be the basis for deferring societal responses to these risks. Many of the actions that would reduce risk are also desirable on other grounds.

3. A major international observational and research effort will be essential to strengthen the knowledge-base on climate processes and human interactions, and to provide the basis for operational climate monitoring and prediction."

7 Conclusion

The purpose of this introductory chapter was to set the scene by describing, in very general terms, the main features of climate change resulting both from natural causes and from human activities. It should provide the background to the subsequent discussion of climate modelling. From that discussion (in chapter 6) we shall see, in turn, that one of the general conclusions about the present knowledge of climate change is that more scientific data are needed. It will be our purpose in this book to examine the extent to which remote sensing techniques can contribute to these data requirements.

References

Barraclough, G., 1989, The Times Atlas of World History (London: Times Books).

Broecker, W. S., and Denton, G. H., 1990, What drives glacial cycles? *Scientific American*, **262**, 43-50.

Canby, T. Y., 1984, El Niño's ill wind. *National Geographical Magazine*, **165**, 144-183.

Charlson, R. J., Lovelock, J. E., Andreae, M. O., and Warren, S. G., 1987, Oceanic phytoplankton, atmospheric sulfur, cloud albedo and climate. *Nature*, **326**, 665-661.

Croll, J., 1867, On the change in the obliquity of the ecliptic, its influence on the climate of the polar regions and on the level of the sea. *Philosophical Magazine*, **33**, 426-445.

Farman, J. C., Gardiner, B. G., and Shanklin J. D., 1985, Large losses of total ozone in Antactica reveal seasonal ClO_x/NO_x interaction. *Nature*, **315**, 207-210.

Foukal, P., and Lean, J., 1990, An empirical model of total solar irradiance variation between 1874 and 1988. *Science*, **247**, 556-558.

Friedli, H., Loetscher, H., Oeschger, H., Seigenthaler, U., and Stauffer, B., 1986, Ice core record of the $^{13}C/^{12}C$ record of atmospheric CO_2 in the past two centuries. *Nature*, **324**, 237-238.

Ginsberg, I. W., and Angelo, J. A., 1990, *Earth Observations and Global Change Decision Making, 1989: A National Partnership* (Melbourne: Krieger).

Gribbin, J., 1978, *The Climate Threat* (Glasgow: Fontana/Collins).

Gribbin, J., 1988, *The Hole in the Sky - Man's Threat to the Ozone Layer* (London: Corgi).

Hall, N. (Editor), 1991, *The New Scientist, Guide to Chaos* (London: Penguin).

Houghton, J. T., (Editor), *The Global Climate* (Cambridge: Cambridge University Press).

Houghton, J. T., 1991, The Bakerian Lecture 1991, the predictability of weather and climate. *Philosophical Transactions of the Royal Society of London*, **A337**, 521-572.

Houghton, J. T., Callander, B. A., and Varney, S. K., 1992, *Climate Change 1992 - The Supplementary Report to the IPCC Scientific Assessment* (Cambridge: Cambridge University Press).

Houghton, J. T., Jenkins, G. J., and Ephraums, J. J., 1990. *Climate change - the IPCC scientific assessment* (Cambridge: Cambridge University Press).

Jäger, J., and Ferguson, H. L., 1991, *Climate Change: Science, Impacts and Policy. Proceedings of the Second World Climate Conference* (Cambridge: Cambridge University Press).

Jones, R., Russell, and Wigley, T. M. L., 1989, *Ozone depletion: health and environmental consequences* (Chichester: John Wiley).

Keeling, C. D., Bacastow, R. B., Carter, A. F., Piper, S. C., Whorf, T. P., Heinmann, M., Mook, W. G., and Roeloffzen, H., 1989, A three dimensional model of atmospheric CO_2 transport based on observed winds.

Kondratyev, K. Ya., 1992, *Global Climate American Geophisical Monograph, No. 55*, edited by D. H. Peterson (Washington D.C: American Geophysical Union), pp. 165-236.

Leggett, J. K., 1990, Global Warning. *The Greenpeace Report* (Oxford: Oxford University Press).

Lorenz, E., 1963, Deterministic non-periodic flow. *Journal of the atmospheric sciences*, **20**, 130-141.

Mandelbrot, B. B., 1982, *The Fractal Geometry of Nature* (Salt Lake City: W. H. Freeman).

Milankovitch, M. M., 1920, Theorie mathématique des phenomènes thermiques produits par la radiation solaire. Academie Yougoslave des Sciences et des Arts de Zagreb (Paris: Gauthier-Villars).

Neftel, A., Moor, E., Oeschger, H., and Stauffer, B., 1985, Evidence from polar ice cores for the increase in atmospheric CO_2 in the past two centuries. *Nature*, **315**, 45-47.

Parry, M. L., 1990, *Climate change and world agriculture* (London: Earthscan).

Pearce, F., 1992, The green diplomat. *New Scientist*, on 21 March 1992 (no 1813), 38-40.

Roan, S. L., 1989, Ozone crisis: the 15-year evolution of a sudden global emergency (New York: John Wiley).

Slingo, A., 1988, Can plankton control climate? *Nature*, **336**, 421.

Tickell, C., 1977, *Climatic Change and World Affairs* (Harvard: Center for International Affairs).
Wilson, R. C., and Hudson, H. S., 1988, Solar luminosity variations in solar cycle 21. *Nature*, **332**, 810-812.

Remote Sensing, Systems and Data

Robin A Vaughan

1 Introduction

To understand global environment change, it is essential to understand and to document how the Earth works as a system. The scientific success of this understanding depends on the integration and management of numerous data sources. Unlike many scientific conundrums of the past, this one will not require the discovery of new laws of nature: the climate works according to the established rules of physics, chemistry and biology. What is needed is the data to make clear what is going on where. The continuous monitoring of the environment and the investigation of both natural and man-induced changes present a very complex and fascinating challenge. No one nation has the resources to generate the comprehensive understanding required, and many international, collaborative programmes have been established to investigate phenomena and processes on a global scale and to coordinate work carried out on a local scale.

One tool now available to environmental scientists is the science - or black art - of remote sensing. The global, continuous and homogeneous information obtained from satellites make them an obvious choice to be used as observation platforms, but their synoptic coverage is offset by their coarse spatial, spectral and temporal resolutions. The interpretation of global phenomena often requires an understanding of local features which can only be monitored from the more proximal viewpoint of aircraft or other platforms. A great deal of work is already being carried out using existing systems, which were probably designed for very different purposes, but programmes such as NASA's "Mission to Planet Earth" and platforms such as those planned for the Earth Observation System (EOS) and instruments such as SeaWiFS have now been designed specifically to contribute to our understanding of climate change on a global scale.

The Earth's environment is a complex system involving interactions between the atmosphere, the land and the oceans, each of which comprises a complex substructure of components, atmospheric chemicals, vegetation, animals, ice, snow, etc. Monitoring from space gives an overview - but of what? An interpretation of the parameters so measured requires an understanding of the phenomena being studied, and this can only be obtained from often long and painstaking fieldwork.

NATO ASI Series, Vol. I 24
Remote Sensing and Global Climate Change
Edited by Robin A. Vaughan and Arthur P. Cracknell

Monitoring the environment implies not only periodic observation but also an attempt at analysing the results. One is in with a reasonable chance of achieving this in the case of local events, those which are caused locally and felt locally over a reasonably short time scale. Examples might be the contamination of a water supply, the emission of toxic fumes from an industrial complex or an oil spillage. A similar appreciation of long term global events requires a completely different approach. On the one hand there is the enormous amount of data that would be required to be analysed if it were available, but on the other hand there is the more important question of what data should be collected and analysed. It is not always clear where the source is, when the event occurred or what intermediate processes may be involved. The *effect* of global warming, which would affect every country and every inhabitant of the world to some extent, would not be difficult to detect. But monitoring, in the context of the environment, carries with it an element of predictive ability. We do not just want to be able to say "it has happened" but we need to be able to detect impending changes in time to take remedial action. We do not require sophisticated satellite monitoring to see the *effect* of algal blooms in the Adriatic - fishermen and those in the tourist industry can point to the 100 million dollars lost in 1989. The 170,000 inhabitants of the Maldives, no part of which is more than 2 m above present sea level, will soon know if global warming raises that sea level. The 20 million people who die each year from malnutrition and dirty water do not need to be reminded of the problem. What they need are solutions to existing problems and prevention of future problems.

The causes and effects of natural events may have completely different scales of both space and time. The ozone hole over the Antarctic is caused not by penguins but by people living in London, Moscow or New York. A volcanic eruption may last only a few minutes but can devastate plant and animal life for decades. An earthquake lasts for a few seconds, but may be the culmination of stresses set up over thousands or millions of years. The El Niño effect causes billions of dollars worth of damage and loss of life along the whole Pacific seaboard.

One well established branch of science that has over many years developed the use of monitoring and prediction is that of meteorology. Great quantities of data from all over the world are analysed using complex mathematical models with a greater or lesser degree of success. There are two problems. One is with the complexity of the data input, and the second is with the ability of the model to predict over time scales greater than a few days. European weather, although felt locally, is only the end product of a series of events extending over vast areas. Heat is exchanged between the atmosphere and the North Atlantic, each square kilometre of which gives off as much energy as a nuclear power station (Rhind 1989). But the heat energy in the North Atlantic was originally absorbed from sunlight in the tropical Pacific and has been carried by ocean currents through the Drake Passage and up into the Atlantic. Even if suitable data were available, the predictive ability of a model depends on its sophistication and on the divergent nature of the phenomena. Small perturbations early on in such an analysis can produce widely varying scenarios. Meteorologists now limit their predictions to only a few days because their longer-range forecasts were unreliable.

One very serious consideration is what data and how much data should be acquired and saved. Is global coverage of all parameters really necessary? Which parameters should be measured and on what spatial and temporal scale? What accuracy is necessary? Since natural events are so complex, perhaps every single parameter should be recorded and used as input to some mathematical model of the Earth's environment (or perhaps of the Universe?). But in the real world, compromise is always required. Unfortunately, one often only finds out retrospectively the value of certain data, so the temptation is to store it just in case - but that probably will not be possible for this situation. Even though great advances in storage media (CD ROM, video disc, etc) and in data compression techniques are continuing to be made, and computer speeds and memories are increasing all the time, perhaps by 50% per year, they are at present still rather insignificant compared to the scale of the problem. By 1995/6, NASA plans to collect up to one Tbyte (10^{12} bytes) of data per day. To hold all of the 1:50000 scale topographic map coverage of Europe would probably require about 3 T of storage. A CD-ROM, even though it may now cost only $2, holds only about 600 Mbytes of data!

The huge quantities of data which will be obtained over the next few years from land, sea and air, and the diversity of their format will require special management systems. Field data, remotely-sensed image data, conventional map data and socio-economic data will need to be coregistered and analysed. Not only are their formats different, but probably so are their accuracy, subjectivity, resolution, classification and stratification. Differences in culture, political regimes, language, population and development will affect the uniformity of data from different parts of the world and provide logistical problems in assimilation into a global database.

Even if the problem of data collection, storage and management could be overcome, there is still the problem of what to do with it. For example, even though it might be possible to *monitor* the development of algal blooms, without local, micro-scale studies we can never really *understand* the *reasons* why the bloom is occurring. Field work and local investigation is still an essential input to any database. But monitoring in itself cannot solve problems. It can point them up, and perhaps point the way, but it does not replace the need for action. Remote sensing may be one of the tools to help us on the road to a better management of our planet, and we would be foolish to ignore the opportunities it presents, but the solution, if there is one, does not lie in the hands of the scientists but of the politicians who need to take their decisions with a sound scientific base.

This chapter will hopefully set the scene for those following in terms of the practicalities of remote sensing. After a brief survey of the principles of remote sensing, an attempt will be made to summarise the types of data sets required in the study of climate, the existing satellite systems and scanners which can provide such data, the Earth Observation (EO) programmes that are planned and the management and manipulation of the data so obtained. The bibliography includes a number of sources of information that the reader may find useful. Just before the Summer School, the British National Space Centre published a report for the Committee on Earth Observation Satellites (CEOS) which sets out the existing and proposed earth observation satellite missions over the next 15 years and describes how these programmes provide information that is crucial to the success of global environmental programmes. This report was produced specifically for delegates attending the UNCED Conference in Rio de Janeiro which took place in June 1992. This is an excellent, comprehensive compilation of data. Subsequent to the Summer School, NASA has published a new edition of their reference handbook on the Earth Observing System (EOS) Programme (Asrar and

Dokken, 1993). This also is a useful compendium of information about platforms and instruments especially designed for long-term global observations of our world. Many articles, newsletters, scientific papers and press reports have been unashamedly used in the preparation of the lectures on which this chapter is based, but the authorship of some of them is unknown and some may not be included in the reference because their origin has been lost with the passage of time. To the authors of these, many apologies. Some of this material was presented at an international workshop held in Leningrad (Vaughan, 1990).

2 Data Requirement

Climate is not just "long term average weather". It is a complex phenomenon which determines the equilibrium state of the Earth's environment. It is governed not only by what happens in the atmosphere, but in the oceans, the cryosphere (glaciers, sea-ice and continental ice caps), the geosphere (the earth's solid surface) and the biosphere (living organisms in the oceans and on the land). The interaction between these various spheres are difficult to predict, not least because their respective processes occur on widely different time scales. The typical equilibrium response times of the climate system's various elements range from a single day to many centuries.

Solar radiation is the only significant source of energy driving the climate system, and the amount of warming depends in part on the nature of the Earth's surface. Ocean and land surfaces warm at different rates, and land covered by vegetation absorbs and reflects solar energy differently from deserts or ice-caps. The radiation reemitted by the surface is absorbed by some of the gases present and heats the lower atmosphere. The oceans are the major regulator of climate. Water circulates within the oceans because of differences in temperature and density. The surface absorbs not only heat but gases such as CO_2 from the atmosphere. When surface currents eventually descend they transport this heat and gas down to the regions of "deep-water formation" where the heat and gases become trapped within deep-water currents for anything up to one thousand years. The oceans therefore modulate the climate by storing heat and CO_2 etc. for long periods of time and changes in circulation patterns and regions of deep-water formation may have a significant influence on climate changes over the centuries.

The atmosphere is very much more mobile than the oceans, and interacts with the surface layers of both land and water. Energy and matter can be exchanged across this boundary, thus coupling the various systems. Perhaps, therefore, it is the atmosphere which plays the most important part in climate changes which take place over human timescales. Although they necessarily have focussed the public mind on the problem, it is not just anthropomorphic changes which may produce climate change. Slow, natural changes in atmospheric content will have concomitant effects, but so also will sudden changes associated with natural disasters. Analysis of measurements of the Earth's radiation budget from satellites revealed that the eruption of Mount Pinatubo in the Philippines in June 1991 resulted in a temporary decrease in global surface temperature of approximately 1°F. The increase in solar radiation reflected from the Earth, responsible for this cooling, continued through to October and is well-correlated with the gradual spread of the Pinatubo dust cloud over the globe. The relative cooling produced by this is 60% greater than any other climate disturbance since systematic global observations of the Earth's radiation balance began in 1976. It has been suggested that volcanic eruptions can have long-term effects on the jet stream patterns, and that we may still be feeling the effects of the Pinatubo eruption. It is not clear, however, whether the effect of volcanic dust and ash in the atmosphere will cause heating or cooling - it depends very critically on the concentration, size, shape etc of the particles in the dust cloud.

The extent and nature of land-cover also determine the amount of energy reflected back into space without heating the Earth or atmosphere. Any changes in ice extent or vegetation cover could therefore have a significant effect on the climate, both directly due to albedo modification and indirectly due to change in greenhouse gases, water vapour and CO_2, resulting from changes in biological activity and evaporation. Clouds can also affect climate, but in two ways. They could act to cool the earth by reflecting solar radiation back into space or to warm it by reducing the outgoing terrestrial radiation.

Ice and snow are of great importance to the world's climate principally because of two physical properties,: their high reflectivities tend to depress daytime temperatures whilst their high emissivities lead to rapid cooling of air over snow- or ice-covered surfaces at night. The result is that cold weather tends to be prolonged in areas which are already snow- or ice-covered. Satellite observations showed a marked increase in the amount of snow cover in the northern hemisphere from 1972 - 74. The theory that a positive feedback mechanism might operate when the extent of snow and ice cover reaches a critical level led to fears at that time of the imminent onset of a new ice-age. Changes in ice-sheet extent will effect the albedo of the Earth's surface and therefore the energy balance. Some regions, particularly in the Antarctic ice sheet, are thought to be unstable and capable of responding to global warming very quickly. Given the influence of ice sheets on global mean sea levels, their growth or shrinkage is of practical as well as scientific interest.

It is therefore necessary to measure the behaviour of the atmosphere, the oceans, changes in vegetation, moisture, snow-cover and the extent of the polar ice pack and to try to understand the complex interactions between them. Only then would it be possible to predict changes which may occur in the climate system due to changes in the various components, but such prediction will require the understanding of the mechanisms of transient climate change to an unprecedented level of accuracy.

The parameters required to be measured are very varied and cover the fields of atmospheric physics (temperature, humidity, winds, rainfall, radiation balance), atmospheric chemistry (ozone, greenhouse gases, chemical reactions in the atmosphere), oceanography (sea-surface temperatures, circulation, heat transport, marine biosphere, topography and roughness), land (biosphere, vegetation structure and function, ice and snow cover, water cycle, radiative properties and soil moisture) and the solid earth (crustal movements, gravity and magnetic fields).

Transient and long term variations in climate depend on the coupling of energy between the various components of the environment. The atmosphere is the "fast component" - changes take place relatively quickly throughout its volume. The oceans react more slowly, but not as slowly as the solid earth. Energy is exchanged between the various components due to interface fluxes of momentum (e.g. wind stress), radiation, other heat forms (convection, circulation) and water (evaporation) (Morel 1991). Energy is transported spatially, and therefore "cause and effect" may have very different time scales and spatial locations. The European climate, for example, although felt locally, is only the end product of a series of events extending over vast areas. Global monitoring of the various parameters is therefore necessary even for local climate modelling. Global modelling also requires local inputs. A consideration of data requirements is therefor a very complex issue (see Chapter 6).

3 Remote Sensing Inputs to Climate Models

Earth observation and environmental monitoring from satellites and aircraft (remote sensing) has been carried out for many years. The satellites which continually orbit the earth, and which carry many different types of detector and recorder, could be the technology which will provide both global and regional observations. Indeed, the technological ability is already available to gather much of the data which would be required to set up a global database - the limitation at present is in our ability to store the data, manipulate it and extract information from it.

Climate models, ranging from simple one-dimensional energy balance models to enormously complex 3-dimensional global models requiring enormous computing power, have been developed over the past twenty years. Until recently, effort was concentrated on developing the model of the global atmosphere coupled to the oceans and cryosphere only through prescribing and updating surface parameters such as temperature and albedo from observations (Mason 1991). However, realistic predictions of long term changes in climate, natural or man-made, must involve the atmosphere, oceans, cryosphere, and eventually the biosphere, treated as a single, strongly-coupled and interactive system.

Predicting global climate change requires understanding of the mechanisms of transient climate variations to an extremely high level of accuracy, and modelling requires a knowledge of a multitude of environmental parameters on a global 3-dimensional grid again to great accuracy. Many of the present data sets have been accumulated through painstaking laboratory and field observations, but remote sensing can provide some of these synoptically and continuously. Parameters such as temperature, moisture and wind velocities can be made from space, as can the radiation fluxes at the top of the atmosphere, cloud extent and properties, atmospheric composition and rainfall. Global measurements of sea surface temperatures, topography, and wind, ocean colour and extent of sea ice can be made, and remote sensing can also be used to monitor the time dependent properties of the land surface vegetation and soil moisture.

It is reasonably easy to detect and describe changes in one or more climate parameters, but it is a completely different matter to try to identify the processes which are the causes of such change. There are so many variables changing in time and space that to try to isolate and detect each one to a satisfactory level of accuracy would be impossible. At present, climate parameters cannot be measured accurately enough nor completely enough using existing operational and experimental systems. Determining the causes of variations and then predicting their future course will require a much more complete and accurate database requiring considerable augmentation of space systems.

The data sets required fall into four main groups; meteorological, energy and water flux data, oceanographic and land surfaces. Meteorological observations are perhaps the most developed applications of remote sensing, and such measurements have been operational for many years. Such data have formed the basis of numerical weather forecast models and also contribute to existing climate models. However, more sophisticated measurements will be required for the more complex models now being developed, and provision will be made during the 1990s to provide such data. High spectral resolution sounders, such as NASA's Advanced Infra Red Sounder (AIRS) and active Doppler lidar sounding for three dimensional wind observations are planned to augment presently inferred data from radiometers and single level observations.

The exchange of energy and water flux with the atmosphere couples it to the other components of the climate system. Measurements of long and short wave radiation from the top of the atmosphere, optical properties of clouds and precipitation are required frequently, at least daily, and will require platforms carrying microwave and infrared radiometers with greater spatial and spectral resolutions than those presently available, together with improved rain radar systems.

Remote sensing provides quasi-instantaneous global coverage of surface properties of the oceans. Since sea water is essentially opaque to electromagnetic radiation, most other (e.g. bulk) observations need to rely, to a great extent, on acoustic remote sensing or *in situ* measurements carried out from ships or other floating or submersible platforms. Radar altimetry will provide information about sea surface topography which will in turn provide information about surface currents. Wind stress can be measured using scatterometry. Surface temperatures have been extensively measured using for example AVHRR, and the Along Track Scanning Radiometer (ATSR) on ERS-1 will continue such monitoring. SeaWiFS will soon be providing information about ocean colour, taking over from the Coastal Zone Colour Scanner (CZCS) which provided such information for many years. Other scanners and imaging spectrometers, such as MERIS (Medium Resolution Imaging Spectrometer) are planned to fly on a number of components of EOS. Other ocean paprameters, such as sea ice, can be monitored both by radar and by passive microwave radiometers.

On land, most of the investigations so far made have used visible and infrared data from the high spatial resolution scanners such as the Thematic Mapper on Landsat. For global mapping of change, however, more frequent coverage is required at a lower resolution. Data from AVHRR has provided most of the global information so far acquired, but this instrument was never designed for such applications. The next generation of medium resolution instruments should provide better quantitative information. The essential consideration will be to ensure the continuity and quality of the data that will be obtained so that a quantitative assessment of changes to the environment can be detected and monitored. Most of the instruments envisaged are to be flown on polar orbiting, sunsynchronous platforms but diurnal cycling of radiation and rainfall needs a non-sunsynchronous orbit. It is for all the above reasons that the EO programme for the next 15 years is so complex.

4 Remote Sensing

4.1 Background

For the purposes of the subject matter of this Summer School, the term "remote sensing" will be limited to the use of satellites, and to some extent aircraft, for the purposes of obtaining environmental data relevant to climate studies. This is a subset, of course, of a much wider field of activity ranging from hand-held monitors to space missions and from aerial photography to altimeter measurements. For a more comprehensive treatment, the reader is referred to the many introductory text books now available, a few of which are listed in the bibliography.

The technology of remote sensing has come a long way since the early days of black and white aerial photography, but even so, aerial photographs still form an important component of remotely-sensed data and modern scanners are regularly carried on aircraft. Although remote sensing from aircraft does not have the glamour that may be associated with satellite systems, it nevertheless has its part to play. The higher resolution, flexibility of revisit and greater number of wavebands all may be advantageous. It can provide a valuable source of data on a local scale and complement those obtained from satellites. The use of modern scanners, however, has extended the spectral range available and can provide data in a form which can be handled on a computer. Multispectral scanners have high radiometric resolution in the narrow and simultaneously recorded wavebands. These wavebands span a relatively large portion of the electromagnetic spectrum from ultraviolet wavelengths to thermal infrared, and the data so collected can be stored in digital form for correction and quantitative analysis. New techniques for data management and interpretation have necessarily been developed which has extended the range of applications considerably.

The carrier of information from the sensed object to the detector is usually electromagnetic radiation. The useful range of wavelengths extends from the ultraviolet, through the visible and infrared, thermal and microwave, and into the long radiowave region. The sources of radiation, as well as the method of detection, are different in different parts of the spectrum. The ways in which the radiation interacts with the atmosphere and surface of the Earth also depends on the energy (hence frequency) of the radiation. The choice of wavelength will obviously be determined by the parameter being studied.

4.2 Satellite systems

Since the launch of Sputnik 1 in 1957, there has been an almost exponential growth in the exploration of space. In October 1959, the first satellite to carry a meteorological experiment, Explorer VII, was launched These two classes of satellites were followed by the Television and Infra Red Operation Satellite (TIROS) in 1960 and the Nimbus experimental satellite series in 1963. The second generation operational meteorological satellites, ESSA, began the fully operational weather satellite system in 1966. Also in 1966, the experimental Applications Technology Satellite (ATS-1) was launched. This carried a spin-scan cloud camera (SSCC) and acquired images of the Earth from about 36,000 km, thus being the forerunner of the present operational geostationary satellites.

ATS-3, launched in 1967, carried a colour camera. The third generation of operational meteorological satellites were those named NOAA in the USA and METEOR in the USSR. These are essentially the precursors of the present-day near-polar orbiting operational weather satellite systems.

A satellite system has a number of components. These are the space segment (spacecraft with its sensor and communication facilities) and the ground segment (stations with their data acquisition, processing and distibution facilities). There is a wide range of complexity of both systems. Space systems can be quite simple and dedicated to observing one particular parameter, or may include a large number of operational and experimental instruments. Similarly, a receiving system for data from a geostationary meteorological satellite can be so simple and cheap that it can even be afforded by some schools, whereas the ground segment for Landsat or ERS-1 is hugely expensive and very complex.

Satellites can be classified as either geosynchronous or (near) polar (low) orbiters. The geosynchronous orbiters appear to hover over the same place on Earth (geostationary) because they have the same period of rotation (24 hours) as the Earth. To maintain this orbit, the satellite is placed over the equator at a height of about 36,000 km. This is a particulary useful orbit because a relatively inexpensive fixed antenna can be used for reception of data and because a large area of the Earth's surface can be observed continuousely. Given the horizon that can be seen from that height, five such satellites are required to achieve complete coverage of the world. The current satellites are: Meteosat, operated by EUMETSAT, GOES-East and GOES-West, operated by NOAA in the USA (but when GOES-West failed a year or so ago, GOES-East was moved to take its place and a spare Meteosat was move to replace that. GOES-Next was scheduled to fly in 1989 but is still suffering delays due to instrument development problems), GMS (Japanese) and INSAT (Indian). Russia plans to launch their geostationary component, GOMS, later in 1993. This geostationary orbit is also the one favoured for communications satellites for similar reasons, namely continuous reception using a fixed antenna. A disadvantage of this orbit is the poor coverage it gives of high latitudes due to the curvature of the Earth and the degradation of the signal from ground "clutter" because the line of sight is so close to the horizon. For this reason, the Russians have experimented with satellites in highly elliptical Molniya orbits which appear to hang in space for up to three quarters of their orbit. ESA is considering using three such satellites to give continuous coverage of Europe, and Motorola is considering a worldwide network using a system of 77 of them (the *Iridium Project*).

The polar orbiting satellites are placed so as to circle the Earth in a nearly polar orbit, 700 - 900 km above the surface, with a period of about 100 minutes. As they orbit, the Earth turns below them, so in a given number of orbits, 364 for SPOT, for example, the satellite can see the whole world (which takes 26 days in this case). The exact speed and orbit inclination are carefully chosen such that the width of the swath viewed enables complete global coverage to be obtained. If the orbital characteristics are carefully planned such that, for example, the northbound, ascending path (which usually occurs in daylight) always crosses the equator at about the same local time, this class is referred to as "sunsynchronous" which provides fairly constant illumination conditions preferred by many groups of scientists. Table 1 lists some of the more widely known current polar orbiting satellites relevant to this subject area. Detailed information about these, their payloads as well as other systems can be found in Kramer (1992) and in Matra (1992)

Other more general orbits are, of course, possible and these have been used for manned space vehicles and for special observations or programmes. All of the relevant parameters must be carefully chosen to provide the required data and suitable resolution and coverage. The medium-inclination orbit of Upper Atmosphere Research Satellite (UARS) was chosen specifically because of the UARP focus on the processes influencing ozone depletion. Topex/Poseidon, a joint mission by NASA and CNES, was put into a high, non-sunsynchronous, circular orbit at an inclination of 66°. This orbit reduces the effects of atmospheric drag, filters out the unevenness in the Earth's gravity field and provides satisfactory coverage of the oceans. On the other hand, high inclination polar orbiting satellites are needed to observe phenomena that require relatively detailed observations on a routine basis, often with a constant solar illumination angle. Geostationary satellites are needed to provide continuous monitoring of high temporal resolution processes. The height and speed of a satellite orbit are constrained by Newton's laws of motion and so these are not independently variable. In an attempt to defy Newton, a none too sucessful experiment has recently been attempted to fly a satellite at the "wrong" height by tethering it to a parent satellite. A mechanical problem involving the deployment of such a long tether caused that particular attempt to be aborted, but this type of system is not completely out of the question and further attempts may be made in the future.

There is another way of classifying remote sensing satellite systems. In very general terms they fall into one of two main categories based on their primary application - meteorology or Earth resources, although the recent trend for using meteorological satellites for non-meteorological applications rather blurs this distinction. Other systems, designed for atmospheric research and radiation budget measurements, also exist, although many meteorological systems carry sounding instruments as part of their payload. Meteorological satellites, since they are monitoring the rapidly changing mobile atmosphere, tend to provide images of much larger areas much more frequently than do the others. Since these satellites are in general operational, the data is effectively "free" and less sophisticated receiving stations are required. Even though the sensors on board were designed for very different purposes, it has been found that they are useful for measuring such parameters as sea surface temperatures, vegetation indices etc on a global scale and fairly frequently . On this scale, the low spatial resolution is quite acceptable.

There are many and varied Earth observation satellite systems. Kramer (1992) lists 72 which are operational or approved (ie flying, in manufacture or to be funded with existing resources). CEOS (1992) lists many more which are planned, but are as yet outside existing resources or beyond the normal timescale of approval. Space does not permit a complete listing here of all these satellites, their instruments and objectives, but a few will be discussed in a following section. It should be noted that there is now a change in philosophy taking place, away from the use of large expensive observation platforms with many different types of sensor to smaller, cheaper spacecraft with very specific objectives. For example Seastar (due to be launched late in 1993) is a single-instrument platform developed specifically to measure ocean colour. Its total cost, including launch and five years of operation, will be about $43M and the

Satellite name	launch date	Height (km)	Inclin- ation	Period (mins)	Repeat (days)	Note
Nimbus-7 (USA)	1978	955	99.3°	104	6	Carried TOMS which is still working. CZCS failed in 1986
NOAA series (USA)	1984-92	850	98.7°	101	12 hrs	NOAA-13 launched in 1993. Operational and always two in orbit giving 4 visits per day. Carries AVHRR, TOVS, HIRS/2, SSU and MSU
METEOR series (Russia/ USSR)	1988-91	950 1200	82.5° 82.5°	104 109		The first was launched in 1975. Present craft carry scanning telephotometer and scanning IR radiometer in a non-sunsynchronous orbit at a greater height, also carries TOMS
Feng-Yun (China)	1998-90	901	99°	103		FY-1A failed after 39 days. FY-2 carries VHRR, designed to be compatible with NOAA series
Landsat 4/5 (USA)	1982-84	705	98.2°	99	16	Power supply problems caused the TM on Landsat-4 to fail in 1983. Landsat-5 is still operating and has far exceeded its 3-year life expectancy.
Landsat 6 (USA)	due 1993	705	98.2°	99	16	Will carry enhanced version ETM with an additional 15m panchromatic band
ERS-1 (ESA)	1991	800	98.5°	100	3	Carries active radar instruments as well as along-track scanning radiometer.
SPOT (France)	1986-90	832	98.7°	101	26	Carries two HRV scanners which can point up to 27° from nadir providing stereoscopic capability and a revisit of 3-7 days
UARS (USA)	1992	600	57°	96	various	Orbiting observatory designed to monitor and record data on upper atmosphere composition, temperature, dynamics and global energy inputs
Topex/ Poseidon (USA/ France)	1992	1334	63°	2hrs	10	Circular, non-sunsynchronous orbit. Seen as successor to Seasat. Heart of WOCE programme, carries altimeters and radiometer

Table 1: Some current polar-orbiting satellites

first data will be received 28 months after NASA signed the contract. In contrast, POEM, which is part of the European contribution to Space Station Freedom, will carry a number of environmental monitoring instruments, will cost over $1000M and will not fly in less than 13 years! In an interesting article Elliott (1992) discusses the cost-effectiveness of environmental monitoring. He maintains that while the supplier (the space agencies) determine what the customer (the environmental scientist) wants and controls the budget to provide it, the result is a vicious circle of large projects requiring large launch vehicles, lack of competition and the need for high reliability and large payloads to justify the large and expensive ground infrastructure (he likens this scenario to paying the Ford Corporation to manage the police service because policemen drive cars!) He then produces a very convincing argument that if the customer were to hold the budget, the spiral would be inward resulting in smaller, cheaper, missions albeit with perhaps greater risk of failure. Applying commercial discounting (accepting that data actually have no financial value), more rapid deployment would result in greater cost effectiveness. Be that as it may, most of the missions at present flying or planned are in the hands of the national or international space agencies, and it is also these agencies which tend to coordinate the international EO programmes in global climate monitoring.

4.3 Sensors

Sensors in operational use include both imaging and non-imaging devices. Non-imaging devices usually make a single spot measurement vertically below the platform (e.g. a radar altimeter) or integrate the average value of a parameter from a large area (e.g. radiometer). Even so, the spatial variation of these parameters can be mapped out from successive measurements, and some instruments, such as the radiometers, can be scanned to build up a coarse image. Such scanning radiometers, which are sensitive to emitted thermal radiation, have been flown on a number of satellites to produce temperature maps and snow and ice extent maps.

Perhaps the earliest imaging system was the camera. This produces an instantaneous, spatially-continuous analogue picture on a film medium. This can be very convenient and has been used for many years for aerial photography. One disadvantage, however, particularly when used in a satellite, is that the film has to be physically transported back to Earth. This has been done with most of the manned space missions, and the Russians have, quite successfully, collected ejected film using high-flying aircraft and large "fishing nets". The more common solution is to digitise the signal and telemeter it back to a ground receiving station. A film or video-type picture can be scanned to produce a digital signal, or a scanning system can be used to produce a similar output directly. The principle of operation of all such scanners is basically similar with subtle differences. At any instant in time the detector records the radiance from the instantaneous field of view (IFOV) of the sensor. In an opto-mechanical scanner, the IFOV is swept across the ground perpendicular to the flight direction by means of an oscillating or rotating mirror. The signal is sampled so that the radiance originating from a ground resolution cell and falling onto a detector represents a pixel on the resulting image. The motion of the platform (satellite or aircraft) provides the second dimension and the speed of the mirror and the speed of the platform are arranged so that successive lines of pixels are contiguous. The signal can be dispersed by means of a grating or prism so that different wavelengths fall on different detectors which enables simultaneous multipectral images can be obtained. An alternative method, which does not involve mechanical scanning and is therefore more robust, is to detect a whole row of image elements simultaneously using a row of detectors for each waveband. This has the advantage of increasing the dwell time for each pixel, but suffers from the problem of intercalibrating perhaps 2000 CCD detectors. This method is favoured for the HRV instrument on SPOT, and also for the new generation of imaging spectrometers now being flown on aircraft. In all cases the detector output is amplified and, at least on board a satellite, the signal is digitised and transmitted to Earth.

In the microwave part of the electromagnetic spectrum, different techniques are used. Neither the sun nor the Earth emit an appreciable amount of energy in this part of the spectrum. The small amount that is emitted by the Earth can be detected using a radiometer with a large IFOV (e.g. 18 $\times$ 23 km on the Scanning Multichannel Microwave Radiometer - SMMR - carried onboard Seasat and Nimbus-7). But in general active systems are used which produce their own signal. In an altimeter, a pulse of microwaves is transmitted vertically downwards and the time taken for the reflected signal to return is used to calculate the height above the surface that the platform is flying (see chapter 15). In the scatterometer, a pulse is sent out obliquely from the instrument which detects the component scattered by the surface. The intensity, depolarisation and pulse degradation give information about the roughness of the surface and can be used, for example, to estimate wind speeds over oceans or to identify different crop types. Imaging systems use a swept beam of microwaves and the time of flight of the returned scattered pulse is used to locate that particular pixel in the image. The spatial resolution of such a "Side Looking Radar" (SLAR) system is limited by the size of the antenna, so to improve this, particularly for use on satellites, a very much longer aerial is synthesized using the Doppler shift of the frequency of the signal returned from various parts of the microwave beam. The relative motion between the platform and an object on the ground varies as the object passes through the fan-shaped microwave beam, and therefore so does the Doppler shift. The *effective* length of the aerial is now the distance travelled by the platform while the object is in the beam, which can be several kilometres. Such a system is called a "Synthetic Aperture Radar" (SAR). Such instruments have been flown on quite a number of aircraft, but only rarely on satellites. Seasat carried a SAR system during its brief 100 day mission in 1978 and proved the feasibilty of the concept. At that time, though, computing powers were barely sufficient to cope with the huge quantity of data received. ERS-1 is now providing SAR data which is being processed almost in real time. Radarsat, a mostly Canadian programme, to be launched in the mid 1990s, will use SAR particulary for sea-ice mapping to aid ship movements - a scheme which is said will pay for itself within three years!

Sensing a target remotely is based upon observing the spectral differences in energy reflected or emitted from the features of the target. Unlike cameras, which can take full colour pictures, scanners record only in a limited number of wavebands. The width and number of wavelength intervals of a sensor define its spectral resolution, i.e. the detail by which the sensor is capable of observing the "colour" of the target. In the visible region, the operational sensor detects solar radiation reflected from the target, and the recorded information is termed the *albedo* of the surface. The albedo is governed by the nature, texture, condition and orientation of the surface. Surfaces cannot be characterised unambiguously by their albedo at one wavelength, but often a "signature" can be defined in terms of the response in a number of wavelengths. But even then quite a range of reflectance values may be obtained from what is ostensibly one ground cover type but which exists in various conditions of growth stage, vigour, aspect, illumination etc. The particular wavebands and bandwidths provided by a particular scanner are chosen for specific applications, but that does not stop them being used for other purposes. For example, AVHRR was designed for meteorological applications, soon found uses by oceanographers for mapping sea surface temperatures, and is now being used for medium-scale vegetation mapping. But, because the wavebands and bandwidths are not the same as those of other systems, care must be taken in correlating the results of any analysis. Different information is being recorded by each instrument, so different interpretations may be required. One cannot directly compare the vegetation index derived from AVHRR with that from TM or MSS. For good spectral discrimination, narrow wavebands are desirable, but narrow wavebands mean that less energy is incident on the detector which in turn means less sensitivity. Larger numbers of wavebands produce greater volumes of data which not only provides problems with transmission back to Earth, but also aggravates the data storage problem.. Multipurpose scanners therefore require compromises to be made. Table 2 shows the spectral characteristics of several of the more well-known scanners which are currently being flown on environmental satellites. Most of these satellites carry a number of other instruments as well. The reader is referred to Kramer (1992) or any of the many textbooks for more detailed information.

Scanners on board aircraft can operate in more wavebands than those on satellites because the problem of data transmission does not occur. The Airborne Thematic Mapper (ATM) from Daedalus, has 12 wavebands, many of them coinciding with those of TM on Landsat. Imaging spectrometers operating in up to 288 wavebands are now being developed, but there is usually a trade off made between spectral and spatial resolution in order to keep data volumes within bounds. Usually one can choose to have fine spatial resolution in a very few wavebands or to have the full spectral resolution but on a very coarse spatial scale, for example in the CASI instrument. An instrument based on a video camera incorporating an interference filter is being developed in Dundee (Sun and Anderson 1992) which collects and stores data with full spatial and spectral resolution which the user can afterwards use software to display in the form he requires. Preliminary trials look very promising, and it means that the decision as to the type of data required does not have to be made before the flight but can be made by any user to suit his own requirements. An added benefit is that it can be flown on any light aircraft and is completely portable.

One great advantage of the digital nature of images from scanners is that they are in a form ideally suited to manipulation by computers. Aerial photographs are often analysed visually, but a great deal more spatial and non-spatial information can be extracted from a digital image. Firstly, digital images usually require preprocessing to restore the fidelity of the image, that is to compensate for geometrical and radiometric distortions that the data may have undergone. Secondly, the visual impression of the image can be improved by various enhancement techniques, and thirdly mathematical or statistical processes can be used to extract useful information. Enhanced images or

Satellite	Scanner	resolution/ swathwidth	wavebands (μm)	image repeat
Landsat	Multispectral Scanner System (MSS)	80m/185 km	0.5-0.6 (green) 0.6-0.7 (red) 0.7-0.8 (red/near IR) 0.8-1.1 (near IR)	16 days
Landsat 4/5	Thematic Mapper (TM)	30m/185km (bands 1-5 & 7) 120m/185km (band 6)	0.45-0.52 (visible) 0.52-0.60 (visible) 0.63-0.69 (visible) 0.76-0.90 (near IR) 1.55-1.75 (near/mid IR) 10.4-12.5 (thermal IR) 2.08-2.35 (mid IR)	16 days
Landsat 6	Enhanced Thematic Mapper (ETM)	15m (panchromatic) 30m (visible/near IR) 120m (thermal IR)	1-7 as for TM (visible/IR) 0.50-0.90 (panchromatic)	
SPOT	High Resolution Visible (HRV) (two, steerable)	10m (panchromatic) 20m (multispeectral)/ 60 km (each)	0.51-0.73 (panchromatic) 0.50-0.59 (green) 0.61-0.68 (red) 0.79-0.89 (near IR)	26 days (nadir)
NOAA	Advanced Very High Resolution Radiometer	1.1 km/3000km (nadir)	0.58-0.68 (visible) 0.725-1.10 (near IR) 3.55-3.93 (mid IR) 10.30-11.30 (thermal IR) 11.30-12.50 (thermal IR)	12 hours
Meteosat	Spin Scan Radiometer	2.5 km (visible) 5.0 km (others)	0.4-1.1 (visible) 5.7-7.1 (IR, water vapour absorpt) 10.5-12.5 (thermal IR)	30 mins
GOES	VAS (VISSR Atmospheric Sounder)	1,2,4,8 km (visible) 8-14 km (IR)	0.55-0.75 (8 channels, visible) 3.7-7.3 (2 channels, mid IR) 7.7-14.7 (2 channels, mid-thermal) 6.7-14.7 (2 channels, mid-thermal)	30 mins

Table 2: Some current scanner systems

spatial imformation can then be displayed on a monitor or passed to some form of printer to produce a hard copy. Colour composites can be made from three bands of data, but computers can handle any number of bands and produce, for example, multispectral classifications or principal components using all seven TM bands. Sophisticated image processing can now be carried out on microcomputers, although large data sets and complex processing still requires the power and speed of mainframe machines or parallel processors.

Different parts of the electromagnetic spectrum interact differently with the components of the atmosphere. Wavelengths which interact strongly can be used to probe these constituents and provide information about the concentration and distribution of gases and chemicals in the atmosphere. Water vapour, ozone and carbon dioxide, for example, can all be studied from space and synoptic maps produced. This, of course, will complement the localised point measurements made from radiosondes, sounding rockets and ground-based spectrometers and provide important input data to global models. In many of the current satellites such instruments form an important part of the payload. Nimbus-7 has been measuring ozone concentrations using TOMS (Total Ozone Mapping Spectrometer) since 1976 and microwave sounders have flown on a number of satellites. Most of the systems planned for the Mission to Planet Earth programme planned for the 1990s (see section 7) will carry sophisticated instruments to study the chemistry and physics of the atmosphere.

The section of the user community which studies the surface of the Earth, however, finds such interactions highly inconvenient. Certain large parts of the electromagnetic spectrum are either not transmitted at all or are modified to a greater or lesser extent during the passage of the radiation through the atmosphere. This limits the usable wavelengths, and complicates the interpretation of the data. Because of multiple scattering, absorption and reemission, the signal received at the satellite is highly contaminated and perhaps only a small portion of the information contained in the data relates to the surface being studied. Atmospheric "windows" can be defined in which the signal is transmitted reasonably well, but even in these preprocessing is usually required to radiometrically correct and to calibrate the data before it can be analysed.

Ultraviolet radiation is strongly absorbed by ozone - hence the worry over the possible effects of an ozone hole - and so little use is made of such wavelengths for terrestrial monitoring. Similarly, blue light is strongly scattered and so that part of the spectrum is usually avoided exept for observations from low-flying aircraft. In the visible and near infrared part of the spectrum, sensors detect mainly the reflected component of the sunlight. Atmospheric absorption is reasonably small unless water is present in the form of cloud or haze. The near infrared signal is particularly important for vegetation monitoring because of the sensitivity of the strong reflection to the health of vegetation. Absorption by water bodies increases with wavelength. Blue light can be transmitted by metres of water, but infrared is absorbed by even a few centimetres. Infrared is useful for delineating water bodies, but blue and green are much more useful for studying the oceans and coastal waters.

The mid infra-red wavelength senses primarily terrestrial radiation along with some reflected solar radiation. Particularly at night, this band, around 3 μm, is useful for detecting wildfires, industrial pollution and for volcano monitoring. In meteorology, certain cloud formations are enhanced in this channel, and the strong absorption by water vapour of wavelengths in the range 6 - 7 μm makes this a very useful band for mapping the global extent of water vapour.

The thermal infrared wavelength range lies in the window region of terrestrial radiation and thus atmospheric absorption is not a significant factor provided the ozone absorption at about 9.5 μm is avoided. Most thermal scanners operate in the range 10 - 12.5 μm. Thus radiation will be directly related to the brightness temperature of the radiating surface and, if the instrument is calibrated, quantitative measurements can be measured. Originally, instruments such as the AVHRR on the NOAA satellites were designed to measure cloud top temperatures. They then found a use for measuring sea surface temperatures and a great deal of work has been done on performing atmospheric corrections using two different frequencies (the so-called split windows of the later AVHRR scanners). More recently, the Along Track Scanning Radiometer on ERS-1 with its helical scanning motion uses a two-look method through different atmospheric path lengths to carry out this correction. Attempts to extend these methods to plot *land* surface temperatures run into the difficulty that the emissivities of land surfaces are very variable and absolute measurements of temperatures have so far been unsuccessful. Some success has, however, been achieved in the field of geology where rock structures, and in particular thermal inertia, can be studied to some extent.

Microwaves are now becoming more widely used in remote sensing. The Earth emits a small quantity of energy in the microwave region of the electromagnetic spectrum, and this can be measured by microwave radiometers, albeit on a spatial scale of 20 - 50 km. Such instruments have been very useful for mapping ice and snow extent. Since microwaves penetrate clouds, and are also not dependent on solar illumination, such instruments can be used night or day, summer or winter, and thus provide useful information from the polar regions all year round. Most other microwave systems are active, sending out pulses and detecting the reflected or scattered returned signal and these again provide all-weather capability. ERS-1 carries an active microwave instrument (AMI) which doubles as a scatterometer and a SAR. Although designed mainly for oceanographic measurements, the data are nevertheless proving useful for land measurements as well. Even radar height measurements over water, ice and land can be a key source of information for the study of the climate and climate change. Surface height features are associated with ocean currents, and these are of special importance to climate research since oceans play a major role in transporting heat from the equator towards the poles, providing those who live at medium to high latitudes with a milder climate than they would otherwise experience. Such studies need height measurements to a precision of a few centimetres and an accuracy of a few decimetres. Seasat proved that such measurements were technically possible, and ERS-1 and Topex/Poseidon form the centerpiece of the World Ocean Circulation Experiment (WOCE), one of the four major experiments within the World Climate Research Programme.

It must be remembered, however, that sensors in the visible and infrared wavebands measure only radiance values, and it requires considerable effort on the part of man and computer to translate those values into useful parameters such as temperatures or land use classes. Automated or semi-automated procedures for calibration and classification are being developed and continually improved, but the quality of the end product depends not only on the sophistication of these analytical techniques but on the inherent limitations of the detecting instruments - on their resolution. The usual understanding

Platform	Image size	Resolution	Repeat Mapping	scale
Geostationary	whole world	2.5 -5 km	$\frac{1}{2}$ hour	global
NOAA (AVHRR)	5000 km	1 km	12 hours	regional
Landsat (TM)	185 km	30 m	16day	1:1,000,000 to 1:100,000
SPOT (HRV)	60 km	20 m (10 m)	27 days (1-5 days)	1:100,000 to 1:25,000
Aircraft	100 m-20 km	< 1 m	as required	1:80,000 to 1:2,500

Table 3: Monitoring systems - resolutions

of the word "resolution" is in terms of the ability to distinguish small objects. In the context of remotely-sensed data, however, there are other connotations. One now has to take account of four distinct dimension: spatial resolution, spectral resolution, radiometric resolution and temporal resolution.

The spatial resolution of a camera or a telescope is limited both by the mechanical perfection of the components and by physical problems produced by the phenomenon of diffraction. Satellite and aircraft-borne scanners, however, suffer from a further limitation set by the scanning system itself which does not produce a continuous picture of the scene but samples small areas and forms an image consisting of a mosaic of picture elements. The signal from the whole of each sampled area is integrated at the detector, and no information about spatial distribution on a scale smaller than each element can be obtained. The spatial resolution of the sensor can be defined as the smallest area that it can distingtuish from areas around it on the image, usually the pixel size. Pixel sizes vary from one satellite to another and from one channel to another on the same satellite. The spatial resolution of some common scanners is shown in Table 3, but it must be remembered that these values refer to the size of the pixel at satellite subpoint, ie vertically beneath the scanner. If the angular field of view of the scanner is significant, the pixel size increases from the nadir point to the edge of the image. AVHRR pixels may be 1 km square at nadir, but represent nearly 4km at the extremities of the scan line. The scanner on board GOES has a resolution of 1 km at the satellite subpoint in the visible channel and 8km in the thermal infrared which decreases to 14 km in the mid infrared. Typically pixel sizes may be of the order of 1 km square for meteorological use, tens of metres for land use, and less than a metre from aircraft. There is a trade-off, of course, between the various resolutions. The higher the spatial resolution the smaller the area imaged and so the less frequent the revisit, as shown in Table 3. The corresponding range of map scales corresponding to some to these images is also shown.

High spatial resolution may not necessarily be an advantage. The smaller the sampled areas, the more pixels that are required to cover a given area and consequently this requires the storage and manipulation of more data. A Landsat thematic mapper scene covers an area of 180 km $\times$ 180 km, and at a resolution of 30 m, contains about 10 million pixels (collected in 24 seconds!). The same area is contained in only about

	Landsat MSS	Landsat TM	AVHRR
Swath width	185 km	185 km	3000 km
Number of pixels/line	2340	5760	2048
Number of lines/image	3600	6928	2577
Number of pixels/image	8.4×10^6	$4x10^7$	5.3×10^6
Acquisition time	24 seconds	24 seconds	6 mins
Number of bands	4	7	5
Total number of pixels	33.6×10^6	28×10^7	26.5×10^6
Number of grey levels	64 (6 bit)	256 (8 bit)	1024 (10 bit)
Total number of bits	$\sim 20 \times 10^7$	22×10^8	26×10^7
Data transmission rate	15 M bps (34 MByte/s)	85 M bps (246 MByte/s)	665 kbps (53 MByte/s)

Table 4: Approximate Comparative Data Volumes

300 pixels in an AVHRR image. Obviously oceanographic and meteorological processes and regional-scale variables with periodicities of hundreds of kilomet are better monitored using AVHRR than Landsat TM, and conversely local phenomena would not show up on an AVHRR image.

In order to telemeter data to ground receiving stations, information is recorded in digital form. Intensities are quantised into a number of grey levels, the number of which determine the radiometric resolution. Landsat MSS (multi- spectral scanner) uses only 64 levels (6 bit data), TM uses 8 bits (256 levels) and AVHRR 10 bits. In its orbit of the Earth a scanner has to image scenes with a wide range of light levels, from bright snowfields to dark evening tundra. Consequently any one scene usually involves only a small range of available intensity levels, and obviously a larger number of levels provides greater sensitivity, but again at the expense of increases in data volume (see Table 4). Image processing on a computer enables a particular small range of values relating to the image in question to be fully utilised. The data transmission rate of these imaging systems should be compared with that for non-imaging systems. The Poseidon altimeter, a single frequency system, has a telemetry rate of only 1 kbps in normal mode (12 kbs in the upload/dump mode). This is achieved by processing much of the data on board and only transmitting the results.

The frequency at which objects are imaged is also an important parameter. Ideally, fluctuating events should be recorded at intervals no greater than one half of their period, otherwise aliasing may occur. The 18-day repeat of Landsat is therefore not adequate for observing short events, but is well suited to seasonal variations. Daily weather patterns can be detected using AVHRR, but more rapid developments require images from Meteosat (see Table 3).

The temporal resolution of the sensor determines its ability to repeat coverage of a certain area. There is a fairly simple relation between spatial resolution and temporal resolution and both are determined by the altitude at which the sensor is used to record images of the Earth. As usual, however, one can never achieve the ideal situation. Higher spatial resolution necessitates a smaller swath width which in turn requires more orbits to cover the entire Earth. Thus any area is revisited less frequently. There is thus a trade-off between spatial and temporal resolution as there is between spatial and spectral and spectral and radiometric resolutions.

5 Satellite Climatology

Until recently, climatology has been seen as a minor, rather specialised area of meteorology. It is now recognised, however, as a science in its own right, concerned as it is with atmospheric processes and their effects over much longer periods of time and often over greater geographical areas. The use of satellite data as input to climate models has accelerated the emergence of this area of research, propelled by the growing concern about environmental issues. Surface-based observations tend to be rather unevenly distributed, being particularly poor in polar regions and over the oceans, the very regions of particular interest to climatologists. Some important fields in which remote sensing can make a particular, direct contribution are in studies of the Earth's radiation budget and the transport of energy around the Earth, measurements of atmospheric moisture content and precipitation, ice and snow monitoring and in some studies of atmospheric pollution. Other land and sea measurements are, of course, necessary for understanding and modelling change.

5.1 The Earth radiation budget

The term "Earth Radiation Budget" (ERB) refers to the balance between the amount of solar energy absorbed by the Earth to the amount of long-wave radiation emitted back into space. In the long term, and taking the Earth as a whole, these two should exactly balance out if the climate system is to be in equilibrium. In practice, at most latitudes and for most of the year, the amount of radiation absorbed and emitted is not in balance. There is always an excess of absorbed over emitted radiation at low latitudes and an excess of emitted over absorbed radiation towards the poles. At intermediate latitudes the balance will vary seasonally. It is the attempt to produce local balance of the budget, through the transport of energy around the globe, which drives the large scale circulation of the atmosphere and the oceans.

The ERB has been the subject of quite intensive satellite monitoring for a number of years. The most notable programme is the Earth Radiation Budget Experiment (ERBE) run by NASA. As part of this programme, a dedicated satellite, the Earth Radiation Budget Satellite (ERBS) was launched in 1984. Other data were obtained from NOAA-9 and NOAA-10. As incoming solar radiation can be absorbed and backscattered into space by the atmosphere, reliable measurements of incident solar energy can only be made from outside the atmosphere. Before the advent of satellites, the only way to do this was by using rockets. The first instrument to be flown on a satellite which could take readings of ultraviolet radiation was carried by Nimbus-4 in 1970, and there have been many others since making measurements in the ultraviolet and visible wavebands.

Such measurements are used to estimate the so-called "solar constant", a term used to describe the Sun's energy output. Estimates of the size of the solar constant vary, and from ground-based measurements it was not possible to conclude whether this represented a genuine variability in the Sun's output or whether it was caused by atmospheric effects. Results from the satellite on the Solar Maximum Mission (SMM) in 1980, however, suggested that the solar constant does indeed vary, by perhaps as much as a few tenths of a percent over a period of weeks, Longer-term variations are smaller and appear to be related to the sunspot cycle. SMM measurements showed a decline in solar output of about 0.08% between 1980 and mid 1985, representing a possible external source of variation in the Earth's climate.

The ERBS also carried radiometers for making direct measurements of the amount of incoming solar and outgoing terrestrial radiation. It must be noted, however, that these instruments can only measure the amount of thermal radiation within their particular wavebands and a profile of the Earth's infrared emission is required in order to estimate the total amount of radiation emitted.

Remote sensing makes other indirect contributions to ERB studies through the monitoring of land and sea-surface temperatures and by making atmospheric temperature soundings. The AVHRR on the NOAA satellites and ATSR on ERS-1 are examples of instruments for measuring land and sea-surface temperatures by measuring the amount of thermal infrared radiation emitted by the surface. Microwave radiometers, such as the Scanning Multichannel Microwave Radiometer (SMMR) flown on Nimbus-7, can also be used to measure surface temperatures as the amount of microwave radiation emitted by a body is related to its temperature. The advantage of the latter is that microwaves can penetrate cloud whereas infrared radiation cannot. One problem of using microwaves for determining temperatures is the complexity of emissivities in this waveband region. This is particularly true when considering the oceans because the emissivity of water is partly a function of its roughness. This can of course be put to use in the determination of surface roughness and hence windspeeds over the oceans. In general, measurements of surface temperatures from satellites have in the past proved to be accurate to better than one dgree, but it is anticipated that current instruments can improve that considerably.

As we saw earlier, ocean currents are responsible for transporting much of the heat from equatorial regions to higher latitudes. Traditional measurements of sea-surface temperatures relied on readings taken from quite a small number of ships and buoys, and so our knowledge of the behaviour of ocean currents and the precise role they play in the global climate is rather incomplete. The advent of satellites has provided the opportunity for adequate monitoring of these currents for the first time.

As well as considering the lateral transport of heat around the world, studies of the ERB must also take into account the vertical transfer of energy through the atmosphere. The temperature and pressure profiles of the atmosphere affect the amount of radiation emitted into space. This is principally because changes in pressure affect the concentration of trace gases such as CO_2 and water vapour which can absorb and re-radiate long wave radiation. The amount of energy radiated by these trace gases is also, of course, related to their temperature. Vertical temperature profiles of the atmosphere can be obtained using the principle of atmospheric sounding. The most frequently used method relies on the fact that CO_2 is known to be uniformly mixed throughout the atmosphere to within 1-2%. The amount of energy emitted in the CO_2 emission spectrum when viewed from above will therefore depend only on the temperature of the layer of atmosphere from which the energy was emitted. The emitted energy can, however, also be re-absorbed by CO_2 in higher layers where the concentration of gas also depends on the temperature at that height. The final measure of the amount of radiation within the CO_2 emission/absorption spectrum is then the result of repeated emission and absorption of energy and could be the result of an almost infinite number of different atmospheric profiles. A first guess, based for example on the climatological average for the place and time of year, has to be made and then, using a mathematical inversion technique, the estimated temperature profile is refined. In clear conditions, the errors can be as low as 1 degK, but are usually nearer to 2 degK because of the presence of aerosols and clouds. Using oxygen emission/absorption frequencies in the microwave wavelengths the problem of absorption by aerosols and clouds can be avoided. TOVS (the Tiros Operational Vertical Sounder), which consists of the HIRS (High Resolution InfraRed Sounder), the SSU (Stratospheric Sounding Unit) and the MSU (Microwave Sounding Unit) has been carried on board a number of NOAA satellites. The next series will carry in addition the MHS (Microwave Humidity Sounder).

The Earth's radiation budget will also be affected by changes in surface albedo, which may alter the amount of energy absorbed. Determining albedos from satellite measurements is actually rather complex as the amount of solar energy reflected depends not only on the nature and state of the surface but also on the direction of reflectance (the angle at which the sensor looks at the target surface) and the solar illumination angle. To calculate the albedo from the reflected radiation also requires a knowledge of the incident radiation which in turn depends on a knowledge of the solar constant. Energy is also reflected strongly by clouds and ice, and the extent of such cover will therefore greatly affect the ERB. Cloud cover monitoring is now a well-tried and tested operation using the several operational meteorological satellites available, and ice extent can now also be mapped from space (see sections 5.2 and 5.3).

5.2 Atmospheric moisture

Atmospheric moisture is obviously of major importance in climatology. Differences in weather largely depend on the amount of water present in the atmosphere at a given place and the state in which it is found. The contributions of remote sensing to the study of atmospheric moisture can be divided into the observation of clouds, observation of water vapour and precipitation estimates. Clouds are particularly easy to sense from satellites and are the dominant features of images in the visible and infrared wavelengths. They are important climatologically for several reasons. They tend to produce lower daytime temperatures because of the amount of solar radiation which they reflect into space, but they do also inhibit the re-radiation of infrared energy from the Earth at night thus leading to higher night-time temperatures, they are reservoirs of latent heat and play a part in transporting energy around the Earth, and they are obviousy related to amounts of precipitation. The classification of clouds into their different types is best done using a combination of visible and infrared data as this combines information concerning the temperature of the clouds, which is related to height, and brightness or reflectivity which depends partly on the depth of the cloud and also on its shape. The inclusion of channel 3, in the mid-infrared, in the AVHRR has helped to make cloud discrimination easier. It is now possible to make estimates of total global cloud cover, not possible from ground-based measurements alone.

In order to understand the atmospheric processes which lead to the formation of clouds and precipitation, it is necessary to know something about the distribution of water vapour through the atmosphere. Satellite sensors which have a water vapour absorption band, such as the 5.7-7.1 μm band on Meteosat, can be used for this purpose. These measurements rely on the principle that the degree of absorption of energy in this part of the electromagnetic spectrum is related to the concentration of water vapour (similar to the principle used in temperature sounding). Remote sensing has also contributed to measurements of rainfall. The most usual methods to estimate the extent and heaviness of rainfall rely on active microwave systems which can be both satellite or ground based. Before these systems were available, many parts of the world had insufficient data coverage for reliable estimates of long-term rainfall averages. Satellite data have also helped in the understanding of precipitation processes.

5.3 Ice and snow

Ice and snow are present to the greatest extent in inhospitable regions of the Earth where ground-based data are sparse. The most permanent components of the global ice and snow cover are the Antarctic and Greenland icecaps. High mountain ranges such as the Himalayas also contain significant amounts of permanent snow and ice cover; also about 3×10^6 square kilometres of the ocean surface is ice-covered. There are huge seasonal variations in the amount of ice cover in each hemisphere, but the global total remains roughly constant. The importance of ice and snow cover to the world's climate were mentioned in section 2, and there is clearly a role for remote sensing in their global measuring and monitoring. Visible wavelengths are of limited use because many of the regions where cover is most extensive are either dark for much of the year or are often cloud-covered, or both. For these reasons, microwave techniques are of great importance. Microwave radiometers are particularly useful because they can discriminate between sea-ice and water, even at the same temperature, because of their different emissivities at these wavelengths. Synthetic aperture radar can produce data in all weather conditions and during darkness. An important but originally unintended application of satellite altimeter data has been that of mapping ice-sheet topography (see chapter 15). Unfortunately, the altimeters flown to date have only covered the latitude range to 72 °N/S, leaving half of Greenland and most of Antarctica unmapped. Also, because these instruments were not designed for operation over topographic surfaces, the sub-metre height accuracies required for mass balance studies were right at the limit of their capabilities and this has led to some controversy over the interpretation of the results obtained so far. The new instruments, including that on ERS-1 will hopefully provide more accurate data.

5.4 Other atmospheric constituents

Any changes in the concentration of gases, dust aerosols etc. in the atmosphere, whether natural or anthropomorphic, could have significant effects on the climate. As well as the other unpleasant effects which continued ozone depletion will produce, it is possible that there will also be climatological consequences, although it is not clear what these might be. Stratospheric ozone levels have been measured continuosly since 1979 by the Solar Backscatter Ultraviolet Spectrometer and the Total Ozone Monitoring System (TOMS) on Nimbus-7. Many other satellite systems are involved in ozone monitoring, amongst these will be the Global Ozone Monitoring Experiment (GOME) on ERS-2. Other pollutants (after all, the absence of a desirable gas is as much a pollutant as the presence of an undesirable one) may also be detected directly or indirectly from satellites. UARS and other satellites have, or will have, sensors specially designed for atmospheric chemistry. Volcanic dust can be monitored, thermal and chemical pollution from cities can be detected, and smoke from wildfires or disasters such as the Kuwait oil field fires can be observed. The so-called greenhouse gases (carbon dioxide and water vapour for example) can, as we saw above, be continuously monitored by sounding systems on satellites. Present and future satellite systems will therefore play a crucial role in providing the frequent, synoptic data required for climate modelling.

6 Data Management

The compilation of a data bank for environmental use is very complex. Not only are the data of different types (field data, satellite data, statistical data, census data, map data etc.) and in different formats (vector, raster, tabular) but also relate to different scales of measurement, both spatially and temporally.

Information systems, using the power of computers, have been developed over a number of years to store and manipulate data. If the data have spatial relationships, such a system is usually referred to as a "Geographic (or Geographical) Information System", (GIS). Perhaps a system relevant to the present discussion should be called an "Environmental Information System" (EIS). Such systems were developed originally for the purpose of resource management and to aid decision-making. They usually contained a limited range of information on a local, or perhaps national, scale. These find use in local government, public utilities, and national agencies such as National Parks etc. They contain such things as a base map, some satellite or aircraft data, population data, infrastructure data (water, gas, electricity, public transport), public amenities, tree distribution, and climate data etc. Techniques have been developed to integrate vector, raster and tabular data and to compress data into manageable forms. Also means of inputting and outputting information have been improved.

Considerable development and sophistication is required to extend the use of GIS into environmental monitoring, particularly on the global scale. In order to be in with a chance of handling the enormous volumes of data involved it is necessary to incorporate the ideas of "intelligent knowledge based systems" (IKBS). Raw data needs to be processed. Satellite images need to be geometrically registered and turned into sea-surface temperature maps, land class maps, vegetation maps etc. Base maps need to be digitised. Image processing and digital cartography therefore play an important part. One then has to overcome the incompatibility of data formats, classification methods and encoding routines, the problem of different kinds and scales of the variables, incompatibility of machines and also the incompatibility of the "hard" scientists and social scientists who both have input and require output from the system. Even before these problems are considered, however, one has to formulate the right questions. Exactly what are the requirements of the system, who will use it, what form should the output take, what inputs are required, who should operate it, what level of expertise is required, and even, dare I suggest, why do it at all?

Another problem is that related to data storage. Magnetic tapes are not only bulky but data retrieval is slow and they have a limited shelf-life of only about ten years. CD ROMs and video discs are an improvement, and extensive use is now being made of these. But one recent development, which at least makes a start at solving the problems involved with the data "floods" of the near future, is that of optical tape. This is a flexible, laser read/write medium based on dye-polymer WORM (write once read many times) technology. It provides a vast increase of storage density, reliability, lifetime and speed. The ICI 1012, for example, holds 1 terabyte of data (which is equivalent to more than 4600 magnetic tape cartridges) on a single 12″ reel. This has a life expectancy of more than 30 years, is immune to magnetic fields and cannot be accidentally erased. Since there is no contact between the read/write head and the tape, optical tape is extremely durable - it can survive more than 40000 long-length rewind passes (nearly more than an order of magnitude better than a magnetic tape) and has an average address time of only 37 seconds and a search of the entire reel can take only 68 seconds. The Canadian Centre for Remote Sensing (CCRS) is currently archiving data on such tape, with a conversion rate of 150 high density magnetic tapes onto a single optical tape. ESA is also looking at it to reduce the size of its archive and to save money, and the Australian Centre for Remote Sensing (ACRES) is well on the way to transcribing its extensive Landsat archive.

Computing speeds will also need to improve in order to make use of all the data that will be generated over the next decade. SAR data can just about be processed in near-real-time. Images from ERS-1 can be produced in about an hour. An example of this was a proof-of-concept programme to provide SAR images to a ship in the ice-strewn Barents Sea to aid its navigation. For SAR processing in real time, computing powers of well over 1000 million floating-point operations per second (Megaflops) will be required. The Cray processor nearly achieved this, and NASA's Massively Parallel Processor (MPP) managed 6×10^9 8-bit addition operations or 1.8×10^9 8-bit multiplication operations per second using 16,384 processors in parallel. It is hoped that by the mid 1990s that computing speeds of 10^{11} or 10^{12} operations per second may be achieved which would make real-time SAR processing feasible.

7 Current and Future Programmes

A great deal of work towards interpreting remotely-sensed data and developing climate models has already been carried out using data from the existing satellites and instruments. Very often, however, these were designed for different purposes and are not ideally suited to providing the information required. Nevertheless, man's ingenuity to adapt has lead to the discovery of unintended applications, such as for the AVHRR on the NOAA series of satellites. NASA is taking a leading role over the next few decades in its Mission to Planet Earth concept which will use space- and ground-based measurements to provide the basis for understanding global climate change. NASA will be working with the scientific communities through research efforts coordinated under the US Global Change Programme (USGCRP), the International Geosphere-Biosphere Programme (IGBP) and the World Climate Research Programme (WCRP). NASA will be contributing satellite missions together with management and analysis of both satellite and *in situ* data. The centrepiece of Mission to Planet Earth will be the Earth Observing System (EOS), a series of polar orbiting and low inclination satellites for long term global observations of the land surface, biosphere, solid Earth, atmosphere and oceans. In tandem with EOS, the polar orbiting and mid-inclination platforms from Europe, Japan and NOAA form the basis of a comprehensive International Earth Observing System (IEOS) intended to operate for at least 15 years. Tables 5 and 6 identify the NASA and international contributions to EOS. More detail can be found in Asrar and Dokken (1993). Two classes of instruments will be carried: facility instruments supplied in response to general mission requirements and Principal Investigator instruments selected through a competitive process and aimed at the focused research interests of the selected investigators. EOS observations will permit assessment of various Earth system processes, including:

Hydrologic processes, which govern the interactions of land and ocean surfaces with the atmosphere through the transport of heat, mass and momentum.

Biogeochemical processes, which contribute to the formation, dissipation and transport of trace gases and aerosols and their global distributions.

Climatological processes, which control the formation and dissipation of clouds and their interaction with solar radiation.

Ecological processes, which are affected by and/or will affect global change, and their response to such changes through adaptation.

Geophysical processes, which have shaped or continue to modify the surface of the Earth through tectonics, volcanism and the melting of glaciers and ice.

The issues identified by USGCRP and the Intergovernmental Panel on Climate Change (IPCC) as key to understanding global climate change include:

The role of clouds, radiation, water vapour and precipitation

The productivity of the oceans, their circulation and air-sea exchange

The sources and sinks of greenhouse gases and their atmospheric transformations

Changes in land use, land cover, primary productivity and the water cycle

The role of polar ice sheets and sea level

The coupling of ozone chemistry with climate and the biosphere

The role of volcanoes in climate change.

In the long term, NASA plans to extend the MTPE programme well into the next century. Small to moderate sized satellites with focused objectives will complement the work of the EOS systems already planned.

Integral with the EOS programme is the Earth Observing System Data and Information System (EOSDIS). This will manage the data resulting from the satellite and field measurement programmes. EOSDIS will supply data archive, distribution and information management services, data product generation, command and control functions and data sets generated by assimilation of applicable observations into global climate models.

The first spacecraft to be launched in the Mission to Planet Earth programme was the Upper Atmosphere Research Satellite (UARS). Its mission is to provide data to improve our knowledge and understanding of the dynamics, chemistry and energy inputs of the stratosphere, including the ozone layer and its chemistry (see chapter 9) and carries 10 instruments. These measure trace gases (especially those involved in ozone chemistry), horizontal winds that transport gases from one region of the stratosphere to another, solar UV radiation and high energy particles inputs into the upper atmosphere and temperature and pressure variations. UARS is in a low-angle (57 ° inclination) orbit at an altitude of 585 km. At this altitude and inclination, instruments can view up to 80 °N and S, providing coverage of more than 98% of the globe and making it possible to sample all local solar times in about 36 days. Prior to UARS, the most comprehensive satellite atmospheric observatory was the polar- orbiting Nimbus-7 satellite, launched in 1978 and still functioning in 1993. This carries the Solar Backscatter Ultraviolet (SBUV) and Total Ozone Mapping Spectrometer (TOMS) instruments, both of which monitor the ozone layer, and the Limb Infrared Monitor of the Stratosphere (LIMS) and the Stratospheric and Mesospheric Sounder (SAMS) which measure related trace gases. Updated versions of TOMS are scheduled to fly on several of the planned EOS platforms (tables 5 and 6).

A number of international environmental programmes related to global climate change have been set up over the past few years, a few of which have already been alluded to. CEOS (1992) describes some of these, particularly their satellite inputs, and further information can be found in other chapters in this volume. The Global Climate Observing System (GCOS) is now accepted by the main international organisations (the WMO, the IOC, UNEP and ICSU) and its data requirements are to upgrade the World Weather Watch (WWW) system, to develop additional facilities for programmes such as Global Atmosphere Watch (GAW), to develop a Global Ocean Observing System (GOOS) and to maintain and enhance operational and research monitoring programmes of changes in terrestrial ecosystems, clouds and the hydrogeological cycle, the ERB and the cryosphere. Satellite data will play a key role in satisfying these requirements.

NASA Satellites (launch status)	Mission Objectives
ERBS (operating) Earth Radiation Budget Satellite	Radiation budget, aerosol and ozone data from 57° inclination orbit
TOMS/Meteor-3 (operating) Total Ozone Mapping Spectrometer	Ozone mapping and monitoring (Joint with Russia)
UARS (operating) Upper Atmosphere Research Satellite	Stratospheric and mesospheric chemistry and related processes
NASA Spacelab series (1992 on)	Atmosphere and solar dynamics (ATLAS) Atmospheric aerosols (LITE) Surface Radar images (SIR C, X-SAR)
TOPEX/Poseidon (operating) Ocean Topography Experiment	Ocean surface topography Ice sheet altimetry
LAGEOS-2 (operational) Laser Geodynamics Satellite	Satellite laser-ranging target for monitoring crustal motions and Earth rotation variations
SeaWiFS (1993) Sea Viewing Wide Field Sensor	Ocean colour for productivity
TOMS/Earth Probe (1993) Total Ozone Mapping Spectrometer	Ozone mapping and monitoring
NSCAT (1995) NASA Scatterometer	Surface wind vectors over the ocean (on Japan's ADEOS)
Landsat-7 (1997) Land Remote Sensing Satellite	High spatial resolution visible and infrared imagery to monitor land surfaces
TRMM (1997) Tropical Rainfall Measuring Mission	Precipitation and related variables Radiation budget in lower latitudes
EOS-AM Series (1998) Earth Observing System -morning crossing	Global change measurements
EOS-COLOR (1998) EOS Ocean Colour Mission	Ocean colour for productivity
EOS-PM Series (2000) EOS afternoon crossing	Global change measurements
EOS-AERO (2000) EOS Aerosol Mission	Global aerosols and stratospheric chemistry
EOS-ALT (2002) EOS Altimeter Mission	Ocean and ice sheet altimetry
EOS-CHEM (2002) EOS Chemistry Mission	Atmospheric chemistry and ocean surface scatterometry

Table 5: NASA contributions to Mission to Planet Earth (From Asrar and Dokken, 1993)

Non-NASA Satellites (launch status)	Mission Objectives
NOAA-9 through J (US - Operational)	Visible and IR imagery, IR atmospheric sounding and ozone measurements
Landsat 4/5/6 (US - Operational) Land Remote Sensing Satellite	High spatial resolution visible and IR imagery
DMSP (US - Operational) Defence Meteorological Satellite Programme	Visible, IR and passive microwave atmospheric and surface measurements
ERS-1 (ESA - Operational) Earth Remote Sensing Satellite	C-band SAR, microwave altimeter, scatterometer and sea surface temperature
JERS-1 (Japan - Pre-operational) Japan's Earth Resources Satellite	L-band SAR and high resolution visible and IR imagery
ERS-2 (ESA - 1994)	As ERS-1 plus ozone mapping
Radarsat (Canada - 1994) Radar Satellite	C-band SAR
NOAA-K through N (US - 1994 on)	As earlier systems plus microwave imagery
NOAA-O through Q (US - 2001 on)	As above, plus microwave sounding
ADEOS (Japan - 1995) Advanced Earth Observing Satellite	Visible and near IR imagery, scatterometry,tropospheric and stratospheric chemistry
POEM-1 (ESA - 1998) Polar Orbit Earth Observing Mission	Global change and meteorological measurements
ADEOS-II (Japan - proposed 1998)	Global change measurements
TRMM -2 (Japan - Proposed 2000) Tropical Rainfall Measuring Mission	Precipitation and related variables Radiation budget in tropics

Table 6: Contributions to Mission to Planet Earth from non-NASA satellites (From Asrar and Dokken, 1993)

The World Climate Research Programme (WCRP) was established in 1979 as a joint undertaking of the ICSU and the WMO to determine to what extent climate can be predicted and the extent of man's influence on climate. Three major projects have been instituted under this programme, namely the Tropical Ocean and Global Atmosphere (TOGA) project, the World Ocean Circulation Experiment (WOCE) and the Global Energy and Water Cycle Experiment (GEWEX). Two further projects are planned: the Climate Variability and Prediction Research Programme (CLIVAR) and the Arctic Climate System Study (ACSYS). GEWEX is particularly demanding in its requirements for satelite data, requiring simultaneous measurements of properties at the top of the atmosphere (cloud top reflectance the ERB) and underlying vertical profiles of temperature, moisture and clouds. Precipitation measurements are also required in conjunction with the profiles of reflectance clouds and atmospheric properties. A number of new instrument developments are required to meet the GEWEX objectives, including a millimetre wave radar imager or radar profiler to measure cloud particle profiles and instruments to measure soil moisture and 3-D winds.

The International Geosphere-Biosphere Project (IGBP), launched in 1986, is a programme to assess and document the effects of global change in an attempt to understand the interactive physical, chemical and biological processes that regulate the total earth system, the changes that are occuring in this system and the manner in which they are influenced by human activities. It is intended to extend over many years, overcoming to some extent the limitations of many short- term observation programmes. Under this programme, global data will be collected *o* stratospheric ozone, sea surface and sea-ice variables and on the Earth's radiation budget. The launch of SeaWiFS (Sea-viewing Wide Field Sensor) on Seastar in 1993 to measure ocean colour is critical to the success of several of the projects in this programme. This will curtail the gap in measurements since CZCS died in 1986.

The Global Environmental Monitoring System (GEMS), part of the UN Environmental Programme has been using NASA data for over a decade. A Global Resources Information Database (GRID) was set up and it is now involved in integrating, storing and exploiting a variety of global environmental databases with regional nodes worldwide each equipped with the same set of hardware, software and local sub-sets of the data. GRID will offer on-line access to a catalogue of environmental data sets and details of the responsible agencies by mid 1990's. Other research programmes such as the World Ocean Climate Experiment (WOCE) and the World Climate Research Programme (WCRP) will also exploit GIS technology to the full.

8 Conclusions

The concept of global environmental monitoring and the setting up of global databases is very attractive, but we should not delude ourselves as to the complexity, expense and logistics of the problem. Much work is required to formulate programmes, to correlate the inputs and to evaluate the results. computing speeds and data storage capacity will need to be improved, as will the management of the data in GIS systems. Most importantly attitudes need to change, both of politician and the users.

The will is there now, as is the technology, to make a start at setting up environmental databases. Until recently, however, the funding has not been available to do much more than try out pilot schemes. Dedicated local systems have been produced on individual bases originally in experimental situations. On regional and national scales, countries such as Canada have made use of GIS, mainly for resource management, and the European Commission has funded several fairly ambitious and comprehensive programmes. NASA, the IGBP and the EOS point the way towards global programmes. Now, however, the environment is firmly on the political agenda, and national governments are not just wanting to be seen to be environmentally aware but can see the cost effectiveness of such awareness. The necessary investment at least to make a start on effective environmental monitoring may at long last be imminent. At present it is a research and development exercise rather than an operational proposition, but the urgency of the situation may soon change that.

By being aware of the changes that are occurring, and by tempering our desires to advance regardless of the consequences, perhaps the Armageddon that is so often prophesied may be averted, or at least delayed. And it so happens that the technology that has brought about the concentration of these problems in our present generation may help us to understand them and perhaps to ameliorate them.

References

Asrar, G., and Dokken, D. J., 1993, *EOS Reference Handbook* (Washington USA: NASA)

Barrett, E. C., and Curtis, L. F., 1992, *Introduction to Environmental Remote Sensing* (3rd ed) (London: Chapman and Hall)

Campbell, J. B., 1987, *Introduction to Remote Sensing* (New York: Guilford Press)

Carlton, A., 1991, *Satellite Remote Sensing in Climatology* (London: Belhaven Press)

CEOS (County Education Offices Society) - The Relevance of Satellite Missions to the Study of the Global Environment (London: British National Space Centre)

UNEP (United Nations Environment Programme), 1992, Climate Change. Dossier of fact sheets produced by UNEP, Geneva.

Cracknell, A. P., and Hayes, L. W. B., 1991, *Introduction to Remote Sensing* (London: Taylor and Francis)

Elliott, C., 1992, Cost Effective Environmental Monitoring. *Earth Space Review*, **1**, **2**, 7-10.

Kramer, H. J., 1992, Earth Observation Remote Sensing (Berlin: Springer-Verlag)

Mather, P. M., 1992, TERRA-1: Understending the Terrestrial Environment. *The Role of Earth Observation from Space* (London: Taylor and Francis)

Matra Marconi Space, 1992, *Earth Observation Spacecraft Directory* (Portsmouth: MMS)

Mason, J., 1991, Models of Global Climate Change. In *The Earth's Environment - an Assesment from Space* held in Venice, on 10-11 April. (Noordwijk: ESA), ESA, SP-337, pp. 47-56.

Morel, P., 1991, Space Observation Requirements for Global Climate Science and Prediction. In *The Earth's Environment - an Assesment from Space* held in Venice, on 10-11 April. (Noordwijk: ESA), ESA, SP-337, pp. 57-59.

Rhind, D. W., 1989, Environmental Monitoring and Prediction Working Report No. 16, South East Regional Research Laboratory, Birkbeck College, London, UK.

Sun, X., and Anderson, J. M., 1993, A spatially variable light-frequency-selective component based airborne pushbroom imaging spectrometer for the water environment. *Photogrammetric Engineering and Remote Sensing*, **59**, 399-406.

Vaughan, R. A., 1987, Remote Sensing Applications in Meteorology and Climatology (Dordrecht: Reidel)

Vaughan, R. A., 1990, Microwave Remote Sensing for Oceanographic and Marine Weather-forecast Models (Dordrecht: Kluwer)

Vaughan, R. A., 1990, Interrelation of Global and Regional Aspects of Environmental monitoring. Presented at an International Workshop on *Key Problems of Global Ecology and Requirements to Observation* held in Leningrad, on 19-24 November.

Satellite Observations in Global Ecology

K. Ya. Kondratyev

1 Introduction

The 20th century is approaching its end under conditions of the intensification of a number of global environmental problems (see "References"). Mankind has entered a phase in its development when the prediction of the great Russian scientist V. I. Vernadsky that human activities become geological-scale forcings capable of changing the world by bringing it to the verge of ecological catastrophe has become quite realistic. This is what determines the absolute urgency of global ecology problems and the need to develop a political concept of priorities in human values, especially in connection with the Second UN Conference on Environment and Development held in June 1992 in Brazil.

During the whole evolution of humanity, the growth of the consumption of natural resources has been accompanied by local ecological disturbances and catastrophes which resulted from forest cuttings, the impact on the water cycle, etc., and led to such consequences as desertification, etc. The growth of consumption increased by an order of magnitude after the industrial revolution started and was mainly the result of population growth. At present, the average annual consumption of raw materials per capita is of the order of 20 tonnes. These raw materials are processed at the expense of consuming 1.5 kW of energy and 800 m^3 of water into final products having a mass of 2 tonnes. The process is accompanied by large-scale chemical transformations of the environment through its pollution and subsequent chemical reactions.

The world's population has reached the figure of 5.3 billion people in 1990 and is expected to grow by another billion to reach 6.3 billion by the year 2000 (Keyfitz and Flieger 1990). About 93% of this population growth will be the contribution of the developing counties. The increase of population at this rate causes a greater pressure on ecosystems and natural resources. This pressure is expected to increase further until the world population figure is stabilized somewhere about 10 billion by the year 2050

NATO ASI Series, Vol. I 24
Remote Sensing and Global Climate Change
Edited by Robin A. Vaughan and Arthur P. Cracknell

or 2070 (the stabilisation can only be achieved, however, if special efforts to control the population growth are pursued today).

As has been revealed earlier (Gorshkov and Kondratyev 1990, Gorshkov 1991, Kondratyev 1990b, Marchuk and Kondratyev 1992), the global biosphere can be sustainable if the utilization of biospheric resources does not exceed 1% of this primary productivity, which was exactly the case before the industrial revolution. Nowadays, the exploitation of biospheric resources has increased by an order of magnitude, now being close to 10%. This is a sure way to the destruction of the global biosphere and thus to a global ecological catastrophe. Since the world population continues growing, the danger of the catastrophe is even more frightening.

Under these circumstances, the Second UN Conference on Environment and Development (SUNCED) has played an important role. The basic objectives of the SUNCED were concerned with the preparation of a number of documents to formulate problems and work out practical recommendations in order to direct the necessary efforts towards a sustainable socio-economic development in the world. The documents can be classified into three categories.

First, there is the "Earth Charter", the principal purpose of which is the substantiation of basic principles concerning the environment and development, and the interrelationships between people and between man and nature.

The second very important document "Agenda 21" contains a programme of actions required to make it possible for humanity to solve the problem of survival and continuation of normal life in the 21st century. In many respects, this programme is similar to relevant documents approved within the UN system. However, "Agenda 21" has an important specific feature; it calls for a multi-disciplinary approach from two points of view, namely multi-disciplinary ecological problems and the proper and efficient coordination of the efforts of numerous institutions.

The third document should have been a combination of three framework conventions on

1. climate change
2. bio-diversity
3. forestry.

Unfortunately the convention on forestry had not been prepared by the beginning of the SUNCED.

The preparation of the above-mentioned documents was based on the United Nations General Assembly Resolution 44/228 of 22 December 1989, which emphasised that the Assembly was "deeply concerned over the continuing deterioration of the state of the environment and the serious degradation of the global life support systems, as well as over the trends that, if allowed to continue, could disrupt the global ecological balance, jeopardise the life-sustaining qualities of the Earth and lead to an ecological catastrophe, and recognising that decisive, urgent and global action is vital to protecting the ecological balance of the Earth".

Climate change is only a part of the problem of global ecological change. Of general interest also are problems of the ozone layer changes, deforestation and desertification, acid rain and its impact on the biosphere and the pollution of the atmosphere and the ocean. The continuous decrease of biodiversity is also very hazardous. These problems are revealed through the biospheric dynamics which determines the closed nature of global cycles of matter due to biodiversity (Gorshkov 1990, Marchuk *et al.* 1992, Marchuk and Kondratyev 1992).

Analysis of the biosphere dynamics requires data from a combined use of both conventional observations (on the land and in the ocean) and satellite observational means. This problem was the central one in the course of the realisation of the International Space Year (Kondratyev 1992b, Marchuk and Kondratyev 1992).

The decision to make the year 1992 an International Space Year has been favourable for a successful solution of the problem of planning and realising a global system of ecological observations. Of the utmost priority are developments and applications of remote sensing techniques to the solution of the following problems: (i) global vegetation dynamics, (ii) the atmospheric greenhouse effect, (iii) the ocean- atmosphere interaction, (iv) the global dynamics of ozone, particularly in polar regions, (v) the bioproductivity of the oceans, (vi) deforestation and desertification processes and (vii) the polar ice-cover dynamics.

Since ecological problems are numerous and diverse, it is necessary to identify the priorities, especially because means for resolving them are rather limited. In this regard, of primary importance are global-scale ecological problems which are sometimes a real threat to human life (Gorshkov and Kondratyev 1990, Gorshkov 1990, 1991, Kondratyev 1990b, Marchuk and Kondratyev 1992).

Life is possible on the Earth due to the interaction of physical, chemical and biological processes taking place in the environment and the biosphere is now under conditions of an ever-growing impact of man's economic activity on nature. There is no doubt, however, that the basic external factor regulating natural processes is the income of solar radiation which is determined by the value of the so-called "solar constant" (Kondratyev 1991a, 1992a), since it is in fact a single source of energy required for the functioning of natural processes. The favourable location of our planet within the Solar System, in terms of its distance from the Sun, has created unique possibilities for the emergence of life on the Earth. This is associated with the existence of a specific heat budget (the difference between the solar radiation absorbed and the loss of heat due to the outgoing long-wave radiation) which is on the average equal to zero (otherwise, a catastrophic heating or cooling of our planet would eventually happen). Of paramount importance for the evolution of life on the Earth are changes in the fine structure of the solar radiation spectral distribution (Kondratyev *et al.* 1990).

With regard to the internal regular features of the functioning of the processes on the planet, they are above all relevant to the matter and energy cycles (Kondratyev 1990b, 1991c).

2 Priorities

2.1 The Earth's heat budget

There are two problems of priority in the Earth's heat budget dynamics, namely (i) the variation in the incoming solar energy, the solar "constant" which is defined as the solar radiation flux integrated over wavelength, at the upper atmospheric boundary level, with an average distance between the Earth and the Sun and (ii) the anthropogenically-induced redistribution of the heat budget components of the surface and the atmosphere, which is above all noticeable as an enhancement of the greenhouse effect of the atmosphere due to the increased concentration of the greenhouse gases (GGs) (Houghton *et al.* 1990, Kondratyev 1992a).

As far as variations in the extra-atmospheric insolation are concerned, which are first of all significant as a climate forming factor, although these are rather considerable (up to a few percent) in the UV spectral region, the solar contact value varies within 0.2% only, as has been revealed by the analysis of satellite observation data. However, even these minor variations should be taken into account in the examination of causes for changes in the global climate. Since some satellite observations are still rather short-term (during about 10 years), it is not yet clear whether one can speak of long-term trends of the solar constant, although there are some grounds to suppose that such long-term trends exist (it is a known fact that during the several billion years of the Earth's evolution, the solar constant increased by 30-35%). On the scale of hundreds, thousands and more years, the role of the redistribution over the globe of extra-atmospheric insolation, caused by variations in the orbital parameters, seems doubtless. More and more observational data are accumulated revealing the influence of solar activity on climate; however, no important progress has been achieved thus far in the understanding of mechanisms determining this influence (Kondratyev 1992a).

From the point of view of monitoring the Earth's heat budget dynamics, most essential still are satellite observations of the extra-atmospheric spectral distribution of solar radiation, the solar constant and the Earth's radiation budget (ERB) components (Marchuk *et al.* 1990). Very important also are observations of the radiation budget (RB) of the surface and the atmosphere making use of ground-based, as well as ship-based, air-borne and balloon-borne observing aids. An exceptionally complicated priority problem is research in the entire set of transformations of the energy and water cycles due to anthropogenic impact. The urgent necessity of identifying the "greenhouse" signal in climate change requires the implementation of complex observational programmes and experiments. The realisation of the Global Energy and Water Cycle Experiment (GEWEX) (Kondratyev 1992a, Marchuk and Kondratyev 1992) will hopefully be an essential contribution to the solution of the above-mentioned problems.

2.2 The biosphere dynamics

As has been emphasised above, climate change is only a part (and not at all the main part, in spite of its high practical importance) of the global ecological change problem. The most acute problems of global ecology are relevant to the anthropogenic impact on the biosphere, namely deforestation, changes in the soil fertility and bioproductivity of the oceans; these result in breaking the closed nature of global biogeochemical cycles of such elements as C, S, N, etc. In the case of carbon, a substantial disturbance of its cycle dates back to the boundary between the last and current centuries.

There is no doubt that the variations of the gaseous composition of the atmosphere influence its greenhouse effect and can definitely affect the global climate. Closely related to this problem is the dynamics of the global ozone layer which is subject to the effect of emissions of chlorofluorocarbons (Kondratyev 1989). Equally important is the problem of tropospheric ozone - that of the content of ozone in the air we breath. One should bear in mind the fact that possible climate warming and the decrease in the total ozone content in the stratosphere will lead to enhanced smog photochemical reactions of the ozone formation near the ground, which will certainly be a threat to human health.

The problem of biosphere dynamics acquires fundamental significance and it should be considered to be the first priority problem of global ecology. An inherent component of global ecology are numerous regional ecological problems (desertification, environmental pollution, the trans-boundary transport, the degradation of forests, soils, etc.).

Of special significance is further progress of the simulation-numerical modelling of the biosphere dynamics (Marchuk *et al.* 1992, Marchuk and Kondratyev 1992), which is still in its initial stage of development and is actually reduced to the setting up and application of mostly simplified climate models. The gradual enrichment of the current biogeochemical cycles (first of all, of carbon, sulphur and nitrogen), and, consequently, of interactive reproduction of changes in the atmospheric chemical composition, the most essential characteristics of the biosphere and climate dynamics, appears to be quite promising. This kind of approach must be fruitful and helpful for achieving a more profound understanding (and, where possible, forecasting) of the regular features of climate change and the biosphere dynamics.

2.3 The observation system

The first priority of the problem of global biosphere dynamics and the disturbance of biogeochemical cycles (above all, the carbon cycle) accounts for the urgent necessity of substantiating and organising an optimal global observation system for monitoring processes taking place in the biosphere and the environment, which can be implemented only on the basis of international cooperation under the auspices of the UN.

The substantiation of an optimal (adequate) observation system which is non-existent today but is rather necessary, can only be made on the basis of the analysis of the simulation numerical modelling data. Its purpose is the formulation of priorities and requirements from the data of observations (Kondratyev and Pokrovsky 1989, Marchuk and Kondratyev 1992). Analysis of the regular features of the biosphere dynamics from observational data requires a combined use of both conventional (land-based and ocean-based) and satellite observational aids. Here it is important to concentrate conventional observational aids (especially ships) in such regions where there are particularly significant ecological changes as well as to use the ecosystematic approach to the planning of the system of observations (Kondratyev 1991b, Marchuk *et al.* 1992). This situation puts forward, as a first priority problem, that of optimal planning of global observational systems, including for example systems planned within the scope of the programme "Mission to Planet Earth".

In this connection (and in the context of priorities in general), of great interest are national global change programmes developed in the USA and in France (Research Strategies..., 1990; Scientific Orientation..., 1990).

3 National and international global change programmes

3.1 The US National Programme

Intensive research which is carried out in the USA under the guidance of the Committee on Earth and Environment Sciences (CEES) and within the scope of the US Global Change Research Programme (USGCRP), which can be regarded as a component part of the World Climate Research Programme (WCRP) and the International Geosphere-Biosphere Programme (IGBP) (Kondratyev and Pokrovsky 1989, Kondratyev 1991b) has the following main objectives:

1) Creation of an integrated, comprehensive and long-term programme of documenting the Earth's system dynamics on a global scale;

2) Implementation of a programme of focused (problem-oriented) studies for a more profound understanding of physical, geological, chemical, biological and social processes determining the Earth's system dynamics, including trends on global and regional scales;

3) Development of integrative conceptual and prognostic models of the Earth's system.

A feature of principal importance of the Programme is the interactivity of its all Sections, the central component of the implementation of the Programme strategy being the combination of adequately balanced conventional and satellite observational aids. Conventional (in situ) observations in combination with theoretical investigations of physical, chemical, biological and geological processes must be complemented with a comprehensive programme of space research which can ensure observations of the key parameters of the environment on a global scale (it is common knowledge that the beginning of the full-scale functioning of a satellite Earth Observing System (EOS) is planned for 1998).

Proceeding from the recommendations of the Intergovernmental Panel on Climate Change (IPCC) (Houghton *et al.* 1990, Kondratyev 1991a, 1992a), the Committee CEES put forward the following main objectives of the USGCRP:

1) Speeding up of research to restrict the uncertainty limits (determined by IPCC) for conclusions about causes and scales of climate change, particularly from the viewpoint of assessing:

1.1 The sources and sinks of greenhouse gases (GGs), the knowledge of which determines the degree of reliability of forecasting the GG concentration in the future:

1.2 The cloud cover and radiation budget dynamics, which determines to a great extent, the climate change on a global and regional scales;

1.3 Processes in oceans and also the interaction between the atmosphere and ocean, which are the key climate forming factors;

1.4 Processes on the land surface, which influence the regional climates and water resources;

1.5 Polar ice dynamics determining potential changes in the World Ocean Levels;

1.6 The ecosystem dynamics which is subject to the effect of climate change.

2) Intensification of complex GG research, considering the following priorities:

2.1 Analysis of GG sources and sinks (not separately, but on the basis of complex studies);

2.2 Development of quantitative indicators of the radiative forcing effects:

2.3 Substantiation and application of economic models aimed at forecasting, taking account of the entire set of GG (as climate change factors) and various branches of economics.

Proceeding from the IPCC recommendations (Houghton *et al.* 1990), four key directions of interdisciplinary research have been defined:

1. Numerical modelling and climate forecasting as a basis for future potentially feasible broader modelling of the Earth's system

In this connection, of decisive significance is the advancement of 4-dimensional assimilation methods, and also those of the parametrisation of climate forming processes, the development of interactive models of the climatic system and their realisation using computers of new generations. Palaeoclimatic studies continue to be very important as an information source for testing the reliability of climate models, but not for using palaeoanalogues for prediction of the climate of the future (Houghton *et al* 1990, Kondratyev 1992a).

2. Global water and energy cycles

Here the main objectives are: a more profound understanding of water cycles (precipitation, evaporation, evapotranspiration, soil moisture, ice and snow cover dynamics) and energy cycles (climate warming/cooling, the radiation budget, latent heat, the solar activity). Attention should be focussed on:

2.1 The cloud cover dynamics (processes of phase transformation, radiative properties of clouds, etc.);

2.2 The study of energy and mass exchange between the ocean (taking account of ice cover) and atmosphere; between the top and bottom ocean layers; the transport in the ocean thickness;

2.3 The study of the energy and mass exchange between the land and atmospheric ecosystems, taking account of anthropogenic impact on the ecosystems;

2.4 Study of changes in the World Ocean level as indicators of the effect of climate variations on the Earth's system (taking account of the ocean thermal expansion, glacier melting and polar ice sheet dynamics).

3. Global carbon cycle

The central task of relevant research should be the quantitative assessment of natural and anthropogenic sources and sinks of basic carbon compounds on the land and in the ocean (for instance, carbon dioxide and such precursors in the processes of the formation of methane and ozone as CO and non-methane hydrocarbons), taking account of chemical reactions in the atmosphere and ocean with the participation of the above-mentioned carbon compounds. Very important are studies of fluxes of these compounds due to physical, chemical and biological processes, as well as flux changes caused by climate variations and the GG dynamics.

4. Ecosystems and the population dynamics.

In this case, the main purpose is the development of methods for more reliable estimations of global change effects on regional scales. The first priority here is given to the analysis of the response to global change on the part of such ecosystems on the land and in the ocean which are subject to strong anthropogenic impact, with special emphasis on the following directions of research:

4.1 The species composition of ecosystems.

4.2 The spatial extension and distribution of ecosystems:

4.3 The productivity of ecosystems.

The key directions of research outlined above do not exhaust the contents of the "Global Change" programme. The studies described above should be complemented, for example, by research on the economics of the use of natural resources, investigations of solar-terrestrial relations and the global dynamics of ozone. As a rule, these developments are envisaged as part of international efforts within various international research programmes, and first of all, WCRP and IGBP. In this connection, very important are such subprogrammes of WCRP as the Tropical Oceans and Global Atmosphere Programme (TOGA) and the Global Energy and Water Cycle Experiment (GEWEX), and also a number of core projects within IGBP, viz.: the Joint Global Ocean Flux Study (JGOFS), the International Global Atmospheric Chemistry Project (IGAC) the Biological Aspects of the Hydrological Cycle (BAHC), the Global Change and Terrestrial Ecosystems (GCTCE), the Past Global Changes (PAGES). It should be mentioned here that all these projects have been discussed in detail in the monograph by Marchuk and Kondratyev (1992).

It is justly stated in the Programme (Research Strategies..., 1990) that political leaders of both national and international ranks must primarily be interested in those global change aspects which are relevant to the problem of providing favourable conditions for life on the Earth. However, one must admit that the contents of the Programme does not reveal an adequate treatment of all basic problems relevant to the USGCRP and possible way of their solution. Since the programmed objective of studying global change are presented elsewhere (Marchuk and Kondratyev 1992) in a sufficiently comprehensive manner, it will be enough here to concentrate on two USGCRP subjects of principal importance, viz.:

1. The thesis that global ecological security (favourable conditions for living on the planet) is determined by the closed nature of global biogeochemical cycles (Gorshkov and Kondratyev 1990, Kondratyev 1990b, Gorshkov 1990, 1991) has not been distinctly

formulated. It is important that among top priority problems in the Programme there is one concerning the global carbon cycle, but the actual essence of the question is far broader. One is surprised by the lack of treatment of such significant problems as those of water resources and desertification.

2. The role of socio-economic "roots" of global ecology problems is underestimated, especially the most alarming of these, namely the growth of population on the Earth (Kondratyev 1990b, Marchuk and Kondratyev 1992) - is obviously underestimated. It goes without saying that without a proper consideration of demographic problems and adequate accounting of the huge contrast in the extent of development between the highly industrialised countries and the Third World, one can hardly speak seriously of analysing the long-term global ecological perspective. However, the problem of the population growth dynamics is mentioned in the Programme only vaguely.

3.2 The French National Programme

Of great interest are similar programme developments in France, a brief survey of which can be found in Report (Scientific Orientation..., 1990). The CNRS (National Centre for Scientific Research) Environment Programme, the implementation of which started in 1991, proceeds for the necessity of complex research integrating the efforts of environmentalists and social scientists, and it will develop along the following lines:

1. law and environmental economy;
2. the historical aspect of environmental problems;
3. analysis of the most important systems and evolution of the systems (land hydrology, coastal environment, urbanisation, etc.);
4. the understanding of the main mechanisms and effects of anthropogenic impact on the environment (reaction of populations and ecosystems to environmental disturbances, biological invasions, etc.):
5. the key aspects of global change (the atmospheric components of global biogeochemical cycles, the global change-induced dynamics of ecosystems, etc.).

All these and other developments must be first of all oriented toward the solution of the problems of human ecology, since the problem of human health is central.

One of the most difficult tasks is that of substantiating the priorities of concrete research developments. With regard to the ecosystem studies, special attention should be given to tropical and temperate forests, steppes and savannas, continental hydrological systems (rivers and lakes), seas and oceans, the interface zones (ecotones), urbanized systems; agro-ecosystems, "stress" ecosystems (deserts, Arctica and Antarctica, high mountain areas, etc.), synthetic ecosystems (e.g. manned orbital stations).

The CNRS Environment Programme emphasises the primary importance of the "human factor" of the ecological problematics and the necessity for the intensification of efforts in this direction. Particularly important is the analysis of the compatibility of demographic and economic growth under conditions of an ever- increasing stress on the environment.

5. Natural and anthropogenic changes in the environment.

Let us consider the main scientific factors that have to be taken into account in any study of global climate change. The evolution of the geosphere and biosphere is characteristic of a more and more intense interaction between their components and this is first of all due to the increasing anthropogenic impact on the environment. As a result of the interaction between geospheric and biospheric components, each is affected by an ever-increasing anthropogenic load. In order to discover the laws of interaction between the geosphere and the biosphere, the following measures appear to be necessary: a systematic approach, interdisciplinary research of unprecedented complexity and consideration of relevant natural and social problems as top priority ones.

Problems of global ecology have become extremely severe by the end of the 20th century. General attention is attracted to the problem of possible climate warming as a result of the ever increasing concentration in the atmosphere of such GGs as carbon dioxide, methane, nitrous oxide, tropospheric ozone, etc. As has been noted, quite to the point, in the Report of the Intergovernmental Panel on Climate change (IPCC), approved by the participants of the Second World Climate Conference (SWCC) (Geneva, 29 October - 7 November 1990) (Houghton *et al.* 1990), for the present moment, it is not at all clear to what extent the global climate warming observed during the last 100 years is due to the natural variability of the climatic system and to what extent it is due to the increasing concentration of GGs. The results of numerical climate modelling that are available are not at variance with the possibility of anthropogenically induced warming, but they cannot be regarded as definitive proof of the possibility either.

Doubtless, the very important fact that water vapour is the basic GG requires special attention. The problem is that water exists on the Earth in three phases and this leads to substantial difficulties in taking account of the phase transformations of water and their consequences in the theory of climate. Thus, for instance, the problem of adequate parametrisation of the processes of formation and evolution of cloud cover, as well as its interaction with radiation, has not so far ben solved. It is very important to remember, as has been pointed out by Lindzen (1991), that the contribution of water vapour and stratus clouds to the downward flux of atmospheric thermal emission at the surface ("gegenstrahlung") reaches about 98% of the total flux, which is equal, on the average, to 340 Wm^{-2}. The rest of the GGs contribute only a little. It is well known for instance that a doubling of the CO_2 concentration might lead to an increase of the "gegenstrahlung" of the order of 3-4 Wm^{-2} (Kondratyev 1992a).

Equally important is the accounting of the climate-forming role of the interaction between the ocean and atmosphere (Houghton *et al.* 1990, Kondratyev 1992a, Losev et al. 1992). In this content, it should be pointed out that it is undoubtedly true that the recommendation of the SWCC on the necessity to reduce GG emissions into atmosphere is most important. It must be remembered, however, that this is not a panacea that will guarantee the ecological safety of the planet. It is equally clear that socio-economic situations in developing countries should not be ignored. In this respect, it seems appropriate to quote here the Beijing Ministerial Declaration (1991), "Changes in the environment are closely related to economic and social activities. Environmental problems must not be viewed in isolation; they need to be addressed together with the process of development, by integrating environmental concerns with the imperatives of economic growth and development. In this connection, the right to development of the developing countries must be fully recognised and the adoption of measures for the protection of the global environment should support their economic growth and development... Each country must be able to determine the pace of transition, based on the adaptive capacity of its economic, social and cultural ethos and capabilities. The environmental problems of the developing countries arise from the conditions of poverty. Sustainable development and steady economic growth constitute a way to break this cycle of poverty and environmental degradation and to strengthen the capabilities of the developing countries for environmental protection... It is therefore imperative to establish a new and equitable international economic order conducive to the sustained and sustainable development of all countries, particularly the developing countries, thereby creating the conditions necessary for global cooperation to protect the environment... The developed countries, in view of their main responsibilities for environmental degradation and their greater financial and technological capabilities, must take the lead in eliminating the dame to the environment, as well as in assisting the developing countries to deal with the problems facing them."

Also of key importance from the viewpoint of climate change is the fact that, under conditions of an unperturbed natural environment, the global biogeochemical cycles of various elements (carbon, sulphur, nitrogen, phosphorus, etc.) are closed with a high degree of accuracy (of the order of 0.01%) (Gorshkov 1990, Gorshkov and Kondratyev 1990, Kondratyev 1990a). During the last two centuries, this closure has decreased by approximately one order of magnitude; if this tendency continues it will constitute a serious threat of a global ecological catastrophe, such as the destruction of the currently existing biosphere in a few centuries from now.

That is why the problem of biosphere dynamics acquires a fundamental significance and should be considered a top priority problem of global ecology.

The Earth's system is an open one in the sense that it is subject to considerable impact of cosmic factors, such as, e.g., the solar activity and galactic cosmic rays. Many aspects of this system are still unknown, which opens up possibilities for sudden events and abrupt changes. What is quite clear, however, is the necessity long-term monitoring of the variations of the extra-atmospheric total and spectral solar radiation with an emphasis on the biospheric aspects of the problem (Kondratyev 1990a, 1991).

Numerous regional ecological problems (desertification, environmental pollution, transboundary pollution transport, forest decline, soil degradation, etc.) are an inherent component of global ecology.

In view of the limited possibilities of conventional observational means (surface observations, balloon and aircraft measurements, ship observations, etc.), it is very important to identify the areas of the global that are of critical environmental significance from various points of view. A concept of the so-called Energy Active Zones (EAZO) (Losev *et al.* 1992) may serve as an illustration of a quantitative approach to the substantiation of environmentally significant regions from the point of view of energy exchange between the atmosphere and the surface.

Another example is the FIFE field experiment First ISLSCP (International Satellite Land Surface Climatology Project) Field Experiment and subsequent similar efforts to study the biosphere as an interactive component of the climatic system under various typical conditions (steppes, forests, deserts, etc.).

Since regional problems have different scales, a necessity arises to identify problems within the range from a continental level (future environments for Europe) to a strictly regional level (the dynamics of the Aral Sea and its environs; the dynamics of the basins of large rivers, such as the Zambezi, Mississippi, Rhine, Neva, Volga, etc.; the dynamics of forests - both tropical and boreal, etc.).

A task of special importance is the further development of three-dimensional numerical models to stimulate relevant regional processes. The monograph by Marchuk *et al.* (1989) contains a description of a 3-dimensional mesoscale model worked out to stimulate the entire succession of cloud development processes with detailed consideration of dynamics, thermodynamics, water phase transformations (including droplet size distribution dynamics) and radiation. One of the purposes of the employment of this model has been substantiation of improved cloud and radiation parametrisations in climate models.

4 Optimal planning of observations

Successful solution of the above-stated problems requires developing a global observational system which would employ both conventional and space-based observation means. The latter must be given priority in view of the global scale of information to be obtained. For this purpose, meterological and Earth resources satellites and also manned orbital stations are generally used nowadays.

4.1 Users' demands with regard to satellite information.

In most cases, users of satellite information (SI) can formulate very precisely their demands concerning the basic characteristics of the data needed (spectral channels, the geometry of the experiment, the spatial-temporal resolution, etc). In studies discussed in what follows, a summary of such demands has been used which is presented by Kondratyev *et al.* (1975) and is relevant to four groups of problems: (i) oceanology (102 requirements), (ii) hydrology (33 requirements), (iii) geology (36 requirements) and (iv) forestry and agriculture (32 requirements). Table 1 illustrates examples of such requirements for several important problems in each of the four groups. The diversity of the requirements makes it necessary to seek optimal conditions for realising the experiment, so that it would be possible to ensure the solution of the entire set of problems under investigation in the best possible way.

The difference between individual requirements is first of all relevant to the choice of spectral sounding channels. comparison of data concerning the choice of channels reveals the fact that the instrumentation available is not at all optimal and thus shows the necessity of solving the problem of optimal planning of the parameters of remote sensing instruments. Table 2 presents requirements of active/passive measurement systems in the microwave region, which are intended for the solution of a number of problems involved in the above mentioned four-groups ((i)-(iv)). The boundaries of spectral intervals are given in Table 3.

Applications	Resolution (m)	View angle (deg.)	Swath width (km)	Spectral Intervals (μm)		
				Visible	Near IR	IR
I Oceanology						
1. Mapping of coastal currents and tides	60-90	45-60	200	0.42-0.44; 0.45-0.47 0.52-0.53; 0.55-0.57 0.57-0.64; 0.64-0.69 0.68-0.70	0.77-0.80	10.9-12.6
2. Turbidity of waters	50-100	0-30	40-100	0.30-0.5; 0.5; 0.65	0.90	
3. Mapping of the ocean surface (colour, salinity, oxygen content, bioorganisms, pollution, temperature)	30-60	0-20	100	0.37-0.40; 0.40-0.43 0.43-0.45; 0.45-0.48 0.48-0.51; 0.51-0.53 0.53-0.55; 0.55-0.57 0.57-0.61; 0.61-0.66 0.66-0.70	0.70-0.82	10.9-12.6
4. Sea Ice	100-200	0-30	100-200	0.45-0.53; 0.53-0.57 0.50-0.58; 0.57-0.64 0.64-0.69	0.73-0.82 0.8 -1.1	10.9-12.6
5. Propagation and migration of sea organisms. Biological pollution	1000	0-10	50-100	0.45-0.53; 0.53-0.57 0.57-0.64; 0.64-0.69 0.50-0.58	0.73-0.82	8-14
6. Coastal pollution	10-30	0-10	50-100	0.45-0.53; 0.53-0.57 0.57-0.64; 0.64-0.69 0.30-0.40; 0.48-0.59 0.50-0.58	0.73-0.82 0.80-0.90	10.9-12.6

Table 1: Requirements for Optical Remote Sensing Systems.

Applications	Resolution (m)	View angle (deg.)	Swath width (km)	Spectral Intervals (μm)		
				Visible	Near IR	IR
II Hydrology						
1. River Basins	5-10	0-30	200	0.68-0.74	0.78-1.2	
2. Flood, high water	5-15	0-10	50	0.67-0.75	0.7 -1.3	10-14
3. Swamp	3-5	0-10	50	0.51-0.57; 0.62-0.67; 0.53-0.57; 0.57-0.64; 0.64-0.69	0.7 -1.3 0.73-0.82	10.9-12.6
4. Snow Cover (compactness, melting)	10-50	0-20	200	0.38-0.51; 0.44-0.52; 0.69-0.88; 0.50-0.67;	0.82-0.84 0.7 -1.3 0.49-1.2	
5. Glaciers and ice cover on rivers and lakes	5-15	0-30	50-100	0.44-0.50; 0.50-0.55; 0.46-0.52; 0.52-0.58; 0.50-0.58	0.79-0.84 0.7 -1.3	8-14
6. Precipitation and soil moistening	5-30	0-10	50	0.49-1.2	0.78-0.84 0.8 -1.3	8-14 10.9-12.6
7. Pollution of water reservoirs	5-10	0-10	200	0.3 -0.4; 0.48-0.59	0.8 -0.9	8-14 10-12

(Table 1 Continued)

Applications	Resolution (m)	View angle (deg.)	Swath width (km)	Spectral Intervals (μm)		
				Visible	Near IR	IR
III Geology						
1. Mapping of the relief	10-20	0-30	100-500	0.51-0.63; 0.63-0.70	0.70-0.89 0.89-1.0 0.8 -1.3	8-14
2. Volcanic activity	10-20	0-20	100-200	0.4 -0.7		10-12 8-14
3. Lithological mapping	10-20	0-30	100-200	0.51-0.63; 0.63-0.70 0.40-0.48; 0.51-0.57 0.62-0.67	0.7 -0.89 0.89-1.0	8-10 10-10.6 10.6-12.9 9.3-12.8
IV Forestry and agriculture						
1. Mapping of soils (compactness, melting)	5-15	0-45	100-200	0.62-0.67	0.74-0.85 1.5 -1.8 2.0 -2.6	8-14
2. Soil Erosion	5-10	0-45	40-50	0.40-0.48; 0.51-0.57; 0.62-0.67	0.73-0.78	
3. Forest inventory	10-30	0-10	50-100	0.55-0.58; 0.70-0.74	1.5 -1.8 2.0 -2.6	
4. Forest fires	10-20	0-70	500-1000	0.4 -0.7	3.5 -5.4	8-14
5. Assessment of crop state	5-10	0-10	50-100	0.55-0.58; 0.66-0.77; 0.70-0.74; 0.51-0.57; 0.59-0.68	0.76-0.83 1.5 -1.8 2.0 -2.6	8-14
6. Forest vegetation disease	10-15	0-10	40-50	0.40-0.44; 0.55-0.58; 0.70-0.74; 0.51-0.57; 0.58-0.68	0.78-0.84 0.83-0.86	8-14

(Table 1 Continued)

Groups of Problems	Economy Problems	Spectral Regions Used	System Nature Active/Passive	Instruments
I Oceanology	1. Monitoring of sea ice drift	K_u, X, C, S, L	Passive	Multichannel radiometer
	2. Control of sea ice return	K_u, X	Passive	Radiometer measuring in directions close to nadir
	3. Compilation of topography map of ocean surface	K_u	Active/Passive	Altimeter-radar directed towards nadir
	4. Temperature distribution in the atmosphere and near the ocean surface	K_a, K, K_u, X	Passive	Multichannel scanning radiometer
	5. Wind near the ocean surface	K_u, C	Active/Passive	2-ray radar-scatterometer operating in 2 azimuthal directions
	6. Mapping of ocean salinity	L	Passive	Radiometer (nadir measurements)
II Hydrology	1. Mapping of water surface	C, L	Passive	Radiometer (nadir measurements)
	2. Precipitation	Q, K_A, K	Passive	Radiometer (nadir measurements)
	3. Atmospheric water vapour	Q, K_a	Passive	Radiometer (nadir measurements)
III Forestry and Agriculture	1. Soil Moisture	L	Passive	Scanning radiometer
	2. Determination of vegetation state (leaf surface index, biomass estimation, bio-productivity)	C	Active/Passive	Radar with 2 receivers conjugated with 2 antennas for horizontal and vertical polarizations
IV Geology	1. Compilation of structure and morphology maps	Z	Active/Passive	Radar-radiometer (nadir measurements)
	2. Mapping of latent peculiar features of relief	L	Active/Passive	

Table 2: Requirements for Microwave Remote Sensing Systems in the atmosphere and near the ocean surface.

Band designations Boundaries(cm)	W 0.13-0.33	Q 0.33-0.63	K_a 0.72-1.0	K 1.0-1.9	K_u 1.9-2.5
Band designations Boundaries(cm)	X 2.5-3.5	C 3.5-7.1	S 7.1-15.8	L 15.8-63	P 63-100

Table 3: Spectral Interval Boundaries in the Microwave Region

4.2 Analysis of the characteristics of instruments used earlier.

The data of Tables 1 and 2 reveal the diversity of requirements for the instrumentation. Analysis of the characteristics of a part of the operational satellite-borne instrumentation (see Table 4) shows considerable overlapping of various parameters. It is interesting to see that spectral sounding channels have often been chosen in a similar way for different instruments, sometimes on board the same satellites. Multispectral Scanner (MSS) data have been widely used by numerous experts in the field of Earth resources studies. Therefore, the limitations of spectral intervals of respective sounding channels have been chosen approximately. Respective characteristics of the Thematic Mapper (TM) have been chosen in the interest of a narrower group of problems. Requirements relevant to a single group of problems in oceanology have been met by the spectral characteristics of the Coastal Zone Colour Scanner (CZCS). The SPOT HRV instrument surveys water basins.

The agreement between the demands of individual information users and the choice of spectral characteristics for satellite borne instruments is illustrated by data borrowed from Kondratyev and Pokrovsky (1977), Kondratyev *et al.* (1975) and Pokrovsky (1984)

4.3 Optimisation of observations

The importance of the problem of optimal planning of systems for remote sensing from space is determined by a number of circumstances, including the costliness of the experiments and limitations on the weight of the instruments, the power supply system and the volume of information stored on board. Besides, one must consider a whole set of scientific aspects of the problem of satellite data interpretation such as the mathematically incorrect inverse problem (the environmental parameters are retrieved by making use of the characteristics of either reflected or absorbed radiation), the fluctuations of atmospheric optical characteristics that cannot be controlled, etc.

When one is busy with solving the problem of planning a remote sensing experiment, one must avoid two cases of extremity, namely excessive or insufficient information as a result of poor choice of observation conditions. In each of the cases, the solution of an inverse problem may greatly deteriorate because of its non- correctness.

Two approaches have been worked out with regard to the solution of the problem of planning a multi-purpose remote sensing system, namely the informative one and the common approach (Pokrovsky 1984). In both cases, it is assumed that the information users are able to formulate accurately their requirements in terms of major characteris-

Instrument	Spectral Channels (μm)	Spacial Resolution (nadir) (m)	Swath Width (km)	Notes
MSS	0.50-0.60			Multi- spectral
	0.60-0.70			Scanner
	0.70-0.80			
	0.80-1.10	80	185	
TM	0.45-0.52			Thematic
	0.52-0.60			Mapper
	0.63-0.69			
	0.70-0.90	30	185	
CZCS	0.43-0.45			Coastal Zone
	0.51-0.53			Colour Scanner
	0.54-0.56			
	0.66-0.68			
	0.70-0.80			
	1.05-1.25	825	1636	
SPOT-HRV	0.50-0.59			Water basin
	0.61-0.68			survey
	0.79-0.89	20	60	
	0.51-0.73			

Table 4: Characteristics of Satellite-borne Instruments in the Visible and near-IR Regions.

tics of the information they need (the spectral channels, the geometry of observations, the spatial-temporal resolution, etc.). As has been mentioned earlier, a summary of the above requirements has been used in some studies discussed below (as described by Kondratyev *et al.* 1975) where four groups of problems relevant to four branches of science can be found, namely (i) oceanology (102), (ii) hydrology (33), (iii) geology (36), (iv) forestry and agriculture (32).

4.3.1 Informative approach.

The idea of this approach (Kondratyev and Pokrovsky 1977) consists in the selection of most important factors and finding respective vectors of factor loads in to formalised space of requirements. The above-mentioned work offers a method of building a space of requirements and a modification of an applied scheme of factor analysis. The results of determining optimal sets of spectral intervals are presented in Table 5 for groups of problems (i-iv). The above publication contains the results of calculations for various complexes of problems. It should be noted that the data in Table 5 (line 1) have been obtained exclusively on the basis of the information users' requirements; therefore, no account has been taken of atmospheric correction in various regions of the electromagnetic spectrum has been made, or of the economic aspects of space-based experiments. Table 5 (line 2) shows the limits of spectral sounding channels specified with the H_2O, O_3, and NO_2 absorption bands excluded in the following intervals: 0.38-0.42 μm, 0.57-0.62 μm, 0.755-0.760 μm, 0.785-0.790 μm, 0.810-0.845 μm, 0.910-0.960 μm, 1.10-1.16 μm. A considerable contribution of molecular scattering on the part of the atmosphere results in the low information content of the 0.30-0.40 μm channel. That is why this channel is not used in the remote sensing instruments (see Table 4).

4.3.2 Analysis of potential possibilites of instrumentation taking account of users' requirements

If sounding channels chosen are organised in the order of increasing wavelengths, a relation between similar channels can be obtained for different groups of problems (Table 6). It can be seen that the short wave channel 0.42-0.44 μm is used for oceanology problems only. One can see from the data of Table 4 that the corresponding channel 0.43-0.45μm is also used exclusively in a specific oceanographic instrument, the Coastal Zone Colour Scanner (CZCS).

The set of channels for oceanology problems presented in Table 6 has much in common with the set of spectral intervals for CZCS (Table 4). The difference in the limits of respective spectral channels is due to the fact that the scheme of the oceanographic scanner does not take into account the users' demands (see Kondratyev et al. 1975) or the spectral optical characteristics of the atmosphere. Thus, for example, the spectral channel 0.62-0.64μm which is present in the requirements for each problem in the oceanology group, is absent in the CZCS scheme.

The data of Table 6 reveals that measurements in the two narrow channels (0.55-0.57 μm and 0.76-0.78μm) are required for the group of problems "forestry and agriculture" only. However, experts interested in this group of problems cannot obtain information in the spectral intervals with any instrument available. The narrow channel 6 is needed for oceanology and hydrology problems.

Specialists in geology and in forestry and agriculture use information obtained in the broad interval 0.62-0.70μm, where there is a combination of channels 5 and 6 for problems of groups (i) and (ii). The spectral region is present in the instruments available as a single spectral interval. This means that the instruments available do not meet the requirements of groups (i) and (ii) as far as the data for channels 5 and 6 are concerned.

Comparison of data in Tables 4 and 6 shows that the information obtained with the TM agrees fairly well with the requirements for group (iii), but not so well with those for group II. The MSS is in poor agreement with the users' requirements. This happens because it yields no information to channel 2 (Table 6). Channels 5 and 6 in its scheme which belong to the optimal set are combined and made into one channel. Information for channels 0.07 - 0.80μm and 0.80 - 1.10μm is heavily distorted due to reflected radiation absorption by the atmosphere. The data reveal the fact that satellite-borne instruments which are presently available are inadequate considering the demands of satellite information users in various fields.

	Group of Problems / Factors	1	2	3	4	5	6	7
I.	Oceanology	0.45-0.53 0.45-0.53	0.64-0.69 0.64-0.69	0.70-0.90 0.70-0.75	0.57-0.64 0.62-0.64	0.82-1.30 0.96-1.10	0.50-0.58 0.50-0.57	0.40-0.44 0.42-0.44
II.	Hydrology	0.70-1.30 0.96-1.10	0.51-0.59 0.51-0.57	0.44-0.52 0.44-0.52	0.67-0.74 0.67-0.74	0.57-0.67 0.62-0.67	0.30-0.40 0.70-0.75	0.38-0.49 0.42-0.49
III.	Geology	0.51-0.63 0.51-0.57	0.62-0.70 0.62-0.70	0.70-0.89 0.70-0.75	0.40-0.51 0.42-0.51	0.8 -1.30 0.96-1.10		
IV.	Forestry and agriculture	0.51-0.57 0.51-0.57	0.59-0.70 0.62-0.70	0.75-0.78 0.76-0.78	0.70-0.74 0.70-0.74	0.40-0.51 0.42-0.51	0.55-0.58 0.55-0.57	0.88-1.30 0.96-1.10

Table 5: Choice of Spectral Intervals (μm) using Factor Analysis

	Groups of Problems	1	2	3	4	5	6	7	8	9
I.	Oceanology	0.42-0.44	0.45-0.53	0.50-0.57		0.62-0.64	0.64-0.69	0.70-0.75		0.96-1.10
II.	Hydrology		0.44-0.52	0.51-0.57		0.62-0.67	0.67-0.74	0.70-0.75		0.96-1.10
III.	Geology		0.42-0.51	0.51-0.57		0.62-0.70		0.70-0.75		0.96-1.10
IV.	Forestry and agriculture		0.42-0.51	0.51-0.57	0.55-0.57	0.62-0.70		0.70-0.76	0.76-0.78	0.96-1.10

Table 6: Optimal Spectral Souunding Channels Relevant to Basic Groups of Problems

4.3.3 Cost-effective approach

Since in the future the development and implementation of space-based remote sensing systems will probably need to be financed by the users of satellite information, there arises the problem of cost-effectiveness of planned instrument complexes intended for long-term operation on board space-based platforms. The multispectral nature of satellite information and multi-purposedness of its application make it possible to analyse the problem of optimal planning and respective observational systems from the viewpoint of maximisation of the integral economic efficiency factor for a multi-purpose observational system V under the condition that the efficiencies for individual problems V_i (i=1, ..., N) are regulated by special limitations of the type of inequalities given above and below.

We shall dwell on the formulation of the problem. Suppose that C_{ij} is the economic efficiency of the j-th measuring channel ($j = 1, \ldots.., M$) for the purpose of the i-th individual problem; d_j is the total cost including expenses for processing information for the j-th channel of the multi-purpose measuring system and b_i^+, and b_i^- are the maximum and minimum values of economic efficiency of the use of observational data for the i-th problem. In this case, for the economic efficiency k_j of the employment of the j-th measuring channel we obtain the following expression:

$$k_j = \sum_i C_{ij} - d_j \quad (j = 1, \ldots\ldots M). \tag{1}$$

By introducing the y_j variable, an indicator of the inclusion into the observational system of the j-th channel $y_j = \{^0_1$, we formulate the basic problem of optimal planning as a standard problem of the Bulev whole number programming:

$$y = \min(y_1, \ldots\ldots, y_M)\{-\sum_j k_j y_j\} \tag{2}$$

$$y_j = \left\{ \begin{array}{c} 0 \\ 1 \end{array} , \quad (j = 1, \ldots, M) \right. \tag{3}$$

$$b_i^- \le \sum_{j=1}^{m} C_{ij} y_j \le b_i^+ \quad (i = 1, \ldots, N) \tag{4}$$

One of the most efficient numerical alogirithms for solving the problem (2) - (4) is Balash's method. This method, realised in a standard set of computer programs, has been used in the calculations for planning multi-purpose remote sensing systems for monitoring the environment from space. As an illustration, we shall consider the results of choosing an optimal set of spectral intervals for a multi-purpose survey intended for application in oceanology (see Table 7) (Marchuk and Kondratyev 1992, Pokrovsky 1984).

The number of spectral channels for the solution of 12 problems (N = 12) relevant to the group "Oceanology" (Table 6) totals 37 (M = 37), the number of particular channels m_i for the i-th specific problem ranging between 18 and 33 ($18 \le m_i \le 33$). This means that the volume of information for every specific problem is sufficiently

1	Sea Surface state
2	Water turbidity
3	Sea ice
4	General near-shore marine survey
5	Mapping of coastal currents and tides
6	Global scale mapping of currents, sea surface survey
7	Mapping of the coastline and shallow waters
8	Bathymetry and toporaphy of ice
9	Distribution of migration of sea organisms
10	Pollution of coastal waters
11	Impact of pollution on marine environment
12	Study of bars, reefs, etc.

Table 7: Optimal Spectral Sōuunding Channels Relevant to Basic Groups of Problems

large. It comprises at least half the total volume of information for the entire groups of problems "Oceanology".

In designing multi-spectral instruments, one proceeds from the fact that the expenses are inverse proportional to the width of the spectral interval and in direct proportion to the spectral resolution. Let β be the cost of a channel with a resolution of 1 nm. Then the cost $d_{\Delta\lambda}$ of a channel with a resolution $\Delta\lambda$ (nm) will be $d_{\Delta\lambda} = \beta/\Delta\lambda$. Let "$b_i$" be the total efficiency of the use of satellite information in the i-th oceanology problem. Then $\sum_i b_i = b$. Let all b_i be identical. Then $\sum b_i = b/N$ is an average efficiency related to a single problem. Introducing a threshold value $\alpha (0 < \alpha < 1)$ of permissible deviations from average efficiencies, we define for equation (4) the constraints:

$$b_i^+ = (1+\alpha)b_i; b_i^- = (1-\alpha)b_i \quad (i = 1, \ldots\ldots, M). \tag{5}$$

Let I_i be a set of channel numbers included in the requirements for an observational system for the i-th individual problem. We introduce an indicator function of the j- th channel, according to the equality:

$$x_j(I_i) = \begin{cases} 1, & \text{if } j \in I_i \\ 0, & \text{otherwise} \end{cases} \tag{6}$$

Then a set of coefficients C_{ij} in the optimization problem (2) - (4) is determined from the equality:

$$C_{ij} = \frac{b_i}{m_i} X_j(I_i) \quad (i = 1, \ldots, M; j = 1, \ldots, N). \tag{7}$$

When certain numerical values were used such as $\beta = 5, a = 31$, all the parameters in (2) - (4) were calculated. Calculations of the economic efficiency factors of measurement channels k_j (j = 1.... ..N) have revealed that some channels are potentially unfavourable (negative values of k_j) for multi-purpose application of the observational system. These are channels 0.40-0.45μm and 0.69 - 0.73μm . Measurements in the following spectral channels are, on the contrary, the most justified economically, namely 0.53-0.55μm,

0.66-0.68μm, 0.59-0.61 μm, 0.61-0.64 μm, 0.55 - 0.57 μm. The maximum values of $\{k_j\}$ are reached here. Thus, even simple calculations make it possible to select from a big initial list of spectral channels those that are most advantageous form the viewpoint of economic criteria.

Thus, out of all channels given in Table 6, channels 1 and 7 appear to be less (or perhaps the least) favourable in the optimal scheme for problems of oceanology. However, this information should only be taken into account when specialised oceanographic instruments are designed. There are similar non - advantageous spectral regions for other groups of problems. For example, for the group of problems "forestry and agriculture", the most favourable channels are 1.5-1.8 μm and the least favourable are 0.96-1.10 μm (channel 9). All channels are economically justified when the "hydrology" group is concerned.

5 Conclusions

5.1 The international scene

The inadequacy of "software" for global ecology problems and the insufficiency of efforts on both national and international scales provoke putting forward a proposal concerning an urgent necessity of discussing the global change problem (above all, from the view point of providing favourable living conditions on the Earth not only in the near future but for a long time ahead) on a level of a Conference called for by the UN Secretary General. A conference on this level might, first having defined the priorities, then stimulate development and implementation (on an international basis) of an adequate programme directed towards the study and solution of the global change problem. The UNCED has only been a step towards such a goal.

5.2 The requirements for satellite data.

1. Basic requirements for studies within the IGBP have been analysed.

2. Requirements of satellite information users for observation systems for basic groups of problems have been systematised.

3. Comparison between these requirements and the characteristics of the instrumentation available today for meteorological and Earth resources satellites has revealed the fact that their features are not optimal.

4. Optimisation using an information maximum has made it possible to determine a set of spectral sounding channels intended for four groups of problems: oceanology; hydrology; geology; forestry and agriculture.

5. Maximisation of the total economic efficiency index, taking account of the multispectral and multi-purpose nature of satellite information, makes it possible to differentiate spectral sounding channels for each group of problems according to their economic benefits.

References

Baker, J. J., 1991, Oceans, Climate Change and Global Observation System. *WMO Bulletin*, **40**, pp. 137-147.

Beijing Ministerial Declaration on Environment and Development, 1991. Ministerial Conference of Developing Countries on Environment and Development, held in Beijing, on 18-19 June 1991.

Corell, R. W., and Anderson, P. A. (Editors), 1991, *Global Environmental Change* (Berlin: Springer-Verlag).

Degens, T., 1989, Perspectives on Biogeochemistry (Berlin: Springer-Verlag) .

Ehrlich, P. R., and Ehrlich, A. H., 1990, The Population Explosion: From Global Warming to Rain Forest Destruction, Famine, and Air and Water Pollution - Why Overpopulation is our No. 1 Environmental Problem (New York: Simon and Schuster).

Ellis, D., 1989, *Environments at Risk* (Berlin: Springer-Verlag).

Gorshkov, V. G., 1990, *Biosphere Energy and Environmental Stability, Itogi Nauki i Tekhniki, Theoretical and General Aspects of Geography, vol. 7* (Moscow: VINITI).

Gorshkov, V. G., 1991, Ten fold Increase of Dissolved Organic Carbon production in the Ocean during the Industrial Era. *Doklady of USSR Academy of Sciences*, **320**, 492-498.

Gorshkov, V. G., and Kondratyev, K. Ya., 1990, The Le Chatelier Principle in the Biosphere. *Ecology*, No. 1, 7-16.

Grigoryev, A. A., 1991, *Lessons in Ecology in the Past and Nowadays* (Leningrad: Nauka).

Houghton, J. T., Jenkins, G. J., and Ephraums, J. J. (Editors), 1990, *Climate Change, The IPCC Scientific Assessment* (Cambridge: Cambridge University Press).

Ittekkot, V., Kempe, S., Michaelis, W., and Spitzy, A., 1990, *Facets of modern Biogeochemistry* (Berlin: Springer-Verlag.

Jacobson, H. K., and Price, M. F., 1990, A framework for Research on the Human Dimensions of Global Environmental Change, ICSU/UNESCO Series 3, Paris.

Keyfitz, N., and Flieger, W., 1990, *World Population Growth and Ageing. Demographic Trends in the late Twentieth Century* (Chicago University of Chicago Press).

Kondratyev, K. Ya, 1987, *International Geosphere Biosphere Programme: the role of observations from space, Studies of the Earth from Space* No. 4, pp. 105-118 (in Russian).

Kondratyev, K. Ya., 1988, *Climate Shocks: Natural and Anthropogenic* (Chichester: Wiley and Sons).

Kondratyev, K. Ya., 1989, Global dynamics of ozone, *Itogi Nauki i Tekhniki. Geomagnetism and Upper Atmospheric. Layers, vol. II*, (Moscow: VINITI).

Kondratyev, K. Ya., 1990a, Global security and ecology, *Geofisica International*, **29**, 123-128.

Kondratyev, K. Ya., 1990b, *Key Problems of Global Ecology, Itogi Nauki i Tekhniki, Theoretical and General Aspects of Geography, vol. 9* (Moscow: VINITI).

Kondratyev, K. Ya., 1990c, The International Space Year: Priorities and Perspectives. In *Studies of the Earth from Space*, No. 1, pp.3-13. (in Russian).

Kondratyev, K. Ya., 1991a, On an International Framework Convention on Climate Change. *Izvestiya of USSR Geography Society*, **123**, 127-136.

Kondratyev, K. Ya., 1991b, The International Geosphere Biosphere Programme: the role of observations from space. *Izvestiya of USSR Geography Society*, **123**, 2-7.

Kondratyev, K. Ya., 1991c, Global ecology priorities and remote sensing problems. In *Earth Studies from Space*, issue 5, 3-9. (in Russian).

Kondratyev, K. Ya. (Editor), 1991, *Aerosol and Climate* (Leningrad: Gidrometeoizdat).

Kondratyev, K. Ya., 1992a, *Global Climate* (Moscow; St. Petersburg: Nauka).

Kondratyev, K. Ya., 1992b, The next century: perspectives for development of space research. In *Earth Studies from Space*, pp. 111-119.

Kondratyev, K. Ya., Grigoryev, A. A., and Pokrovsky, O. M., 1975, *The information content of remote sensing data on environmental parameters and natural resources obtained from satellites* (Leningrad: Leningrad University Press).

Kondratyev, K. Ya., and Pokrovsky, O. M, 1977, Planning of multipurpose experiments on remote sensing of environmental parameters and natural resources. *Izvestiya of USSR Academy of Sciences, Series. Geography*, **3**, 83-89.

Kondratyev, K. Ya., and Pokrovsky, O. M., 1989, The International Geosphere Biosphere Programme. Key aspects in requirements to Earth observing data from space. *Izvestiya of USSR Academy of Sciences, Series. Geography*, **1**, 20-28 (in Russian).

Kondratyev, K. Ya., Telegina, T. A., and Fedchenko, P. P., 1990, The effect of radiation in Mg lines on the process of inverse transcription, *Doklady of USSR Academy of Sciences*, **314**, 988-990,

Lahsof, D. A., 1990, The role of the biosphere in the global carbon cycle: Evaluation through biospheric modelling and atmospheric measurement. PhD. U.M.I. Dissertation Information Service, Ann Arbor.

Lindzen, R. S., 1991, Response to AMS Policy Statement on Global Climate Change. *Bulletin of the American Meteorological Society*, **72**, 515.

Losev, K. S., Kondratyev, K. Ya., Gorskov, V. G., 1992, The ecology of Russia, *Itogi Nauki i Tekhniki, Fundamental and General Aspects of Geography*, **12** (Moscow: VINITI).

Lubchenko, J., Olson, A. M., Brubaker, L. B. *et al.*, 1991, The Sustainable Biosphere Initiative: An Ecological Research Agenda. *Ecology*, **72**, 371-412.

Marchuk, G. I., and Kondratyev, K. Ya., 1992, *Priorities of Global Ecology* (Moscow: Nauka).

Marchuk, G. I., Kondratyev, K. Ya., and Kozoderov, V. V., 1990, *Earth Radiation Budget: Key Aspects* (Moscow: Nauka).

Marchuk, G. I., Kondratyev, K. Ya., Kozoderov, V. V., 1992, *Space and the Biosphere* (Moscow: Nauka).

Marchuk, G. I., Kondratyev, K. Ya, Kozoderov V. V., Lappo, S. S., Sarkisian, A. S., and Khvorostyanov, V. I., 1989, *Energy-Active Zones: Conceptual Aspects, Itogi Nauki i Tekhniki, Atmosphere-Ocean-Space, the "Razrezy" Programme, 10 and 11* (Moscow: VINITI).

National Research Council, 1990, Research Strategies for the U.S. Global Change Research Program, National Research Council, Washington.

Perry, J. S., 1992, Global change: from rhetoric to reality, U.S. National Report to IUGG, 1987-1990 (Washington: AGU).

Pokrovsky, O. M., 1984, *An Optimisation of Meteorological Sounding of the Atmosphere from space* (Leningrad: Gidrometeoizdat).

Pokrovsky, O. M., 1991, Determination of information characteristics of sounding systems from space, *Optika Atmosfery*, **4**, 227-241.

Polunin, N., and Barnett, J. H., 1990, *Maintenance of the Biosphere* (New York: St. Martin's Press).

Ponting, C., 1992, Historical perspectives on sustainable development. *Environment*, **32**, 4-9, 31-33.

Schimel, D. S., Kittel, T. G. F., and Parton, W. J., 1991, Terrestrial biogeochemical cycles: global interactions with the atmosphere and hydrology. *Tellus*, **43A-B**, 188-203.

Schlesinger, W. H., 1991, Biogeochemistry (New York: Academic Press)

Scientific Orientation and Programming, 1990, Interdisciplinary Research Environment Programme (Paris: CNRS).

Sorokhtin, O. G., and Ushakov, S. A., 1991, *Global Evolution of the Earth* (Moscow: Moscow University Press).

Telichenko, M. M., and Ostroumov, S. A., 1991, *Introduction to Biogeochemical Ecology Problems* (Moscow: Nauka).
U.S. Global Change Research Program. Our Changing Planet, 1992, The FY 1992. A Report of the Committee on Earth and Environmental Sciences, A Supplement to the U.S. President's Fiscal Year 1992 Budget, Washington.
Woodwell, G. M. (Editor), 1991, *The Earth in Transition* (Cambridge: Cambridge University Press).
World Commission on Environment and Development, 1987, *Our Common Future* (Oxford: Oxford University Press).

Satellite Data Archiving, Retrieval and Distribution

Maurizio Fea

1 Introduction

Any satellite system is constituted in general of two main components, namely the Space Segment and the Ground Segment. The Space Segment is basically the spacecraft itself, one or more, and its payload: it includes all the hardware and software tools to make the satellite(s) operate as planned and commanded by the ground control centre, from the trusters to perform the orbit and attitude manoeuvres to all the equipment to execute embarked instrument operations and to ensure the data transmission to the ground. The Ground Segment encompasses all the ground facilities required to manage and control the space segment, and to carry out all the operations related to the satellite data handling: data acquisition, processing, archiving, retrieval and distribution; consequently it also includes the user interfaces and services.

Since the launch of the first meteorological satellite TIROS-1 on 1 April 1960, it has been clear that a satellite programme for Earth observation from space is in fact a multidisciplinary tool, which initially is meant to serve a specific community, but which in fact provides useful information to many other disciplines. Over the last three decades major progress has been made in the understanding of the potential of such space systems for local and global observations and in building powerful facilities for the utilisation of such data. Satellite observations allow the monitoring of the Earth's environment: atmosphere, oceans, ice caps and land. The corresponding instrument data complement the usual conventional ground observations and are a very important additional input to the numerical models for the analysis of the state of the environment. It is worth pointing out that, under equal forecasting model conditions, the accuracy of the forecast is directly dependent upon the accuracy and completeness of the analysis: the more detailed and complete the description of the initial status of the observed parameters (analysis), the more accurate the final parameter values generated by the model (forecast) will be.

From the above considerations, it is also clear that a user has in general a big advantage in making use of the information available from various sources, and in particular from all those satellite missions which can provide the data of interest. For many application disciplines, the so-called multi-approach is very useful:

NATO ASI Series, Vol. I 24
Remote Sensing and Global Climate Change
Edited by Robin A. Vaughan and Arthur P. Cracknell

- multi-mission (polar and geostationary satellite data)
- multi-sensor (optical and microwave, passive and active instruments)
- multi-temporal (several re-visits by the same system).

As mentioned above, the multi-approach obviously includes other datasets when available, like surveys from airborne instruments, ship measurements and ground observations. In effect, at the time of designing an Earth Observation mission one of the key activities pursued is the survey and analysis of the users' requirements. Remote sensing users in most disciplines require for their applications suitable:

geographical coverage
temporal sampling
geometric and radiometric accuracy
data continuity
easy access to data
timely delivery of data.

Consequently, any user-oriented ground segment should include skilled and diversified data handling facilities and ultimately user-friendly data manipulation tools. In this context, data archiving, retrieval and distribution are fundamental issues which can determine the success of a space mission, through an easy availability of data products and friendly user services. In addition, a critical element is the long-term continuity of the data types. A prerequisite for a sensible use of remote sensing data in many environmental domains and in climate studies is the availability of long time series of the same parameter, the prime condition for its use in monitoring its evolution and detecting its trend. Consequently, a well-structured long-term archive has a real value on its own for any space mission and requires much attention in both the development and operational phases of the mission.

2 Ground Segment Aspects

The ground segment of the most common space Earth observation missions nowadays supports two main groups of user needs, first on-line and quasi-real-time services and secondly the off-line requirements. The former are meant to serve all those operational activities for which the time response is critical to the good end of the application, like in the case of meteorological offices, ship traffic authorities, hydrological institutions, environmental emergency task forces and last, but certainly not least, the overall satellite system performance control centres. The off-line facilities are charged with ensuring the long-term availability of the data (archiving and retrieval) and the generation of precision products through additional auxiliary information not available for the real-time operations, like in-situ measurements and calibration and validation parameters.

The high-level scheme of an Earth observation ground segment data flow is shown in Figure 1, where the main steps to ensure that the satellite data are saved and the products reach the users according to their time requirements are indicated.

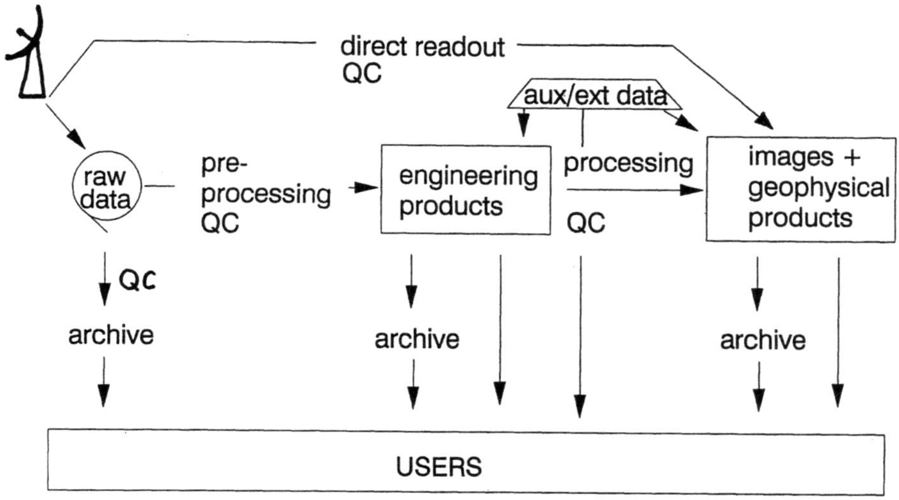

Figure 1: Scheme of an Earth Observation Ground Segment Data Flow

The acquisition of satellite telemetry is the prerequisite for any further ground operation. Meteorological satellite systems, like NOAA (USA), GOES (USA), Meteosat (ESA/Eumetsat), GMS (Japan), Fengyun-1 (China), Meteor (ex-USSR) allow direct readout of pre-processed image data in a reasonable user-accessible way. Until now, such satellite images are freely available to everybody and a very large community of users has been formed world-wide; as a consequence the derived image data have become very popular, particularly through television weather reports. A large variety of user acquisition tools exists, from the simple and small omni-directional antenna to receive the NOAA low-resolution image data in the visible and thermal infrared channels, to the large 5 m parabolic antenna required to acquire the Meteosat digital full-resolution images in the visible, thermal infrared and infrared water vapour absorption spectral bands. An additional consideration concerns an important difference between polar orbiting and geostationary weather satellites. Due to their very high altitude (36,000 km) the latter provide image data of the whole Earth region observed by the spacecraft to each user station in the region of visibility, whereas the polar-orbiting satellites, flying at much lower altitudes (700 to 1500 km) provide direct read-out image data only for an area of some two thousand km width around the user station as a strip from horizon to horizon during the satellite pass.

Earth observation polar-orbiting satellites like Landsat, SPOT, ERS-1 and JERS-1 are designed for a much higher ground resolution (few tens of metres), required by their mission objectives for land resources and environmental monitoring. Therefore they need much more sophisticated and complex ground acquisition facilities, and direct read-out by an unqualified and unauthorized user is not possible. In addition, a commercial interest exists in the image products of the latter satellites, therefore data distributor firms and consortia have been established through which a user can have access to the satellite products.

The growing user requirements in terms of ground resolution and spectral resolution has forced an increasing data rate in the satellite down-link to be implemented. This situation has made it very difficult for many satellite systems to support the global on-board data recording, particularly in the case of synthetic aperture radar (SAR) instruments. The SAR image data of the currently operational ESA ERS-1 mission, in fact, are not stored on-board and can be acquired only if there is a suitable ground station in view of the spacecraft. Consequently, in the case of polar orbiters a world-wide network of ground stations is required to ensure global coverage, even if some of them like SPOT and JERS-1 (and ERS-1 for the non-imaging instruments) carry on-board recorders to allow some imaging of areas not assisted by ground stations. This situation makes the management of the ground segment more challenging, since all stations must acquire the same satellite down-link with the same accuracy and standard to allow a global service to be offered to the users; this adds some complications for the establishment, maintenance and updating of a homogeneous and complete global mission archive.

The other main blocks of a ground segment for an Earth observation mission are strictly associated with the key functions planned for the space programme in the areas of processing, archiving, retrieval, and distribution. In view of the large variety of implementations and programmes operated by various space agencies, we confine ourselves in this chapter to a few examples related to the experience of the European Space Agency (ESA) in these domains. They are related to the processing, archiving, retrieval and distribution of data of some missions currently handled at ESOC and ESRIN, the ESA establishments respectively in Darmstadt (Germany) and Frascati (Italy).

2.1 The Technological View

As for other human ventures, the performance and success of a satellite mission are always associated with the technological progress which permits not only the development of advanced on-board instrumentation, but also the availability of more powerful and cheaper processing and data-handling tools. In the last decade a drastic improvement has occurred in the field of computing devices; from a world of huge mainframes and relatively slow and not very user-friendly terminals, the market has moved very rapidly towards the availability of personal computers and versatile work stations, with an overall trend of more powerful and less expensive tools. As an example, during the development life of the ERS-1 programme in the 1980-1990 decade, the market in computing equipment went through two major steps: one in 1980 with the diffusion of the 16-bit machines and the second in the years 1988-1989 with the 32-bit machines, although the delay of the software market in catching up with the hardware progress often creates problems to users. Since the approval and funding of any programme is based on a proposal addressing also the technical implementation of the hardware and software configurations, it is easy to understand that the risk of implementing an architecture with tools which can become obsolete, or in the worst case not maintainable, at the start of operations many years after is real and requires a high degree of management attention.

Without entering into an endless analysis of these tools, which in any case would be very rapidly obsolete because of the fast evolution of the market, it is important to point out that the applications of remote sensing data have received an enormous push ahead by the progress in the technology. However, it is also worth noting that a serious problem for the users in the fields of communication and exchange of data is due to the lack of agreed common standards in some key industrial sectors, for instance in the archiving and distribution media.

One important consideration is concerned with the need for the long-term duration and accessibility of an archive, and the use of common easily useable distribution media. The frequent technological progress has a contradicting impact on the ground segment activities, since on the one hand it makes available faster and more capable media, while on the other it makes them rapidly obsolete and proposes different standards. That is the reason why computer compatible tapes (CCTs) are still commonly used, though their density has increased from 800 to 6250 bpi (bits per inch). Another typical example of this problem is the issue of optical disks. Big technological promise of the early 1980s, the optical disks have not reached the goal of a common and standard high capacity medium yet; only the CD-ROM technology is nowadays well consolidated, primarily for mass distribution of data and information.

Consequently, the long-term management of an accessible and updated archive must address the possible transfer of data onto different media as an important topic in the system planning. Any decision on this issue has to be made as a trade-off between technical advantages, financial considerations, operational impact and improvements to the data retrieval and user services. Adapting to the technological trend and at the same time maintaining a coherent implementation is certainly a challenging task for a user service facility, as it is quite often heavily affected by the conflict between the market situation and the operational constraints.

The issues associated with the archiving, retrieval and distribution of space Earth observation data address the following functions:

- archiving
- product generation
- cataloguing
- retrieval
- distribution
- user services.

As previously indicated, these functions will be analysed briefly in the following section making reference to existing and planned ESA programmes:

- Meteosat, operated by ESA on behalf of Eumetsat, to give an idea of the functions associated with a geostationary satellite mission
- AVHRR, to describe a service from a non-ESA polar orbiting mission

- ERS-1, the ESA world-wide environmental polar orbiting mission
- GENIUS, an ESA study for a possible European network for environmental satellite data.

3 The Meteosat Mission

Meteosat is the European contribution of a geostationary satellite to cover the position over the Gulf of Guinea for the Global Observing System of the World Meteorological Organisation. The Meteosat system has been developed by ESA and is now operated by ESA on behalf of Eumetsat, the European Organisation for Meteorological Satellites. Under this programme five satellites have been launched, nominally hovering at 35,800 km in the geostationary position at 0 degrees latitude and longitude:

Meteosat – 1 launched on 23 November 1977

Meteosat – 2 launched on 19 June 1981

Meteosat – 3 launched on 15 June 1988; it was moved in August 1991 to 50°W to help observing the Atlantic, and in February 1993 to 75°W at the US GOES East position, to support observations over Americas until GOES -7 will be launched in 1994

Meteosat – 4 launched on 6 March 1989, first of the improved series

Meteosat – 5 launched on 3 March 1991

Meteosat – 6 will be launched in November 1993.

The Meteosat mission includes:

- imaging the Earth disk every 30 minutes in the three spectral channels
- deriving and disseminating geophysical products
- disseminating image data and selected products via the spacecraft
- archiving and retrieval of the image data and products
- handling data from automatic data collection platforms.

The Meteosat Second Generation (MSG) programme is being developed for the year 2000 and beyond and its objectives include, besides the follow-on of the present Meteosat programme, higher spatial and spectral resolution imaging, more frequent imaging (especially for Europe), temperature and humidity sounding, improved data circulation, and the inclusion of a scientific package. Currently, a Meteosat Transition Programme (MTP) is being implemented in order to form a bridge between today's programme and the first MSG satellite, planned for launch in 1998.

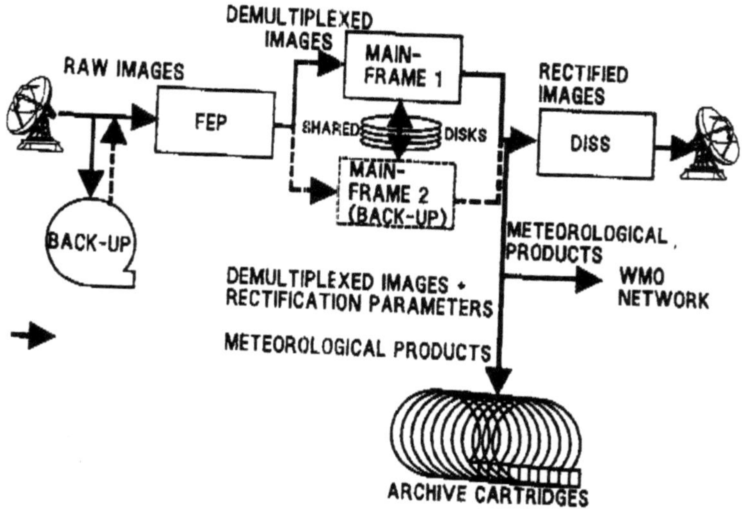

Figure 2: Simplified Diagram of the Meteosat Ground Computer System

The Meteosat radiometer images the Earth in the visible, thermal infrared and infrared water vapour absorption bands. The satellite data stream is acquired at the Michelstadt ground station, located near Darmstadt in Germany. Through high-speed links the data are transferred to the Meteosat Ground Computer System (MGCS, Figure 2) at ESOC, the European Space Operations Centre of ESA, where the data are pre-processed and the image data rectified and prepared for the further dissemination to users in various formats via the same spacecraft. Then the three spectral channel image data are further processed to derive geophysical products. The six types of products generated on-line and interactively quality controlled at the ESOC central facility are:

- cloud motion vectors (CMVs) for deriving wind fields
- sea surface temperatures (SSTs)
- cloud analysis (CA)
- upper tropospheric humidity (UTH) data
- Precipitation Index (PI) data
- International Statellite Cloud Climatology Project (ISCCP) images
- cloud top height (CTH) maps
- climate data sets (CDSs).

The first four products are formatted as WMO SATOB bulletins and disseminated through the WMO Global Telecommunication System (GTS) to the world's meteorological services, whereas the CTH maps are disseminated as images via the spacecraft.

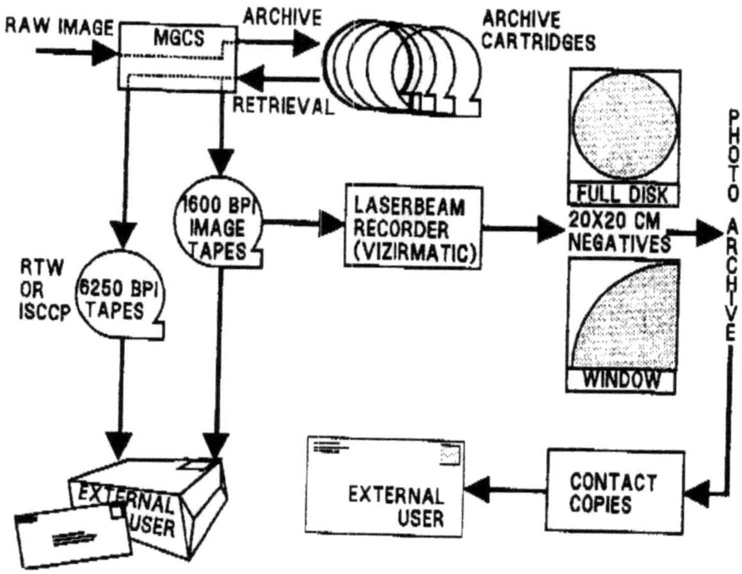

Figure 3: Flow of Data through the Archiving System

The CDSs, which are not disseminated, contains the results of the image interpretation and form the basis for extracting all the other products. Sampled image data are routinely sent to the World Data Center for the global archive of the International Satellite Cloud Climatology Program (ISCCP). All the products, except the CTH, are archived in the form of digital files on cartridge at the MGCS (Figure 3). For many years, the image data have been stored on High Density Digital Tapes, each tape containing the hundred images received daily. However, in order to improve the archive maintenance and accessibility, a major exercise was carried out to copy the original data onto IBM cartridges, which then became the Meteosat archive supporting media. A photographic high resolution 20 cm × 20 cm negative archive is also maintained by the Meteosat Data Service, which provides the programme interface for the distribution of the archived products (Table 1).

4 A Network for AVHRR Products

Originally, the image data of the polar-orbiting NOAA meteorological satellites were meant to be used only in real-time for operational meteorological applications. But it rapidly became obvious that the data had a much larger informative potential and many other disciplines started to take advantage of such data, in particular of the Advanced Very High Resolution Radiometer (AVHRR) visible and thermal infrared image data. However, the direct readout approach of the mission created some difficulties, not only because the archiving of such data was left to individual initiatives, but also because the availability of data at each station was limited to the visibility area covered by that station.

Item	Product Description
10	One Computer Compatible Tape, 1600 BPI, containing one image file in one channel or containing one file of Meteorological Products
11	Each additional File on the above CCT, if more that 1250 lines long
12	Each additional window of less than 1251 lines
13	Each additional file of METEOROLOGICAL Products on the above CCT
20	One CCT 6250 BPI with ISCCP or Real Time Window data
1	One Photographic Contact Positive (20 × 20 cm) from an existing Negative
2	One Photographic Contact Positive (40 × 40 cm) from an existing Negative
3	One Photographic Contact Positive (20 × 20 cm) requiring the production of a Negative
15	Image Bulletin - Monthly Issue
16	Catalogue (per volume) of Digital, Photographic or Meteorological data
30	Documentation (per volume - List available from the Meteosat Data Service)

Table 1: Meteosat Products

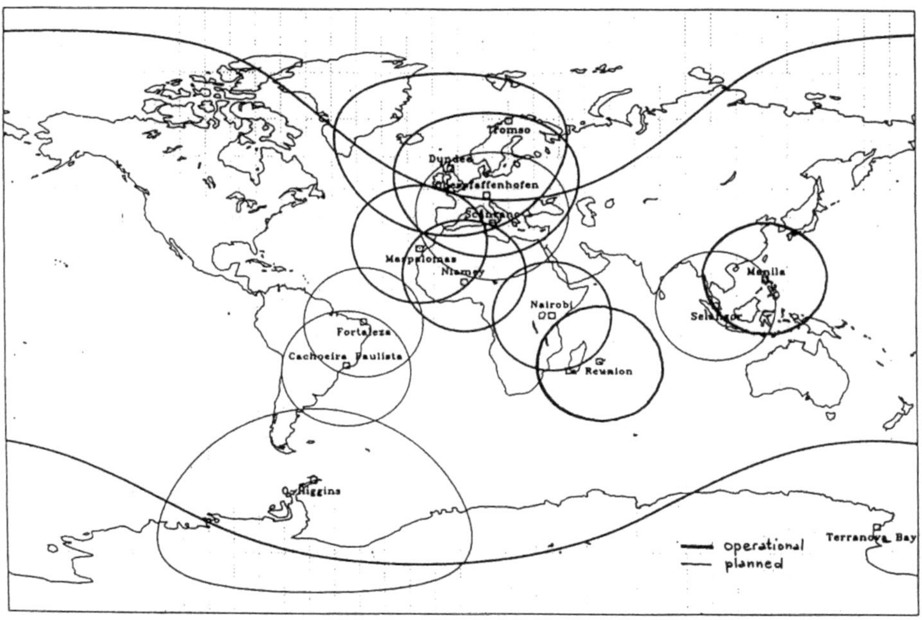

Figure 4: The Earthnet Coordinated TIROS Network

Data Rate	HDR vis/ir	HDR SAR (mw)	MDR vis/ir	LDR ocean mw
Instrument examples	TM MSS OPS	ERS JERS RADARSAT	AVHRR CZCS SeaWiFS	RA,SCAT WAVE TOPEX/POS.
ACQUISITION CHAIN				
antenna	---------- common --------		common	
recei/synchr/decom	---------- common --------		no need	
tempor. recording	HDDT	HDDT	direct on disk	OB. Tape
Quick look	---------- common --------		common	no need
ARCHIVING				
long term media	HDDT/OT	HDDT/OT	OD	OD
CATALOGUE				
user interface	------------------------- c o m m o n -------------------------			
inventory	---------- common --------		common	common
PROCESSING				
product definition	---------- common --------		common	per inst.
product format	CEOS	CEOS	CEOS	CEOS

Table 2: Examples of instruments type versus common functions in multi-mission environment

Considering both the interest in an extensive long-term European archive and the need for a wide coverage of derived geophysical products, to complement individual stations operations ESA started in the mid-1980s the Earthnet Coordinated TIROS Network (ECTN), which nowadays includes also non-European stations (Figure 4). TIROS AVHRR High Resolution Picture Transmission (HRPT) and TOVS data are acquired and archived at the ECTN stations, where data products are generated and delivered in response to user requests addressed to the Central User Service at ESRIN. Data acquired by a station where a processing facility is not available are sent to ESRIN, where they are processed to generate quick-looks and a certain number of products, as described in Table 2, and archived. In order to allow easy international use, the data are stored at the stations and at ESRIN in the CEOS SHARP (Standard-family HRPT Archive Request Product) format. SHARP products are defined and processed according to internationally agreed levels and specifications (Table 3).

A centralized catalogue is maintained within the on-Line Earthnet Data Archive (LEDA) catalogue system at ESRIN; this catalogue

- contains HRPT Acquisitions Records from ECTN stations (about 30,000 entries)
- contains AVHRR SHARP Level 1B entries (to date some 50,000), as archived within ECTN, and quality parameters from the corresponding quick-looks
- is available for world-wide access since 1987
- handles Committee of Earth Observation Satellites (CEOS) Inventory Exchange Format files
- is planned to include world-wide station inventories and guide information, like historical ephemeris messages (TBUS), calibration data, etc.

LEVEL	A	B	C	
0 ACQUISITION				DATA STREAM
1 RAW DATA	RAW +ANCILLARY	RAW CALIBRATION/ NAVIGATION APPENDED	RAW GEOMETRIC SYSTEM CORRECTIONS	SCAN ORIENTED
2 CALIB/GEOPHYS	RADIOMETRIC UNITS 1 SCENE	GEOPHYSICAL UNITS 1 SCENE		SCAN ORIENTED
3 REMAPPED	REMAPPED 1 SCENE	REMAPPED GEOPHYSICAL 1 SCENE		SINGLE PASS
4 REDUCED	MULTI-TIME/ MULTI-SCENE			COMPOSITE

Table 3: International Ciassification of Optical Instrument Products

An on-line bulletin board is maintained, and data products are distributed to users, normally via public mail services. An interesting feature of this system is the digital fax communication standard (Gammafax boards). An example of SHARK AVHRR telefax is given in Figure 5. A browser software package is being developed at ESRIN, which will permit among other functions the distribution of quick-look products on CD-ROM. Finally, it is worth noting that the Earthnet Coordinated TIROS Network is involved in different global environmental projects to which it is contributing by providing either standard or special products.

5 The ERS-1 Programme

The European Remote Sensing satellite programme launched its first satellite, ERS-1, on 17 July 1991. The data continuity will be ensured through the launch of ERS-2 by January 1995. The ERS-1 payload is composed of various equipment: an Active Microwave Instrument(AMI), consisting of a wind scatterometer and a synthetic aperture radar (SAR), which in turn can be operated in image or wave mode, a radar altimeter (RA), Along-Track Scanning Radiometer (ATSR), a Laser Retro-Reflector device (LRR) and a Precise Range and Range Rate Equipment (PRARE), the only equipment out of order. The AMI SAR operated in image mode generates a high bit rate downlink of 105 Mbps, which cannot be stored on board and therefore requires an in-sight ground station. All other instruments generate a low bit rate (LBR) data flow, which is stored on board along the orbit through tape recorders and dumped to an ESA ground station once every orbit.

28 JUN 1991	NOAA-10	Archive ID: TRT91176S11A
Start: 07:35:58	DESCENDING	Scene ID: TR109106280737
Stop: 07:39:58		Acquired: TSS TROMSOE

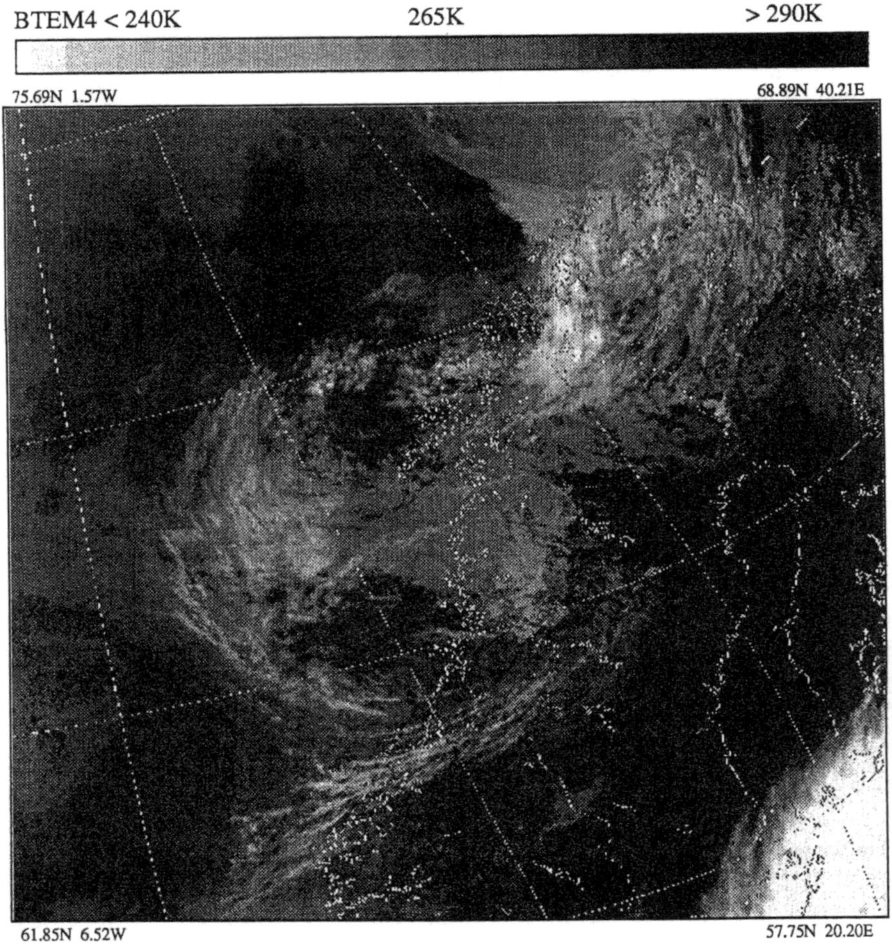

European Space Agency / Earthnet Programme Office
ESRIN, Via Galileo Galilei, I00044 Frascati, Italy
Telephone: +39 (6) 941801 Telex: 610637 ESRIN I
Fax: +39 (6) 94180361

Percentage of each cover type:

sea	0	cloud	0
sunglint	0	snow/ice	0
land	0	unclass.	100

Figure 5: A SHARK AVHRR Telefax

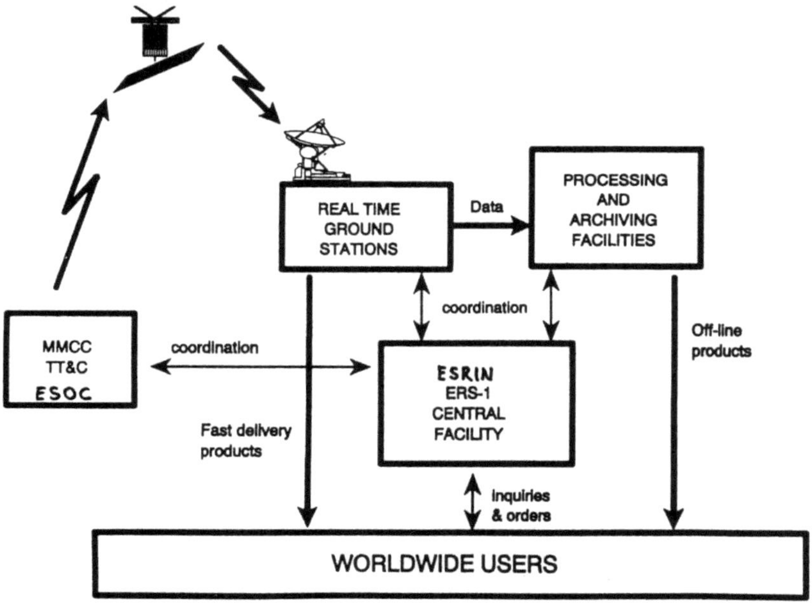

Figure 6: The ERS-1 Overall Ground Segment Concept

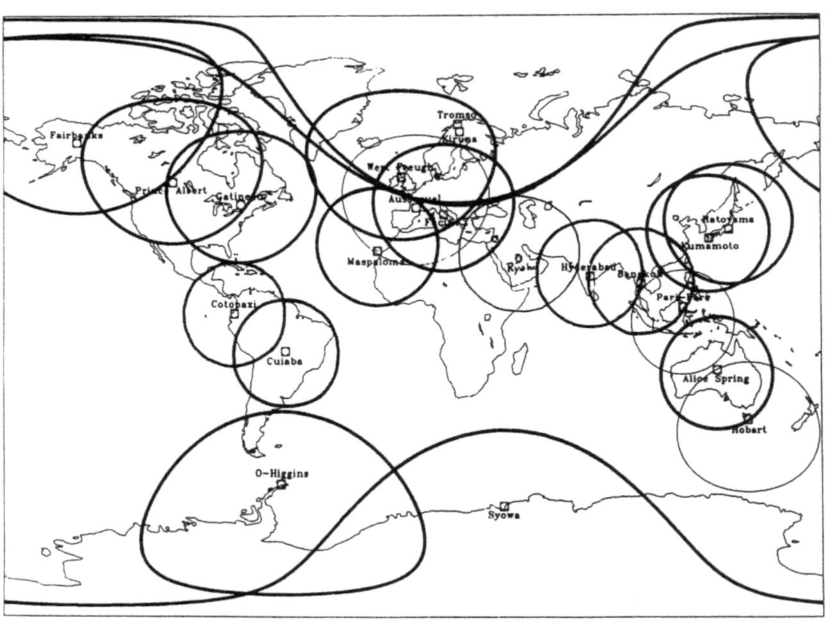

Figure 7: ERS-1 Ground Stations Coverage

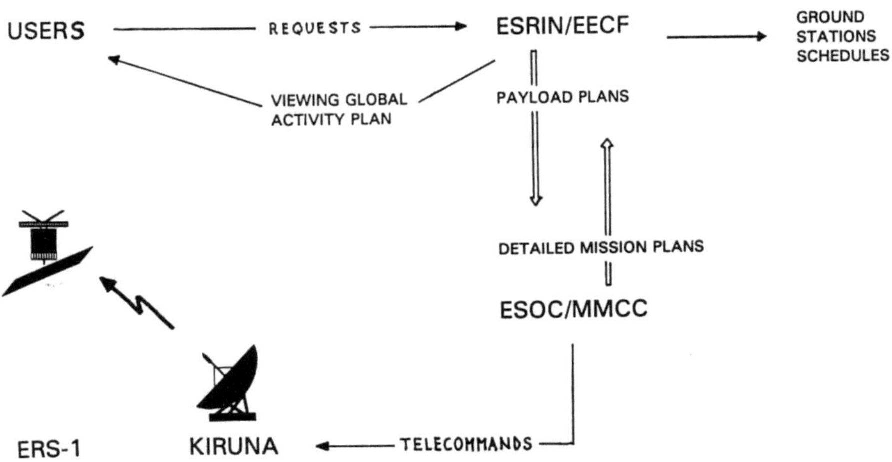

Figure 8: ERS-1 Mission Planning

Multidisciplinary in nature, the ERS-1 mission is supported by a complex distributed ground segment, whose main components are shown in Figure 6. This includes:

- a number of ESA, national and foreign ground stations (to date 18 in total and more are being built (Figure 7))
- centralised ESA management, control and user service facilities
- four Processing and Archiving Facilities (in France, Germany, Italy and the United Kingdom)
- large network facilities for data distribution.

The ERS-1 ground segment is operated to handle high rate SAR data for regional coverage and global low bit rate instrument data, and to generate, archive and distribute fast delivery and off-line precision products. The complexity of the management and of the data flow aspects is at least of an order of magnitude larger than has been the case for previous missions. This is well depicted in Figure 8 and Figure 9, where the data flow associated with the payload; operations planning to respond to user requests and the interaction among the various components of the system are schematically indicated. In fact, the ERS-1 system serves a wide range of users with very different requirements. Operational entities are supported through the dissemination of the fast delivery (FD) products, which are generated at the ESA ground stations and rapidly distributed through land lines and satellite links (Figure 10 and Figure 11). In particular, the LBR FD data are centralised at the ESRIN ERS Central Facility in Frascati, reformatted into WMO BUFR bulletins and injected into the GTS links through the nodes of Rome (Italy) and Bracknell (UK). The SAR image data, instead, are uplinked from the ESA stations of Kiruna and Fucino to the Eutelsat satellite, through which they are

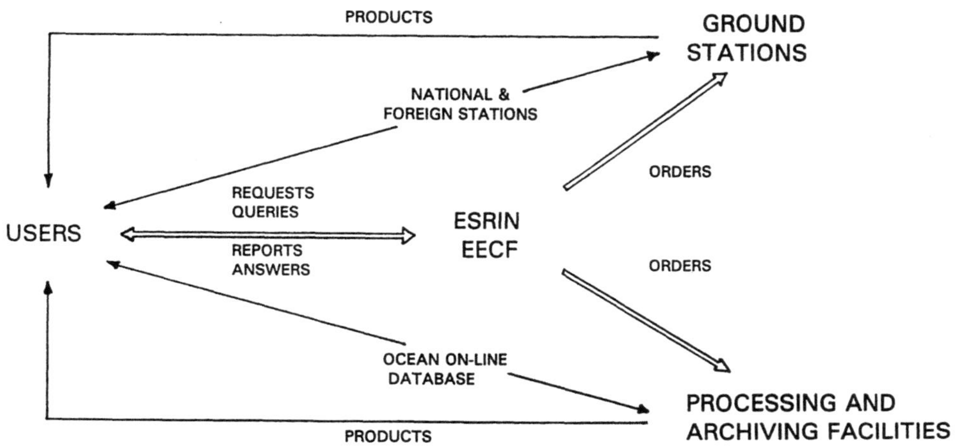

Figure 9: Interaction with ERS-1 Users

downlinked to the national nominated centres, where they are stored at the disposal of the final users.

In view of the wide range of potential applications and of its multidisciplinary nature, the ERS-1 archiving functions are shared among various facilities. Primarily, the four Processing and Archiving Facilities (PAFs) have been established by cooperation between ESA and national agencies to put at the disposal of the programme existing national skill and expertise in technical and application domains, reflected in the harmonised sharing of functions (Figure 12). Highlights of this scheme are:

- the prime archive of LBR data is maintained on optical disks at the French PAF, which is dedicated to ocean data
- the archive of SAR image data is split as follows:
  1. at the UK PAF for the data acquired in Kiruna and datasets of European interest acquired by foreign stations and stored on Penny & Giles HDDTs (high density digital tapes)
  2. at the Italian PAF for the data acquired in Fucino and Maspalomas
  3. at the German PAF for the data acquired in O'Higgins (Antarctica) and datasets of European interest acquired by foreign stations and stored on Honeywell HDDTs
- the generation of off-line SAR products is carried out as follows:
  1. by the ground stations for their national customers and investigators, if processing facilities are available (Alaska SAR Facility, Brazil, Canada, Ecuador, France, India, Japan, Norway, UK)

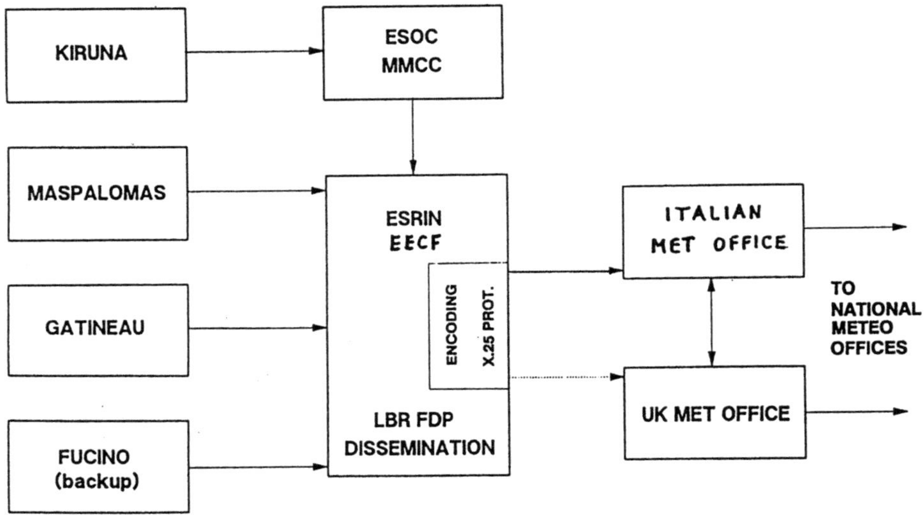

Figure 10: ERS-1 Low Bit Rate Fast Delivery Product Dissemination

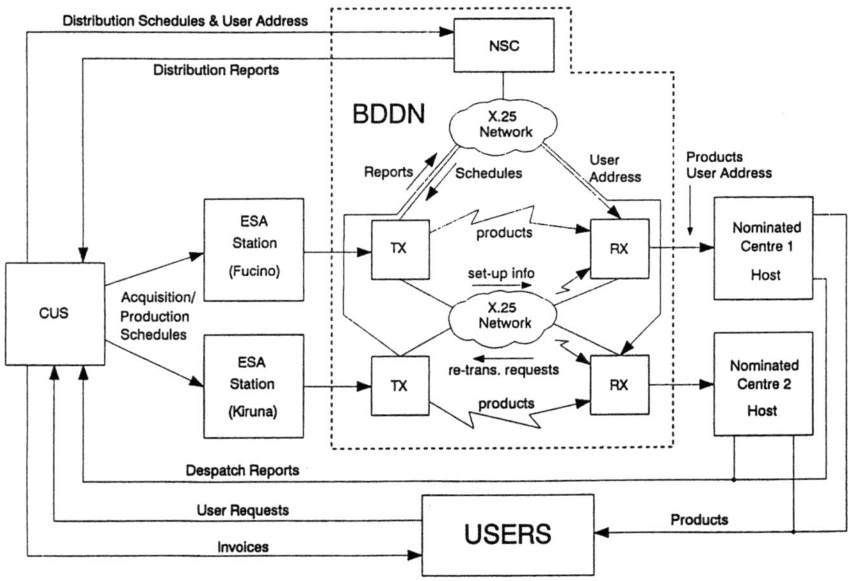

Figure 11: ERS-1 High Bit Rate (SAR) Fast Delivery Product Dissemination

2. by the PAFs according to their archived data, in particular by the German PAF if a geocoded product is required

- the generation of off-line LBR products is done as follow:

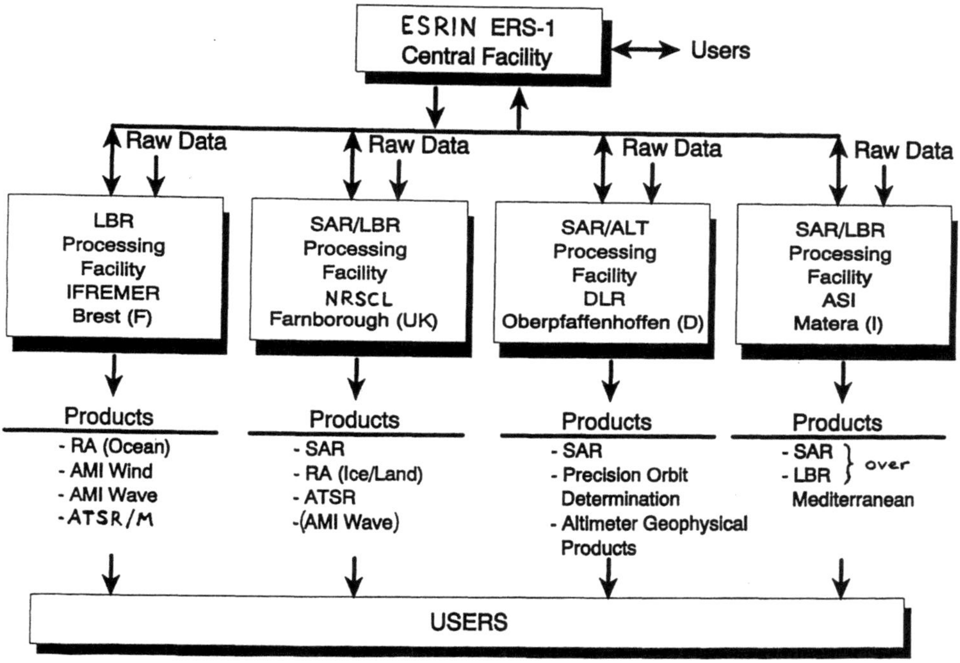

Figure 12: ERS-1 Processing and Archiving Facilities

1. over the ocean at the French PAF
2. over the Mediterranean sea at the Italian PAF (by the end of 1993)
3. over sea-ice, ice and land at the UK PAF
4. precise orbit calculation, geodetic and high level topography products at the German PAF

- a working HDDT archive of those data used for routine quality control and performance assessment is maintained at the ESRIN ERS-1 Central Facility (EECF), where the SAR Verification Mode Processor and a copy of the ESA FD station processing chains are used.

The EECF hosts the ERS-1 world-wide central catalogue, which contains the record of acquired data and processed products and which can be accessed on-line. The management of the archiving, retrieval, processing and distribution functions is carried out by the EECF User Services. They ensure also the management of the user requests and the planning of the ERS-1 payload accordingly, in close operational cooperation with the Mission Management and Control Centre at ESOC. The ESA PAFs do not interface the users directly for collecting the user orders; they receive the production schedules from EECF and deliver the requested products to the users. It is worth noting that an ERS (Eurimage, Radarsat International, Spot Image) Consortium has been appointed as the ERS-1 data distributor world-wide, with national representatives in most countries. Further details about the aspect of product ordering and distribution can be obtained from the ERS-1 help desk at ESRIN. In summary, an ERS-1 user will forward his request for products either to the data distributor or, in the case of ESA Principal Investigators and Pilot Projects, to EECF and he will receive the products directly either on-line through land-lines or satellite links, or by mail from the PAFs.

6 A Possible Way towards a European Network for Environmental Data

The ERS-1 ground segment was designed as a distributed set of specialised facilities in support of a world-wide multidisciplinary polar mission. In fact, the basic idea behind the ERS-1 concept was to develop an infrastructure which would also constitute the long-term core structure for the Earth Observation missions for the year 2000 and beyond. Consequently, the ERS-1 ground segment has been implemented for the immediate need of supporting ERS-1 operational tasks, but also as a test bed for a larger future configuration. In effect, the latest decisions of the European ministers to ensure the funding for the future ESA Earth Observation missions, after the ERS-1 and ERS-2 series, have given credit to that concept. Envisat, the new generation of environmental satellites, will require a ground segment at least one order of magnitude more complex than that of ERS-1! In addition, the need for a structured multi-purpose data distribution network, where the user can be served by several missions and would not be troubled and confused by the individual features of each of them, has grown consistently and has become the objective of several initiatives. For instance, one example is the study for the Centre for Earth Observation (CEO), managed by the Joint Research Centre of the Commission of European Communities (CEC), focusing on the applications aspects of future ground infrastructure.

One of these initiatives is a study funded by ESA and addressing these issues; it has been called GENIUS (Global Environmental space data Network Information and User System) and its objectives are:

- safeguarding the historical archives
- improving the maintainability and efficiency of the existing infrastructure
- extending to new missions
- creating a networking capability
- vastly improving user services
- fostering applications expertise at the European and national levels
- providing a focus for industrial development, These are specifically related to the topics discussed in this chapter, namely payload data acquisition, archiving and cataloguing, processing and quality control and data distribution and user interfaces. This ESA study will be complemented by the above-mentioned CEC activities focused on high-level products generation, calibration and validation, development of algorithms and models and handling of ground measurements.

The GENIUS strategy considers the following issues:

- multi-mission system environment at local, regional and global scale (historical, current and future missions)
- distributed and decentralised approach, to optimise the provision of data where needed and in the required time frame and to make maximum use of national expertise
- evolutionary development, using as baseline the ERS-1 ground segment architecture
- matching of end-user requirements, as regards availability of user services, information and data products.

In terms of user communities, GENIUS aims to serve in near real-time the users for operational environmental monitoring, the European research and application users requiring access to available environmental data, the international research and application users accessing European data holdings, the commercial and value-added industries and other related users. Consequently, the high-level functionality entails

- an integrated data and information management system to satisfy real-time and off-scales user needs
- end-to-end networks, to meet the required time-scales, data quality and data volumes
- architecture maximising function unification across the missions
- multi-mission grouping of instruments in a common processing environment (Table 3)
- long-term continuity of services.

In the context of a multi-mission environment, the ERS-1 architectural scheme has been retained (Figure 13), with the addition of the Application User Centres, which are discipline-oriented entities which would be the centres of excellence for specific instruments (algorithms, calibration, validation, long-term instrument performance), and major networks and interconnectivity improvements. In this enlarged scheme, the Payload Data Coordination Centres, managing and coordinating all payload data ground activities, would have responsibility for the following functions:

- catalogue and user services
- network and data distribution services
- product control service
- international missions interface services

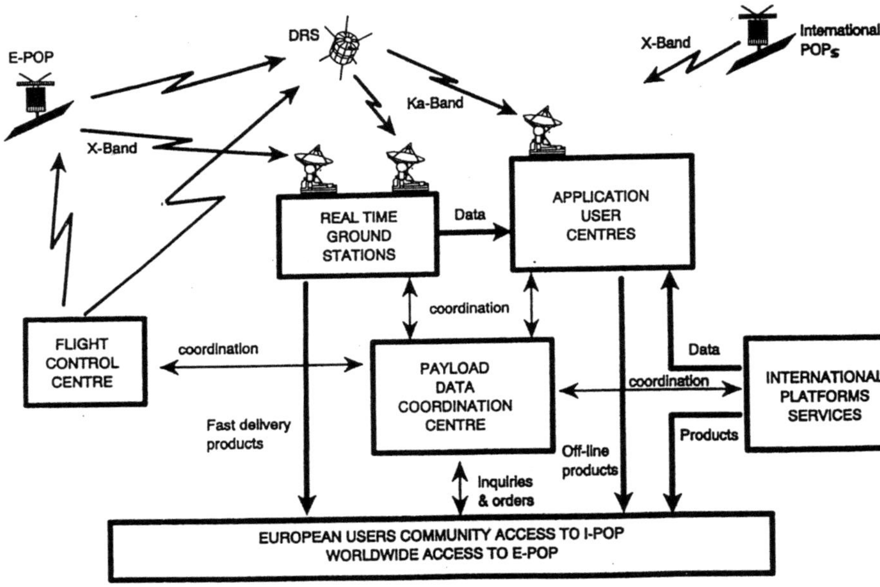

Figure 13: GENIUS Multimission Ground Segment Concept

- high level global products archive
- system engineering services.

In the domain of data archiving and retrieval, some basic considerations need attention, related to the fact that the long-term mission archive has an actual intrinsic value and is critical for several reasons. These include the following considerations: historical data are important and need to be preserved at an affordable cost, the cost of archival support media is becoming a sizable fraction of Earth observation exploitation costs and re-processing of historical data is often needed as better models or calibration parameters are generated.

As indicated earlier, the issue of archiving media is very much affected by technological progress. As an example, the ESA Earth observation missions archive at ESRIN (Nimbus-7 CZCS, HCMM, Landsat, MOS, JERS, ERS mission data for European interest) is held on some 40,000 HDDTs, with a monthly increase of more than 300 HDDTs! However, the effort of transferring historical data from one type of archive support medium to another more modern type is a very costly and time consuming exercise. The GENIUS study has addressed this issue and has indicated a possible strategy, where the envisaged advantages of a new approach to archiving and retrieval is that it will be possible to have a large volume of data processed (or re-conditioned) without human intervention. That would ensure the migration of large historical volumes from one type of medium to others more modern at a reasonable cost. Taking into

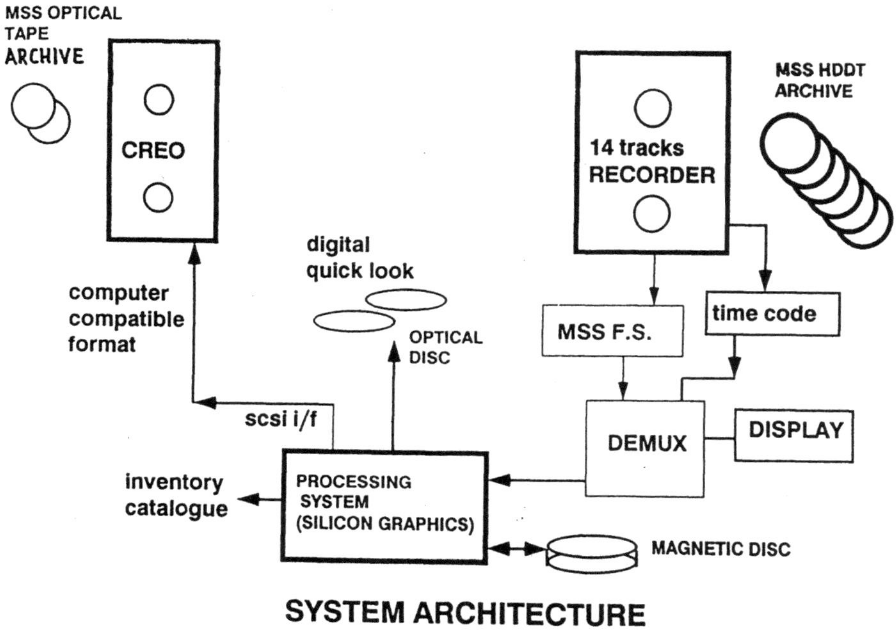

Figure 14: Landsat MSS Archive Transcription

account the fact that the ERS-1 SAR generates one Tbyte of data every week, a test exercise has been carried out at ESRIN, whereby a prototype activity has demonstrated the potential utilisation of optical tape as a standard computer peripheral to store a very large volume of data (about one Tbyte per reel, equivalent to a hundred of the currently-used HDDTs). The exercise is being run to transfer the Landsat MSS data archive from HDDT onto optical tapes (Figure 14). Other promising technologies exist, but still require consolidated results from operational and endurance tests. The basic requirement for the future is that large volumes of data are made accessible for direct retrieval by the user community permitting large use of satellite data and the analysis of long time series.

Another useful objective is the generation of a mission archive of digital quick-looks, as a browse service on-line; that would allow the quick-look data to be transferred quickly and economically through the network to users, who would be in a position to look at the data and select what they need before ordering a product. A possible user approach sequence is depicted in Figure 15.

Finally, it is worth mentioning the issue of the proliferation of data catalogues, which quite often becomes a discouraging element for the user. International initiatives have started to streamline the user access to catalogues, in particular to the so-called metadata, which is the information describing the content of the catalogue and the characteristics of the data. International compatibility and interconnectivity aspects have been addressed by the space agencies and good progress has been achieved. Under the auspices of the Committee for Earth Observation Satellites (CEOS), NASA, NASDA

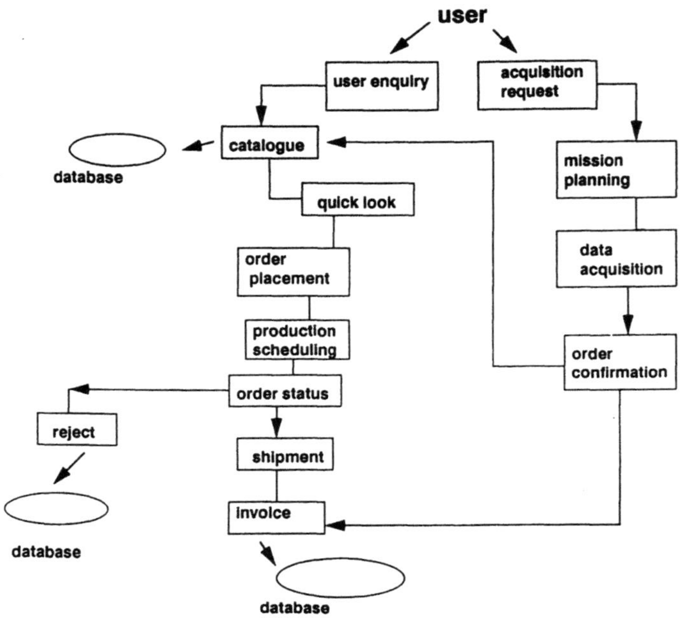

Figure 15: User Approach Sequence

and ESA have implemented and operate the International Directory Network (IDN, Figure 16), a high level interface to very distributed Earth observation data and services providers. The IDN is an international effort to give global change researchers access to information on datasets available around the world; it is free, is an on-line directory which provides access to scientific Earth and Space observations information world-wide, it requires no special training and no account is necessary for the use of it. It has automatic LINKs, with no telecommunication charges; users can LINK directly to other on-line systems, and this immediate connection allows them to access to more information on a directory entry and, in some cases, to browse, manipulate or order data of interest. In Asia the IDN node is operated by NASDA in Saitama (Japan) and in America by NASA/GSFC in Greenbelt (Maryland, USA). The European node, still called Prototype International Directory (PID), is operated by ESA at ESRIN in Frascati (Italy). National sub-nodes contribute to coordinating the IDN population and its utilisation. It is worth noting that through IDN a user can at present access information around the world on many historical and current Earth observation satellite missions, like Landsat, SPOT, MOS, NOAA AVHRR, Seasat SAR, Nimbus CZCS and GOES.

7 Conclusions

The large variety of technical and architectural solutions for the archival, retrieval and distribution of data from Earth observation remote sensing missions from space makes

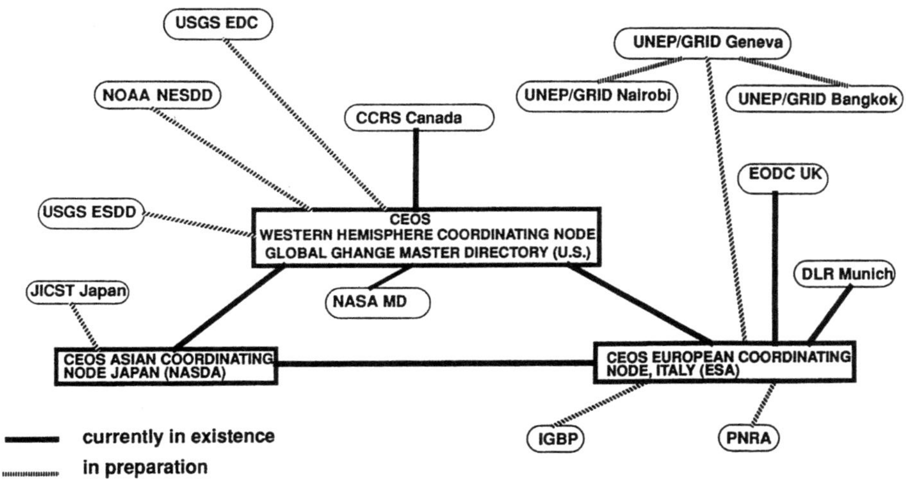

Figure 16: International Directory Network

it very difficult to discuss those topics in a exhaustive way. However, from the limited number of cases presented, it is clear that these issues are of great importance both for the satellite operators and for the users. The international effort in coordinating procedures and formats, interconnecting catalogues and directories and integrating networks has triggered a major step towards real progress for user benefit in the domains of data access and availability. In addition, the requirement for accessible long-term updated archives has become an essential function of the Earth observation missions. Similarly, data retrieval and timely distribution to different user communities are definite parts of most such space programmes.

Up to now, besides the agencies responsible for specific missions, two main international organisations have been deeply involved in producing standards for space data handling, the Consultative Committee for Space Data Systems (CCSDS) and the Committee on Earth Observations Satellites (CEOS), which has already been mentioned. Recently, the Co-ordination Group for Meteorological Satellites (CGMS) started to discuss these issues also for data from the geostationary meteorological satellites. Consequently, despite the number of remote sensing programmes and their different historical traditions, the space missions of the last decade of this millennium are characterised by a bigger emphasis on data continuity, long-term updated archives and internationally integrated user services.

Bibliography

ERS-1 System - ESA SP-1146, 1992 (Paris: European Space Agency)

ERS-1 User Handbook - ESA SP-1148, 1992 (Paris: European Space Agency)
ESA ERS-1 Product Specification - ESA SP-1149, 1992 (Paris: European Space Agency)
METEOSAT System Guide - Issue 3, METEOSAT Data Service, ESOC, Darmastdt (D)

Special Sensor Microwave/Imager (SSM/I): Instrument, Data Analysis and Applications

J. Noll, J.P.V. Poiares Baptista, S. Buonomo and A. Rognes

1 Introduction

The increasing concern for our environment has promoted interest in using remote sensing of the Earth's surface and atmosphere and microwave measurements from space are particularly important because they can provide all-weather monitoring on a global, day and night basis.

Radiometric techniques using microwaves were developed in the 1930s to measure electromagnetic energy of extraterrestrial origin. Terrestrial observation started much later (1960) and since then passive microwave sensors were employed in meteorological, hydrological and oceanographic applications. Since the first Cosmos satellite (U.S.S.R.) was launched in 1968, a large number of systems have been designed and flown to exploit the advantages of microwave spaceborne radiometers.

Since then the complexity of systems has increased and multi–frequency, multi–polarisation radiometers are currently in operation in space. The Special Sensor Microwave/Imager (SSM/I) first launched in 1987 on the American DMSP-F08 satellite is an example of such a radiometer, operating at 19, 22, 37 and 85 GHz with both horizontal and vertical polarisations. In Europe, an even more advanced radiometer (the Multi–frequency Imaging Microwave Radiometer — MIMR) is programmed to fly in 1998.

This chapter deals with the objectives of the SSM/I and with the user requirements. It will discuss the environmental parameters that can be obtained. As several organisations are starting to work with data, available data formats and data processing software are discussed. Remote sensing applications require direct and inverse algorithms to be applied to geophysical and electromagnetic data. Applications for the atmosphere, sea and land are presented and we shall show how microwave radiometers also used in conjunction with other instruments to provide valuable data for studies on global energy exchange and other areas.

NATO ASI Series, Vol. I 24
Remote Sensing and Global Climate Change
Edited by Robin A. Vaughan and Arthur P. Cracknell

2 Requirements for spaceborne radiometers

Observation of the Earth in the last decade has shown that local events may often have an impact on a worldwide scale and that several sets of environment variables or measurables are closely inter-related. The Pinatubo eruption which brought sizable amounts of dust into the higher atmosphere is an example of this, as the event is thought to have delayed global warming (greenhouse effect) by some years.

2.1 User requirements for global Earth observation

Due to global interaction researchers do not concentrate on one unique parameter but cover a whole set of observables e.g. even if the requirements for meteorologists might look different from those for studies of the global energy exchange the basic parameters of the atmosphere, ocean and land have to be known since they are all inter-related (see chapter 6). Very often the difference in requirements in two different fields (e.g. meteorology and climatology) are mainly to do with space and time resolution of the observations.

For the atmosphere the basic parameters required are water vapour (WV), liquid water (LW) and precipitation contents. For the ocean the parameters are sea surface temperature (SST), sea wind and salinity, whereas for land the soil moisture content, the land surface temperature and the plant water content are required. Snow and ice observations are also very important as they store a large amount of energy and water.

2.2 Instrument requirements

Radiometers are the most important sensors for passive remote sensing of the above mentioned parameters. Starting with the Cosmos 243 satellite in 1968, microwave radiometers have been carried aboard several spacecraft (Table 1).

To fulfil multiple requirements, multi–channel radiometers have been designed, starting with the SMMR (Scanning Multi–Channel Microwave Radiometer) on Nimbus–7 and Seasat. The SSMR was able to measure the sea surface temperature and wind speed together with the atmospheric water vapour, the liquid water and rain rate, the sea ice concentration and snow coverage. At present, the SSM/I is monitoring the Earth's surface (Hollinger *et al.* 1990). It provides high–performance microwave measurements of geophysical parameters. As the number of frequencies is no longer a very difficult requirement for instrument designers, the user requirements are more and more important for the definition of the frequencies.

As can be seen from Table 2, geophysical parameters can be related to microwave frequencies due to their physical absorption/emission characteristics. The SSM/I radiometer covers a wide spectrum of these parameters, especially those related to sea ice, snow and atmospheric features. In Europe the Multifrequency Imaging Microwave Radiometer (MIMR) is programmed to fly on the polar platform, which is to be launched in 1998. This instrument will increase the possibilities of SSM/I, as it has a higher resolution (see Table 1) and with the additional 6.8 and 10 GHz channels covers the required ocean parameters (Menard and Reynolds 1991). In ESA SSM/I data are

Sensor, Spacecraft, Country	year	frequency and polarisation	inc. $\angle^{\circ}$	resolution [km]
Cosmos 243, USSR	1968	3.5, 8.8	0	37
Cosmos 384, USSR	1970	22.2, 37	0	
ESMR, Nimbus 5, USA	1972	19.35	var	25
NEMS, Nimbus 5, USA	1972	22.2, 31.4, 53.6, 54.9, 58.8	0	200
S-193, Skylab, USA	1973	13.9	var.	16
S-194, Skylab, USA	1973	1.4	0	115
Meteor, USSR	1974	37 V H	35	
ESMR, Nimbus 6, USA	1975	37 V H	var	20 43
SCAMS, Nimbus 6, USA	1975	22.2, 31.6, 52.8, 53.8, 55.4		150
SSMR, Seasat, USA	1978	6.6, 10.7, 18.0, 21.0, 37.0 V H	50	22 35
SSMR, Nimbus 7, USA	1978	6.6, 10.7, 18.0, 21.0, 37.0 V H	49	18 27
SAMIR, Bhaskara-II, India	1981	19.4, 22.2, 31.4		125
SSM/I, DMSP F08, USA	1987	19.3 V H, 22.2 V, 37.0 V H, 85.5 V H	50	70x45 – 16x14
MIMR, ESA	future	6.8, 10.7, 18.7, 23.8, 36.5, 90 all V H	50	39x60 – 3x5

Table 1: Spaceborne microwave radiometers for Earth observation

			1.4	6.8	10	18	23	37	50-60	90	160	180
Land	Soil moisture		●		O			O				
Ocean	SST			●	◒	◒	◒	O				
	Wind speed			O	●	◒	O	O				
	Salinity		●	◒	◒	◒	◒					
Sea Ice	Boundary					O	O	●		O		
	Concentration, Type, Temperature			O	O	●	O	●				
	Dynamics			O	O	●	O	●		O		
	Surface melt			O	O	●	O	●				
Ice Sheets	Accumulation				O	●	O	●				
	Surface melt				O	●	O	●				
Snow	Snow cover extent					O	O	●		O		
	Water equivalent				O	●	O	●				
	Water content				O	●	O	●				
Atmosphere												
	Liquid Water	land								●	●	◒
		sea				◒	◒	●				
	Water vapour	land								◒	◒	●
		sea				◒	●	◒				
	Rain	land						●		●		◒
		sea			O	●	◒	●				
	Temperature	profile					O	O	●	O		
	Humidity	profile								O	◒	●

● Key channel
◒ Necessary
O Helpful

Table 2: Relation between microwave frequencies and geophysical parameters as guiding rules for radiometer designer (Gudmandsen and Skou 1991)

being used for testing and development of retrieval algorithms for MIMR since it uses frequencies similar to those of SSM/I.

3 Special Sensor Microwave/Imager (SSM/I)

This section focus on the SSM/I sensors that are flown on board several spacecraft in the Defense Meteorological Satellite Program (DMSP). The SSM/I are designed to provide high-performance microwave measurements for the determination of global weather and critical atmospheric, oceanographic and land parameters.

The first SSM/I was launched on the DMSP-F08 spacecraft on 19 June 1987 and numbers two and three followed on 1 December 1990 on the DMSP-F10 and on 28 November 1991 on the DMSP-F11.

3.1 Sensor description

The SSM/I (Figure 1a) is a seven-channel four-frequency linearly polarised, passive microwave radiometric system which measures brightness temperatures at 19.3, 22.235, 37.0 and 85.5 GHz. It receives both vertically (V) and horizontally (H) linearly polarised radiation for all channels except for the 22 GHz channel for which it receives only vertically polarised radiation. The advantages in terms of power consumption and mass for passive instruments compared to active instruments is seen for this sensor which consumes only 45 W of power by a weight of 48.5 kg.

The DMSP-F08 spacecraft flies in a circular Sun–synchronous orbit with an inclination of 98.8° and an altitude of 835 km. At this height the orbital period is 102.0 minutes, so that approximately 14.1 orbits per day are performed. With a swath width of 1400 km an almost global coverage is obtained within one day. Only the grey shaded areas of Figure 1b are not covered within one day. Due to the 98.8° inclined orbit small circular areas with a radius of 280 km at the north and south poles are not covered at all. The instrument is mounted on top of the spacecraft and consists of an offset parabolic reflector (61 cm × 66 cm), a corrugated, broad–band seven port-horn antenna and the radiometric subsystem.

Calibration is performed using cold and hot references. A small mirror mounted on top of the drum, off–axis, provides the cold (2.7 K) cosmic background temperature reference. The hot reference is obtained using a heated absorber. Measurements of these two references are performed for each channel once every scan. This scheme provides an overall end–to–end absolute high quality calibration which includes the feed horn. A total–power radiometer configuration is employed providing a sensitivity that is twice that which can be obtained by a conventional Dicke switched radiometer. When including the data preprocessing that corrects for spillover, cross–polar coupling and antenna pattern effects, the performance of the SSM/I is significantly improved compared to previous spaceborne microwave radiometric systems.

3.2 SSM/I environmental parameters

With the calibration procedure of the SSM/I channels the accuracy for a subset of the user-required environmental products (see Table 1) can be achieved. The subset summarized in Table 3 reflects the channel sensitivity of the 19, 22, 37 and 85 GHz channels in terms of geophysical parameters.

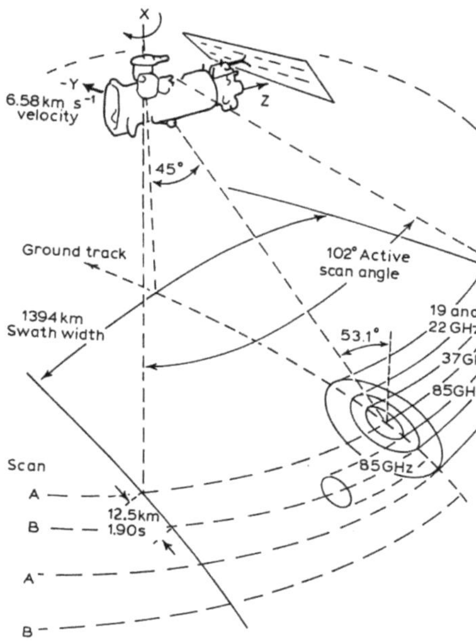

a) Scan geometry

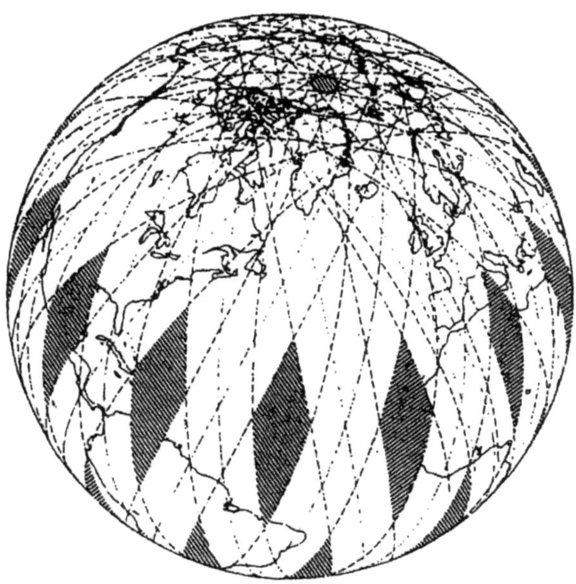

b) 24 h coverage

Figure 1: Scan geometry and global coverage of the SSM/I aboard the DMSP-F08 spacecraft (Hollinger *et al.* 1990)

Parameter	Spatial Resolution [km]	Range of Values	Quantization Level	Absolute Accuracy
Ocean surface wind speed	25	3 – 25	1	±2 m s^{-1}
Ice				
- area covered	25	0 – 100 %	5	±12%
- age	50	1st year, multi-year	1 yr, > 2 yr	none
- edge location	25	N/A	N/A	±12.5 km
Precipitation over land	25	0 – 25	0, 5, 10, 15, 20, > 25	±5 mm h^{-2}
Cloud water	25	0 – 1	0.05	±0.1kg m^{-2}
Integrated water vapour	25	0 – 80	0.1	±2.0kg m^{-2}
Precipitation over water	25	0 – 25	0, 5, 10, 15, 20, > 25	±5 mm h^{-1}
Soil moisture	50	0 – 60 %	1	none
Land surface temperature	25	180 – 340 K	1	none
Snow water content	25	0 – 50 cm	1	±3 cm
Surface type	25	12 types	N/A	N/A
Cloud amount	25	0 – 100%	1	±20%

Table 3: SSM/I environmental products (Hollinger 1990)

As an example from Table 3 the specified results show that measurements of liquid water, water vapour and rain over sea using the 19, 22 and 37 GHz channels are quite good. However, it should be noticed that even the relatively low frequency of 19 GHz may saturate in some occasions (e.g. tropical rain storm). On the other hand, requirements for land use like soil moisture are not well established with SSM/I and will require a lower radiometric frequency (see Table 2).

4 Availability of SSM/I data

As discussed before (section 2) the instrument and platform characteristics define the way that the Earth surface is scanned and the data are collected. Looking in the aft direction of the spacecraft, the scan direction is from left to right with a spin rate of 31.6 rpm of the sensor. On each scan (scan B) 128 uniformly-spaced samples of the 85.5 GHz scene are taken over the 102.4° scan sector. The sampling interval of 4.2 ms is equal to the time required for the beam to travel 12.5 km in the cross track direction. As the footprint of the 19–37 GHz channels is larger, sampling for these channels is done every 2nd scan (scan A) with 64 uniformly spaced samples collected with an 8.4 ms interval. The characteristics of the two scans (A and B) are summarised as follows. Scan A denotes scans with all channels sampled (from 19 to 85 GHz), while scan B collects only 85 GHz data (see Figure 1a). Start and stop times of the integrate–and–dump filters are designed to maximize the radiometric integration time and achieve concentric beams for all sampled data.

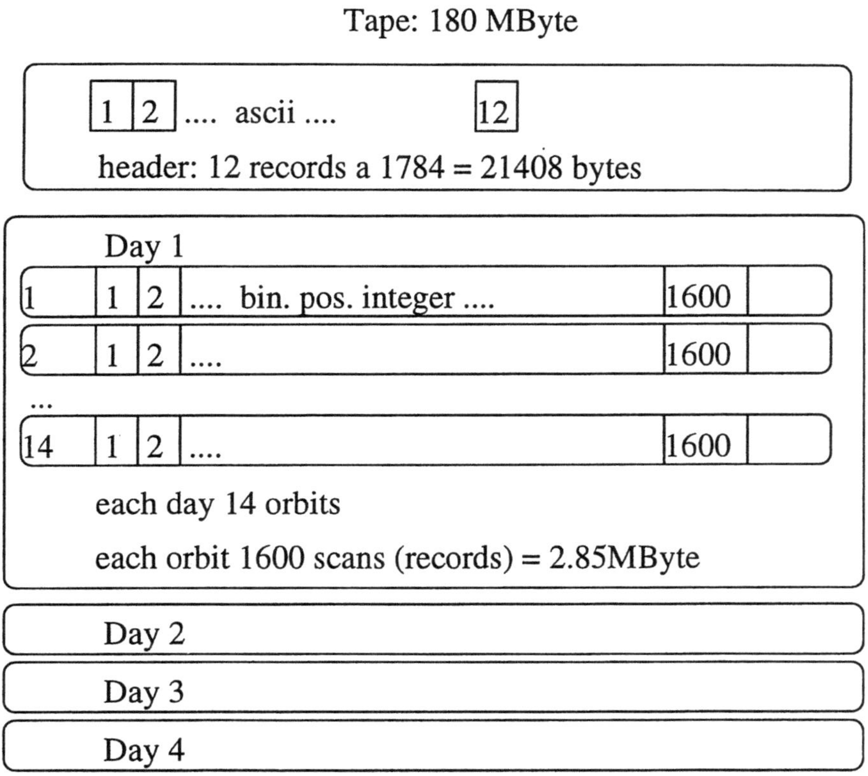

Figure 2: Data storage of the Global coverage data on 6250 bpi tapes

4.1 Raw data format ('Global coverage data')

On board the spacecraft the measured antenna temperature T_A is stored together with instrument specific parameters like reference voltages and temperatures of the components in the receiving chain. This data stream, containing the 19–37 GHz T_A data for the A–scan and the 85 GHz T_A for both scans, is transmitted to a ground station and stored at the site. Organisations like ESA Earthnet keep copies of these data that are then made available to users.

As the data format of the SSM/I is similar to that of other spaceborne missions, it will be described here in more detail. The raw data, usually delivered on 6250 bpi tapes, contains a header and four days of data in four files (Figure 2). The header contains general information stored as ASCII characters. The first line identifies the tape according to year, month, day and revision. The following lines of the header give specified begin and end times of the tape. Because of data gaps, the actual begin and end time may differ from the specified ones. Some additional information conclude the header that contains in total 21,408 bytes. The header is divided into 12 logical records to match the record length of the data files.

The remaining files are used for the T_A data, 4 days per tape, 14 orbits per day and

1600 scans (one record per scan) per orbit. Each logical record corresponds to a pair of A and B scans and contains 1784 bytes. With this, one orbit holds about 2.85 Mbyte and one day about 40 Mbyte, assuming no data gaps during that period.

Each record contains 23 fields of different length, where the first lines contain information for calibration purposes like hot load temperatures, reference voltages and coefficients for the gain control. The fields 19–21 contain the longitude and latitude of the A and B–scans. Finally the 19–37 GHz channels are stored in field 22 as 320 items, which are the five T_A values for the 64 points per scan. Similarly the 85 GHz channels are stored as 384 items in field 23 (Wentz 1991).

A 'predecode' FORTRAN program is delivered together with the tapes, so that extraction of the antenna temperature and a computation of the brightness temperature T_B can be performed.

4.2 Polar data on CD–ROM

As mentioned earlier several organisations already work with SSM/I data to retrieve geophysical parameters. One of these groups is the World Data Center A for Glaciology. In this Center predecoding of the data is focussed on the polar regions. As a longitude/latitude display is useless for the polar regions, a transfer of the satellite scans into a stereographic projection is performed (Figure 3a). The brightness temperatures of data inside the rectangle are stored as 16 bit unsigned integer in an x–y grid, beginning with the top left to bottom right corner (NSIDC 1992).

Different file names are used to indicate the stored information (Figure 3b). Data of the northern area (Nxx) and the southern area (Sxx) are stored differently for the 19–37 GHz channels (xxB) and the 85 GHz channels (xxA). As can be seen from the data format, the xxA files contain four times the number of cells due to the double resolution of the 85 channels. In addition to the data files the coastal outlines as well as documentation are stored on each 650 Mbyte CD–ROM.

Several users have already applied these data to map geophysical parameters like the cloud liquid water content (Alishouse *et al.* 1990).

5 Algorithms for direct modelling and inversion

Modelling is an important part of the data analysis as it combines the measured T_B data with geophysical parameters and vice versa. In principle, modelling uses two types of algorithms, the direct and the inverse algorithms (Figure 4).

Direct algorithms are the ones that, from a given geophysical environment, allow for the calculation of T_B for a given frequency and polarisation. These algorithms are, for example, used by instrument designers when it is fundamental to know what can be expected, in terms of measurement using a particular instrument, from a certain environment.

On the other hand, users need to retrieve geophysical parameters (e.g. wind speed, sea surface temperature) out of the observed brightness temperatures. This relation is expressed by inversion algorithms. Modelling is completes if both kind of algorithms are

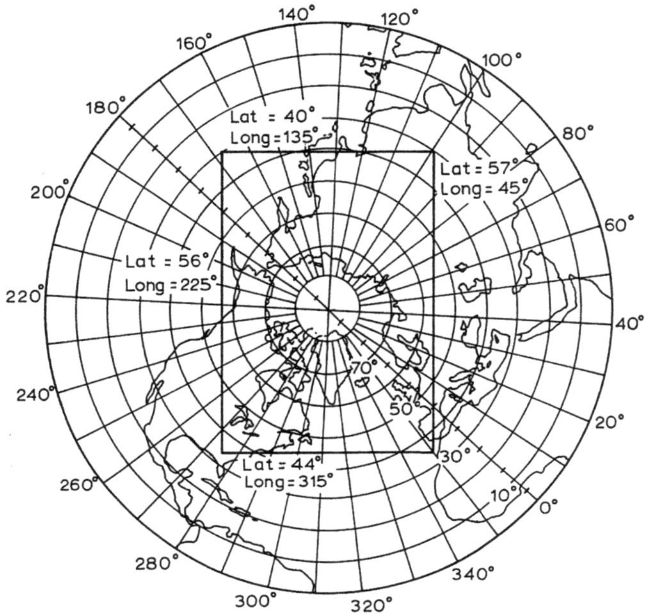

a) Stereographic projection (NSIDC 1992)

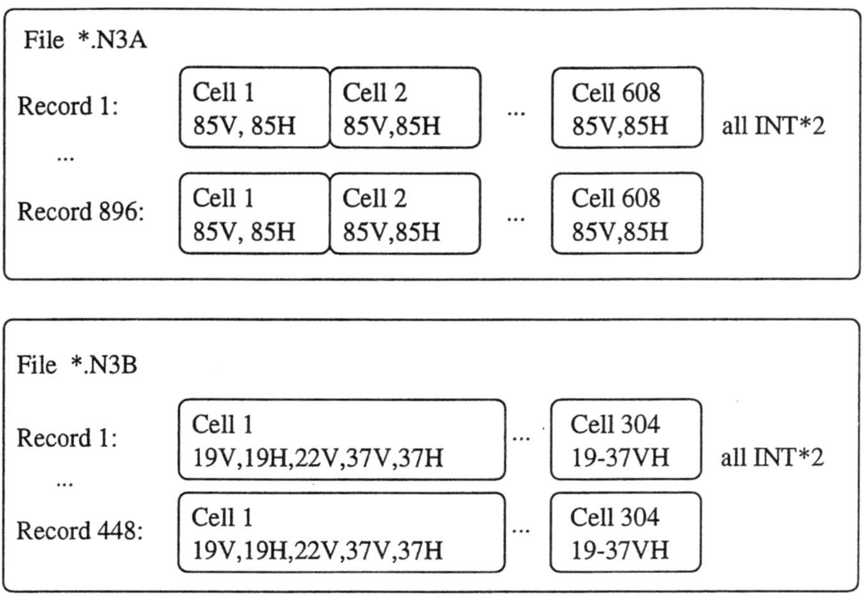

b) Data format

Figure 3: Stereographic projection of the north polar area and file structure as used for SSM/I polar data files

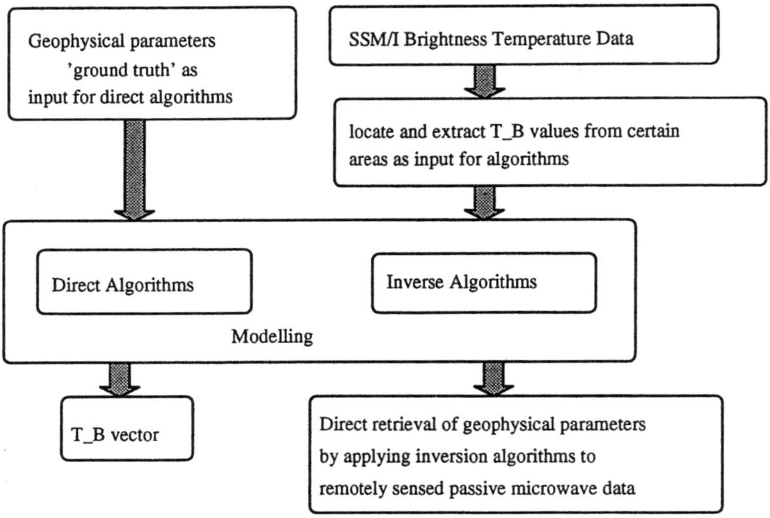

Figure 4: Connection of SSM/I T_B data and geophysical parameters by modelling

available and allow one to close the loop from geophysical parameters to temperatures and back. In such a case, sensitivity analysis can be performed to give the accuracy of the observed results. It is possible to derive the increase in accuracy or number of geophysical parameter retrieval by increasing the number of radiometer channels.

Radiative transfer theory can be used to model the influence of the physical scene that includes the solid Earth and the atmosphere. The latter can be defined by a vertical temperature profile. Absorption due to gas and water vapour as well as scattering due to water droplets and ice crystals are also taken into account by the radiative transfer approach. However, a general inversion of this theory is, in most cases, very difficult to perform. Instead, most inversion algorithms used in an operational environment are based on empirical or statistical formulae that are very efficient in terms of computing time. Several have been proposed in the literature to retrieve geophysical parameters from the measured brightness temperatures. Radiative transfer models, however, give a better insight into the physical processes and for this reason are used in research and development work.

Most algorithms are quite complex and their detailed descriptions go beyond the scope of this chapter. Only some example empirical or statistical algorithms are presented in the following to demonstrate the use of direct and inverse modelling.

5.1 Water vapour over the sea

Recently, research programmes like the Tropical Ocean and Global Atmosphere(TOGA) and the World Ocean Circulation Experiment (WOCE) have been initiated to understand the dynamic processes based on the energy exchange between the oceans and the atmosphere. Passive remote sensing with ESMR, SMMR, SSM/I and MIMR can provide global information about sea surface temperature, sea surface wind speed and other ocean parameters (Goodberlet *et al.* 1990).

For water vapour (WV), Lojou has proposed algorithms for SSMR, SSM/I and MIMR (Scire' Scappuzzo 1992). Using the MIMR channels (see Table 1) WV (in g cm^{-2}) can be expressed as a function of the 18.7, 23.8 and 37 GHz channels as follows

$$\begin{aligned} WV[\mathrm{g/cm^2}] &= 8.45 + 3.71 ln(280 - T_{H19}) + 1.05 ln(280 - T_{V18}) - 6.52 ln(280 - T_{H23}) \\ &\quad -2.69 ln(280 - T_{V23}) + 3.32 ln(280 - T_{H37}) - 0.417 ln(280 - T_{V37}). \quad (1) \end{aligned}$$

The above algorithm is valid only for $T_B < 280$ K and assumes an emissivity with an averaged homogeneous distribution that can only be found when observing ocean surfaces. After applying this formula to measured data, the water vapour is obtained and its accuracy proved by comparing the results with measurements taken from other sensors. Similar formulae can be applied for other parameters like rain rate and liquid water content.

Most inversion algorithms that are used in an operational environment are based on formulations as simple as the one given above. These type of formulae are obtained through parameter fitting to either existing measurements or to calculated values obtained from previously tested direct models.

5.2 Sea surface temperature

Brightness temperature as a function of ocean parameters is presented as an example for direct algorithms. Three major physical properties influence the emission characteristics, namely the dielectric constant, the surface geometry and the foam cover.

The permittivity ε_w of sea water is a function of frequency, salinity and water temperature and can be calculated by the modified Debye equation (Ulaby *et al.* 1986). At frequencies higher than 6 GHz (all SSM/I channels) the effect of salinity is negligible. The surface geometry is regarded as a rough surface with partial foam cover depending on the wind speed W. Wilheit (1979) has proposed an algorithm to model this behaviour where the reflectivity $\Gamma_{h,v}$ given by the Fresnel reflections coefficients with $\Gamma_{h,v} = R^2_{h,v}(f, \Theta)$ is averaged over the surface slope distribution of a wind roughened sea and one uses a geometrical optics (GO) approach to the sea surface. A partial foam cover is taken into account for $W > 7$ m s^{-1}, as it reduces the surface reflectivity. An empirical expression for the foam reduction is given as

$$K_{foam} = \begin{cases} \alpha(1 - \exp) - f/f_0(W[\mathrm{m\ s^{-1}}] - 7) & \text{for } W \geq 7\mathrm{m\ s^{-1}} \\ 0 & \text{for } W < 7\mathrm{m\ s^{-1}} \end{cases}, \quad (2)$$

where $\alpha = 0.006$ and $f_0 = 7.5$ GHz and W in is m s^{-1}.

Finally the emissivity e_{sea} is defined by the foam-reduced averaged reflectivity as

$$e_{sea}(f, \Theta, W, SST) = 1 - [\Gamma_{h,v;ave}(f, \Theta, W, SST)\,(1 - K_{foam}(f, W))] \quad (3)$$

and has a typical behaviour shown Figure 5.

As can be seen from the direct modelling, T_B increases with SST and frequency. The 23 GHz channels show the highest sensitivity that can be explained by the influence of the water vapour absorption around 22.2 GHz.

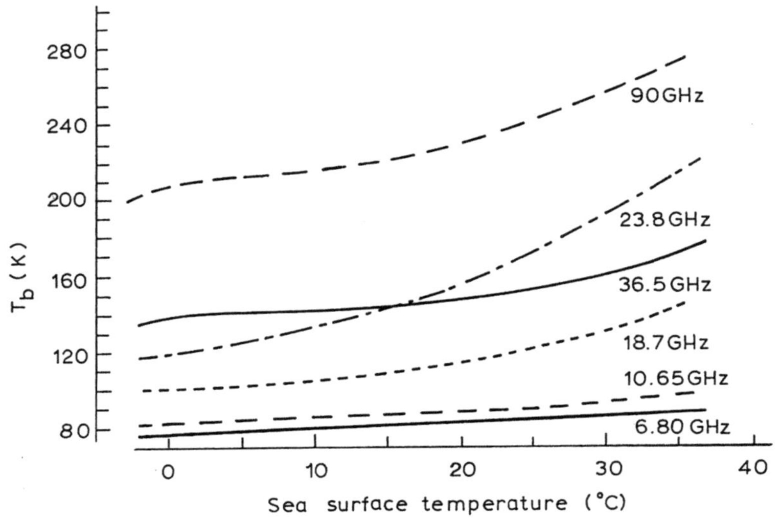

Figure 5: Brightness temperature as function of sea surface temperature (SST) using Wilheit's model for partial foam covered sea ($W = 0$ m s^{-1}, $\Theta = 0°$)

It should be noted that the model, was tested only for frequencies lower than 37 GHz. However, due to the theoretical basis of the model the results at 90 GHz seem to be acceptable. Further investigations have to be undertaken to describe the polarisation dependence of foam.

5.3 Land surface type classification

This section will describe a completely empirical algorithm for the classification of land surface types. The algorithm can be considered as a combined physical/statistical method, where frequency and polarization differences are used to set up the classification rules.

The classification is in terms of topography, natural vegetation, land use, climate, soils and water resources. A sizeable number of formulae/relations with all the channel measurements are used to perform this classification. An example using the set of formulae for the identification of natural high density forests, as found in the Amazon, Congo jungles or the Appalachian forest *Dense Vegetation* is presented below.

$$T_{22V} - T_{19V} \leq 4 \quad (4)$$
$$0.5\,(T_{19V} + T_{37V} - T_{19H} - T_{37H}) \leq 1.9 \quad (5)$$
$$T_{85V} - T_{37V} < 0.5 \quad (6)$$
$$T_{85H} - T_{37H} < 6.0 \quad (7)$$
$$T_{19V} > 262 \quad (8)$$

Each of these equations is related to the influence of a different physical parameter. Equation 5 represents the polarisation behaviour of the vegetation covered soil. In the

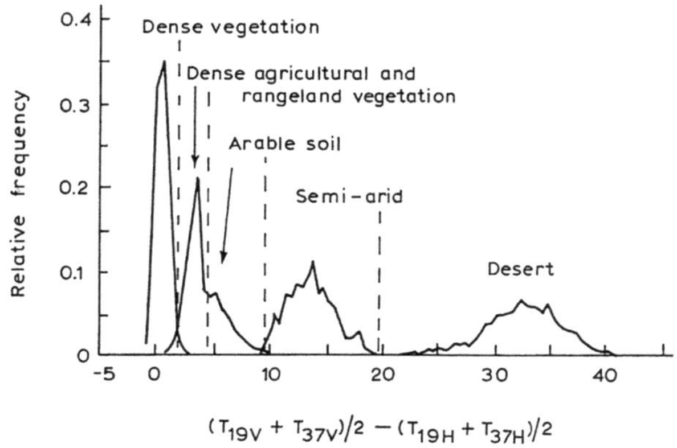

Figure 6: Vegetation classification using the polarimetric difference of the 19 and 37 GHz channels (Neale *et al.* 1990)

case of a lossy dense canopy layer emission is isotropic showing very small polarimetric differences. As the density decreases (jungle → desert) polarisation differences caused by the soil become significant, thus classifying the density of vegetation (Figure 6).

Condition 6 is a precipitation flag showing precipitation clouds over vegetation. Rule 8 ensures that no ice or snow is present and that what the instrument is seeing is vegetation for this class of land type. The two remaining conditions 4 and 7 check the presence of large quantities of water on the surface or, on the other hand, differentiate between dry and moist soil, depending on the vegetation coverage. This check is based on the influence of water in an essentially bare soil as it decreases the brightness temperature and increases the polarisation difference.

All these rules can be explained by the physical behaviour of the soil types. However, classification is useful only for relatively large areas, with sizes greater than at least a couple of footprints (pixels).

6 Conclusions

In this chapter, mm–wave spaceborne radiometric remote sensing has been described using the SSM/I sensor. This multi–polarisation, multi–frequency sensor allows the observation of geophysical parameters on a global scene, consuming less power and weight than active instruments. The retrieval of environmental parameters can be simplified by using frequencies (*key channels*) that are closely related to the physics of the parameter. With such a configuration, a spaceborne radiometer can supply a wide set of environmental parameters from atmosphere, ocean and land.

The SSM/I sensor on board the DMSP F08 spacecraft and its instrument and orbital parameters were presented in terms of scan geometry, footprint, radiometric parameters, spatial resolution and radiometric sensitivity.

As the relation between the observed brightness temperatures and the ground parameters is required by users, direct and inversion algorithm have been presented by example for atmospheric (water vapour), oceanic (sea surface temperature) and land (type classification) retrieval applications. However, due to the large number of geophysical parameters to be taken into account, it was shown that modelling is still incomplete and the complexity of the algorithms has to be increased to fulfil the user requirements in passive remote sensing from space.

Acknowledgments

The authors would like to thank Frank Wentz at the University of Colorado (World Data Center A), Gerhard Triebnig at ESA–ESRIN for providing us with SSM/I data and Maurice Borgeaud at ESA–ESTEC and Francesca Scire' Scappuzzo at the University of Catania for supporting us with the basic decode and display software. All those who were involved in ESA contracts during the last years and delivered various direct and inverse algorithm are gratefully acknowledged. The authors wish also to thank Yvon Menard at ESA–ESTEC for comments on future radiometer missions.

References

Alishouse, J. C., Snider, J. B., Westwater, E. R., Swift, C. T., Ruf, C. S., Snyder, S. A., Vongsathorn, J., and Ferraro, R.R., 1990, Determination of Cloud Liquid Water Content Using the SSM/I, *IEEE Transactions on Geoscience and Remote Sensing*, **28**, , 817-822.

Goodberlet, M. A., Swift, C. T., and Wilkerson, J. C., 1990, Ocean Surface Wind Speed Measurements of the SSM/I, *IEEE Transactions on Geoscience and Remote Sensing*, **28**, 823-828.

Gudmandsen, P., and Skou, N., 1991, User Wishlist for future Spaceborne Microwave Radiometry, *ESTEC Contract No. 8968/90/NL/US(SC)*, Vol. **2**, ESA ESTEC.

Hollinger, J. P., 1990, Introduction, *IEEE Transactions on Geoscience and Remote Sensing*, **28**, 779-780.

Hollinger, J. P., Peirce, J. L., and Poe, G. A., 1990, SSM/I Instrument Evaluation, *IEEE Transactions on Geoscience and Remote Sensing*, **28**, 781-790.

Menard, Y., and Reynolds, M., 1991, The design of the ESA multiband imaging microwave radiometer MIMR, *Proceeding of IGARSS '91 Symposium, Espoo, held in Finland*, 2359–2363.

NSIDC, 1992, User's Guide: DMSP SSM/I Brightness Temperature and Sea Ice Concentration Grids for the Polar Regions on CD–ROM, *National Snow and Ice Data Center,* NSIDC Special Report, 1, Boulder, Coralado.

Neale, C. M. U., McFarland, M. J. and Chang, K., 1990, Land–Surface–Type Classification Using Microwave Brightness Temperatures from the Special Sensor Microwave/Imager, *IEEE Transactions on Geoscience on Remote Sensing*, **28**, 829-838.

Scire' Scappuzzo, F., 1992, Study of Passive Remote Sensing Emission/Inversion Models and Use of SSM/I Data for MIMR Analysis, *ESTEC working paper no. 1639*, ESA ESTEC.

Ulaby, F. T., Moore, R. K., and Fung, A. K., 1986, *Microwave Remote Sensing: active and passive. Volume III: From Theory to Applications*, (Dedham: Artech House).

Wentz, F. J., 1991, User's Manual SSM/I Antenna Temperature Tapes, *RSS Technical Report 120191*, Remote Sensing Systems, Santa Rosa.

Wilheit, T., 1979, A Model for the Microwave Emissivity of the Ocean's Surface as a function of Wind Speed. *IEEE Transactions on Geoscience Electron*, **GE-17**, 244-249.

Basis and Structure of Climate Models

Arthur P Cracknell

1 Introduction

In Chapter 1 we discussed many natural phenomena and human activities which affect the weather and the climate. It would be an interesting and useful scientific activity if we were able to model the climate, taking into account all the natural phenomena, and predict the future behaviour of the climate in the absence of any effects of human intervention. Given the present public concern, it would be particularly useful if we were then able to model the effects of human activities, as well as the natural phenomena, and make reliable predictions of the effects of human activities on the future climate. Unfortunately, the system is so complicated, our present knowledge is somewhat uneven and our historical knowledge is very little indeed. However, in spite of all the difficulties, a great deal of effort has gone into climate modelling in recent years and some very useful results have been obtained. Given what was said in sections 3 and 4 of chapter 1 about weather forecasting, the difficulties of producing weather forecasts for more than a few days ahead, and the whole idea of chaos versus determinism, it is perhaps both surprising and encouraging that some success in climate modelling can be achieved. There are various books on climate modelling, but the most up-to-date text at present is the volume edited by Trenberth (1992); it contains a comprehensive list of references to earlier work.

In 1988 the World Meteorological Organisation (WMO) and the United Nations Environment Programme (UNEP) set up an Intergovernmental Panel on Climate Change (IPCC). This was charged with assessing the scientific aspects of the climate change issue, assessing the environmental and socio-economic impacts of climate change and formulating response strategies. The IPCC established several working groups and, of these, Working Group I given the task of assessing the scientific information on climate changes. That Working Group's report was published (Houghton *et al.* 1990); the production of that report involved contributions from 175 scientists from 25 countries and a further 200 scientists were involved in the peer review of the draft report. Subsequently a supplementary report was also produced recently (Houghton *et al.* 1992).

NATO ASI Series, Vol. I 24
Remote Sensing and Global Climate Change
Edited by Robin A. Vaughan and Arthur P. Cracknell

It is also relevant to mention Prof. K.Ya. Kondratyev's book (Kondratyev 1992) which unfortunately has not yet been translated into English. In this chapter I shall follow the IPCC report (Houghton *et al.* 1990) quite closely, occasionally making some verbatim quotations of extracts from it. The original IPCC report (Houghton *et al.* 1990) comes with a *Policymakers Summary* of about 20 pages which was prepared "to meet the needs of those without a strong background in science who need a clear statement of the present status of scientific knowledge and the associated uncertainties". It is really quite remarkable that from the contributions of so many scientists, who surely must have many disagreements among themselves on matters of details, there have emerged some very clear conclusions. The two page *Executive Summary* of the *Policymakers Summary* is a model of scientific writing and is reproduced here.

EXECUTIVE SUMMARY

We are certain of the following:

- there is a natural greenhouse effect which already keeps the Earth warmer than it would otherwise be.
- emissions resulting from human activities are substantially increasing the atmospheric concentrations of the greenhouse gases: carbon dioxide, methane, chlorofluorocarbons (CFCs) and nitrous oxide. These increases will enhance the greenhouse effect, resulting on average in an additional warming of the Earth's surface. The main greenhouse gas, water vapour, will increase in response to global warming and further enhance it.

We calculate with confidence that:

- some gases are potentially more effective than others at changing climate, and their relative effectiveness can be estimated. Carbon dioxide has been responsible for over half the enhanced greenhouse effect in the past, and is likely to remain so in the future.
- atmospheric concentrations of the long-lived gases (carbon dioxide, nitrous oxide and the CFCs) adjust only slowly to changes in emissions. Continued emissions of these gases at present rates would commit us to increased concentrations for centuries ahead. The longer emissions continue to increase at present day rates, the greater reductions would have to be for concentrations to stabilise at a given level.
- the long-lived gases would require immediate reductions in emissions from human activities of over 60% to stabilise their concentrations at today's levels; methane would require a 15-20% reduction.

Based on current model results, we predict:

- under the IPCC Business-as-Usual (Scenario A) emissions of greenhouse gases, a rate of increase of global mean temperature during the next century of about 0.3°C per decade (with an uncertainty range of 0.2°C to 0.5°C per decade); this is greater than that seen over the past 10,000 years. This will result in a likely increase in global mean temperature of about 1°C above the present value by 2025 and 3°C before the end of the next century. The rise will not be steady because of the influence of other factors.
- under the other IPCC emission scenarios which assume progressively increasing levels of controls, rates of increase in global mean temperature of about 0.2°C per decade (Scenario B), just above 0.1°C per decade (Scenario C) and about 0.1°C per decade (Scenario D).
- that land surfaces warm more rapidly than the ocean, and high northern latitudes warm more than the global mean in winter.
- regional climate changes different from the global mean, although our confidence in the prediction of the detail of regional changes is low. For example, temperature increases in Southern Europe and central North America are predicted to be higher than the global mean, accompanied on average by reduced summer precipitation and soil moisture. There are less consistent predictions for the tropics and the Southern Hemisphere.
- under the IPCC Business as Usual emissions scenario, an average rate of global mean sea level rise of about 6cm per decade over the next century (with an uncertainty range of 3 - 10cm per decade), mainly due to thermal expansion of the oceans and the melting of some land ice. The predicted rise is about 20cm in global mean sea level by 2030, and 65cm by the end of the next century. There will be significant regional variations.

There are many uncertainties in our predictions particularly with regard to the timing, magnitude and regional patterns of climate change, due to our incomplete understanding of:

- sources and sinks of greenhouse gases, which affect predictions of future concentrations.
- clouds, which strongly influence the magnitude of climate change.
- oceans, which influence the timing and patterns of climate change.
- polar ice sheets which affect predictions of sea level rise.

These processes are already partially understood, and we are confident that the uncertainties can be reduced by further research. However, the complexity of the system means that we cannot rule out surprises.

Our judgement is that:

- Global - mean surface air temperature has increased by 0.3°C to 0.6°C over the last 100 years, with the five global-average warmest years being in the 1980s. Over the same period global sea level has increased by 10-20cm. These increases have not been smooth with time, nor uniform over the globe.
- The size of this warming is broadly consistent with predictions of climate models, but it is also of the same magnitude as natural climate variability. Thus the observed increase could be largely due to this natural variability; alternatively this variability and other human factors could have offset a still larger human-induced greenhouse warming. The unequivocal detection of the enhanced greenhouse effect from observations is not likely for a decade or more.
- There is no firm evidence that climate has become more variable over the last few decades. However, with an increase in the mean temperature, episodes of high temperatures will most likely become more frequent in the future, and cold episodes less frequent.
- Ecosystems affect climate, and will be affected by a changing climate and by increasing carbon dioxide concentrations. Rapid changes in climate will change the composition of ecosystems; some species will benefit while others will be unable to migrate or adapt fast enough and may become extinct. Enhanced levels of carbon dioxide may increase productivity and efficiency of water use of vegetation. The effect of warming on biological processes, although poorly understood, may increase the atmospheric concentrations of natural greenhouse gases.

To improve our predictive capability, we need:

- to **understand** better the various climate-related processes, particularly those associated with clouds, oceans and the carbon cycle.
- to **improve** the systematic observation of climate-related variables on a global basis, and further investigate changes which took place in the past.
- to **develop** improved models of the Earth's climate system.
- to **increase** support for national and international climate research activities, especially in developing countries.
- to **facilitate** international exchange of climate data.

There are two approaches which one can take to climate prediction. One is the palaeo-analogue approach, which is based on attempting to estimate future climatic change from reconstructions of past climates using palaeo-climatic data. The other is to use a numerical model, a general circulation model (GCM). Such models are derived from numerical weather forecast models; they include representations of additional elements of the climate system (using ocean models, land surface models, etc) with varying degrees of sophistication.

2 Predictability of Climate Change

In our earlier discussion of weather forecasting (see section 3 of chapter 1) it was commented that weather forecasts are reliable for a period of a few days ahead, that the temporal extent of their reliability is improving but that the ultimate limit of their reliability may be of the order of 2 to 4 weeks. This was discussed in terms of determinism versus chaos. One is therefore entitled to ask whether there is any hope of ever being able to make meaningful climate forecast predictions over the scale of a century or two if one has no hope of being able to make reliable weather forecasts for a time such as only 6 or 12 months ahead. This question is particularly pertinent since we are saying that if one takes the numerical modelling approach to climate prediction one is using general circulation models of the atmospheric circulation which have been derived from weather forecast models and which are based on the same physical mechanisms and equations as are used in the weather forecast models. General circulation models are based on the physical conservation laws which describe the redistribution of momentum, heat and water vapour by atmospheric motions. All of these processes are formulated in the "primitive" equations, which describe the behaviour of a fluid (air or water) on a rotating body (the Earth) under the influence of a differential heating (the temperature contrast between equator and pole) caused by an external heat source (the Sun). These governing equations are non-linear partial differential equations, whose solution cannot be obtained except by numerical methods. These numerical methods subdivide the atmosphere vertically into discrete layers, wherein the variables are "carried" and computed. For each layer the horizontal variations of the predicted quantities are determined either at discrete grid points over the Earth, as in grid point (finite difference) models, or by a finite number of prescribed mathematical functions as in spectral models. Climate models can be tested by running them to predict historical climatic conditions from an earlier historical starting point and considerable success has been obtained; undoubtedly there has been some fine tuning of the models in this work, but that is fair enough. If it is surprising that any success can be achieved with climate models then it may help to appreciate this if we recall that in looking at climate we are looking at an average of the weather over a reasonably long time, e.g. one year, or ten years for example. The chaotic elements, or the statistical fluctuations, in the climate are the weather systems. In looking at climate change we are looking at long-term changes in this average. Viewed over a century or a millennium we see that climatic parameters (temperature, rainfall, etc.) are basically stable and vary only slowly. It is the nature of these slow long-term variations that are of concern to us in climate studies after the local short-term fluctuations have been smoothed out. It is the stability of the long-term components that makes climate prediction possible. In a climate model we consider the effect of various forcing conditions on a given mean state. Analogies are always dangerous and should not be pushed too far. However, let us draw an analogy with the behaviour of a gas. We can use very successfully an equation of state for a gas to determine the effect of a change in pressure, volume or temperature on the gas. But if one were to seek to study the behaviour of the individual molecules in the gas and try to make predictions of the path of any given molecule in the gas one would be faced with an extremely difficult problem. Moreover, even if one could find the solution it would probably not be very interesting or useful.

In spite of randomness or of chaos at the small-scale level one can nevertheless make useful and meaningful calculations at the macroscopic level based on a transition from one mean (or quasi-equilibrium) state to another. Thus we commonly assume in climate prediction that the climate system is in equilibrium with its forcing. That means, as long as its forcing is constant and the slowly varying components alter only slightly within the time-scale considered, the mean state of the climate system will be stable and that if there is a change in the forcing, the mean state will change until it is again in balance with the forcing. The time-scale of the transition period until an equilibrium state is re-established is determined by the adjustment time of the slowest climate system component, i.e. of the ocean. The stable ("quasi-stationary") behaviour of the climate system gives us the opportunity to detect changes by taking time averages. Because the internal variability of the system is so high, the averaging interval has to be long compared to the chaotic fluctuations to detect a statistically significant signal which can be attributed to the external forcing.

Studies of the completed change from one mean state to another are called equilibrium response studies. Studies of the time evolution of the climate change due to an altered forcing, which might also be time dependent, are called transient response experiments.

The weather systems in the ocean have much smaller horizontal scales (less than one hundred kilometres) than those in the atmosphere; this leaves the large-scale features of the world ocean circulation to be non-chaotic. The success of classical dynamical oceanography depends on this fact. Observations of the penetration of transient tracers into the ocean show that the large-scale ocean currents are stable over periods of several decades. Palaeo-oceanographic evidence shows that the currents and gyres adjusted smoothly to the ice age cycle, see figure 10 of chapter 1. That evidence and theoretical understanding of the large-scale ocean circulation suggests that we are indeed dealing, in the ocean, with a predictable system, at least on time-scales of decades. The question is whether the existence of predictability in the ocean component of the Earth's climate system makes the system predictable as a whole. However, this seem to be a reasonable working hypothesis, which receives some support from the smooth transient response simulated by coupled ocean-atmosphere models (see section 9.2 below).

The main interest in climate modelling seems to be in predicting the effects of human activities, principally the release of large amounts of CO_2 into the atmosphere, over time-scales that, in terms of climatic studies, are quite short, i.e. decades or one or two centuries. This does not mean that climatologists are not interested in other causes of climate change. It simply means that CO_2 and the reasonably short-term effects are the aspects which are of great interest to society at large and, therefore, to politicians. As a consequence it tends to be for work in these areas that research grants are made to climatologists.

I have already indicated that I shall follow quite closely the IPCC report already mentioned (Houghton *et al.* 1990). The first two chapters of that report give a rather more detailed discussion of what we have considered already in section 5 of chapter 1 of the present volume. The climate system consists of five components

- atmosphere
- the oceans
- the cryosphere
- the biosphere
- the geosphere.

The ideas of numerical modelling of the climate may be regarded as a development

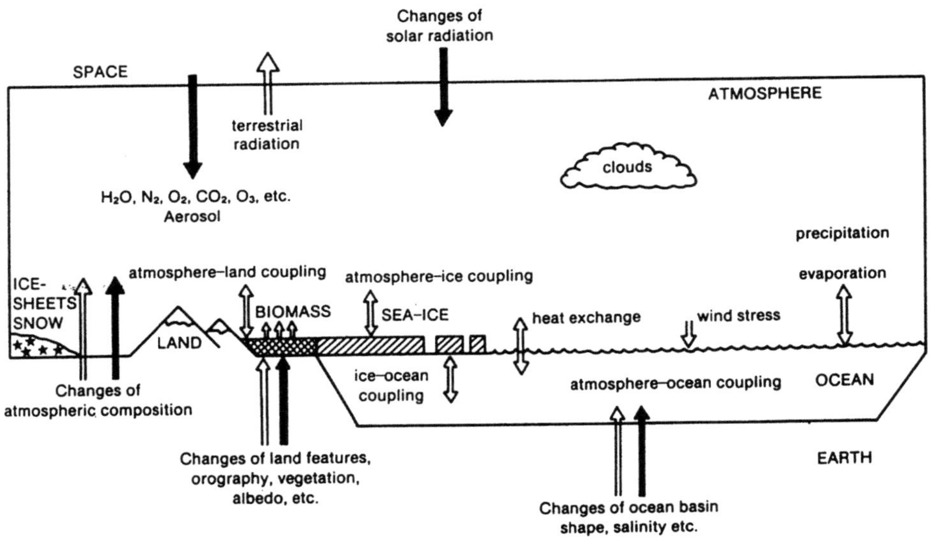

Figure 1: Schematic of the climate system (Houghton 1984).

from numerical weather forecast models. While following the discussion in chapter 3 of Houghton *et al.* (1990), it should be mentioned that this can be supplemented, if one needs a more detailed description, by the books by Gates (1975), Houghton (1984) and Washington and Parkinson (1986). Figure 1 is a much-reproduced diagram illustrating the atmosphere-ocean-ice-land climatic system. The solid arrows in this figure indicate external processes while the open arrows represent internal processes. The fundamental processes driving the global climate system are the heating by incoming electromagnetic radiation of rather short wavelength and the cooling by the loss of longer wavelength radiation into space. The heating is strongest in tropical latitudes and the cooling is predominantly in winter in the polar regions. The temperature gradients between equatorial and polar regions drive both the atmospheric circulation and the oceanic circulation; these circulation mechanisms then transport large quantities of heat from equatorial to polar regions. We shall consider briefly the five components of the climatic system which were listed above so that we can see, in general terms, the processes that need to be included in a climatic model.

3 Factors Involved

3.1 The Atmosphere

Most of the incoming solar radiation passes through the atmosphere and is absorbed by the Earth's surface (ocean, ice, rock, soil, vegetation, etc.). The evaporation of moisture and the direct heating of the surface lead to heat transfer between the surface and the atmosphere in the form of latent and sensible heat. The atmosphere transports this heat meridionally, mainly via transient weather systems, with a time-scale of the order of days.

The following processes are important in determining the behaviour of the atmospheric component of the climate system

- Turbulent transfer of heat, momentum and moisture at the surface of the Earth
- The surface albedo (i.e. its reflectivity) which depends on the nature of the surface; this determines the proportion of reflected to incoming solar radiation
- The release of latent heat when water vapour condenses; clouds, which play an important role in reflecting incoming solar short-wave radiation and in absorbing and emitting long-wave radiation
- The radiative cooling and heating of the atmosphere by CO_2, water vapour, ozone and other trace gases
- Aerosols (such as volcanic dust), the orbital parameters, mountain ranges and the land-sea distribution.

These processes, however, are not all independent and, as we shall see later on, feedback between some of these processes and between these processes and the surface of the Earth is very important. The last two sets of items in this list of atmospheric processes include components that are strongly influenced by human activity.

3.2 The Oceans

Over half of the solar radiation reaching the Earth's surface is first absorbed by the oceans, where it is stored and redistributed by ocean currents before escaping to the atmosphere, largely as latent heat through evaporation, but also as long-wave radiation. The currents are driven by the exchange of momentum, heat and water between the ocean and atmosphere. They have a complicated horizontal and vertical structure determined by the pattern of winds blowing over the sea and the distribution of continents and submerged mountain ranges. The major features of the horizontal distribution of the surface currents on the oceans were shown in figure 9 of chapter 1.

The vertical structure of the oceans comprises three layers

- The seasonal boundary layer, mixed annually from the surface, is less than 100 m deep in the tropics and reaches hundreds of metres in the sub-polar seas (other than the North Pacific) and several kilometres in very small regions of the polar seas in most years
- The warm water sphere (permanent thermocline), ventilated (i.e. exchanging heat and gases) from the seasonal boundary layer, is pushed down to depths of many hundreds of metres in gyres by the convergence of surface (Ekman) currents driven directly by the wind

- The cold water sphere (deep ocean), which fills the bottom 80% of the oceans' volume, ventilated from the seasonal boundary layer in polar seas.

There are physical and biological mechanisms in the oceans which are important in controlling the concentration of CO_2 in the climate system. CO_2 is transferred from the atmosphere into the interior of the ocean by the physical mechanism caused by differences in the partial pressure of CO_2 in the ocean and the lowest layers of the atmosphere. The rate of gas exchange depends on the air-sea difference in partial pressure of CO_2 and a coefficient which increases with wind speed. The ocean branch of the carbon cycle involves a flux of CO_2 from the air into the sea at locations where the surface mixed layer has a partial pressure of CO_2 lower than the atmosphere and vice versa. The annual ventilation of the seasonal boundary layer from the surface mixed-layer controls the efficiency of the biological mechanism by which ocean plankton convert dissolved CO_2 into particulate carbon, which sinks into deep water. These two pumps are responsible for extracting CO_2 from the global carbon cycle for periods in excess of a hundred years. The mixed-layer partial pressure of CO_2 is depressed by enhanced solubility in cold water and enhanced plankton production during the spring bloom.

The following processes control the climate response of the ocean

- The small-scale (of the order of 50km) transient eddies inside the ocean influence the structure of permanent gyres and streams and their interaction with submerged mountain ranges. The eddies also control the horizontal dispersion of chemicals (such as CO_2) dissolved in seawater

- The small-scale (tens of kilometres) patches of deep winter convection in the polar seas and the northernmost part of the North Atlantic, which transport heat and dissolved CO_2 below one kilometre into the deep reservoir of the cold water sphere, and the slow currents which circulate the newly implanted water around the world's oceans.

- The more extensive mechanism of thermocline ventilation by which some of the water in the surface mixed layer flows from the seasonal boundary layer into the warm water sphere reservoir of the ocean, which extends for several hundreds of metres below most of the oceans' surface area

- The global transport of heat, freshwater and dissolved chemicals carried by ocean currents, see figure 9 of chapter 1, which dictate the global distributions of temperature, salinity, sea-ice and chemicals at the sea surface. Fluctuations in the large-scale circulation have modulated these patterns over years and decades, see figure 10 of chapter 1. They also control the regional variations in sea surface properties which affect climate at this scale

- The biological pump in the seasonal boundary layer by which microscopic plants and animals (the plankton) consume some of the CO_2 dissolved in the seawater and sequester the carbon in the deep ocean away from the short term (up to a hundred years) interactions between ocean and atmosphere.

3.3 The Crysophere

The cryosphere can be classified as follows

- Seasonal snow cover, which responds rapidly to atmospheric dynamics on time-scales of days and longer. In a global context the seasonal heat storage in snow is small. The primary influence of the cryosphere comes from the high albedo of a snow-covered surface
- Sea ice, which affects climate on time scales of seasons and longer. This has a similar effect on the surface heat balance as snow on land. It also tends to decouple the ocean and the atmosphere, since it inhibits the exchange of moisture and momentum. In some regions it influences the formation of deep water masses by salt extrusion during the freezing period and by the generation of fresh water layers in the melting period
- Ice sheets of Greenland and the Antarctic, which can be considered as quasi-permanent topographic features. They contain 80% of the existing fresh water on the globe, thereby acting as long-term reservoirs in the hydrological cycle. Any change in size will therefore influence the global sea level
- Mountain glaciers are a small part of the cryosphere. They also represent a freshwater reservoir and can therefore influence the sea level. They are used as an important diagnostic tool for climate change since they respond rapidly to changing environmental conditions
- Permafrost affects surface ecosystems and river discharges. It influences the thermohaline circulation of the ocean.

3.4 The Biosphere

The biosphere on land and in the oceans (which has already been mentioned) controls the magnitude of the fluxes of several greenhouse gases, including CO_2, CH_4, H_2S and DMS (dimethyl sulphide), between the atmosphere, the oceans and the land. The processes involved are sensitive to climatic and environmental conditions, so that any change in the climate or the environment (e.g. increases in the atmospheric abundance of CO_2) will influence the atmospheric abundance of these gases. A detailed description of the feedbacks and their respective magnitudes can be found in chapter 10 of Houghton *et al.* (1990).

3.5 The Geosphere

We have already referred to the role of the land in terms of its exchange of energy with the atmosphere through the reflection, absorption and emission of radiation. The land also plays an important part in the hydrological cycle. The processes involved concern the amount of fresh water stored in the ground as soil moisture (thereby interacting with the biosphere) and in underground reservoirs, or transported as run-off to different locations where it might influence the ocean circulation, particularly in high latitudes. The soil interacts with the atmosphere by exchanges of gases, aerosols and moisture, and these are influenced by the soil type and the vegetation, which again are strongly dependent on the soil wetness. Present knowledge about these strongly interactive processes is rather limited.

3.6 Feedback Mechanisms

An important feature of the climate system is the presence of feedback. Feedback may, of course, be either positive or negative, that is, it may amplify or diminish the climate response to a given forcing. We have already mentioned one example of feedback when we mentioned what happened as the surface area of the Earth covered by ice expands (see section 5.2.2 of Chapter 1), i.e. the expansion of the area covered by ice leads to conditions which cause the area of ice to expand still further - until some other process becomes more significant and finally causes the expansion to come to a halt. We can look at it from the opposite viewpoint - if there is a warming of the Earth, leading to less snow and ice cover, then the surface of the Earth becomes (on average) less reflecting and more absorbing. This causes further warming and a further reduction in the snow and ice cover, i.e. this is an example of positive feedback. The real situation is, of course, more complicated and eventually the amplification of the warming by positive feedback (hopefully) comes to a halt. There are two other important feedback mechanisms, one involving water vapour and the other involving clouds. We shall consider these in turn.

Let us suppose that an increase in concentration of CO_2, a greenhouse gas, occurs. This will lead to some global warming. As a result of this warming, the amount of water vapour in the atmosphere will increase. The water vapour is also a greenhouse gas and so we see that an increase in the concentration of one greenhouse gas (CO_2) leads to an increase in another greenhouse gas (water vapour) and thus provides a positive feedback mechanism. Calculations quoted in section 3 of chapter 3 of Houghton *et al.* (1990) indicate that a doubling of the concentration of CO_2 in the atmosphere would lead to an increase of 1.2°C in the global-mean surface temperature without feedback, but that if one includes water vapour feedback this would be increased to 1.9°C.

The question of cloud feedback is very complicated. Let us consider first the effect of clouds in the present climate before we consider changes in the cloud situation. Table 1 indicates the effect of clouds on the global climate system for annual mean conditions. The presence of clouds heats the climate system by 31 Wm^{-2} through reducing the infrared emission at the top of the atmosphere (TOA). Although clouds contribute to the greenhouse warming of the climate system, they also produce a cooling because they reflect incoming solar radiation and reduce the intensity of the solar radiation reaching the surface of the Earth; this produces a cooling by 44 Wm^{-2} (or a heating by -44 Wm^{-2}). The cooling dominates and so the net effect of clouds on the annual global climate system is a 13 Wm^{-2} radiative cooling.

Although clouds produce net cooling of the climate system, this must not be construed as a possible means of offsetting global warming due to increased concentrations of greenhouse gases. Let us suppose that we start from a certain dynamic equilibrium situation in which the cooling mechanisms and the heating mechanisms balance exactly. Cloud feedback arises from changes in the net cloud radiative forcing. However, this

Mechanism	Forcing(Wm^{-2})
Infrared	31
Solar	-44
Net	-13

Table 1: Annual mean values of infrared, solar and net cloud-radiative forcing (CRF).

feedback arises as the net result of several separate effects and it is far from obvious whether to expect the feedback to be positive or negative. Cloud feedback is not a simple matter; it depends on a variety of factors which we shall now enumerate.

- **Cloud Amount:** If the amount of cloud decreases because of global warming, as occurs in typical general circulation model simulations, then this decrease reduces the infrared greenhouse effect attributed to clouds. Thus as the Earth warms it is able to emit infrared radiation more efficiently, moderating the warming and so providing a negative climate feedback mechanism. But there is a related positive feedback; the solar radiation absorbed by the climate system increases because the diminished cloud amount causes a reduction of reflected solar radiation by the atmosphere. Thus there is no simple way of determining the sign of this feedback component.

- **Cloud Altitude:** A vertical redistribution of clouds will also induce feedbacks. For example, if global warming displaces a given cloud layer to a higher and colder region of the atmosphere, this will produce a positive feedback because the colder cloud will emit less radiation and thus have an enhanced greenhouse effect.

- **Cloud Water Content:** There has been considerable recent speculation that global warming could increase cloud water content, thereby resulting in brighter clouds and hence a negative component of cloud feedback. However, it has also been demonstrated that this negative solar feedback induces a positive infrared feedback. The net result of these two feedback mechanisms of opposite sign may be positive or negative.

The question of cloud feedback is clearly very complex and what actually happens in any given situation will be the result of the various factors we have just outlined.

4 The Palaeo-analogue Method of Climate Prediction

As we have already mentioned, research work in climatology is very largely directed towards predicting the effects of increased emissions of CO_2 on the climate. The palaeo-analogue method has two distinct and rather independent parts. The first involves the derivation of an estimate of global temperature sensitivity to atmospheric CO_2 concentration based on estimates of CO_2 concentration at various times in the past and the corresponding global average temperatures; adjustments are made to allow for past changes in albedo and solar constant. In the second part regional patterns of climate are reconstructed for selected past epochs and they are regarded as analogues of future climates under enhanced greenhouse conditions. For a further discussion of the method (see for example Budyko 1982, Budyko and Izrael 1987 and Budyko *et al.* 1987).

In the derivation of the sensitivity of mean global temperature to changes in CO_2 concentration there are three stages

(i) determining the global mean changes for past palaeo-climates.

This is done for four periods (Early Pliocene, Early and Middle Miocene, Palaeocene-Eocene and the Cretaceous)

(ii) subtracting the temperature change attributed to changes in the solar constant, which is assumed to have increased by 5% every billion years, and to changes in surface albedo

(iii) relating the residual warming to the estimated change in atmospheric CO_2 concentrations. The CO_2 concentrations are derived from a carbon cycle model.

The result is a sensitivity of 3.0°C ± 1.0°C for a doubling of CO_2; this is very similar to that obtained on the basis of numerical simulations.

In their construction of analogue patterns Budyko *et al.* (1987) used the mid-Holocene (5,000-6,000 years bp (before the present)), the Last Interglacial (Eemian or Mikulino, 125,000 years bp) and the Pliocene (3-4 million years bp) as analogues for future climates. January, July and mean annual temperatures and mean annual precipitation were reconstructed for each of the above three epochs. Estimates of the mean temperatures over the Northern Hemisphere exceeded the temperature at the end of the pre-industrial period (the 19th century) by approximately 1°, 2° and 3-4°C during the mid-Holocene, Eemian and Pliocene, respectively. These periods were chosen as analogues of future climate for 2000, 2025 and 2050 respectively. For further information see sections 3.5.1 and 7.2 of Houghton *et al.* (1990).

When reconstructions of past climate conditions are accurate and thorough, they can provide relatively reliable estimates of self-consistent spatial patterns of climatic changes. Weaknesses in developing these relations can arise because of uncertainties

(i) in reconstructing past climates

(ii) in extending limited areal coverage to global scales

(iii) in interpreting the effects of changing orography and equilibrium versus non-equilibrium conditions

(iv) in determining the relative influences of the various factors that have caused the past climate changes.

5 Use of General Circulation Models

5.1 Introduction

We have already indicated in general terms what is involved in an atmospheric general circulation model (GCM) (see section 3 of chapter 1) when discussing weather forecasts. It will be recalled that in such a model the atmosphere is divided into layers and the parameters studied are specified on a grid of points on each layer. The spatial resolution of a GCM is constrained for practical reasons by the speed and memory capacity of the computer used to perform the numerical integrations. Increasing the spatial resolution not only increases the memory required (linearly for vertical resolution, quadratically for horizonal resolution), but also generally requires a reduction in the integration time step. Consequently, the computer time required increases rapidly with increasing resolution. In using an atmospheric general circulation model for weather forecasting purposes the surface temperature over the surface of the Earth (land, sea, ice and snow) provides a very important set of boundary conditions on the model. It is therefore very important to have accurate values of surface temperatures, especially sea surface temperatures (given the large area of the oceans) as input data for the model. While the atmosphere reacts very rapidly to changes in its forcing (on a time-scale of hours or days), the ocean reacts more slowly on time-scales ranging from days (at the surface layer) to millennia in the greatest depths. The ice cover reacts on time-scales of days for sea ice regions to millennia for ice sheets. The land processes react on time-scales of days up to months, while the biosphere reacts on time-scales from hours (plankton growth) to centuries (tree-growth). Because oceanic conditions change very much more slowly than atmospheric conditions, one does not make elaborate provision to take into account the circulation of heat within the oceans in a weather forecast model. However, in making calculations of predicted climate change one is dealing with much longer periods of time and the circulation of water and of heat in the oceans is important. Temporal changes (but not circulation) in the land, ice and snow also need to be taken into account. For climate modelling, therefore, one needs to use general circulation models for both the atmosphere and the ocean and to allow these models to be coupled together by exchanging information about the sea surface temperature, the ice cover, the total (latent, sensible and net longwave radiative) heat flux, the solar radiation and the wind stress. Because this is so much more complicated than using only an atmospheric GCM for weather forecasting, the spatial resolution and temporal resolution used in practice in climate modelling tends to be coarser than those used in weather forecasting. Grid spacings of 90 km or even 40 km in the horizontal surface were quoted in section 3 of chapter 1 for numerical models for weather forecasting. For climate modelling typical models have a horizontal resolution of 300 to 1000km and between 2 and 19 vertical levels. The horizontal resolution of a typical atmospheric model used for climate studies is shown in figure 2. Such resolution is sufficient to represent large-scale features of the climate, but it allows only a limited interpretation of results on the regional scale. As far as time-scales are concerned, 24 hours is a long time in weather forecasting terms whereas the time-scales of climate models range from hours to 10,000 years. High resolution models, which can resolve eddies, are enormously demanding in terms of computer power. Synchronously coupled atmosphere-ocean models are extremely time consuming and limited computer resources prohibit equilibrium being reached

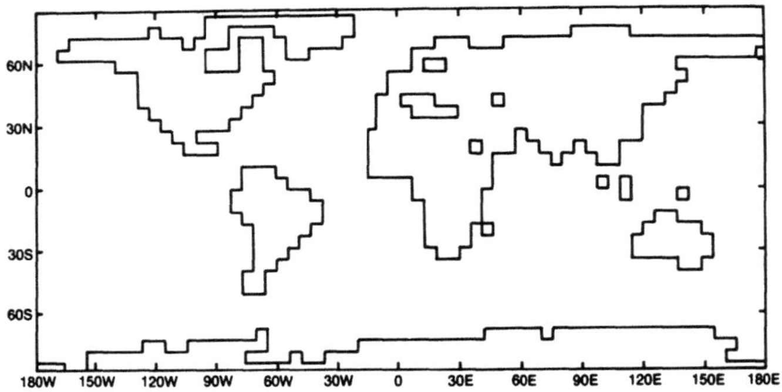

Figure 2: The model land-sea mask for a typical climate model (after Cubasch 1989, Cubasch *et al.* 1990).

except with mixed-layer models. Various asynchronous coupling techniques have been suggested to accelerate the convergence of a coupled model. However, the problem is far from being solved and can only really be tackled successfully by using faster computers. A second basic problem that arises through such coupling is model drift (for details see section 3.5.6 of Houghton *et al.* 1990). Table 2 is taken from Houghton *et al.* (1990). It gives a useful summary of many important numerical GCMs; these models were used in an experiment to determine the effect on the global mean surface temperature if the concentration of CO_2 in the atmosphere were to be doubled. Some characteristics of these models are described in the table itself and its footnotes; some further discussion will be found later in this chapter. One of the problems which we had previously noted with weather forecast models is that because of the rather course spatial resolution there are many processes or features that are not allowed for because they are on a scale that is very small relative to the grid spacing. However, the statistical effects of these sub grid-scale processes on the scales resolved by the GCM have to be incorporated into the model by relating them to the resolved scale variables (wind, temperature, humidity and surface pressure) themselves. Such a process is called parametrisation and it is based on both observational and theoretical studies. Figure 3 shows the physical processes parametrised in a typical GCM, and their interactions.

5.2 Radiation and the effect of clouds

The parametrisation of radiation is possibly the most important issue for climate change experiments, since it is through radiation that the effects of the greenhouse gases are transferred into the general circulation. A radiation parametrisation scheme calculates the radiative balance of the incoming solar radiation and the outgoing terrestrial long-wave radiation and, as appropriate, the reflection, emission and absorption of these fluxes in the atmosphere. Absorption and emission are calculated in several broad spectral bands (for reasons of economy) taking into account the concentration of different absorbers and emitters like CO_2, water vapour, ozone and aerosols.

One sensitive part in any radiation scheme is the calculation of the radiative effect

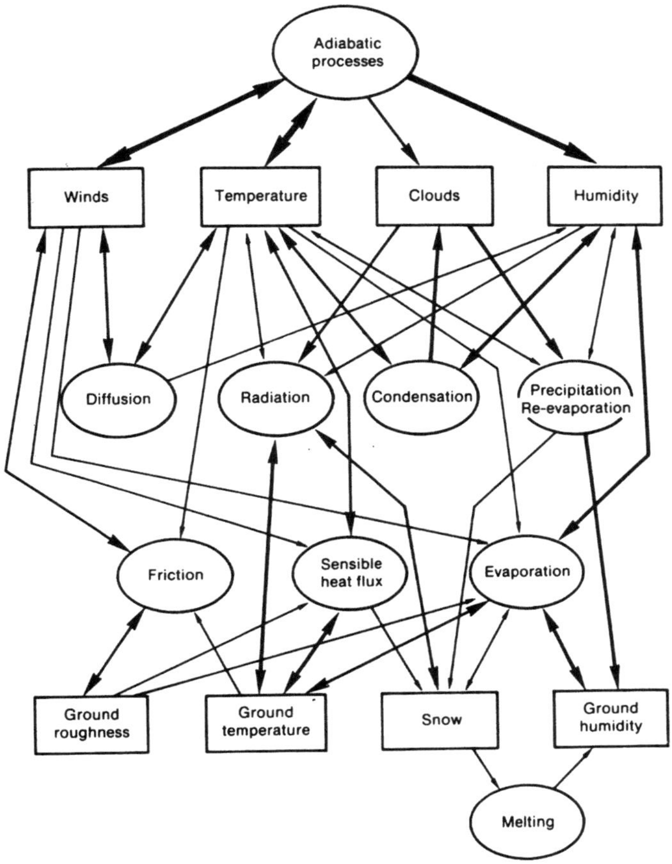

Figure 3: The processes parametrised in a numerical atmosphere model (ECMWF) and their interaction. the thickness of arrows indicates the strength of the interaction (from Houghton 1984).

of clouds. We have already mentioned that determining the effect of clouds is very complicated. In early work with numerical GCMs clouds were prescribed using observed cloud climatologies (fixed cloud (FC) experiments, see part A of table 2), and were not allowed to alter during the experiments with (for example) changed CO_2 concentration. Later schemes contained interactive cloud parametrisations of various sophistication, but mostly based on an estimate of the cloud amount from the relative humidity (RH) experiments, see table 2. Only the most advanced schemes involve calculating the variation of cloud optical properties by the cloud water content (CW) experiments, see table 2. Later chapters will give considerable attention to the question of clouds, see chapters 8 and 12.

The seasonal variation of the solar insolation is included in almost all experiments, but a diurnal cycle is omitted in many simulations. Climate experiments run without a

ENTRY	Group	Investigators	Year	RESOLUTION No. of waves, or °lat. x °long.	No. of Vertical Layers	Diurnal Cycle	Conv-ection	Ocean Heat Trans-port	Cloud	Cloud Prop-erties	ΔT (°C)	ΔP (%)	COMMENTS
A.	**Fixed, zonally averaged cloud; no ocean heat transport**												
1.	GFDL	Manabe & Stouffer	1980	R15	9	N	MCA	N	FC	F	2.0	3.5	Based on 4 x CO_2 simulation
2.		Wetherald & Manabe	1986, 8	R15	9	N	MCA	N	FC	F	3.2	n/a	
B.	**Variable cloud; no ocean heat transport**												
3.	OSU	Schlesinger & Zhao	1989	4° x 5°	2	N	PC	N	RH	F	2.8	8	
4.			1989	4° x 5°	2	N	PC	N	RH	F	4.4	11	As (3), but with revised clouds.
5.	MRI	Noda & Tokioka	1989	4° x 5°	5	Y	PC	N	RH	F	4.3 *	7 *	* Equilibrium not reached .
6.	NCAR	Washington & Meehl	1984	R15	9	N	MCA	N	RH	F	3.5 *	7 *	* Excessive ice. Estimate ΔT = 4°C at equilibrium.
7.			1989	R15	9	N	MCA	N	RH	F	4.0	8	As (6), but with revised albedos for sea-ice, snow.
8.	GFDL	Wetherald & Manabe	1986, 8	R15	9	N	MCA	N	RH	F	4.0	9	As (2), but with variable cloud.
C.	**Variable cloud; prescribed oceanic heat transport**												
9.	AUS	Gordon & Hunt	1989	R21	4	Y	MCA	Y	RH	F	4.0	7	
10.	GISS	Hansen et al.	1981	8° x 10°	7	Y	PC	Y	RH	F	3.9	n/a	
11.		Hansen et al.	1984	8° x 10°	9	Y	PC	Y	RH	F	4.2	11	
12.		Hansen et al.	1984	8° x 10°	9	Y	PC	Y	RH	F	4.8	13	As (11), but with more sea-ice control.
13.	GFDL	Wetherald & Manabe	1989 †	R15	9	N	MCA	Y	RH	F	4.0	8	
14.	MGO	Meleshko et al.	1990	T21	9	N	PC	Y	RH	F	n/a	n/a	Simulation in progress.
15.	UKMO	Wilson & Mitchell	1987	5° x 7.5°	11	Y	PC	Y	RH	F	5.2	15	
16.		Mitchell & Warrilow	1987	5° x 7.5°	11	Y	PC	Y	RH	F	5.2	15	As (15), but with four revised surface schemes.
17.		Mitchell et al.	1989	5° x 7.5°	11	Y	PC	Y	CW	F	2.7	6	As (16), but with cloud water scheme.
18.			1989	5° x 7.5°	11	Y	PC	Y	CW	F	3.2	8	As (17), but with alternative ice formulation.
19.			1989	5° x 7.5°	11	Y	PC	Y	CW	V	1.9	3	As (17), but with variable cloud radiative properties.
D.	**High Resolution**												
20.	CCC	Boer et al.	1989	T32	10	Y	MCA	Y	RH	V	3.5	4	* "Soft" convective adjustment.
21.	GFDL	Wetherald & Manabe	1989 †	R30	9	N	MCA	*	RH	F	4.0	8	* SSTs prescribed, changes prescribed from (13).
22.	UKMO	Mitchell et al.	1989	2.5°x3.75°	11	Y	PC	Y	CW	F	3.5	9	As (18), but with gravity wave drag.

All models are global, with realistic geography, a mixed-layer ocean, and a seasonal cycle of insolation. Except where stated, results are the equilibrium response to doubling CO_2.

R, T = Rhomboidal/Triangular truncation in spectral space;
N = Not included;
PC = Penetrative convection;
FC = Fixed cloud;
F = Fixed cloud radiative properties;
GFDL = Geophysical Fluid Dynamics Laboratory, Princeton, USA;
MGO = Main Geophysical Observatory, Leningrad, USSR;
AUS = CSIRO, Australia;
ΔT = Equilibrium surface temperature change on doubling CO_2;
Y = Included;
CA = Convective adjustment;
RH = Condensation or relative humidity based cloud;
† = Personal communication.
NCAR = National Center for Atmospheric Research, Boulder, CO, USA;
CCC = Canadian Climate Center;
ΔP = Percentage change in precipitation;
MCA = Moist convective adjustment;
CW = Cloud water;
V = Variable cloud radiative properties;
n/a = Not available
MRI = Meteorological Research Institute, Japan;
UKMO = Meteorological Office, United Kingdom;

Table 2: Summary of results from global mixed-layer ocean-atmosphere models used in equilibrium $2 \times CO_2$ experiments.

seasonal cycle are limited in scope and their reliability for climate change experiments is therefore doubtful. The inclusion of the diurnal cycle improves the realism of some feedback mechanisms and therefore the quality of the climate simulations.

5.3 Sub Grid-Scale Transports

Most of the solar radiation absorbed by the climate system is absorbed at the surface. The energy becomes available for driving the atmospheric general circulation only after it has been transferred to the atmosphere through the planetary boundary layer (PBL), primarily by small-scale turbulent and convective fluxes of sensible and latent heat, but also by net long-wave radiative exchange. On the other hand, the general circulation is slowed down by frictional dissipation which basically takes place in the PBL through vertical transport of momentum by turbulent eddies.

In most GCMs the turbulent fluxes of heat, water vapour and momentum at the surface are calculated from empirical "bulk formulae" with stability-dependent transfer coefficients. The fluxes at the PBL top (at a fixed height generally) are either neglected or parametrised from simple mixed-layer theory. In GCMs that resolve the PBL, the eddy diffusion approach is generally employed. Considerable efforts are made to incorporate into the PBL parametrisations the effects of cloud, vegetation and sub-grid-scale terrain height.

Cumulus convection in a vertically unstable atmosphere is one of the main heat-producing mechanisms at scales which are unresolvable in GCMs. A common procedure is to adjust the temperature and water vapour profile to a conditionally stable state (Moist Convective Adjustment, MCA). The second class of cumulus parametrisations often employed in GCMs is based on a moisture convergence closure. Other GCMs use Penetrative Convection schemes to mix moist conditionally unstable air from lower model layers with dry air aloft. The question of how sophisticated convective parametrisations in GCMs need to be, and how much the sensitivity of climate change experiments depends on their formulation, is still an open question.

5.4 Land Surface Processes

Another important parametrisation is the transfer of heat and water within the soil, for instance the balance between evaporation and precipitation, snow melt, storage of water in the ground and river runoff. This parametrisation is of extreme relevance for climate change predictions, since it shows how local climates may change from humid to arid and vice versa depending on global circulation changes. It furthermore reflects, in some of the more sophisticated schemes, the changes that could occur through alterations in surface vegetation and land-use.

Most soil moisture schemes used to date are based either on the so-called bucket method or the force-restore method. In the former case, soil moisture is available from a single reservoir, or thick soil layer. When all the moisture is used up, evaporation ceases. In the latter method, two layers of soil provide moisture for evaporation, a thin near-surface layer which responds rapidly to precipitation and evaporation and a thick deep soil layer acting as a reservoir. If the surface layer dries out, deep soil moisture is mostly unavailable for evaporation and evaporation rates fall to small values. However, in the presence of vegetation, realistic models use the deep soil layer as a source of moisture for evapotranspiration.

At any given grid-point over land, a balance between precipitation, evaporation, runoff and local accumulation of soil moisture is evaluated. If precipitation exceeds evaporation, then local accumulation will occur until saturation is achieved. After this, runoff is assumed and the excess water is removed. The availability of this runoff as fresh-water input to the ocean has been allowed for in ocean models only recently (Cubasch et al. 1990). Most models differ in the amount of freshwater required for saturation, while few treat more than one soil type.

The force-restore method has recently been extended to include a range of soil types (Noilham and Planton 1989).

5.5 Boundary conditions and ocean models

To determine a unique solution of the model equations, it is necessary to specify a set of upper and lower boundary conditions. These are

- input of solar radiation (including temporal variation) at the top of the atmosphere
- orography (mountain topography) and land sea distribution
- albedo of bare land
- surface roughness
- vegetation characteristics.

The lower boundary over the sea is either prescribed from climatological data or, as this is not very appropriate for climate change experiments, it has to be calculated from an oceanic global circulation model (GCM) as we have already mentioned above. The typical oceanic GCM used for climate simulations follows basically the same set of equations as the atmosphere if the equation defining the water vapour balance is replaced by one describing salinity. As with atmospheric GCMs, one specifies appropriate surface boundary conditions (i.e., fluxes of heat, momentum and fresh-water) either from observations (uncoupled mode) or from an atmospheric GCM. The vertical and horizontal exchange of temperature, momentum and salinity by diffusion or turbulent transfers is parametrised. Solutions are obtained by using numerical methods.

The most commonly used ocean model coupled to atmosphere models is the mixed-layer model. This model describes the uppermost layer of the ocean where the oceanic temperature is relatively uniform with depth. It is frequently modelled as a simple slab for which a fixed depth of the mixed layer is prescribed and the oceanic heat storage is calculated; the oceanic heat transport is either neglected, or is treated as being carried only within the mixed layer, or is prescribed from climatology. Variations of mixed-layer depth, oceanic heat flux convergence, and exchanges with the deep ocean, which would entail an additional storage and redistribution of heat, are all neglected as well. Sea ice extents are determined interactively, usually with a variant of the thermodynamic sea ice model due to Semtner (1976). However, some models already include dynamical effects such as sea ice drift and deformation caused by winds and ocean currents. The simple mixed-layer model has strong limitations for studies of climate change, particularly as it does not allow for the observed lags in heat storage of the upper ocean to be represented. A number of more complicated ocean models have been developed (for references see section 3.5.3 of Houghton *et al.* 1990). Moreover, some work has been done on trying to incorporate carbon cycle models to describe the exchange of CO_2 between the atmosphere and the ocean and, more recently, people have begun to develop models to cover the inclusion of trace gases in the general circulation.

6 Use of Models

It should be apparent from the above discussion that there are many uncertainties and problems associated with any given model. It would be rather risky to rely on absolute predictions made with one of these models. We should therefore consider the manner in which these models are used and this is basically done by attempting to eliminate the effects of the uncertainties or errors. Thus a model is normally used to investigate the sensitivity of climate to internal and external factors and to predict the climate change by first carrying out a control integration with the parameters set for the present climate in order to establish a reference mean model climatology and the necessary statistics on modelled climatic variability. The climate change (perturbation) run is then carried out by repeating the model run with appropriately changed parameters (a doubling of atmospheric CO_2 concentration for example) and the difference between this and the parallel control run examined. The difference between the control and the perturbed experiments is called the response to the change in the values of the parameters used. The significance of the response must be assessed against the model's natural variability (which can be determined from the control run) using appropriate statistical tests. These enable an assessment to be made of the confidence that the changes obtained represent an implied climatic change, rather than simply a result of the natural variability of the model.

The period for which the integration in a model is carried out ranges typically from 5 to 100 years, depending on the nature of the investigation. Until now, the greatest effort has been devoted to studying the response to increased levels of greenhouse gas concentrations by determining the equilibrium response of climate to doubling of CO_2, using atmospheric models coupled to slab ocean models. In an equilibrium response experiment both simulations, the control experiment and the perturbation experiment, are run for a sufficiently long time to achieve the respective equilibrium climates. A review of such experiments is given by Schlesinger and Mitchell (1987). For a mixed-layer ocean the response time to reach equilibrium amounts to several decades; such calculations can be handled with present computers. A summary of work done for a doubling of the concentration of CO_2, using atmospheric models coupled to slab ocean models, is given in table 2, to which we have referred to earlier. The results in table 2 give ΔT, the calculated change in equilibrium surface temperature, and DP, the calculated percentage change in precipitation, resulting from a doubling of the concentration of CO_2 in the atmosphere. All the models show a significant increase in

Group	Year	RESOLUTION No. of Spectral Waves	Atmos. Levels	Diurnal Cycle	Conv-ection	Ocean Levels	Cloud	COMMENTS
GFDL	1989	R15	9	N	RH	12	MCA	100 Years, 1% CO_2 increase compounded.
NCAR	1989	R15	9	N	RH	4	FC	30 Years, 1% CO_2 increased linear.
MPI	1990	T21	19	Y	PC	11	CW	25 Years, instantaneous CO_2 doubling.
UHH	1990	T21	19	Y	PC	9	CW	25 Years, instantaneous CO_2 doubling.

All models are global, with realistic geography and a seasonal cycle of insolation.

R, T = Number of waves in spectral space;
N = Not included;
MCA = Moist convective adjustment;
CA = Convective adjustment ;
FC = Fixed cloud;
Y = Included;
PC = Penetrative convection;
CW = Cloud water;
V = Variable cloud radiative properties.

GFDL = Geophysical Fluid Dynamics Laboratory,Princeton, USA;
MPI = Max Planck Institut für Meteorologie, Hamburg, FRG;
UHH = Met Institute, University of Hamburg, FRG;
NCAR = National Center for Atmospheric Research, Boulder, Co, USA.

Table 3: Summary of experiments carred out with global coupled ocean-atmosphere models.

global mean temperature ranging from 1.9° to 5.2°C, with most results between 3.5° and 4.5°C.

These equilibrium response studies for given CO_2 increases are valuable as standard benchmark calculations for model intercomparison. The results may, however, be misleading, if they are applied to predict actual climate change caused by human activities; this is because the atmospheric CO_2 concentrations do not change abruptly but have been growing by about 0.4% per year. Moreover, the timing of the atmospheric response depends crucially on the ocean heat uptake which might delay the CO_2 induced warming by several decades. Thus, for realistic climate scenario computations, one has to simulate successfully not only the atmospheric processes but also the oceanic heat transport which is largely governed by ocean dynamics. In other words one needs to use a coupled atmospheric and ocean circulation model. For a fully coupled atmosphere-ocean model the equilibrium response time would be several thousand years; such calculations cannot be achieved with present computers. A few attempts have been made to determine the transient (i.e. time-dependent) climate response to anthropogenic forcing using coupled atmosphere and ocean circulation models, see table 3; these will be discussed further in section 9.2.

A second calculation used for the intercomparison of various models was carried out by a perpetual July simulation in which it was supposed that the climate was changed by imposing a 4°C perturbation on the global sea surface temperature while holding sea ice fixed. Since a perpetual July simulation with a general circulation model results in very little snow cover in the Northern Hemisphere, this effectively eliminates snow-ice albedo feedback. The results of this simulation using 17 different models are described briefly in section 3.6 of Houghton *et al.* (1990).

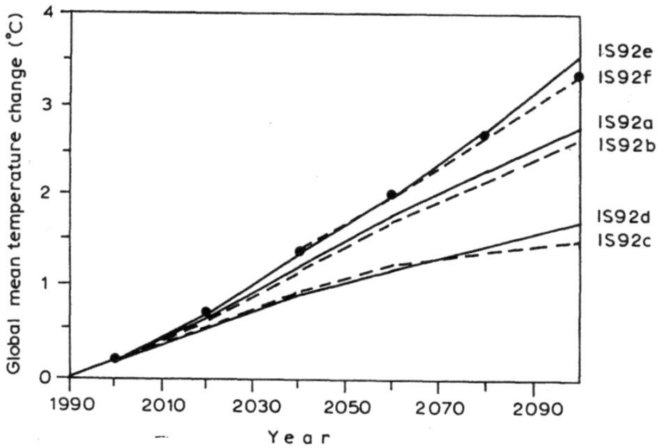

Figure 4: Estimates of global mean temperature change for the IPCC 1992 scenarios (IS92a-f) (see table 4) assuming the IPCC "best estimate" climate sensitivity (Houghton *et al.* 1992). The effects of sulphate aerosol and ozone depletion have not been taken into account. SA90 is represented by solid circles.

7 Consensus

We have quoted very extensively in this chapter from the IPCC report (Houghton *et al.* 1990). This report summarises the work to which nearly 200 climatologists contributed and the work was checked by another 200-climatologists. There is now also a supplementary report (Houghton *et al.* 1992) which updates the earlier report. In this later report, six alternative scenarios are studied. These are outlined in table 4 and the consequent global mean temperature change over the next century for each scenario is shown in figure 4. The (original) IPCC report is widely, but not universally, accepted; there is a very good *New Scientist* article by Gribbin (1990) on the dissenting view which is held by a few Americans. This dissenting view will not be discussed here in detail but it seemed only fair to mention its existence and to cite a reference; I can quote the opening sentences of John Gribbin's article.

"At the same time as scientists meeting at the World Climate Conference in Geneva agreed that global warming was likely, an alarming report came in from Hawaii. US President George Bush, asked to comment on the consensus that the world faces a potentially disastrous global warming linked to the greenhouse effect, replied "that's not what my scientists are telling me"."

Gribbin then gives an account of the views of two scientists from the Marshall Institute in Washington, DC, and finally concludes, after 3 pages, "Small wonder that one eminent US climatologist has referred to the Marshall study as "noisy junk science". Perhaps somebody should tell the President."

Scenario	Population	Economic Growth		Energy Supplies ††	Other †††	CFCs
IS92a	World Bank 1991 11.3 B by 2100	1990-2025: 1990-2100:	2.9% 2.3%	12,000 EJ Conventional Oil 13,000 EJ Natural Gas Solar costs fall to $0.075/kWh 191 EJ of biofuels available at $70/barrel	Legally enacted and internationally agreed controls on SOx, NOx and NMVOC emissions.	Partial compliance with Montreal Protocol. Technological transfer results in gradual phase out of CFCs also in non-signatory countries by 2075.
IS92b	World Bank 1991 11.3 B by 2100	1990-2025 1990-2100	2.9% 2.3%	Same as "a"	Same as "a" plus commitments by many OECD countries to stabilize or reduce CO_2 emissions.	Global compliance with scheduled phase out of Montreal Protocol.
IS92c	UN Medium Low Case 6.4 B by 2100	1990-2025 1990-2100	2.0% 1.2%	8,000 EJ Conventional Oil 7,300 EJ Natural Gas Nuclear costs decline by 0.4% annually.	Same as "a"	Same as "a"
IS92d	UN Medium Low Case 6.4 B by 2100	1990-2025 1990-2100	2.7% 2.0%	Oil and gas same as "c" Solar costs fall to $0.065/kWh 272 EJ of biofuels available at $50/barrel	Emission controls extended worldwide for CO, NO_x, NMVOC, and SO_x. Halt deforestation. Capture and use of emissions from coal mining and gas production and use.	CFC production phase out by 1997 for industrialized countries. Phase out of HCFCs.
IS92e	World Bank 1991 11.3 B by 2100	1990-2025 1990-2100	3.5% 3.0%	18,400 EJ conventional oil Gas same as "a" Phase out nuclear by 2075	Emission controls (30 % pollution surcharge on fossil energy).	Same as "d"
IS92f	UN Medium High Case 17.6 B by 2100	Same as "a"		Oil and gas same as "e" Solar costs fall to $0.083/kWh Nuclear costs increase to $0.09/kWh	Same as "a"	Same as "a"

† The assumptions for the 1990 Scenario A are described in IPCC (1990) Annex A, pp.331-339.

†† All scenarios assume coal resources up to 197,000 EJ. Up to 15% of this resource is assumed to be available at $1.30/gigajoule at the mine.

††† Tropical deforestation rates (for closed and open forests) begin from an average rate of 17.0 million hectares/year (FAO, 1991) for 1981-1990, then increase with population until constrained by availability of land not legally protected. IS91d assumes an eventual halt of deforestation for reasons other than climate. Above-ground carbon density per hectare varies with forest type from 16 to 117 tons C/hectare, with soil C ranging from 68 to 100 tons C/ha. However, only a portion of carbon is released over time with land conversion, depending on type of land conversion.

Table 4: Summary of assumptions in the six IPCC alternative scenarios.

8 Validation of Climate Models

If climate models are to be used to make predictions of climatic change, whether natural change or change induced as the result of human activities, then it is obviously important to attempt to assess the accuracy of the results produced by models.

To carry out the validation of climate models one needs appropriate observed historical climatic data. Hopefully we shall see in this book that satellite remote sensing systems are expected to provide an important new source of data for climatic model validation. Chapter 4 of the IPCC Report (Houghton *et al.* 1990) is devoted to the question of the validation of climatic models and further details will be found therein.

9 Examples of Climate Change Predictions

We have noted above that two main types of climate modelling calculations have been performed. One is equilibrium response experiments and the case which has been widely studied using a large number of climate models has been to consider the effect of doubling the concentration of atmospheric CO_2 and supposing that equilibrium is re-established. The second type of calculations are time-dependent response experiments. We shall describe briefly the present state of these two types of work.

9.1 Equilibrium response experiments

In table 2 the calculated increases in the global mean surface temperature and precipitation obtained from a large number of models were given. However, the climate models do not just produce estimates for global means; neither does a global mean necessarily convey useful information for everyone. There may well be important regional variations and for many purposes it is the local rather than the global climate effects which are important. Chapter 5 of the IPCC Report (Houghton *et al.* 1990) describes some of the rather more detailed results of the use of the various models to attempt to determine the consequences of an equilibrium situation with double the present concentration of CO_2 in the atmosphere. We quote a few examples of results and then an extract from the executive summary of that chapter.

Figure 5(a) shows the changes in the ten-year mean surface air temperature for the three-month period December-January-February obtained from 3 different models for a doubling of the concentration of CO_2 in the atmosphere. There are many differences of detail between the results from different models. However many of the general features of the results are very similar for all the models. Figure 5(b) shows similar results for the three-month period June-July-August. Figures 6(a) and (b) show changes in the ten-year mean precipitation from the same three models for a doubling in the concentration of CO_2 in the atmosphere and for the same two three-month periods. Figures 7(a) and (b) show changes in the ten-year mean soil moisture concentration obtained from the same three models for a doubling of the concentration of CO_2 in the atmosphere and for the same two three-month periods. Coloured versions of figures 5, 6 and 7 are also given in the IPCC Report (Houghton *et al.* 1990). Chapter 5 of the IPCC Report also includes the results of calculations of the values of some other parameters in the scenario of a doubling of the concentration of CO_2 in the atmosphere and the re-establishment of equilibrium.

The important thing is, perhaps, not to concentrate too much on the finer spatial details of the results for one time and place with any one particular model, but rather to try to concentrate on what appears to be common to the results of all, or at least most, of the models. Remember, of course, that what has been illustrated in figures 5 - 7 is only for 3 models and there are many more models for which we have not attempted to reproduce any detailed results. The general conclusions are, perhaps, best taken from the Executive summary of chapter 5 of the IPCC Report (Houghton *et al.* 1990) and I shall quote part of that summary verbatim.

1. All models show substantial changes in climate when CO_2 concentrations are doubled, even though the changes vary from model to model on a sub-continental scale.

2. The main equilibrium changes in climate due to doubling CO_2 deduced from models are given below. The number of * indicates the degree of confidence determined subjectively from the amount of agreement between models, our understanding of the model results and our confidence in the representation of the relevant process in the model. Five * indicate virtual certainties, one * indicates low confidence.

Temperature:

***** the lower atmosphere and Earth's surface warm;

***** the stratosphere cools;

*** near the Earth's surface, the global average warming lies between +1.5° and +4.5°C, with a "best guess" of + 2.5°C;

*** the surface warming at high latitudes is greater than the global average in winter but smaller than in summer. (In time-dependent simulations with a deep ocean, there is little warming over the high latitude southern ocean);

*** the surface warming and its seasonal variation are least in the tropics.

Precipitation:

**** the global average increases (as does that of evaporation), the larger the warming, the larger the increase;

*** increases at high latitudes throughout the year;

*** increases globally by 3 to 15% (as does evaporation);

** increases at mid-latitudes in winter;

** the zonal mean value increases in the tropics although there are areas of decrease. Shifts in the main tropical rain bands differ from model to model, so there is little consistency between models in simulated regional changes;

** changes little in subtropical arid areas.

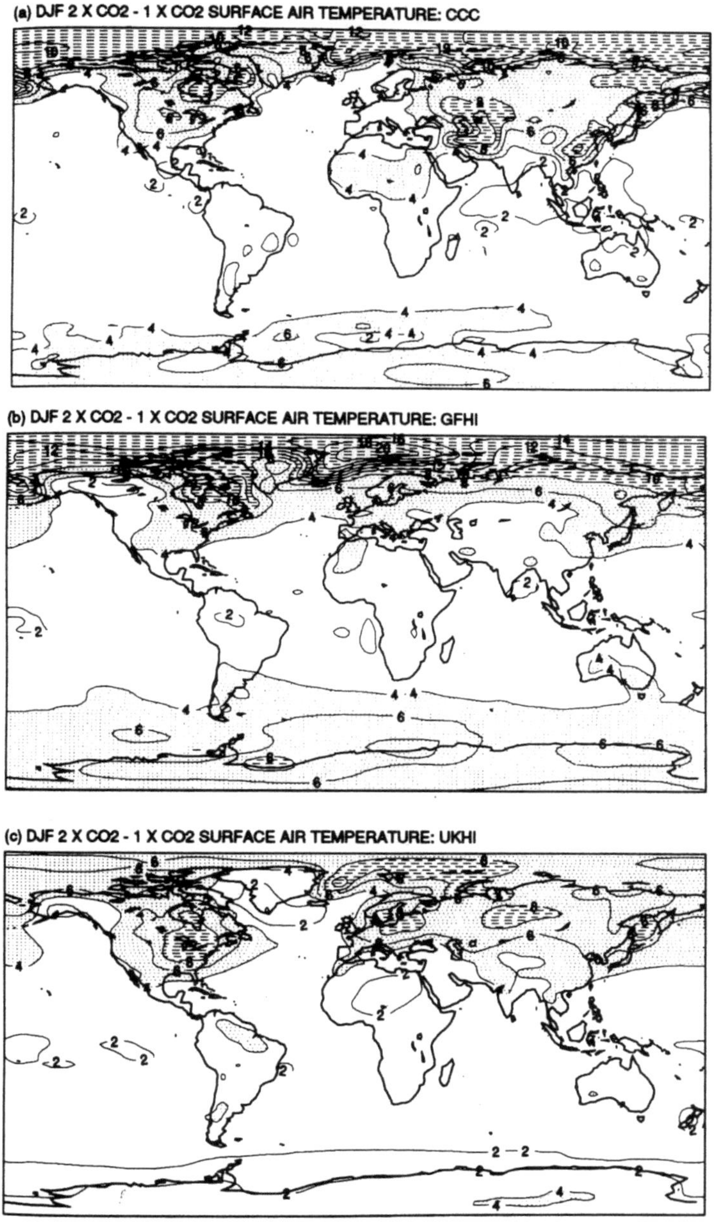

Figure 5: Change in surface air temperature (ten year means) due to doubling CO_2, for months December/January/February, as simulated by three high resolution models: (a) CCC: Canadian Climate Centre, (b) GFHI: Geophysical Fluids Dynamics Laboratory and (c) UKHI: United Kingdom Meteorological Office and for months June/July/August for the same three models (d) CCC, (e) GFHI and (f) UKHI. Contours every 2°C, light stippling where the warming exceeds 4°C, dashed shading where the warming exceeds 8°C (Houghton *et al.* 1990).

(d) JJA 2 X CO2 - 1 X CO2 SURFACE AIR TEMPERATURE: CCC

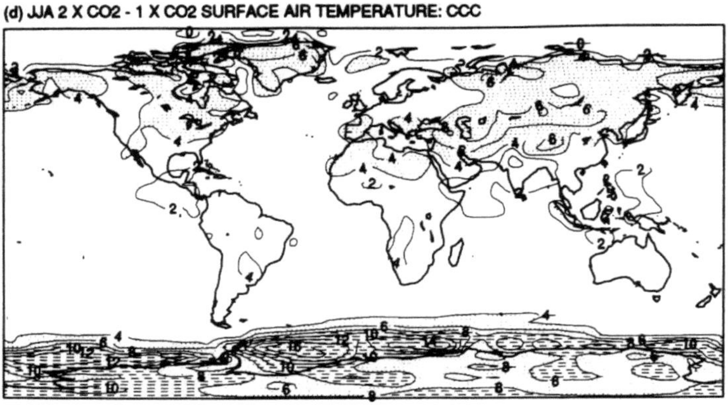

(e) JJA 2 X CO2 - 1 X CO2 SURFACE AIR TEMPERATURE: GFHI

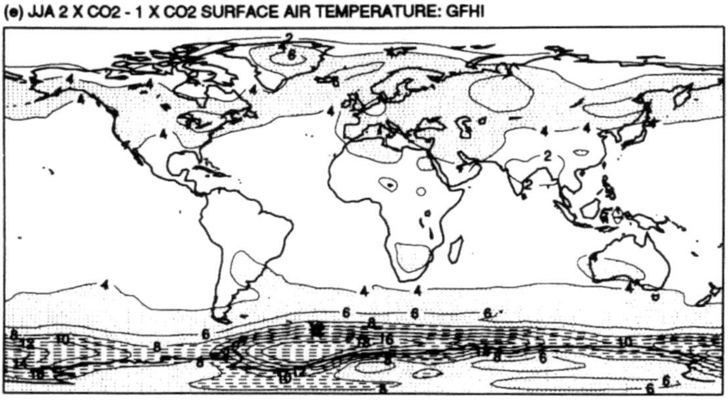

(f) JJA 2 X CO2 - 1 X CO2 SURFACE AIR TEMPERATURE: UKHI

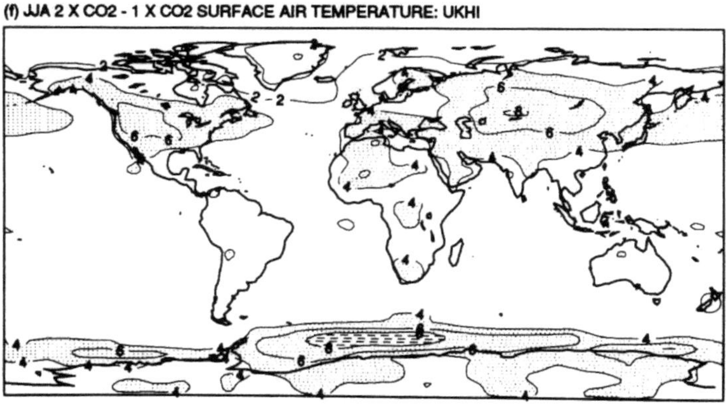

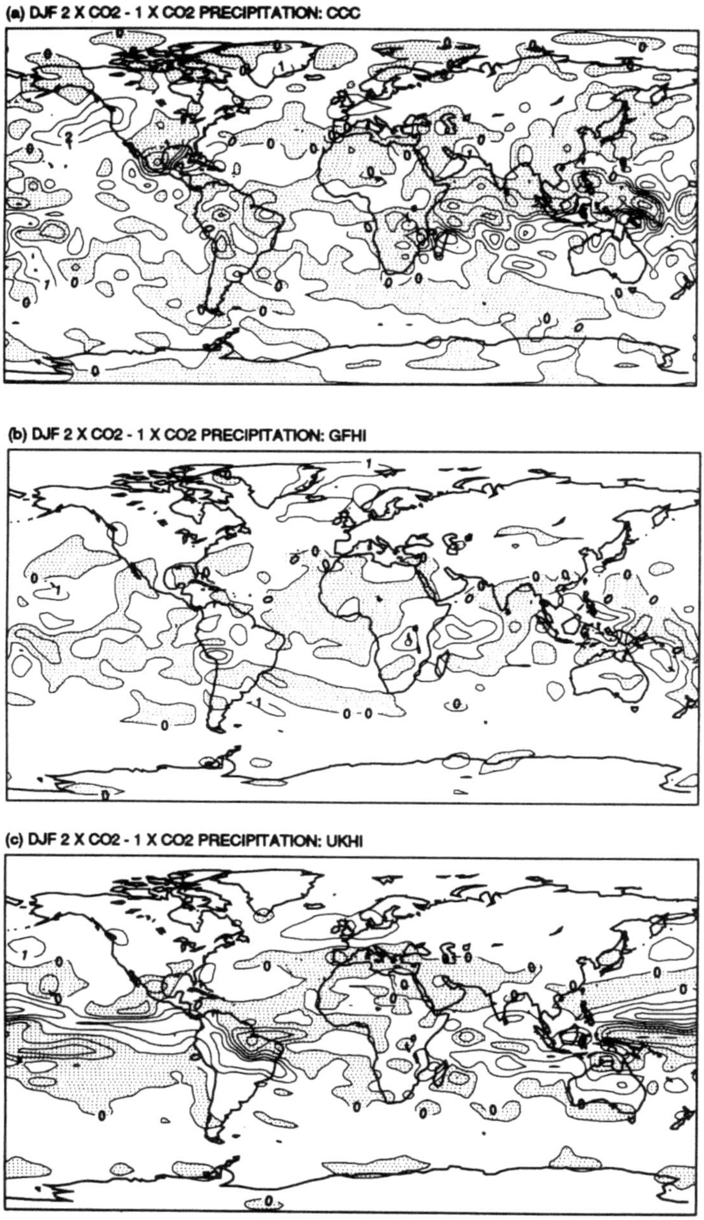

Figure 6: Change in precipitation (smoothed 10-year means) due to doubling CO_2, for months December/January/February, as simulated by the same three high resolution models as in figure 5: (a) CCC, (b) GFHI and (c) UKHI and for months June/July/August (d) CCC, (e) GFHI and (f) UKHI: Contours at ±0 1, 2, 5 mm day^{-1}, areas of decrease stippled (Houghton *et al.* 1990).

(d) JJA 2 X CO2 - 1 X CO2 PRECIPITATION: CCC

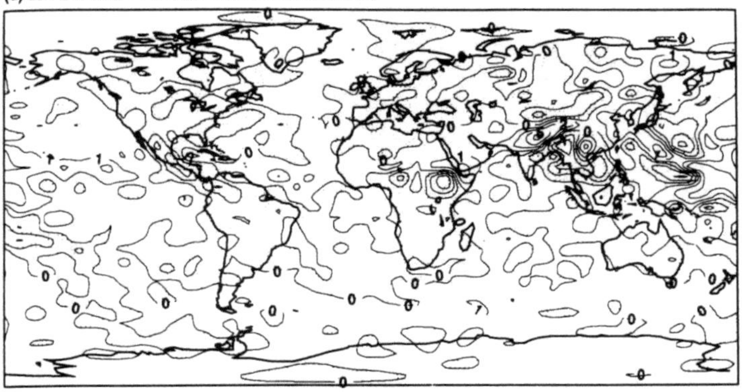

(e) JJA 2 X CO2 - 1 X CO2 PRECIPITATION: GFHI

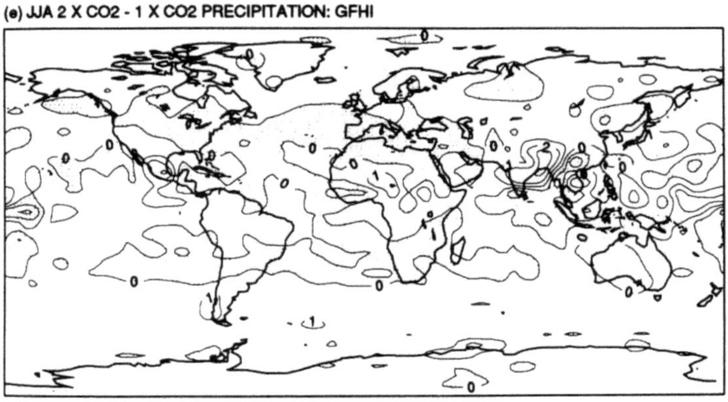

(f) JJA 2 X CO2 - 1 X CO2 PRECIPITATION: UKHI

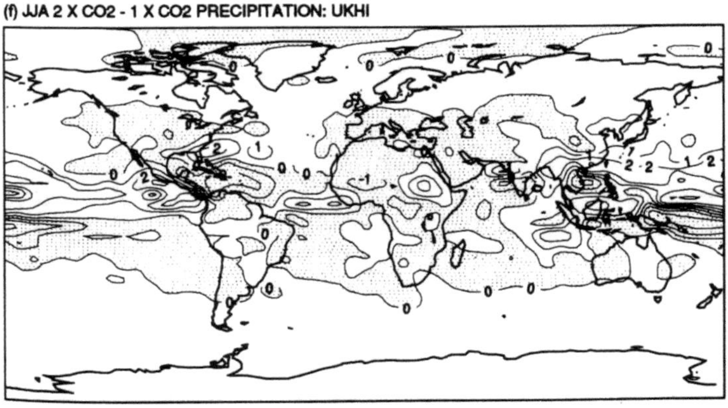

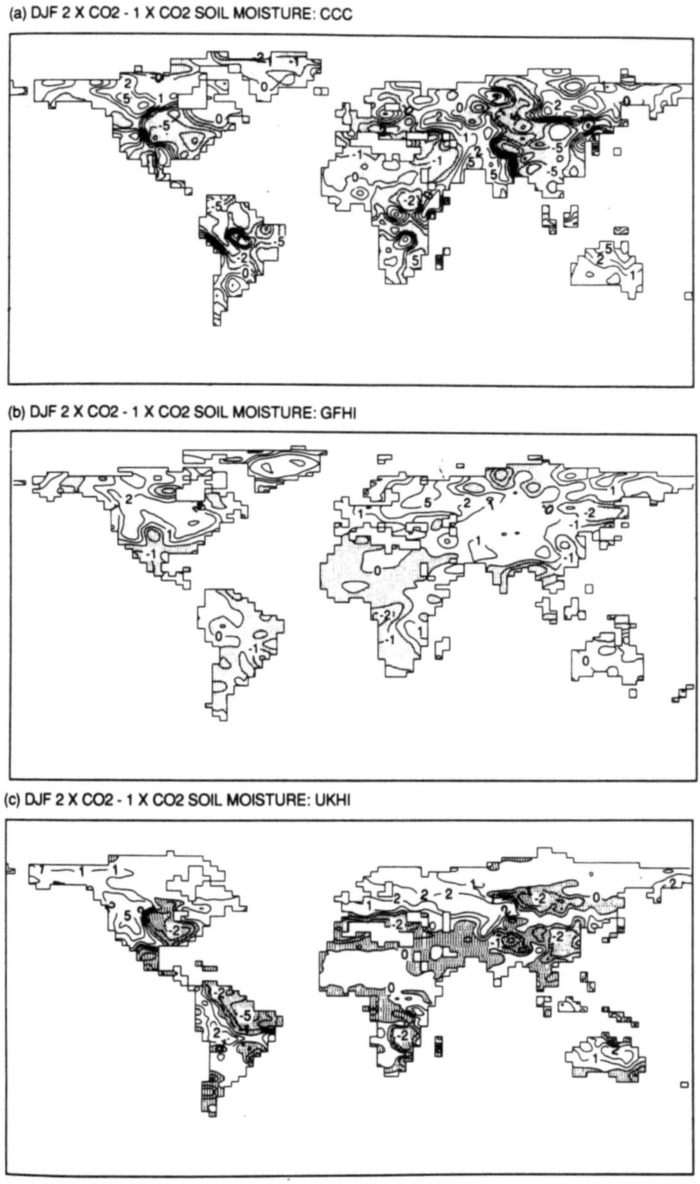

Figure 7: Change in soil moisture (smoothed 10-year means) due to doubling CO_2, as simulated by the same three high resolution models as in figure 5 for months December/ January/ February: (a) CCC, (b) GFHI and (c) UKHI and for months June/July/August (d) CCC, (e) GFHI and (f) UKHI (Houghton *et al.* 1990). Note that (a) has a geographically variable soil capacity whereas the other models have the same capacity everywhere. Contours at ±0, 1, 2, 5 cm, areas of decrease stippled.

(d) JJA 2 X CO2 - 1 X CO2 SOIL MOISTURE: CCC

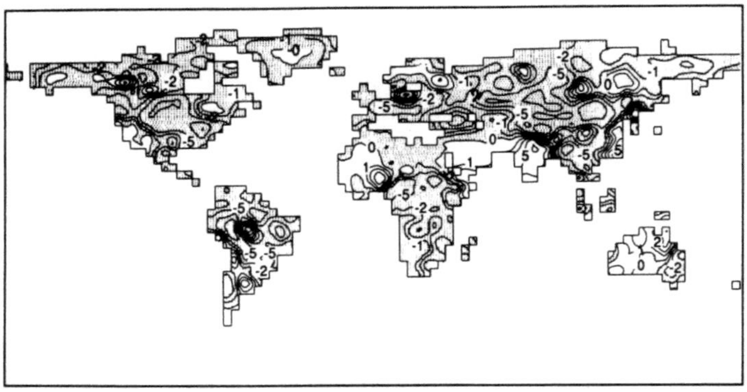

(e) JJA 2 X CO2 - 1 X CO2 SOIL MOISTURE: GFHI

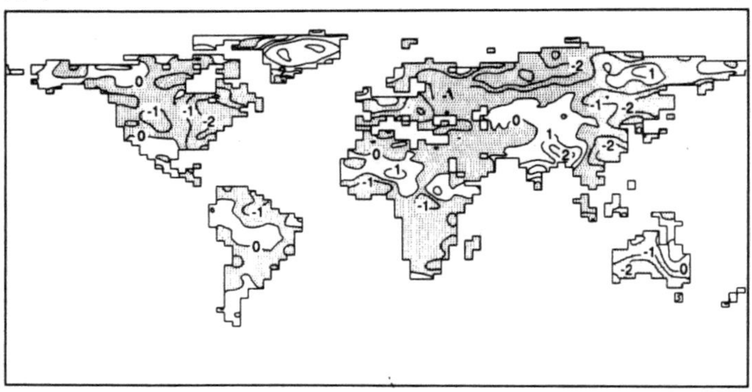

(f) JJA 2 X CO2 - 1 X CO2 SOIL MOISTURE: UKHI

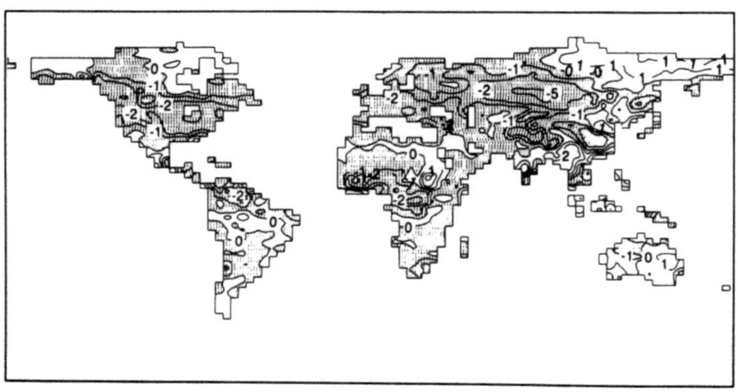

Soil moisture:

*** increases in high latitudes in winter;

** decreases over northern mid-latitude continents in summer.

Snow and sea-ice:

**** the area of sea-ice and seasonal snow-cover diminish.

The results from models become less reliable at smaller scales, so predictions for smaller than continental regions should be treated with great caution. The continents warm more than the ocean. Temperature increases in southern Europe and central North America are greater than the global mean and are accompanied by reduced precipitation and soil moisture in summer. The Asian summer monsoon intensifies.

3. Changes in the day-to-day variability of weather are uncertain. However, episodes of high temperature will become more frequent in the future simply due to an increase in the mean temperature. There is some evidence of a general increase in convective precipitation.

4. The direct effect of deforestation on global mean climate is small. The indirect effects (through changes in the CO_2 sink) may be more important. However, tropical deforestation may lead to substantial local effects, including a reduction of about 20% in precipitation."

One can ask, quite reasonably, what is the value of carrying out equilibrium calculations such as those just mentioned. Climate is in equilibrium when it is in balance with the radiative forcing. Thus, as long as greenhouse gas concentrations continue to increase the climate will not reach equilibrium; moreover, even if concentrations of greenhouse gases are eventually stabilised it would be a long time before equilibrium would be established. Thus equilibrium calculations cannot be used directly to provide forecasts of climate changes. The first reason given for carrying out equilibrium simulations (usually with double the concentration of CO_2 in the atmosphere) is that they are done with atmospheric-oceanic mixed-layer models which ignore both the deep ocean and changes in the ocean circulation; these require less computer time than time-dependent simulations which include the influence of the deep ocean. Secondly, because these simulations are relatively inexpensive (in terms of computer time and therefore of money) to carry out, they can be used fairly easily for intercomparison of different models and for studying the effects of different parametrisations (e.g. of cloud). Thirdly, it appears that apart from areas where the oceanic thermal inertial is high, equilibrium solutions can be scaled and used as approximations to the time-dependent response.

9.2 Time-Dependent Response Experiments

In the models used in the equilibrium response calculations described in section 9.1 one supposes that an abrupt change has been made in the atmospheric CO_2 concentration and that this leads to a change in the radiative forcing; this is allowed to operate in the model for a period corresponding to a few years until a new equilibrium is established.

The atmospheric models used are detailed and highly developed but the treatment of the ocean, as we have already mentioned, is quite primitive. However, in reality, greenhouse gas concentrations are changing continuously, not abruptly. The climate's reaction is not instantaneous because of the thermal capacity of the oceans. At any given time the response corresponds to only part of the equilibrium change for the corresponding instantaneous value of the forcing. For the remainder, part of the response is delayed by storage in the stably stratified layers of the upper ocean and takes several decades, or even a century, to occur. However, there is also a part of the response which remains invisible for a long time (many centuries) until the deep ocean begins to influence the surface temperature. Apart from the thermal inertia of the oceans there may also be other oceanic effects, e.g. the ocean currents can lead to a spatial redistribution of greenhouse warming. Also even in the absence of changes in the radiative forcing the ocean-atmosphere interactions lead to fluctuations which may temporarily mask long-term climate changes.

In order to make reliable predictions of climate change under realistic scenarios of slow changes in the radiative forcing, it is necessary to take the oceanic behaviour properly into account. To do this means that one has to make use of coupled atmosphere-ocean general circulation models (GCMs). We should not be surprised that this is necessary. If we are to be surprised at anything, perhaps it should be at the fact that such a considerable degree of success can be obtained in the equilibrium response experiments when using only a very simple oceanic model. There is discussion of time-dependent model calculations in chapter 6 of the IPCC Report (Houghton *et al.* 1990) and in chapter 20 of Trenberth (1992). The use of a coupled atmosphere-ocean GCM means that the computing requirements for a calculation are very high. The experiment which is described involved a 100-year simulation. Three runs of a model were carried out with (a) no change in atmospheric CO_2 concentration, (b) a 1% per annum (compounded) increase in atmospheric CO_2 concentration and (c) a 1% per annum (compounded) decrease in atmospheric CO_2 concentration. The choice of 1% increase in the CO_2 concentration is based on the fact that this has the equivalent effect, in terms of radiative transfer, to the combined effect of the increases in all the various greenhouse gas concentrations at present. Figure 8 shows some of the results. Figure 8(a) shows the mean surface air temperature in years 60-80 for the 1% per annum increase in the concentration of CO_2. Figure 8(b) shows the equilibrium response to a doubling of the concentration of CO_2 in the atmosphere. Figure 8(c) shows the ratio of the air temperature increases presented in figures 8(a) and 8(b). By the years 60-80 of the model calculation the concentration of atmospheric CO_2 has built up at a (compound) rate of increase of 1% per annum to approximately double the initial value. Thus, both figures 8(a) and 8(b) correspond to a doubling of the concentration of atmospheric CO_2, but in figure 8(a) the concentration has been built up slowly and equilibrium has not yet been reached whereas in figure 8(b) the concentration of CO_2 was suddenly doubled and equilibrium was supposed to be re-established rather quickly.

The use of models in time-dependent response calculations is still in its early days; the conclusions from the results that have been obtained so far will be found in the IPCC Report (Houghton *et al.* 1990); part of the Executive Summary of chapter 6 is quoted verbatim.

"(a) For a steadily increasing forcing, the global rise in temperature is an approx-

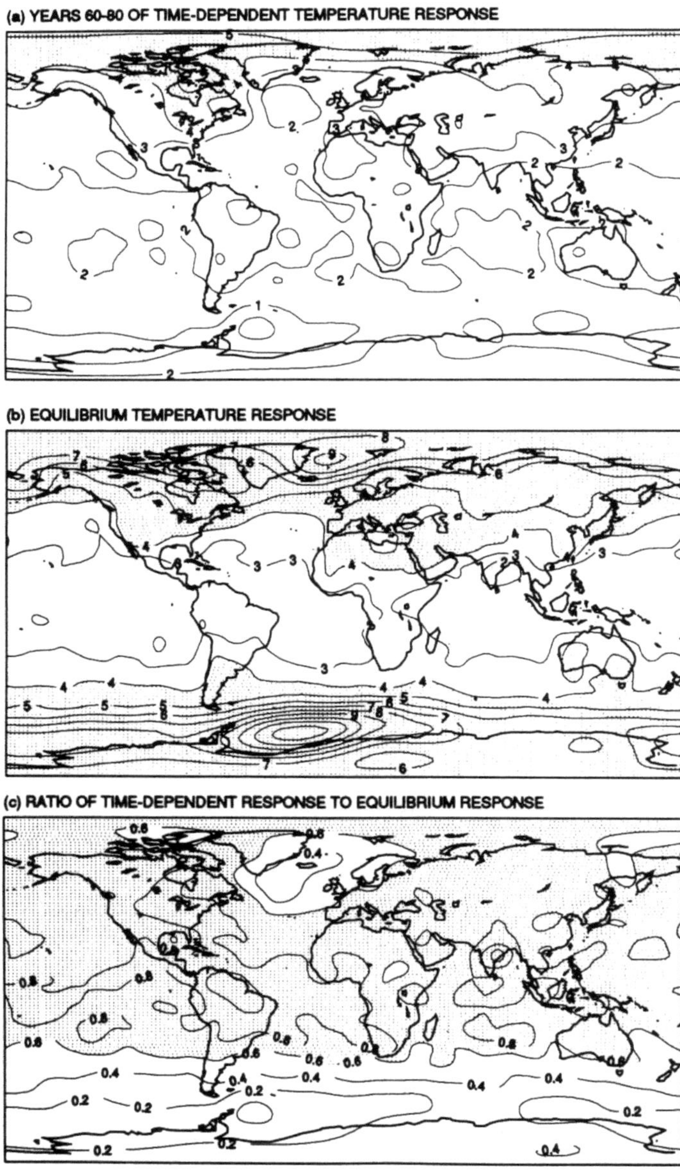

Figure 8: (a) The time-dependent repone of surface air temperature (°C) in the coupled ocean-atmosphere model to a 1% yr^{-1} increase of atmospheric CO_2.The difference between the 1% yr^{-1} perturbation run and years 60-80 of the control run when the atmospheric CO_2 concentration approximately doubles is shown, (b) the equilibrium response of surface air temperature (°C) in the atmosphere-mixed-layer ocean model to a doubling of atmospheric CO_2 and (c) the ratio of the time-dependent to equilibrium responses shown above (Houghton *et al.* 1990).

imately constant fraction of the equilibrium rise corresponding to the instantaneous forcing for a time that is earlier by a fixed offset. For an atmospheric model with temperature sensitivity 4°C for a doubling of CO_2, this fraction is approximately 66% with an offset of 11 years. In rough terms, the response is about 60% of the current equilibrium value. Extrapolation using robust scaling principles indicates that for a sensitivity of 1.5°C the corresponding values are 85%, 6 years and 80% respectively.

(b) The regional patterns of temperature and precipitation change generally resemble those of an equilibrium simulation for an atmospheric model, though uniformly reduced in magnitude. Exceptional regions are around Antarctica and in the northern North Atlantic, where the warming is much less.

(c) These results are generally consistent with our understanding of the present circulation in the ocean, as evidenced by geochemical and other tracers. However, available computer power is still a serious limitation on model capability, and existing observational data are inadequate to resolve basic issues about the relative roles of various mixing processes, thus affecting the confidence level that can be applied to these simulations.

(d) The conclusions about the global mean can be extended, though with some loss of rigour, by using a simple energy-balance climate model with an upwelling-diffusion model of the ocean, similar to that used to simulate CO_2 uptake. It is inferred that, if a steadily increasing greenhouse forcing were abruptly stabilized to a constant value thereafter, the global temperature would continue to rise at about the same rate for some 10-20 years, following which it would increase much more slowly, approaching the equilibrium value only over many centuries.

(e) Based on the IPCC Business-as-Usual scenarios, the energy-balance upwelling diffusion model with best judgement parameters yields estimates of global warming from pre-industrial times (taken to be 1765) to the year 2030 between 1.3°C and 2.8°C, with a best estimate of 2.0°C. This corresponds to a predicted rise from 1990 of 0.7-1.5°C with a best estimate of 1.1°C."

Part (e) of the above talks of the IPCC Business as Usual scenarios, that is under the assumption that few or no steps are taken to limit the emissions of greenhouse gases.

10 Conclusion

We have quoted extensively from the earlier parts of the IPCC Report (Houghton *et al.* 1990). It is also useful to comment briefly on what is in the rest of the Report. There are more detailed discussions of recorded climatic variations in the past and of evidence in the recent records for a human-induced increase in global mean temperature, sea-level rise and effects on ecosystems. These all make fascinating reading but we have no space to consider these in any detail here. Figure 9 shows historical evidence on sea-level changes over the last century, while Figure 10 illustrates the retreat of some selected glaciers. Figure 11 shows projections of global sea-level rise over the next century for the IPCC Business-as-Usual scenario, A, and for the other three policy scenarios B, C and D. The final chapter of the IPCC Report describes an Action Plan; this is a scientific action plan to enable us in improve our ability to make effective predictions of the climate. It talks about the various major international scientific programmes that can be expected to contribute to this. Some of these major programmes are mentioned elsewhere in this book; an example of the consideration of satellite data input to one of these programmes is provided by Townshend (1992). In addition to these major international programmes there is also something that is most relevant to this book, namely some discussion of the role of satellite observations for gathering data about the atmosphere, the land, the oceans and the cryosphere. Once more, we quote *verbatim* from the IPCC Report, this time from Chapter 11. There is a discussion of the importance both of modelling and of observations and the need of both of them to be brought together to develop an understanding of the climate system. The symbiosis between observations and modelling is illustrated in figure 12. The argument then proceeds:

"11.3.1 Improvement of the Global Atmosphere and Land Surfaces Observing System

Basic information on climatic processes, climate variations and systematic trends originates from the operational meteorological observing systems of the World Weather Watch, complemented by various operational hydrological networks and environmental measurements of the Global Atmospheric Watch. In addition to the maintenance of the basic meteorological systems, specific improvements are needed, in particular:

(a) improved infrared and microwave atmospheric sounding instruments on meteorological satellites, to obtain more accurate temperature/moisture information with better vertical resolution in the troposphere (e.g., high spectral resolution infrared spectrometer/radiometer);

(b) improved tropospheric wind observations from geostationary satellites (cloud-drift winds) and platforms in low Earth-orbit (Doppler wind lidar);

(c) rain-radars and passive multiple-frequency imaging radiometers in the microwave spectrum, to estimate rainfall over the whole globe; and

(d) radiometers to determine more accurately the Earth's radiation budget.

Temporal and spatial patterns in key vegetation properties can, when calibrated and interpreted in the context of ground-truth data, be efficiently studied with satellite sensors.

These data are critical in detecting regional shifts in ecosystem form and function in response to climate and land use changes. Sustaining these activities at levels appropriate to the study of climate change requires:

(e) improved commitment to the acquisition and archival of AVHRR data for determining rates of change (for example, those arising from desertification and land clearing), shifts in seasonal vegetation cycles, and rates of plant production needed for carbon cycling modelling; and

(f) improved high spectral resolution sensors, like those being defined for EOS, to determine vegetation characteristics (including physical and chemical properties) needed as input for global biogeochemical models.

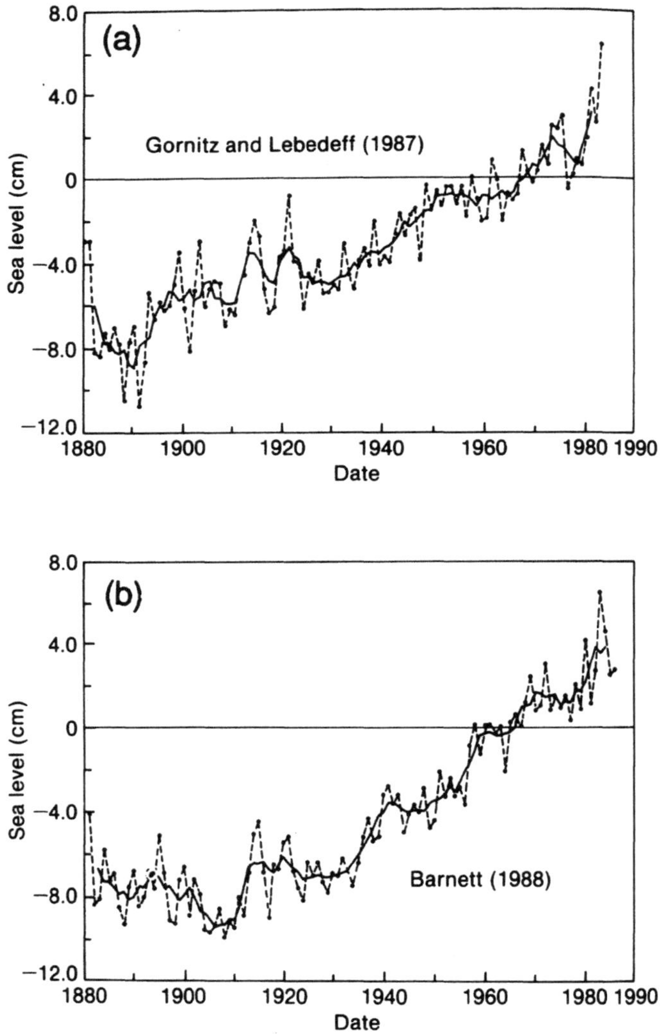

Figure 9: Global-mean sea level rise over the last century. The baseline is obtained by setting for the period 1951-1970 to zero. The dashed line represents the annual mean and the solid line the 5-year running mean. (Houghton *et al.* 1990).

11.3.2 Development of a Global Ocean and Ice Observing System

From the perspective of global climate change, systematic ongoing observations of the global ocean are needed for several purposes. The key to predicting the rate of change of the global system is to be found in observations of the ocean circulation and heat storage. Predictions of future climate change will eventually be carried out starting

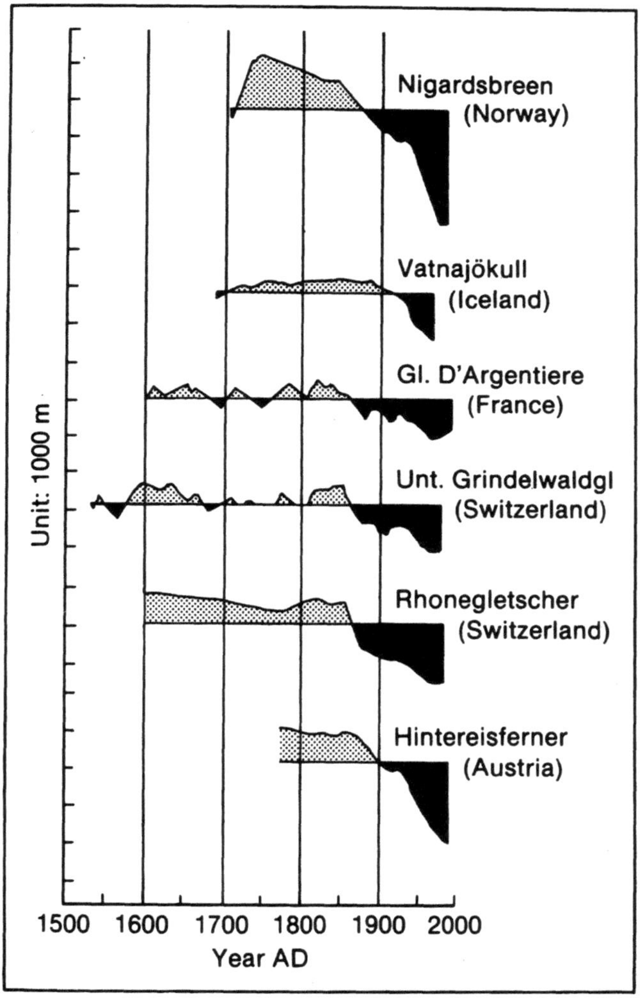

Figure 10: Variations of some selected glaciers as measured by their length (Houghton *et al.* 1990).

from the observed state of the combined atmosphere and ocean system. An example of this approach is the forecast of El Niño and other tropical climate disturbances by the WCRP Tropical Ocean and Global Atmosphere (TOGA) Programme.

A comprehensive ocean and ice observing system requires:

(a) satellite observations of the ocean surface temperature, wind and topography, sea-ice concentration and chlorophyll content (ocean colour) and of the topography of the Antarctic and Greenland ice sheets, by an international array of space platforms in suitable orbits around the Earth;

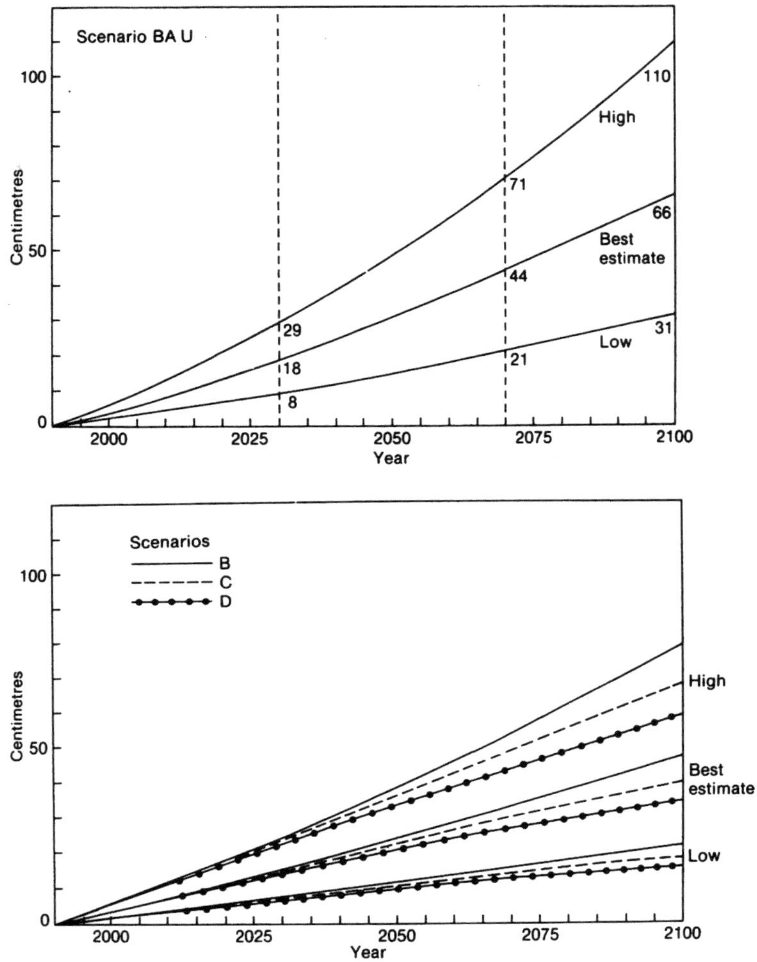

Figure 11: Global sea-level rise, 1990-2100, for (a) Policy Scenario Business-as-Usual and (b) Policy Scenarios B, C and D of the IPCC Report (Houghton *et al.* 1990).

(b) an international operational upper-ocean monitoring programme, to determine the time and space dependent distribution of heat and fresh water in upper ocean layers, seasonal variations and long-term trends; and

(c) an international programme of systematic sea-level and deep-ocean measurements, at suitable time and space intervals, to determine the state of the ocean circulation, ocean volume and transport of heat.

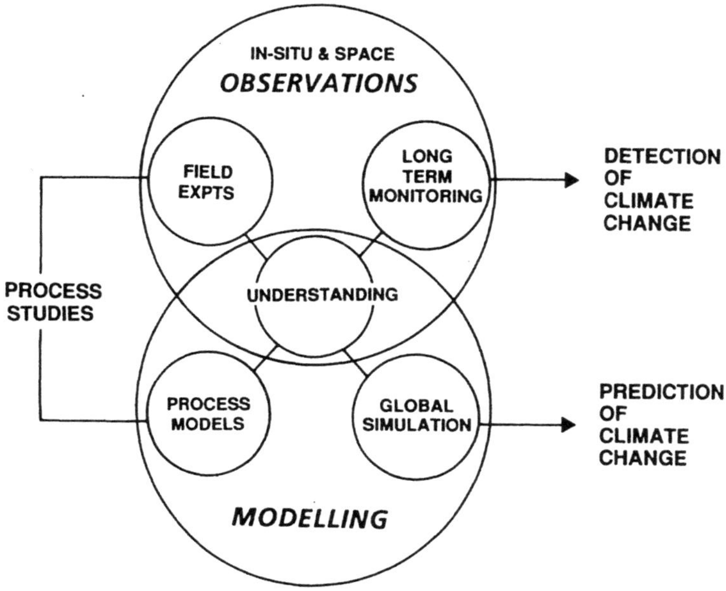

Figure 12: The symbiosis between observations and modelling (Houghton *et al.* 1990).

11.3.3 Establishment of a Comprehensive System for Climate Monitoring

In the previous two sections (11.3.1 and 11.3.2), improvements to the global atmosphere and land surface observing system and development of a global ocean and ice observing system were discussed. These observing systems must be coupled with existing observing systems to establish a comprehensive system for climate monitoring. It must be recognised that the requirements for climate monitoring are different from those for weather prediction. Failure to recognise this in the past means that there are a number of uncertainties which have been introduced into long-term climatological time series. A long-term commitment is now needed by the world's national weather services to monitor climate variations and change. Changes and improvements in observational networks should be introduced in a way which will lead to continuous, consistent long-term data sets sufficiently accurate to document changes and variations in climate. Some observing systems have and will be established for research purposes. These research systems are usually of limited duration and area and may have different emphases from a climate monitoring system (Figure 12). However, it is important that these observations also be integrated with the information from long-term monitoring systems.

Satellite observations are not yet of long enough duration to document climate variations. In order for these data to be most useful, it is very important that they be analysed and interpreted with existing in situ data. High priority should be given to the blending or integration of space-based and in-situ data sets in such a way as to build upon the strengths of each type of data. Examples of ongoing projects that have used mixed satellite data sets, careful cross calibration and coordinated data processing to produce global data sets are the WCRP's International Satellite Cloud Climatology Project (ISCCP) and the Global Precipitation Climatology Project (GPCP). For some applications, it is necessary to assimilate the observations into operational numerical weather prediction models for which purpose the observations need to be available in near real time. All satellite instruments should be calibrated both prior to launch and in flight. For those instruments now in operation, techniques should be developed to prevent or detect instrument drift. In planning new satellite observations for long-term

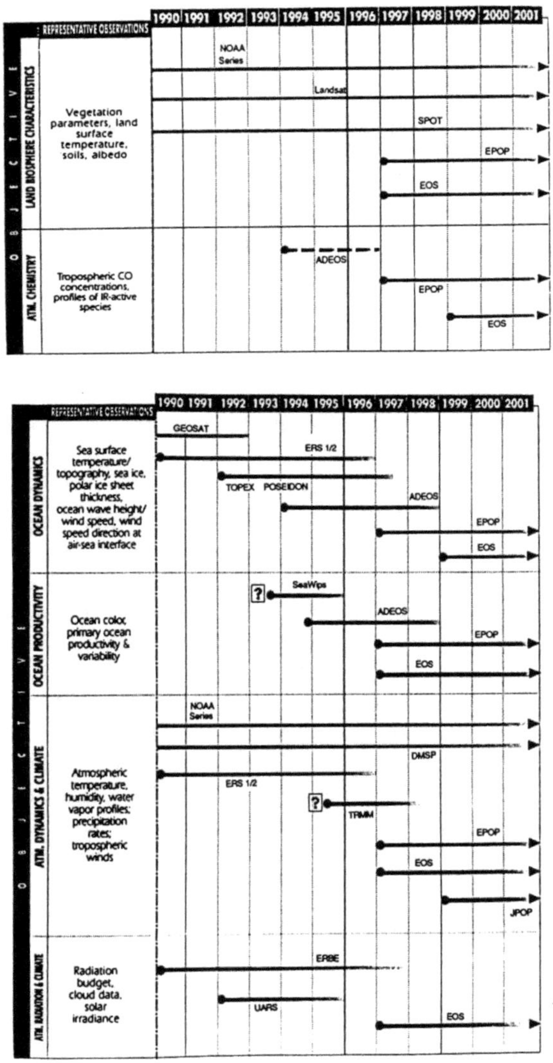

Figure 13: Illustrative examples of important satellite measurements (Houghton *et al.* 1990).

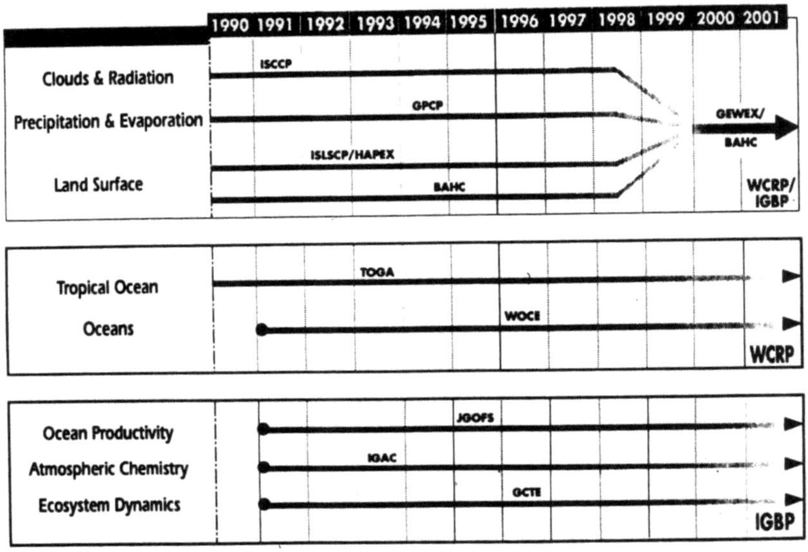

Figure 14: Schematic time schedules of major scientific programmes (Houghton *et al.* 1990).

monitoring, special attention should be paid to continuity of calibration and processing, archival and access to the data."

An indication of the role to be played by space-borne instruments is indicated in figure 13. A highly expanded version of this figure will be found in the BNSC report produced for the recent Rio conference (CEOS 1992). A comparison figure illustrating the time-scale of the major international scientific programmes is also given in the IPCC Report, see figure 14.

References

Broecker, W. S., and Denton, G. H., 1990, What drives glacial cycles? *Scientific American*, **262**, 43-50.

Budyko, M. I., 1982, *The Earth's climate, past and future* (English translation) (London; New york: Academic Press)

Budyko, M. I., and Izrael, Y., 1987, *Anthropogenic climate change* (Moscow; St. Petersburg: Gidrometeorizdat)

Budyko, M. I., Ronov, A. B., and Yanshin, A. L., 1987, *History of the Earth's atmosphere* (English translation) (Berlin: Springer-Verlag)

CEOS (Committee on Earth Observations Satellites), 1992, *The relevance of satellite missions to the study of the global environment. UNCED Conference Rio de Janeiro 1992* (London: British National Space Centre)

Cubasch, U., 1989, Coupling a global atmospheric model with a global ocean model using the flux correction method, in "Aspects of coupling atmosphere and ocean models", Report, No. 6. Meteorological Institute, University of Hamburg, FRG.

Cubasch, U., Maier Reimer, E., Santer, B., Mikolajewicz, U., Roeckner, E., and Boettinger, M., 1990, The response of a global coupled O-AGCM to CO_2 doubling. MPI report, MPI für Meteorologie, Hamburg, Germany.

Gates, L., 1975, The physical basis of climate and climate modelling, GARP Publication Series no. 16, WMO, Geneva.

Gribbin, J., 1990, An assault on the climate consensus. *New Scientist*, **on 15 December 1990**, 26-31.

Houghton, J. T., 1984, (Editor), *The Global Climate* (Cambridge: Cambridge University)

Houghton, J. T., Callander, B. A., and Varney, S. K., 1992, Climate Change 1992. The Supplementary Report to the IPCC Scientific Assessment (Cambridge: Cambridge University Press).

Houghton, J. T., Jenkins, G. J., and Ephraums, J. J., 1990. Climate change - the IPCC scientific assessment. (Cambridge: Cambridge University Press).

Kondratyev, K. Ya., 1992, *Global Climate* (Moscow; St. Petersburg: Nauka)

Noilham, J., and Planton, S., 1989, A simple parametrization of land surface processes for meteorological models. *Monthly Weather Reviews*, **117**, 536-549.

Schlesinger, M. E., and Mitchell, J. F. B., 1987, Climate model simulations of the equilibrium climatic response to increased carbon dioxide. *Reviews of Geophysics,* **25**, 1760-1798.

Semtner, A. J., 1976, A model for the thermodynamic growth of sea ice in numerical investigations of climate. *Journal of Physical Oceanography*, **6**, 379-389.

Townsend, J. R. G., 1992, Improved global data for land applications. A proposal for a new high resolution data set. IGBP Report No. 20. Stockholm, IGBP Secretariat.

Trenberth, K.E., 1992, *Climate System Modeling* (Cambridge: Cambridge University Press).

Washington, W. M., and Parkinson, C. L., 1986, *An introduction to three dimensional climate modelling* (Mill Valley: University Science Books).

The Use of Satellite Data in Numerical Weather Prediction

A.P. McNally

1 Introduction

The basic ideas of numerical weather prediction (NWP) are introduced and it is shown that due to the complexity of the physical systems we are trying to model, frequent and accurate observations of the global atmospheric state are required. In particular, the need to supplement so called conventional observations with remotely-sensed (satellite based) measurements is demonstrated. The true nature of satellite observations will be discussed, highlighting the important fact that they are not direct measurements of the quantities that we require in current NWP systems. The problem of converting the measurements (radiances) to useful quantities will be introduced and a number of retrieval approaches described. The variational approach to the NWP analysis problem will be introduced; this makes the explicit conversion of satellite measurements to products unnecessary. The potential benefits and extensions of such an approach are described. Finally, a number of satellite observation systems currently being investigated for future operational uses are described.

2 Observations in numerical weather prediction.

2.1 What is numerical weather prediction ?

The current demand for accurate weather forecasts is such that some of the most powerful computers in the world are devoted to process called numerical weather prediction (NWP). The basic approach involves the construction of a computer model of the atmosphere that is allowed to evolve in time and space under the influence of a number of different physical processes (see section 3 of chapter 1). The state of the atmosphere is usually described in terms of temperature, pressure, humidity and wind values specified on some (global) 3-dimensional grid (Figure 1) and may be represented by some multi-dimensional vector $\mathbf{X}$. The thermodynamic processes that drive the atmospheric motions and changes of state may be parametrised (Figure 2) such that from some initial state X_o a state X_t at some time in the future can be predicted.

NATO ASI Series, Vol. I 24
Remote Sensing and Global Climate Change
Edited by Robin A. Vaughan and Arthur P. Cracknell

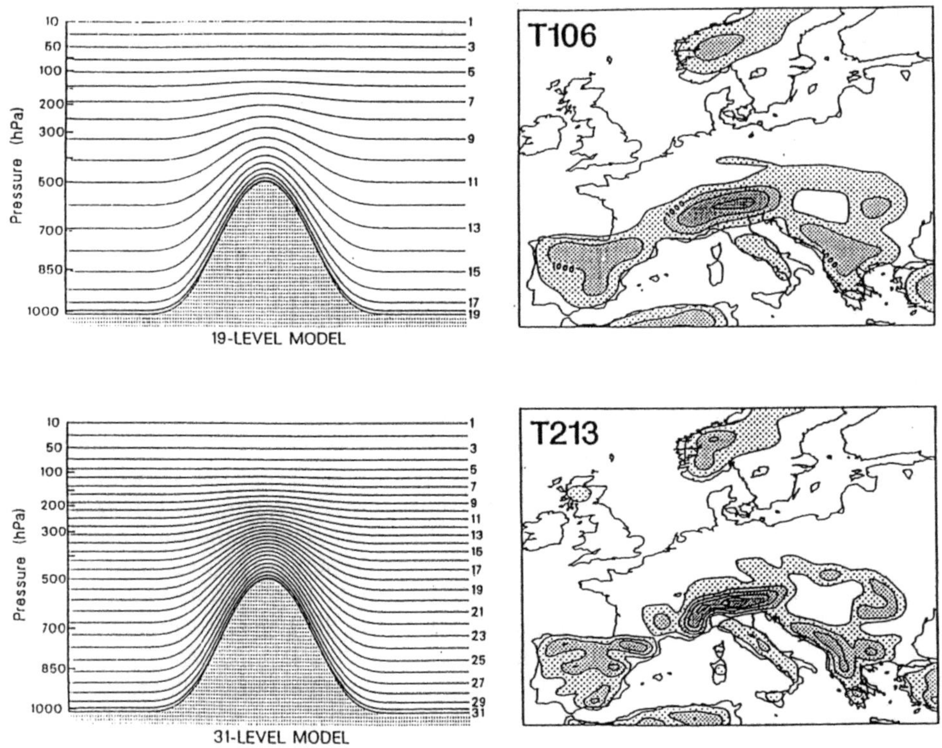

Figure 1: The ECMWF model grid is shown (T106-L19 pre Autumn 91 and T213-L31 after). The vertical resolution is presented in terms of the model levels mapped onto pressure coordinates. It can be seen that the hybrid system has sigma levels which vary with surface pressure in the troposphere (and thus avoid intersecting orography) and fixed pressure levels above 50mb. The horizontal resolution of the model is demonstrated by its representation of orography in the European area.

A simple analogy to the NWP process would be heating a block of metal at some known rate. If we know the initial temperature of the block and the thermal properties of the block (and its environment), we could predict the temperature of the metal at some time in the future. Similarly we could predict the position and speed of some projectile released from some known initial conditions. Though the atmosphere and the physical processes which drive it are much more complex than the analogies given, the principles of modelling the system and predicting its time evolution are the same.

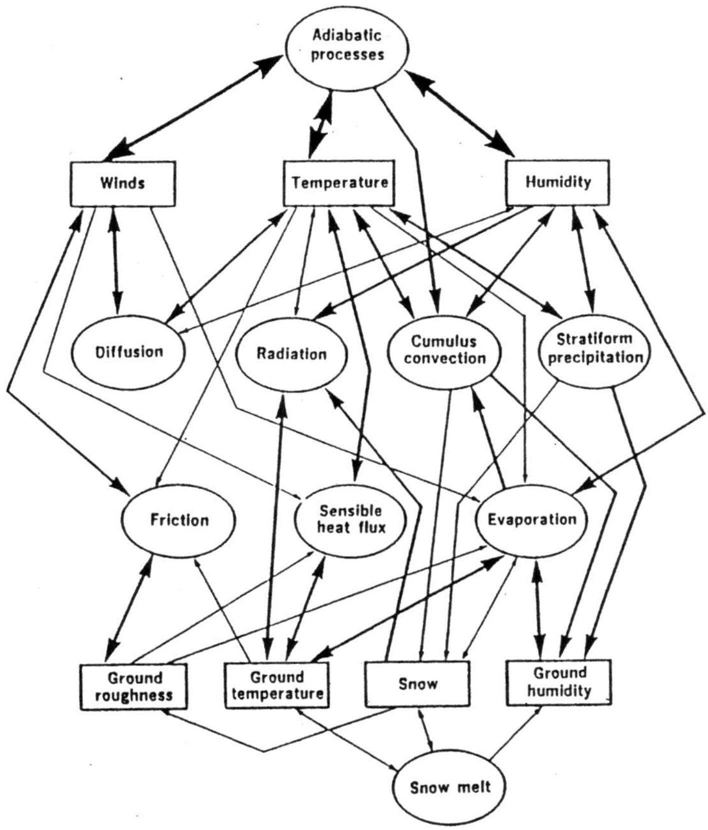

Figure 2: A schematic representation of the various physical processes which drive the ECMWF computer model of the atmosphere.

2.2 Why do we sometimes get it wrong ?

There are two clearly identifiable sources of error in the NWP process. First, our model of the atmosphere (i.e. representation) may be inappropriate or an over-simplification, in that even if the true state were known, it could not be represented accurately by a set of 'true' model variables. Also our knowledge, or more importantly our parametrisations of the thermodynamic processes will be inexact simplifications or approximations of real processes. These errors will be collectively known as model errors.

The second source of error arises since we do not know, and have little chance of ever knowing, the initial state of the atmosphere. While the initial temperature of a block or the position of a projectile may be reasonably well measured, to measure the state of the atmosphere is clearly very difficult, if not impossible, due to its size and complexity. This source of uncertainty will be termed initial state errors. While a small error in the initial temperature of our metal block should not be too serious in terms of our prediction accuracy, it is unfortunate that even the smallest errors in specifying the initial state of the atmosphere tend to grow very quickly (see also the discussion of chaos in chapter 1). Figure 3 shows how parallel forecasts from eight apparently very similar initial states can diverge after a period of about 7 days such that similarities between the eight fields are difficult to recognise.

The combination of these sources of error and the rate at which they grow mean that numerical models of the atmosphere cannot be run for long periods of time without constantly updating the initial state on the basis of observational information.

2.3 What observations do we have ?

Many thousands of so called conventional or in-situ observations of the atmosphere are collected throughout the day, every day, and so far as is possible all over the globe. These include aircraft, drifting buoys and surface and balloon (radiosonde) measurements from ground-stations and ships (Figure 4). While the coverage would appear to be reasonably comprehensive, for the purposes of NWP it is insufficient in both its spatial and temporal resolution and coverage. Large areas of the globe which are crucial to medium range weather forecasting remain unobserved at any time of the day or night.

The problem of specifying a unique global atmospheric state from insufficient measurements is a task known as the analysis or data assimilation. It is achieved by supplementing the insufficient measurement data with prior or background information from a previous short-range forecast. The measurement and background information are combined as we might combine two imperfect measurements of a particular quantity, (the background information essentially being treated as a 'virtual measurement'). The combination is optimal in the sense that each source of information is inversely weighted by its expected error. In this way we avoid drawing too close to poor observations or too close to the background in areas where the short-range forecast is known to be inaccurate.

Though the estimate of the atmospheric state is optimal on the basis of the observations and background state, it is usually not the best state from which to launch a forecast. The analysis may be unbalanced such that small-scale gravity waves are initiated as the state is evolved in time. While such modes do occur in real atmospheric motions and would be damped given time by the forecast model, they are undesirable in the short-range forecast used in data assimilation. Thus the analysis undergoes an initialisation step which essentially filters the atmospheric motions corresponding to gravity modes.

The details of the data assimilation procedure (shown schematically in Figure 5) will not be discussed here (see Hollingsworth 1989) but an important restriction of such schemes is that the observations have to be the same as, or at least a linear transformation of, the model variables to be analysed. Observations that do not conform to this constraint (i.e. remotely-sensed data) have to be converted to appropriate quantities before input to the analysis.

By using the forecast background (which contains information from previous observations evolved in time and space) the analysis is to some extent able to bridge the gaps in the measurement network and transport information. However if the gaps for

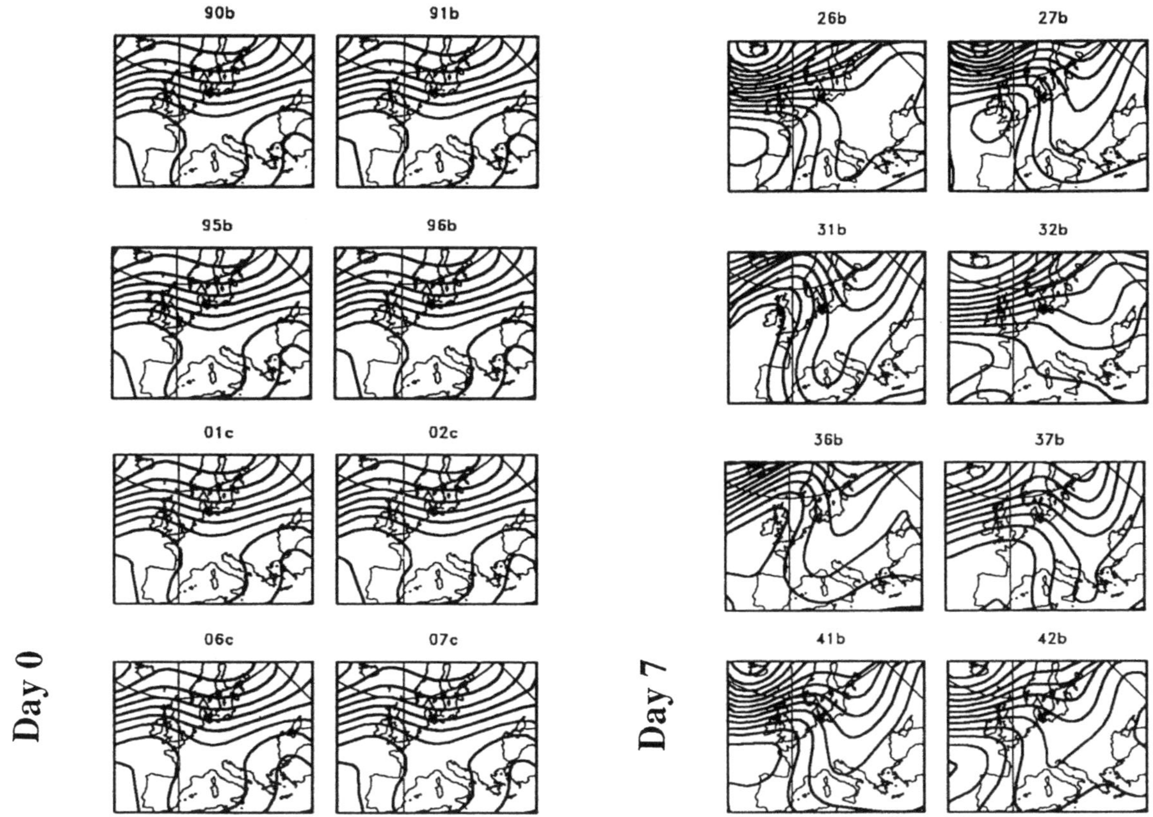

Figure 3: A demonstration of how small differences in the initial state used to launch a forecast grow very quickly in time.

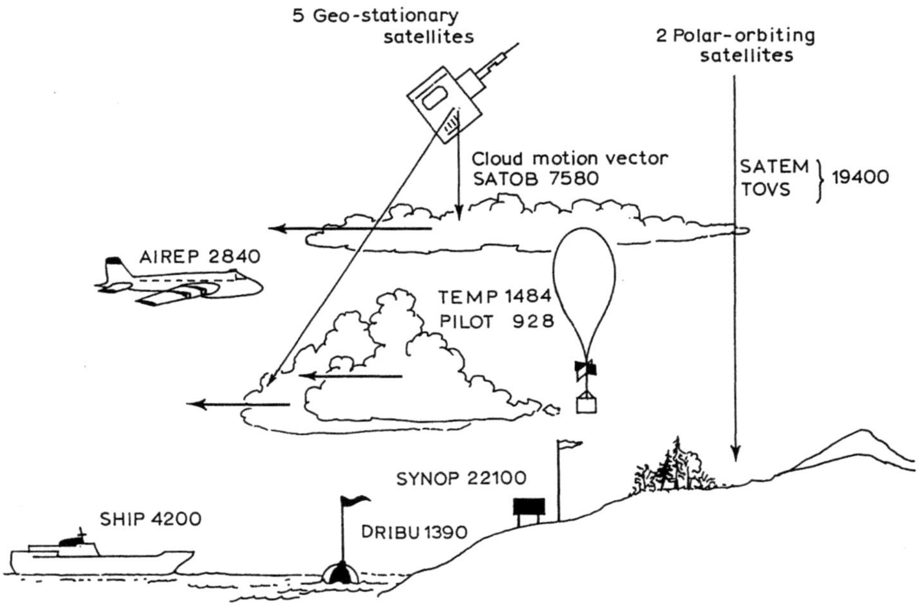

Figure 4a: The various types of observational networks.

example happen to be the Pacific and Atlantic oceans or indeed the southern hemisphere it is very difficult for the analysis to produce a sufficiently accurate estimate of the atmospheric state.

3 Remotely-sensed (satellite) observations

The frequency of coverage and the spatial and temporal resolution obtainable with a pair of polar-orbiting satellites (such as those operated by NOAA) is superior to that which could be reasonably achieved by any conventional system. If we consider the additional observations provided by the current network of geostationary spacecraft it can be seen that satellites are of extreme significance in NWP. Some of the most important instruments and their associated meteorological products currently used at NWP centres are described below.

3.1 Polar-orbiting spacecraft

On board the NOAA polar-orbiting spacecraft (currently NOAA-11 and NOAA-12) is the TIROS Vertical Operational Sounder TOVS which is actually three instruments in

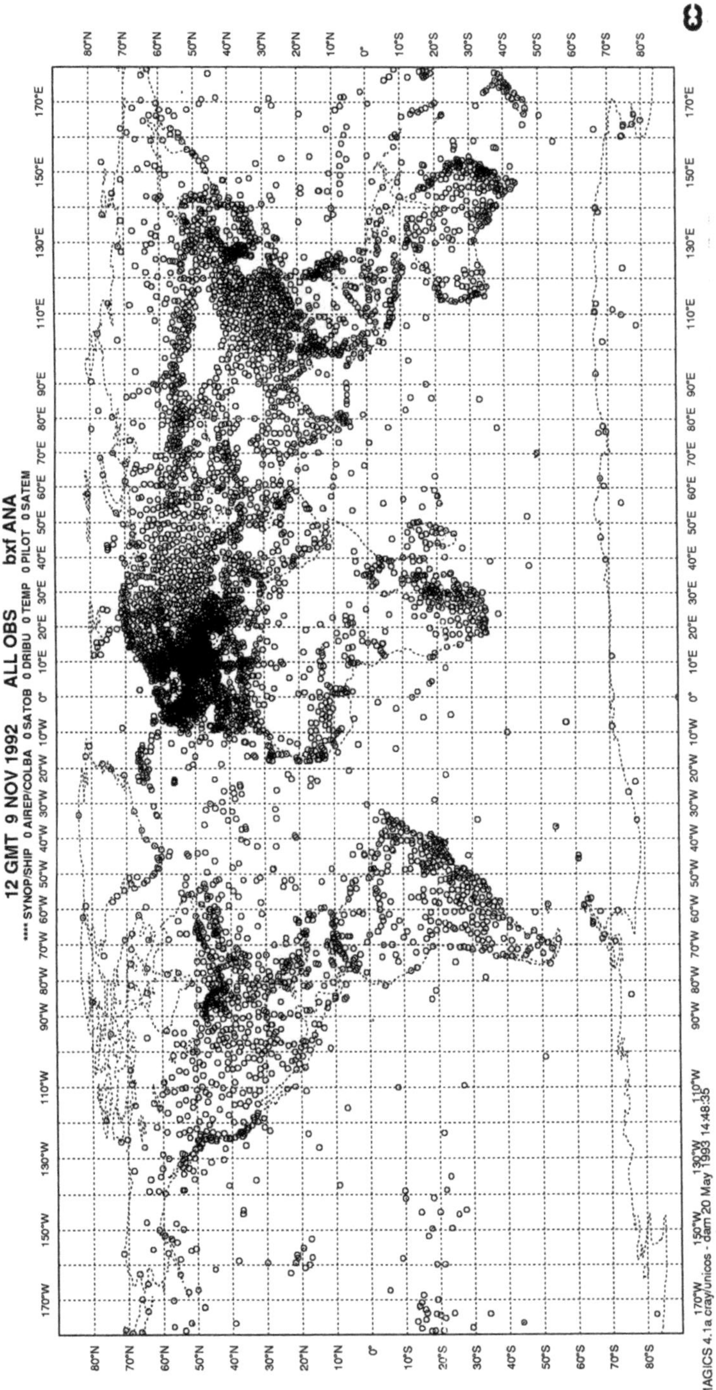

figure 4b: The spatial distribution of measurements described as 'conventional' or 'in situ' observations.

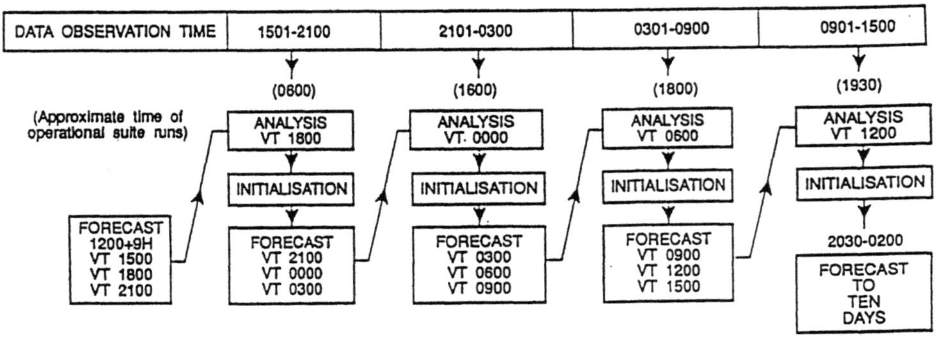

Figure 5: A schematic representation of how the short range forecast is used in the data assimilation scheme at ECMWF. It can be seen that every six hours there is an updated estimate of the current atmospheric state. The operational 10 day forecast is launched from the estimate valid for 12Z.

one. The HIRS, MSU and SSU are sounding radiometers and from their measurements of atmospheric radiation are generated profiles of temperature and humidity. TOVS and DMSP coverage patterns for a six hour period are shown in Figure 6. The TOVS products are routinely distributed to a number of meteorological centres by NESDIS at a resolution of approx 120 km (which is about 1 sounding in every 9). The typical accuracy of the these temperature sounding products is about 2 degK (rms) but can incorporate significant local bias errors. TOVS products are probably the best established source of remotely-sensed data used in NWP.

Also carried by the NOAA polar-orbiting satellites is the Advanced Very High Resolution Radiometer AVHRR which is an imager. Infra-red and visible channels sense the surface of the Earth or (if they are in the line of viewing) the top of clouds. Products of land and sea surface temperature in cloud-free areas are routinely generated and are the primary use as far as NWP is concerned. However many centres are using the data to generate cloud parameter products (such as amount height and type), and to supplement HIRS sounding data which are very sensitive to cloud contamination.

Some centres also make use of the ERBE (Earth Radiation Budget Experiment) measurements to provide observations of outgoing longwave radiation (OLR) from the top of the atmosphere.

3.2 Geostationary spacecraft

A number of space agencies operate satellites which by virtue of their orbital characteristics maintain a fixed position above the Earth (e.g. ESA-Meteosat, NOAA-GOES E/W, JAPAN-GMS, INDIA-INSAT). These are able to monitor almost a hemisphere of the Earth continuously but at the expense of spatial resolution (due to their altitude of approx 36,000 km). Such satellites are best known for providing visible and infra-

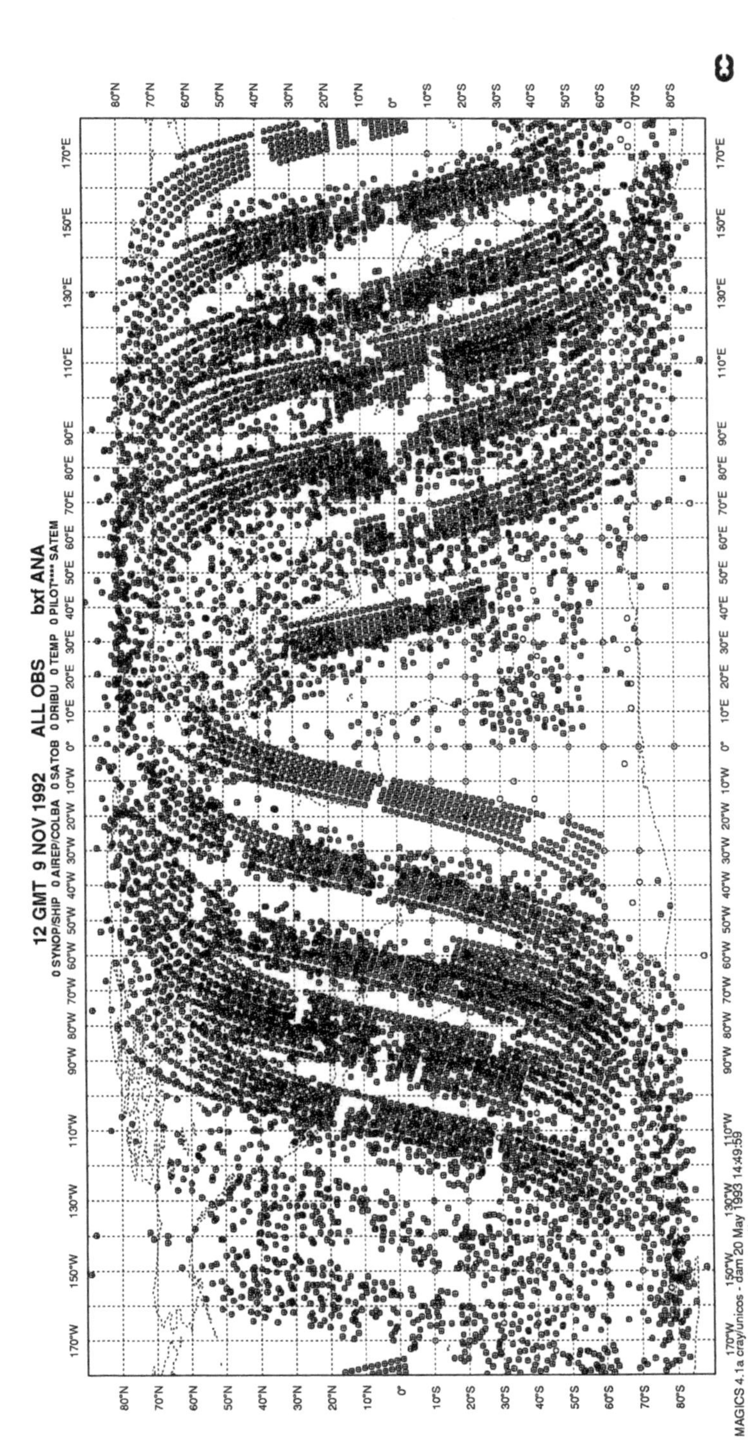

Figure 6: The observation coverage from polar orbiting spacecraft received at ECMWF.

red imagery which is used extensively in synoptic fore/nowcasting. This imagery can be used to confirm or otherwise the location of frontal systems and imminent storms. However the primary use of these satellites in NWP is the generation of cloud-track wind measurements. As their name suggests they are produced from observing the cloud motions (assumed to be following the wind) assigned to a particular level in the atmosphere. The combined geostationary wind (SATOB) wind measurement network is shown (Figure 7). Obviously such wind measurements are restricted to areas containing identifiable cloud cover. The typical accuracy of such wind measurements is around 5-10 ms^{-1}, the primary sources of error being incorrect height assignment of the cloud and inappropriate selection of tracer (i.e. using things not moving with the true wind).

3.3 Is the problem solved ?

If we examine the additional coverage provided by satellite-based measurements it appears that the globe is almost completely observed and so we have a wealth of information from which to uniquely specify the state of the atmosphere. There is however a major problem with using remotely-sensed observations in NWP since they are not direct measurements of the quantities we really need to know about (i.e. pressure, temperature, etc.).

It is often not appreciated (or at least the implications are not well understood) that satellite instruments do not measure temperature, humidity, wind, surface pressure or any of the many quantities we often see quoted as satellite products. Satellite instruments (passive) measure the electromagnetic radiation leaving the top of the atmosphere and extremely complicated conversion or retrieval techniques may be required to extract information about the quantities we are interested in (i.e. that we require for the analysis in NWP). Invariably this conversion introduces significant and very complicated errors into the final products, sometimes to such an extent that their use has little or no positive impact upon the NWP problem (Andersson *et al.* 1991). This situation is most acute in parts of the northern hemisphere which are well covered by conventional measurement systems (e.g. Europe and N. America). In these areas the state of the atmosphere can be specified accurately without satellite data, the addition of which has actually been found to degrade the quality of forecasts.

Such results have led to a great deal of scepticism regarding the use of (particularly) TOVS sounding data. It will hopefully become clear that much of the blame for this is due to the way the data are used and the methods traditionally used to convert the raw data to products.

4 Treatment of Satellite Data in NWP

4.1 What is measured by the satellite ?

This section concentrates on the use of satellite radiance measurements in NWP. Depending on the region of the electromagnetic spectrum in which a particular radiometer operates, the upwelling radiation measured at the top of the atmosphere results from a number of different processes (see Figure 8). It is essential to understand fully the

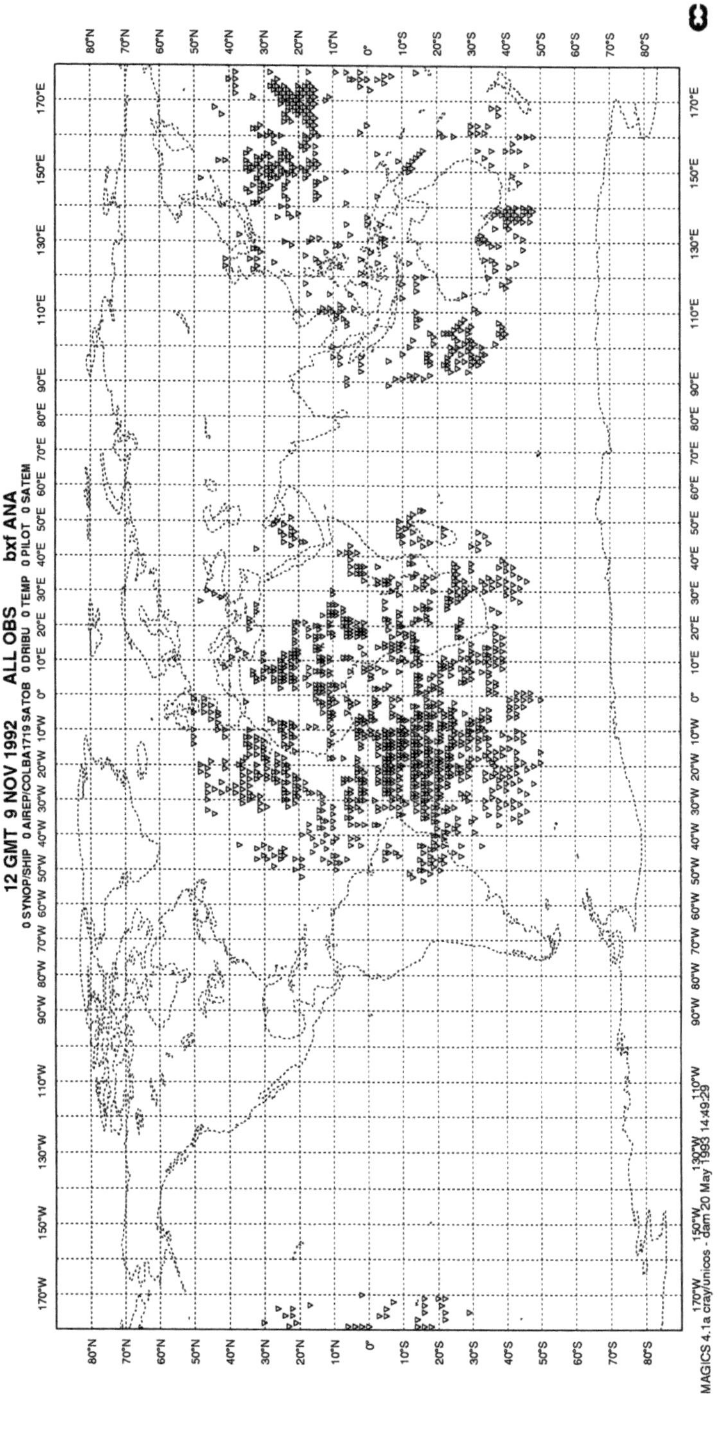

Figure 7: The observation coverage from geostationary spacecraft (only GOES Meteosat and GMS used at ECMWF).

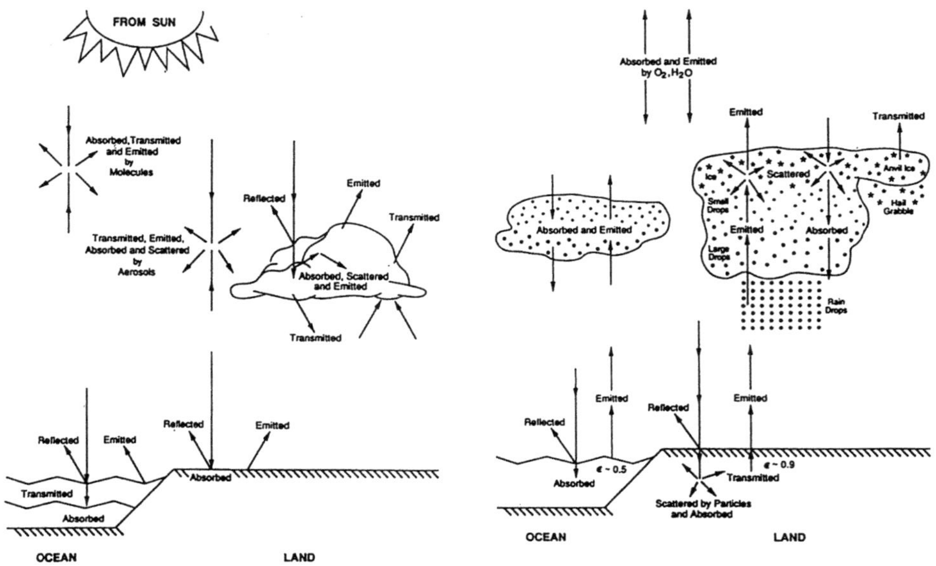

Figure 8: A schematic representation of how radiation is modified in the atmosphere by various processes. It is these processes which must be described by the forward model.

various sources and modification mechanisms if we are to make best use of the radiance measurements and extract useful information.

We formulate our knowledge of the radiative mechanisms in the atmosphere in a forward model which is given the state of the atmosphere viewed by the satellite (i.e. a profile of temperature, humidity, etc.) and predicts the radiances that could be measured in a given part of the spectrum. We may consider this as a mapping from state (or profile) space to measurement space

$$\underset{(\text{ profile })}{X} \xrightarrow{\text{forward model}} \underset{(\text{ radiances})}{Y}$$

The basic assumption in the the use of remote sensing is that if a particular atmospheric parameter influences the radiances that may be measured by the satellite (i.e. it appears in the forward model), we should in principle be able to extract information about that parameter from the radiances that are actually measured. However in practice we may not be able extract information uniquely about this parameter as its effect on the measured radiances may be (to some extent) inseparable from that of other parameters. The question posed in the processing of satellite measurements for use in NWP is, given some measured radiances, to determine the state of the atmosphere from which they originated. That is we require the inverse of the forward model, to perform the mapping

$$\underset{(\text{ measurements })}{Y} \xrightarrow{\text{inverse model}} \underset{(\text{ profile })}{X}$$

As an illustration we consider the problem of retrieving values of the atmospheric temperature profile from radiance measurements. We simplify the situation by assuming a radiometer measuring in part of the spectrum where the only source of radiation is emission from the atmosphere (and surface) and the only modification process is attenuation due to absorption by well mixed gasses (this is actually a good approximation for infra-red measurements around 14 μm in the CO_2 region when there is no cloud). In this case

$$Y = \tau_0 B(T_0) + \int_0^\infty B[T(z)] \frac{d\tau(z)}{dz} dz \tag{1}$$

if we furthermore consider only channels for which the atmospheric transmittance from the surface to space is zero, it can be seen that our radiance measurements correspond to weighted averages of the Planck function (which in turn is directly related to the temperature profile). The weighting function in this case is the vertical derivative of the transmittances and it is shown in Figure 9 a for real HIRS channels of the TOVS. While we can (at least with careful selection of channels) perform the forward mapping, extracting the temperature profile for use in NWP is a fundamentally ill -posed problem. The radiances are very deep vertical averages and do not contain sufficient information to specify uniquely the temperature profile T(z). Even if the spectral selection could be done in such a way that the weighting functions were very sharp (i.e. essentially being sensitive to a single layer) we would still only obtain a limited sampling of the temperature profile. There are areas or components of the atmospheric state which cannot be uniquely resolved by the radiance measurements and these are often called null space components.

Considered in this way we see that the problem of retrieving useful information from radiance measurements is analogous to the analysis problem discussed earlier. That is to say we are trying to construct the state of the atmosphere (in this case a profile) from an insufficient finite number of observations (radiances). We should not then be surprised that the approach adopted to solve the problem involves supplementing the observed radiances with some sort of prior information to extract the quantities we require. Two approaches will be described that are clearly distinguished by the source of prior information used.

4.2 Regression methods

This is an extremely common approach to the problem of extracting required information from satellite radiance measurements. It uses a statistical prior knowledge to

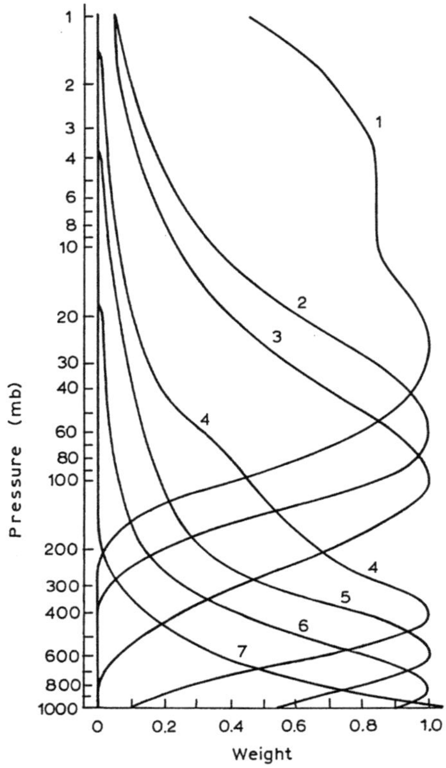

Figure 9: Weighting functions showing the contribution from various atmospheric levels to the radiances measured in the HIRS longwave channels.

relate the occurrence of a particular set of radiance measurements to one or a set of atmospheric state parameters. In its simplest form, a set of temperature profiles (e.g. observed by radiosondes) is collected together with coincident satellite-measured radiances and coefficients are generated (using the usual multivariate regression equations) to map deviations from the mean (of the samples) in measurement space to profile space. In their most complicated form regression methods employ elaborate schemes of air-mass selection to decide which of a variety of different sets of coefficients to use (each generated by

$$X - \overline{X} = [REGCOF] \cdot (Y_m - \overline{Y}) \tag{2}$$

to apply to various meteorological situations), but the basic principle of supplementing the measurements with purely statistical prior information is the same. A method based on this principle is currently employed by NESDIS to produce their global 120/500km TOVS products, which are currently distributed to and used by a large number of meteorological centres all over the world.

While such methods are reasonably simple and convenient there are two major drawbacks limiting the usefulness of regression-generated products in NWP. First, the prior information used (i.e. the statistical relation) in many situations is inaccurate compared to the quality of the NWP model itself (at least in conventional data dense areas) producing retrievals with corresponding poor quality and excessively smooth characteristics. It is true that because of their poor vertical resolution the radiances cannot suggest anything else, but the assimilation of these smooth retrievals has been found to attenuate sharp features present in the NWP model, which however statistically anomalous with respect to the regression sample, are perfectly justifiable from the physics of the situation. While it is true to say that the radiances cannot introduce small-scale vertical structure, they should not be allowed to contradict such features captured well in the NWP model. This can be the case with regression-based products due simply to the nature of the prior information used.

A second drawback to regression products is that to assimilate the products correctly in NWP (i.e. to give them an appropriate weight in the analysis) we must have a good specification of the expected retrieval error structure. To do this correctly requires a good knowledge of the prior information used on each occasion (i.e. the variance of the regression samples) and if a complex air-mass selection scheme is employed a large amount of auxiliary information must be distributed together with the products themselves. This becomes logistically very difficult if a centre aims to produce and distribute global satellite products.

4.3 Use of the forecast as prior information

A possibly obvious remedy to the problems associated with regression schemes is to supplement the radiances with prior information from a previous short-range forecast in the form of a guess profile. The quality of such prior information is generally better than in the worst case is still as good as, the NWP model into which the products will be assimilated. The error characteristics of these products should also be better known such that they may be treated better the subsequent analysis (i.e. as optimally as possible).

An optimal variational approach to this method has been adopted at ECMWF (the European Centre for Medium-range Weather Forecasting). We essentially search for a profile X which is most probable on the basis of the measured radiances obtained and our prior or background knowledge of the atmospheric state. If we assume that random errors in the measurements and forward model are gaussian and described by the combined covariance $[O+F]$, the probability that the measurements originated from an atmospheric state X is given by

$$P(Y_m|X) \propto exp\ [\ -\frac{1}{2}[Y_m - Y(X)]^T(O+F)^{-1}[Y_m - Y(X)]\ ]. \tag{3}$$

The corresponding prior probability for X is

$$P(X) \propto exp[-\frac{1}{2}(X - X_0)^T B^{-1}(X - X_0)] \tag{4}$$

where the matrix B describes the error covariance of the forecast background. The probability that we actually require is that of X conditional upon measurements Y_m

having been observed and this is found using Bayes Theorem

$$P(X|Y_m) = \frac{P(Y_m|X)P(X)}{P(Y_m)}. \tag{5}$$

If we examine the components of this expression we see that maximising the required probability with respect to X is achieved by minimisation of a so-called cost function defined by

$$J(X) = (X - X_0)^T B^{-1}(X - X_0) + [Y - Y(X)]^T (O + F)^{-1}[Y - Y(X)] \tag{6}$$

Thus the solution which is most probable may be regarded as that which simultaneously minimizes the distance between the observations (in radiance space) and the prior estimate of X. It can easily be shown that a solution satisfying the above criteria has an expected error covariance given by the expression

$$S = B - (KB)^T[KBK^T + (O + F)]^{-1}(KB) \tag{7}$$

where the matrix K contains the partial derivatives of the radiance in each channel with respect to each profile element (essentially the discretised weighting functions).

Apart from producing better quality products (essentially by the use of better quality prior information) both of the drawbacks associated with regression based schemes are removed. The subsequent analysis of the products (if performed correctly) adjusts the NWP model only on vertical scales consistent with those of the radiances, leaving components unseen by the radiances unchanged. Also we have good knowledge of the errors in the prior information from which to give an appropriate weight to the observations.

Another advantage of such an approach over regression schemes is that it is easily extended to highly nonlinear situations such as the treatment of cloud contaminated radiances (although the numerical minimisation of the cost function may be more problematic). Most centres currently use radiances in some cloud-cleared form (i.e. pre-processed to remove the effect of cloud), but this pre-processing (as well as others such as limb-correction) introduce additional errors into the radiances. Thus the variational method is able, at least in principle, to by-pass these pre-processing stages and use radiances of intrinsically higher quality. In practice, however, there are still problems with this extension since clouds must be effectively retrieved (with all the other atmospheric parameters) from the radiances and background information on clouds from the NWP model is inappropriate.

A small number of centres now use NWP prior information in this way for the treatment of satellite radiances (e.g. LASS in the UK Meteorological Office and 1DVAR at the ECMWF) or intend to in the near future (ABOM and NMC).

Experiments carried out at ECMWF suggest that the 1DVAR scheme produces better sounding products as verified against collocated radiosondes (see Figure 10), but the most encouraging results are those of forecast impact studies (various viewgraphs). However the problem is far from being solved and a number of outstanding difficulties exist; these include:

Specification of radiance error: This must allow not only for radiometric noise but also for any errors introduced by the various pre-processing of the measurements (e.g. cloud clearing or limb correction). The matrix O+F must also contain a contribution from errors in the process of simulating radiances Y from a profile X.

4.4 Summary

The most important point addressed in this section is the fact that satellite instruments in general do not measure the many quantities they are often quoted as providing. These quantities are generated by what can be a series of very complicated conversion and extraction steps. While for stand-alone uses the origin of the products may not be important, a complete understanding of the information contained in remotely-sensed observations is required if they are to be useful in today's extremely demanding NWP models.

The extraction of information from remotely-sensed observations has been shown to be analogous to the NWP analysis problem itself, attempting to reconstruct atmospheric parameters from incomplete data. Two examples of how prior information is used to supplement the measurements have been shown with the latter preferred because it uses background information consistent with that used in the subsequent NWP analysis.

5 Future methods and observing systems

5.1 Variational analysis of observations

The analysis of observations in NWP has traditionally been performed using optimal interpolation. Observations linearly related to the variables of the model we wish to specify are essentially combined in an optimal way (i.e. inversely weighted by their errors) with a forecast background treated as a virtual measurement. It was explained earlier that the constraint of a there being a linear mapping between the prior and measurement information requires observations such as satellite radiances to be converted to retrieved products before they can be assimilated.

The principles behind the variational approach to the retrieval of profiles from satellite radiances (1DVAR) may be applied to the general analysis problem. We search for the most probable state of the X (now a 3-dimensional vector) which is equivalent to that which simultaneously minimises the distance from some background state X_0 (prior estimate) and the distance from all the observational data. That is to say we minimise a cost function defined by

$$J(X) = (X - X_0)^T B^{-1}(X - X_0) + J_{sonde} + J_{synop} + J_{TOVS} + \cdots \tag{8}$$

where for any observation type its contribution to the cost function is

$$J_{obs}(X) = [OBS - OBS(X)]^T (O + F)^{-1} [OBS - OBS(X)] \tag{9}$$

The matrices O and F describe as before the errors in the measurements (and any pre-processing) and errors in the prediction of the observation form the state vector

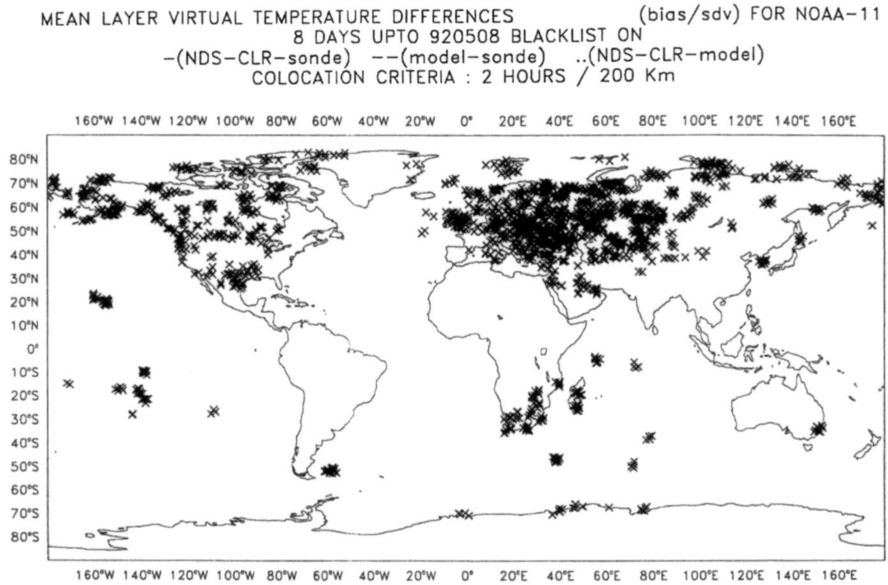

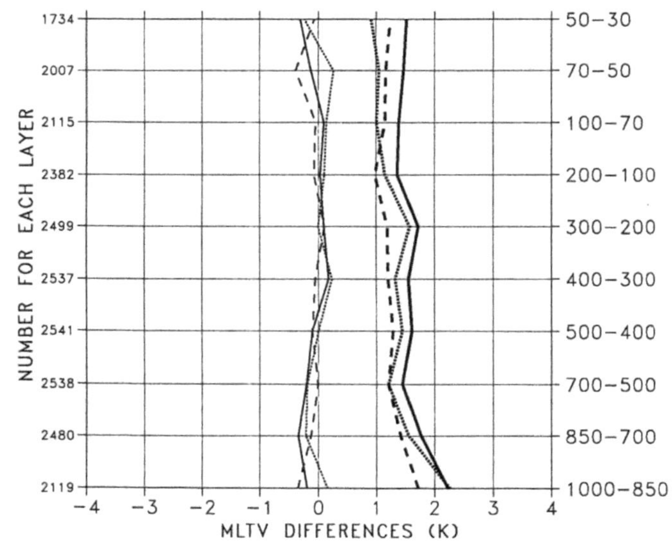

Figure 10a: A comparison between RMS statistics of TOVS retrieved profiles against radiosondes for NESDIS (using a regression based scheme) and the ECMWF 1DVAR (which uses forecast prior information). The dashed lines represent the RMS accuracy of the forecast measured against the same radiosondes. The maps show the locations where coincident TOVS/radiosonde comparisons are possible.

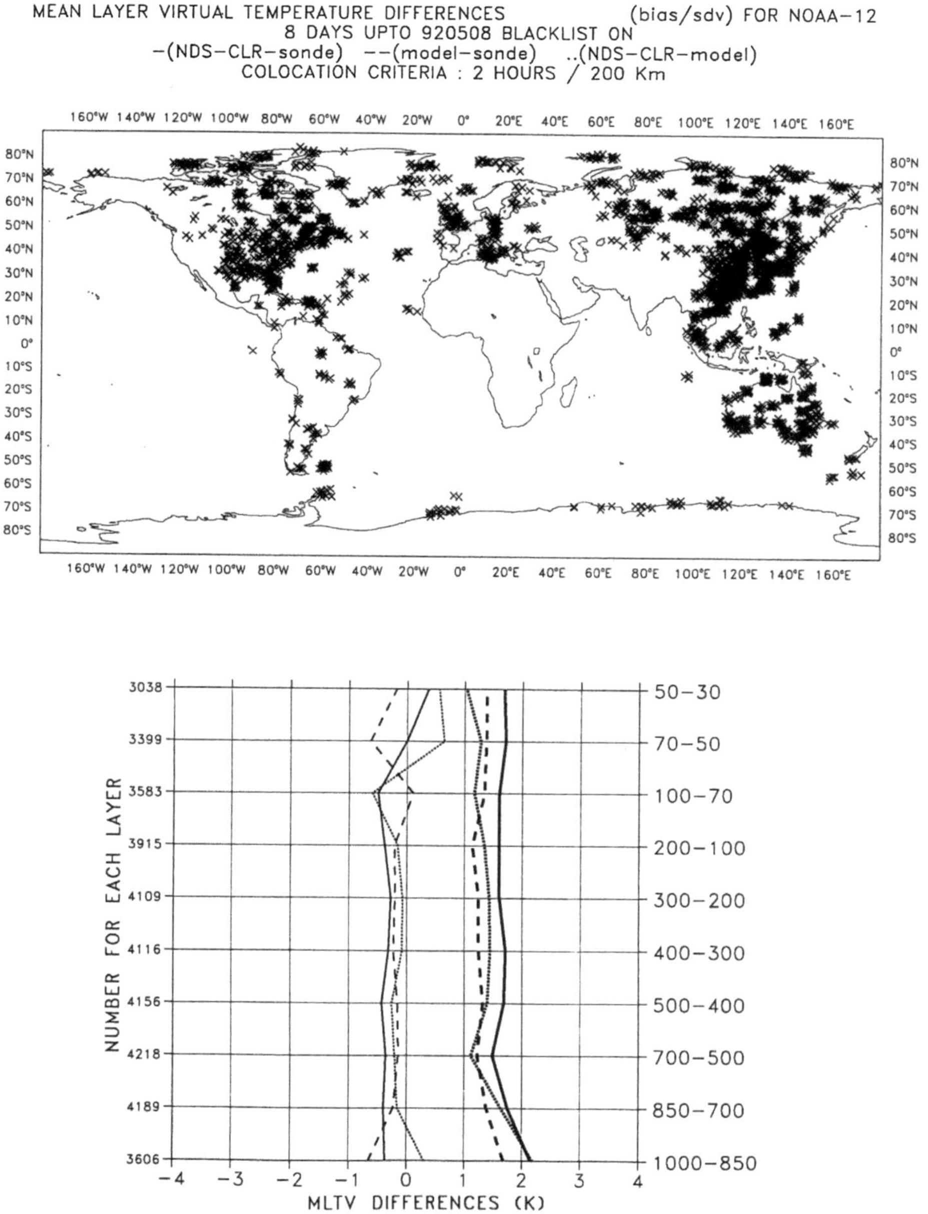
MEAN LAYER VIRTUAL TEMPERATURE DIFFERENCES (bias/sdv) FOR NOAA-12
8 DAYS UPTO 920508 BLACKLIST ON
-(NDS-CLR-sonde) --(model-sonde) ..(NDS-CLR-model)
COLOCATION CRITERIA : 2 HOURS / 200 Km
160°W 140°W 120°W 100°W 80°W 60°W 40°W 20°W 0° 20°E 40°E 60°E 80°E 100°E 120°E 140°E 160°E
80°N 70°N 60°N 50°N 40°N 30°N 20°N 10°N 0° 10°S 20°S 30°S 40°S 50°S 60°S 70°S 80°S
NUMBER FOR EACH LAYER
3038
3399
3583
3915
4109
4116
4156
4218
4189
3606
50-30
70-50
100-70
200-100
300-200
400-300
500-400
700-500
850-700
1000-850
-4 -3 -2 -1 0 1 2 3 4
MLTV DIFFERENCES (K)

Figure 10b

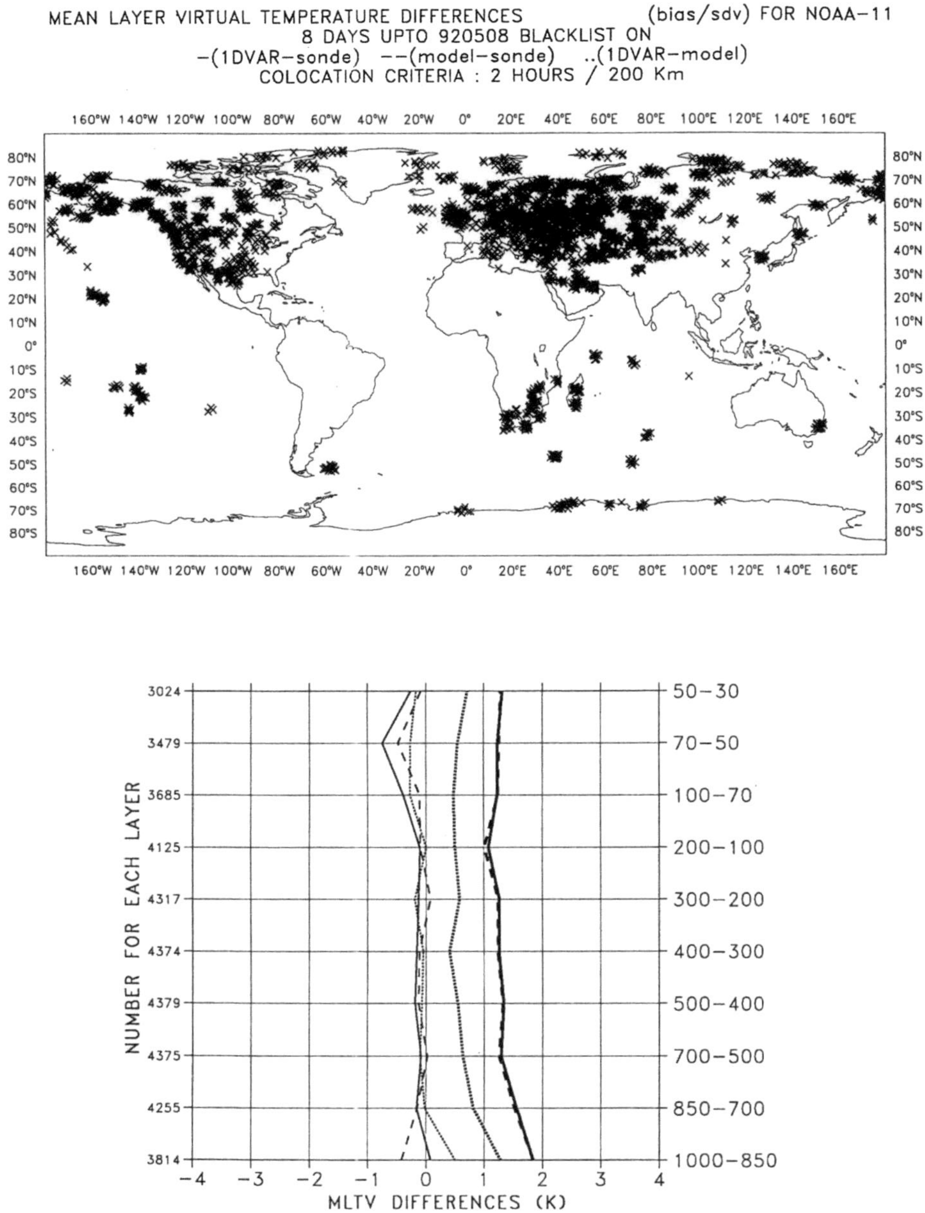
MEAN LAYER VIRTUAL TEMPERATURE DIFFERENCES (bias/sdv) FOR NOAA-11
8 DAYS UPTO 920508 BLACKLIST ON
-(1DVAR-sonde) --(model-sonde) ..(1DVAR-model)
COLOCATION CRITERIA : 2 HOURS / 200 Km
160°W 140°W 120°W 100°W 80°W 60°W 40°W 20°W 0° 20°E 40°E 60°E 80°E 100°E 120°E 140°E 160°E
80°N 70°N 60°N 50°N 40°N 30°N 20°N 10°N 0° 10°S 20°S 30°S 40°S 50°S 60°S 70°S 80°S
NUMBER FOR EACH LAYER
3024
3479
3685
4125
4317
4374
4379
4375
4255
3814
50-30
70-50
100-70
200-100
300-200
400-300
500-400
700-500
850-700
1000-850
-4 -3 -2 -1 0 1 2 3 4
MLTV DIFFERENCES (K)

Figure 10c

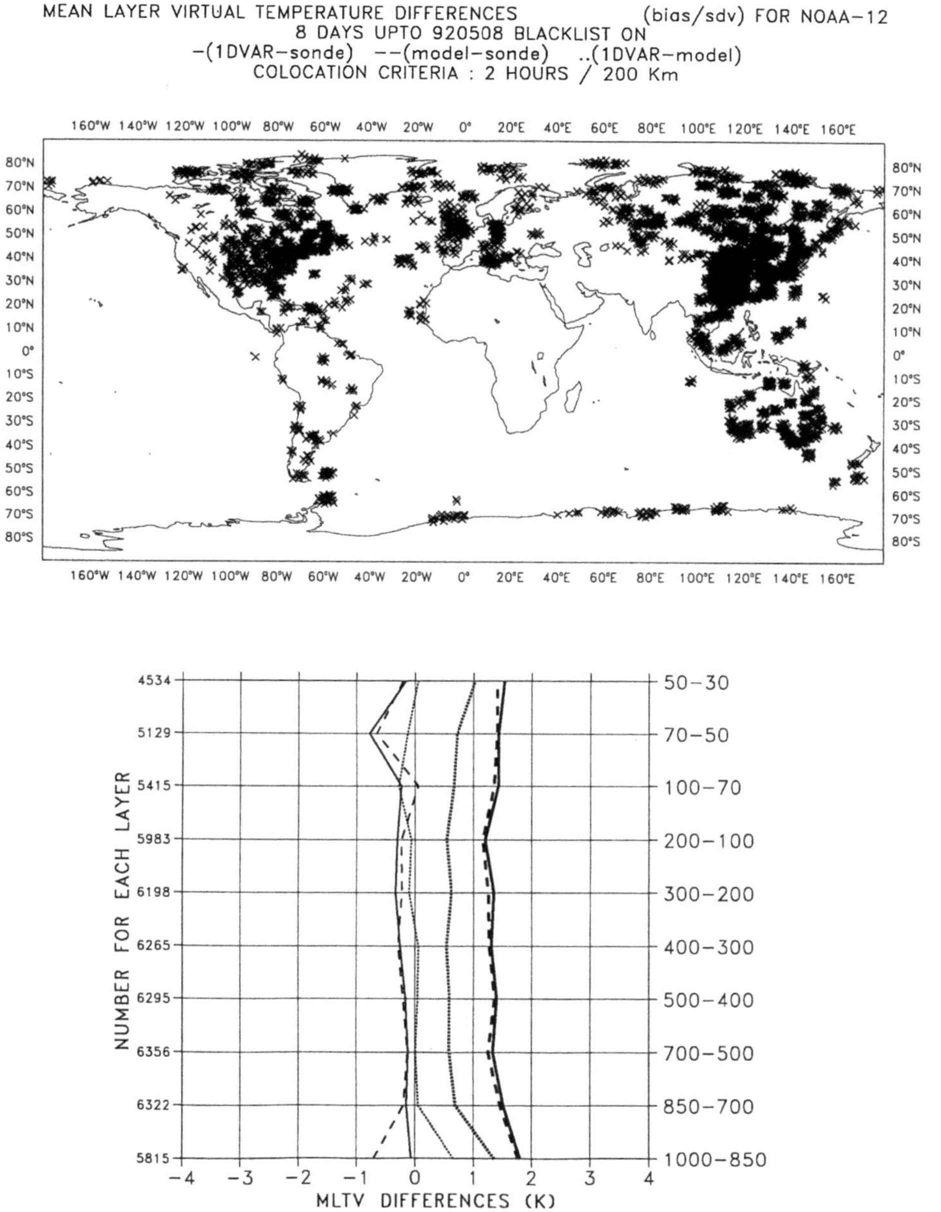
MEAN LAYER VIRTUAL TEMPERATURE DIFFERENCES (bias/sdv) FOR NOAA-12
8 DAYS UPTO 920508 BLACKLIST ON
-(1DVAR-sonde) --(model-sonde) ..(1DVAR-model)
COLOCATION CRITERIA : 2 HOURS / 200 Km
160°W 140°W 120°W 100°W 80°W 60°W 40°W 20°W 0° 20°E 40°E 60°E 80°E 100°E 120°E 140°E 160°E
80°N 70°N 60°N 50°N 40°N 30°N 20°N 10°N 0° 10°S 20°S 30°S 40°S 50°S 60°S 70°S 80°S
NUMBER FOR EACH LAYER
4534
5129
5415
5983
6198
6265
6295
6356
6322
5815
50-30
70-50
100-70
200-100
300-200
400-300
500-400
700-500
850-700
1000-850
-4 -3 -2 -1 0 1 2 3 4
MLTV DIFFERENCES (K)

Figure 10d

(i.e. forward model errors), and the matrix B now describes the 3-dimensional error covariance of the forecast background.

In the 1DVAR we essentially adjusted a background profile on the basis of the radiance information, the 3DVAR method (as it is known at ECMWF) adjusts a 3-dimensional background state using the radiances together with all the other conventional observations.

The restriction that a linear mapping exists between the observation and analysis variables is no longer required, only that we have a forward model which can simulate the particular observation from the atmospheric state vector. The obvious and most important implication of this is that radiances can be assimilated directly without the need to convert them to retrieved products. Indeed the extension to the analysis of nonlinearly related observations with the variational approach is analogous to the extension of 1DVAR to cloudy radiances which was mentioned very briefly in section 4.

An examination of the terms in the above equation shows that the 1DVAR has essentially been absorbed into the analysis itself and an explicit retrieval step is redundant. Clearly then any problems that would be associated with allowing for the prior component in a subsequent optimal interpolation analysis of satellite retrievals also disappear.

Another advantage of the 3D approach is that the adjustments applied to the forecast by the radiances are automatically constrained to be horizontally and vertically consistent in a single step, rather than a split 1D retrieval followed by an analysis.

The potential benefits of a full variational approach to the analysis problem are considered sufficient to warrant a large research effort and it is hoped that a 3DVAR scheme will be operational by 1993 at the ECMWF. The areas still requiring a significant amount of work are:

1. Development of the adjoint techniques actually required to perform the minimisation of the cost function in 3 dimensions. This is obviously a much bigger numerical problem than for the 1DVAR which can be solved by simple Newtonian methods.

2. Specification of background error covariance in 3 dimensions to represent horizontal and vertical (inter-level) correlations of error.

3. Specification of measurement and forward model errors (in particular for radiances) allowing for horizontal correlations due to, for example, cloud clearing.

4. Introduction of additional constraints into the cost function to ensure physical balance between analysed variables and to produce solutions which are suitable initial states for integration in time by the forecast model (i.e. absorb the initialisation step into the analysis).

5.2 Further in the future-4DVAR ?

As a further extension of the variational approach it is possible to search not for a single state X which is simultaneously consistent with a set of observations and a background state, but a time series or trajectory of states which over some assimilation time period

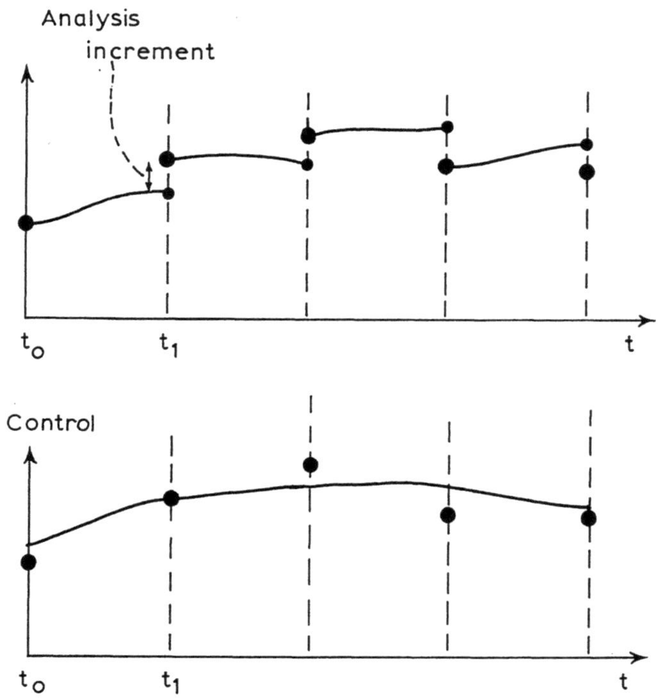

Figure 11: Schematic representation of 4DVAR obtaining a trajectory of atmospheric states which evolve according to the model physics in time. This compared to intermittent adjustment (say every six hours) schemes such as 3DVAR or Optimal interpolation.

does the same (Figure 11). In the same way as 3DVAR improves over 1DVAR in its consistent horizontal adjustment of the model, 4DVAR offers the extra benefit of a model adjustment by the observations which are consistent with the physical time evolution of the atmosphere (or at least our model of it).

In the same way that 3DVAR is able to make better use of the observed data (i.e. without imperfect conversions or retrievals), the 4DVAR does better still, since it can use information from observations at the time they were actually made. In intermittent or 3D schemes observations made far from the time at which we require the analysis have to be down-weighted (i.e. given bigger errors) even though they may be high quality measurements in their own right. Such an extension is part of the basic variational approach in trying to use remotely-sensed data for what they actually are without conversion or down-weighting for reasons which are not intrinsic to the measurements, but consequences of how we wish to use them.

The major problem associated with such a scheme is the numerical expense. The forward model in 1D and 3D variational schemes essentially involves radiative transfer calculations, but in a 4D scheme we must simulate the observation from the model state at a particular time, so the forward model actually requires time integrations of the forecast model itself. The minimisation of the cost function will of course also involve adjoint calculations of the forecast model to be performed. If we consider the number of observations we are likely to have and the possibility that a significant number of iterations (adjustments) will be required before an acceptable solution is found, we have a potentially enormous numerical problem.

Research into a future implementation of OICVAR is well under way at the ECMWF and initial controlled experiments suggest that with expected advances in computing power over the next few years, the numerical cost of such a scheme will not be prohibitive.

5.3 Future observing systems in NWP

The move towards the direct assimilation of remotely-sensed observations using variational schemes reduces or completely removes many of the problems associated with the use of retrieved products in NWP. However observation systems must try to keep pace with corresponding advances in data assimilation and numerical modelling. The remainder of this section will briefly describe the most recent systems being considered.

5.4 High resolution sounders

While the variational assimilation of radiance measurements exploits the available information to the full, the usefulness of satellite sounders is still fundamentally limited by their vertical resolution. This limitation is primarily due to the technological difficulties of obtaining fine spectral separation and selection of the outgoing atmospheric radiation. Traditionally various types of filters have been used to do this but new instruments are planned (and to some extent already exist) which make use of interferometers. Instead of measuring filtered radiation they obtain an interferogram from which the whole measured spectra (over some range) can be extracted using fourier techniques.

The spectral resolution of such instruments is far in excess of conventional radiometers and weighting functions (introduced in section 4.1) are shown for the HIRS instrument in Figure 12 (Revercombe *et al.* 1988). It can be seen that are huge number of channels exist with very much sharper characteristics than for example the HIRS channels. The enhanced spectral (and therefore the effective vertical resolution) has been investigated by a number of groups (Huang *et al.* 1991) and initial results suggest that interferometer measurements have a significant potential in NWP.

A great deal of effort has been spent on the technological aspects of instrument design and flight feasibility, but there are a number of uncertain areas associated with exactly how the data will be used in NWP. First, it is not clear how the great volume of data available from the interferometer measurements will be handled, both in transmission and processing. There are proposals to reduce the data to a set of pseudo-channels (i.e. combinations of channels) but performing the data compression in an optimal way (i.e. preserving as much of the original information as possible) is difficult.

Secondly, however the data is compressed (if at all) the method that should be used to extract the atmospheric information from the radiances is the source of considerable

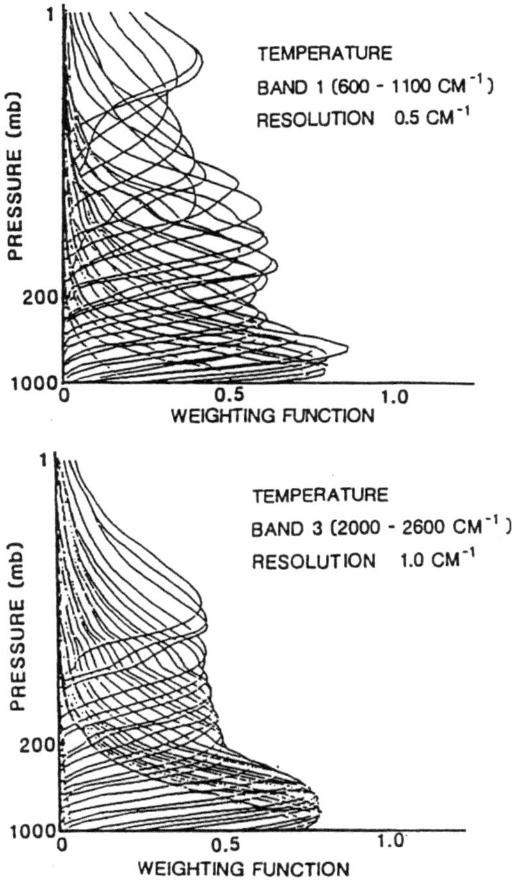

Figure 12: The many thousands of channels (shown in terms of the weighting functions) that are obtained from the simultaneous spectral measurements of an interferometer. Weighting functions of a normal radiometer may be considered as enveloping several interferometer channels due to their poor spectral selection.

debate. The advantages of variational techniques (which are computationally very expensive) are less clear since retrieved products generated explicitly (and easily) from interferometer measurements would require considerably less prior information due to the enhanced vertical resolution in the radiances. Thus many of the problems associated with assimilating the retrieved products in NWP should be reduced. Of course we may reasonably suppose that NWP models will also advance in complexity and sophistication such that what is currently considered a small degree of contamination by prior information will be unacceptable in the future, and variational or at least forecast-guess-based methods could be still be required.

5.5 Lidar wind measurements

Like temperature and humidity, observations of the 3-dimensional atmospheric wind field from conventional sources are poor in resolution and coverage. While numerous ground-based measurements of the surface wind are available, the only reliable source of vertical wind profiles are radiosondes due to the obvious requirement of suitable clouds for cloud-track wind observations (SATOBS). Space-based laser measurements of the atmospheric wind are currently being investigated, relying on the principle that a volume of air moving (i.e. with the wind) will cause a doppler shift of radiation backscattered from it (Bilbro *et al.* 1986). It can be shown that a typical shift of about 200kHz can be observed in the pulse from a CO_2 (10 μm) laser targeted upon a volume of air moving at $1 \mathrm{m\ s}^{-1}$.

It is proposed that the NASA EOS project should carry a Lidar instrument called LAWS. The conical scan pattern of such an instrument is shown in Figure 13. Observing frequency shifts in the backscattered radiation implies a value for the motion of an air parcel along the line of sight (V_{LOS}) which is related to the vector wind components (u, v, w) by the expression

$$V_{LOS} = u\ SIN(\theta)\ COS(\varphi) + v\ COS(\theta)\ SIN(\varphi) + w\ SIN(\varphi) \tag{10}$$

where θ and ϕ are the azimuth and elevation angles respectively (Figure 14). Even if we assume no vertical atmospheric motion (i.e. $w = 0$) it can be seen that to solve uniquely for u and v the wind must be sampled form two different viewing angles. This is achieved by the conically scanning Lidar for a number of points towards the extremes of the swath as the following aft-beam resamples some of the atmosphere sampled by the fore-beam. However the single sampled points still provide useful information on the V_{LOS} component of the wind.

There are number of problem areas in the development of an operational Lidar wind measurement system, both technological and theoretical.

It is questionable if the motion of the volume of air sampled by the Lidar is representative of the atmospheric motions we are actually interested in for NWP. Care must also be taken to ensure the signal due to the atmospheric motions is clearly separated from that due to the relative motion of the platform.

It is not clear to what extent the data obtained form Lidar measurements must be averaged and smoothed to obtain reliable estimates of the wind vector.

It is impossible to obtain Lidar measurements in the region of or below clouds which may leave large areas unobserved.

It is impossible to obtain Lidar measurements in regions of the atmosphere where the aerosol concentration is very low.

Technological difficulties exist with such instruments, for example their large power requirements (in general a problem with all active remote sensing).

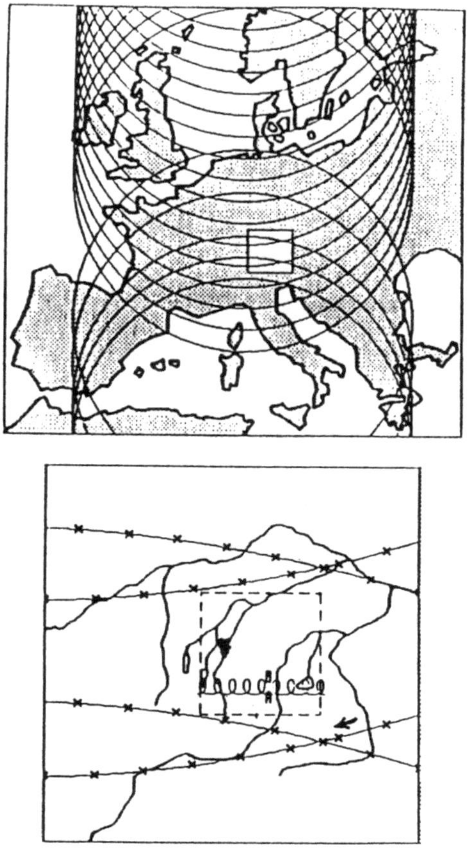

Figure 13: The scanning and sampling of LIDAR wind measurements. Top, surface map of Europe with conical scan pattern overlayed. Bottom, the highlighted section of the map the dashed box is 100km by 100km.

5.6 Active ocean sensors

The coupling of NWP models to corresponding models of the ocean state and circulation is currently under investigation at a number of meteorological centres, as are the possible sources of observational data which might act as input to such schemes. Two such instruments which are both carried by the ERS-1 satellite are mentioned very briefly here.

The radar altimeter targets the ocean surface (at nadir) with a carefully controlled pulse of radiation. By observing the time delay before the pulse returns there is a direct measurement of the mean height of the sea surface below the satellite. Local variations in the height of the sea surface may then be translated into information about the low-level winds causing such changes. This may be done explicitly by some retrieval technique similar to those described earlier, or the information may be assimilated di-

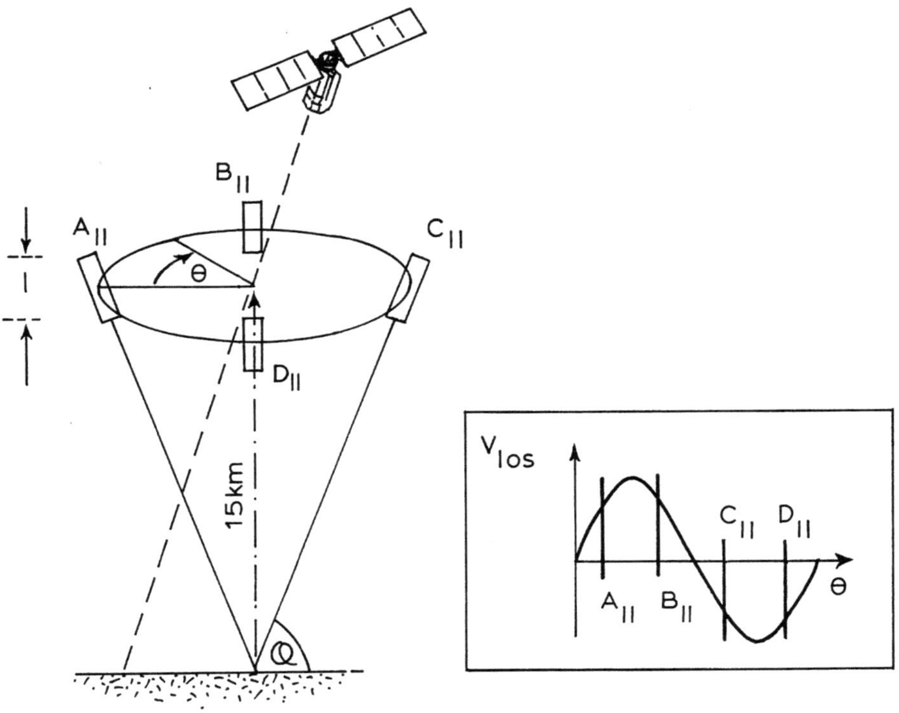

Figure 14: Geometry of LIDAR wind measurements.

rectly into some coupled atmosphere-ocean model. The altimeter has other important uses since the shape of the radar pulse returned from the surface provides a signature of the surface characteristics. Thus the altimeter also provides useful information regarding the sea state (i.e. its roughness) or accurate mapping of sea ice edges.

Also on board ERS-1 is the wind scatterometer which measures a backscattered signal (also radar) from the sea surface (Figure 15). It also relies on the principle that returned signal power is related to the roughness of the ocean surface which is in turn driven by the low level atmospheric wind speed. Measurements are taken simultaneously by three separate beams (fore, mid and aft) in an attempt to determine wind direction as well as speed from the data. Work at ECMWF has been directed towards predicting the measured signal (sigma noughts as they are known) from parameters of the current atmospheric model, that is to say the development of an accurate forward model. This is because the primary use of such data is seen to be in a variational context and preliminary experiments suggest a positive impact upon the forecast quality in some cases.

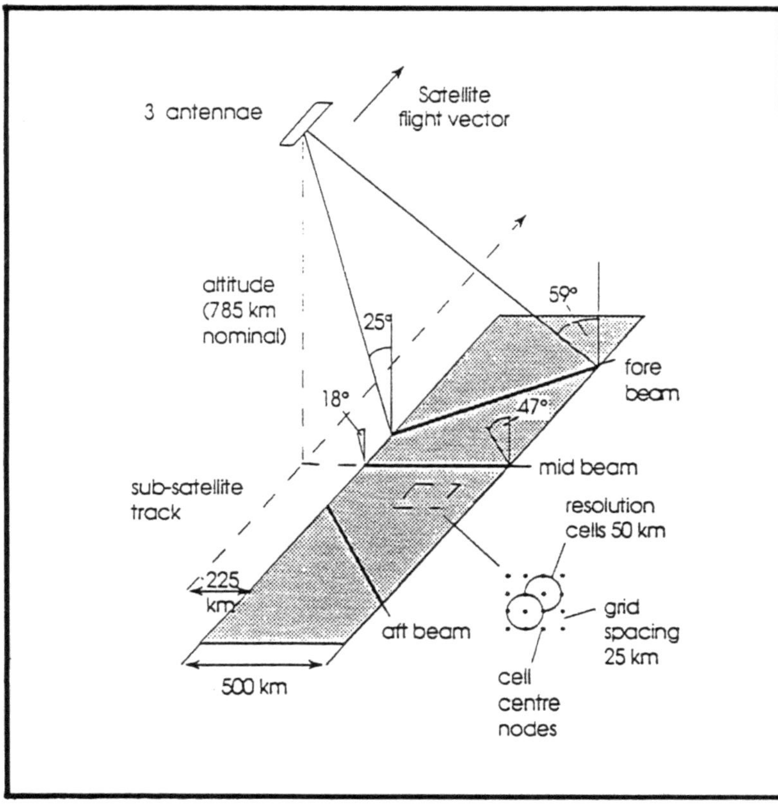

Figure 15: The scanning of the 3 beam scatterometer on the ERS-1 spacecraft.

References

Andersson, E., Hollingsworth, A., Kelly, G., Lonnberg, P., Pailleux, J., and Zhang, Z., 1991, Global observing system experiments on operational statistical retrievals of satellite sounding data. *Monthly Weather Review* , **119** 1851-1864.

Bilbro, J. W., Dimarzio, C., Fitzjarrald, D., Johnson, S., and Jones, W., 1986, Airborne Doppler wind lidar measurements. *Applied Optics*, **25**, 3952-3952.

Hollingsworth, A., 1989, The role of real time four-dimensional data assimilation in the quality control and, interpretation and synthesis of climate data. In *Oceanic Circulation Models: Combining Data and Dynamics* (Dordrecht: Kluwer Academic Publishers), pp. 303-342.

Huang H. J., Smith, W. L., and WOOLF, H. M., 1991 Vertical resolution and accuracy of atmospheric infrared sounding spectrometers. *Journal Applied Meteorology,* **31**, 256-274.

Lorenc, *et al.*, 1985, The analysis of high resolution satellite data by the UK Meteorlogical office. *Proceedings European Centre for Medium-range Weather Forecasting Workshop "High-Resolution Analysis"*, held in Reading, June 1985, 69-87.

Pailleux, 1990, A global variational assimilation scheme and its application for using TOVS data. WMO symposium on *Assimilation of observations in meteorology and oceanography; Clermont-Ferrand*; July 1990. WMO report, 501-505.

Revercombe, H. E., Buijs, H., H. B., Laporte, D. D., Smith, W. L., and Sromovsky, L. A., 1988, Radiometric calibartion of IR fourier spectrometers; Solution to a problem with a high resolution interferometer sounder. *Applied Optics*, **27**, 3210-3218.

Remote Sensing Input to Climatological Datasets

Kendal McGuffie

1 Introduction

Earth-orbiting satellites have provided a wealth of data which has spawned a revolution in the sciences of meteorology and climatology (Rao *et al.* 1990). Through the use of satellites we are able to monitor many aspects of the surface and atmosphere of the Earth. Meteorological satellites have enhanced our understanding of the synoptic processes and now form a routine part of weather information which is distributed to the general public. These satellites have also provided increased understanding of many smaller-scale processes which were not resolved by the surface synoptic network. Aside from these obvious implications for meteorology, satellites have broader implications for the study of large-scale climate dynamics (Ohring *et al.* 1989). Satellites also provide essential information for climate modelling. As global climate models include increasingly complex treatments of the land surface, detailed information on soil moisture distribution, snow cover etc. are required for model validation. Satellite data on clouds provide an important method for evaluating model dynamics together with measured top of the atmosphere radiative fluxes. To enhance our understanding of the climate system, satellite techniques must address many areas of climate research. Some topical examples might be (i) the dynamics of drought (e.g. El Niño), (ii) the monitoring of deforestation, (iii) the monitoring of stratospheric ozone concentrations and the delicate chemical balance of the Antarctic stratosphere and (iv) the potential microphysical changes induced in clouds by the injection of anthropogenic and natural aerosols (such as those produced by volcanoes or dimethyl sulphide (DMS) released from the oceans (Twomey *et al.* 1984 and Somerville and Remer 1984, Charlson *et al.* 1987).

Satellites were first launched to monitor the Earth's environment around 30 years ago but in the past decade, the importance of satellite data in global climate monitoring has increased. There are a number of reasons for this growth. Satellite data is now available on time scales which are long enough to permit realistic evaluation of climatological values and to permit at least preliminary evaluation of likely magnitudes of perturbations which occur presently within the climate system. In parallel with the increased quality of satellite data significant advances in data storage technology have meant that the large amounts of data storage required to analyse satellite data are

NATO ASI Series, Vol. I 24
Remote Sensing and Global Climate Change
Edited by Robin A. Vaughan and Arthur P. Cracknell

now widely available. In addition, data are now routinely distributed on Compact Disc (CD-ROM), usually at little cost to investigators.

The climate problem is being increasingly addressed by current and planned satellite systems. Carleton (1991) examines in detail many of the satellite systems (past and future) which make a contribution to climatology. The future holds great potential for satellite climatology but there remains a wealth of data already gathered, which has yet to be used to its full potential. This chapter examines some of the available data and considers the characteristics which make a satellite-derived dataset of value for climatological studies. Some of the available datasets are considered and their qualities examined although this chapter does not contain an exhaustive study of available datasets.

When satellite data are to be used over an extended period, it is important to recognise the impact of changes in instrumentation between different satellite platforms, likely drift in instrument calibration and errors which are inevitably present in retrieval algorithms. Varying time sampling and spatial sampling biases will also have an influence on the quality of the resulting archive. This chapter concentrates primarily on data for outgoing longwave radiation and of cloud amount although the data from the Microwave Sounder Unit (MSU) are discussed briefly since they provide a relatively long-term dataset on global temperatures suitable for the examination of short term features such as ENSO (EI Niño Southern Oscillation) events. The two top of the atmosphere quantities were chosen because of the potential to compare them with output from global climate models.

2 Microwave Sounder Unit (MSU)

The Microwave Sounder Unit (MSU) (e.g. Spencer *at al.* 1990) has operated on a range of NOAA satellites since 1978. Throughout this time there have been either one or two satellites monitoring the atmosphere. Over the tropics around 40 observations are collected in a 2.5° x 2.5° grid box in each month from each satellite. The number is greater at the pole (around 100) and there are only a few periods over the history of the satellite where the data are from one satellite only. As a consequence of the long time interval of data collection, which includes two important ENSO events the dataset is of considerable value for understanding the global climate system. Stability of the channels between satellite platforms has been determined to be good, with hemispheric brightness temperatures agreeing to within 0.01 °C. On a monthly basis Spencer and Christie found that the reproducibility was better than 0.1 °C making the dataset well suited to studies of phenomena such as ENSO, where typical surface temperature anomalies are of the order of 1-2 °C on a regional basis and 0.2-0.5 °C on a hemispheric or zonal average. Figure 1 shows temperature anomalies from the MSU from Hurrell and Campbell (1992). The surface temperature anomalies associated with the ENSO events in 1982/83 and 1986/87 are clearly visible in Figure 1(a). Figure 1(b) shows the geographical distribution of the temperature anomalies for January 1983. The anomaly in the Tropical East Pacific which is associated with the 1982/83 ENSO event is clearly visible.

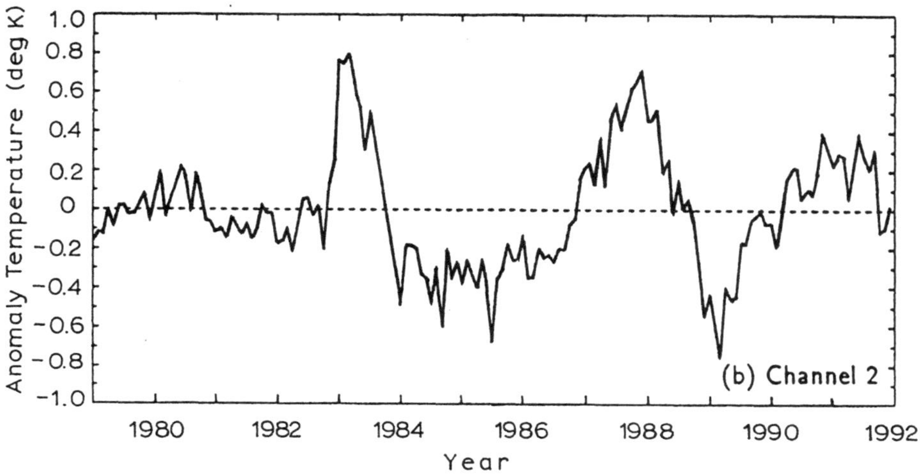

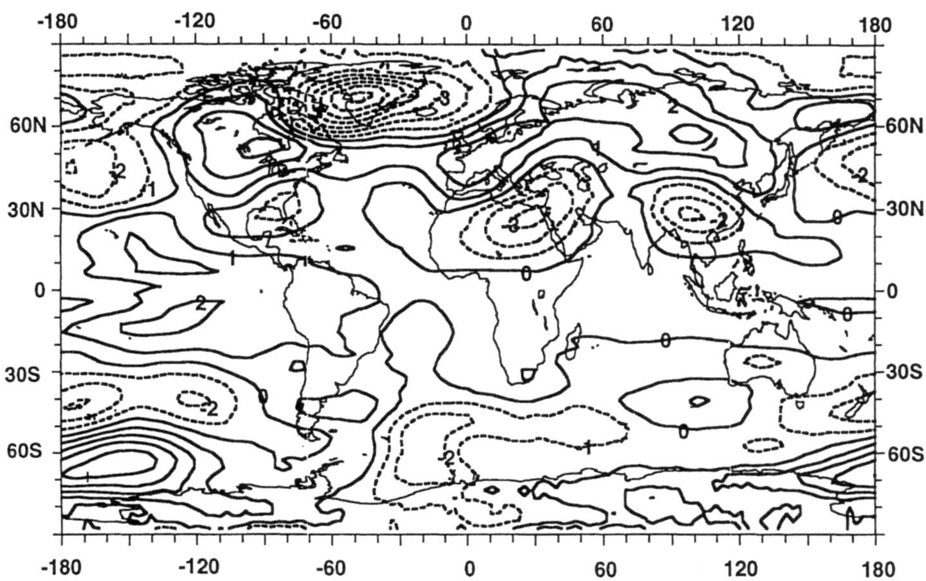

Figure 1: Top, time series of monthly MSU brightness temperature averaged over 20 °N–20 °S for MSU channel 2. Bottom, geographical distribution of temperature anomalies derived from MSU channel 2R for January 1983, showing temperature anomalies in the tropical east pacific Ocean associated with the 1982/83 ENSO event (after Hurrell and Campbell 1992).

3 Cloud Climatologies

Cloudiness is perhaps the most elusive part of the climate system which is monitored from satellites. The short lifetime of clouds requires that observation be frequent, while the three-dimensional structure of the cloud field makes this observation both difficult and crucial for effective analysis of the atmospheric radiation budget. Although it appears straightforward to recognise clouds in satellite imagery, the detailed (automated) analysis of satellite imagery to produce reliable statistics of global cloudiness is fraught with difficulty. Cloud retrieval is difficult over snow and ice areas, where the background brightness in the visible wavelength region is close to the reflectivity of the cloud field. Also, in polar regions, the temperature structure of the atmosphere can be such that the thermal contrast between clouds and surface may be reduced or even reversed, so that automated cloud detection algorithms give unreliable results. Although some satellite-based archives have attempted to overcome some of the drawbacks of satellite cloudiness retrievals by blending in surface observed cloud amounts, this process is also fraught with difficulty because of the differing views of the 3D structure of the global cloud field and the uneven distribution of surface observing stations.

The International Satellite Cloud Climatology Project (ISCCP) was begun in 1983 as the first project of the World Climate Research Program (WCRP). The principal aim of the project was to collect an archive of cloud data (originally over a 5-year period, although this has now been extended) at a resolution which would make the final dataset comparable with the results from climate models. The dataset derived could be used for the validation of climate model results and for the testing of cloud parametrisation schemes within models. The cloud algorithm chosen for ISCCP was the result of a detailed investigation and comparison of the then available cloud retrieval algorithms; it was designed to operate entirely from satellite-derived information (other long term analysis archives such as the USAF RT Nephanalysis use surface-derived data to assist in the satellite analysis. The ISCCP design involved a heirarchy of products which could provide, as well as cloud products, a consistent set of radiances which could provide additional information on cloud optical properties and cloud top height.

ISCCP cloud detection is based on visible and infrared thesholds which are based on time composite imagery of the location in question. A radiative transfer model is then used to derive information on the optical properties and cloud top height. Problems which exist with ISCCP are due to the different temporal sampling of different parts of the globe associated with using both geostationary and polar-orbiter data, spectral differences between different satellites and problems associated with differing view angles of the satellites.

Polar cloud retrieval is a significant problem for all cloud algorithms, but this will potentially be reduced in ISCCP with the inclusion of a 1.6 μm channel data to provide a 3-channel retrieval over polar regions during daylight. Figure 2 shows how the coverage of ISCCP has varied over the time of the project.

The ISCCP is not the only cloudiness product derived from satellite data. The Nimbus-7 climatology covers the period 1979-1985 (Stowe *et al.* 1988). Figure 3 shows the mean cloud amount for 00Z during July 1983 derived from ISCCP and the preliminary Nimbus-7 based ISCCP. The differences are a result of both differing satellite

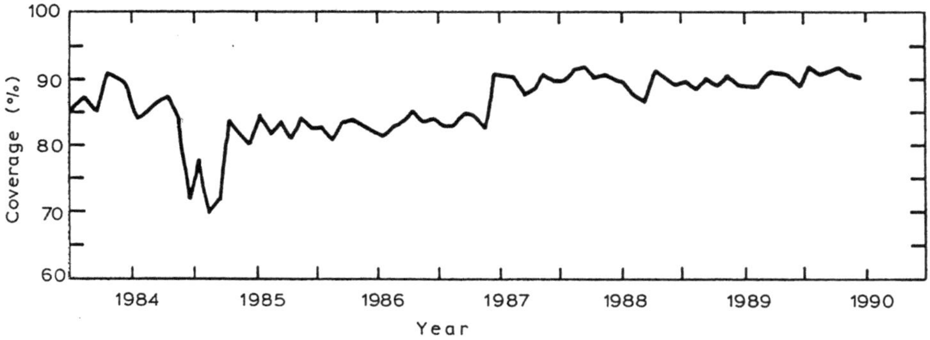

Figure 2: ISCCP coverage of the globe as a percentage of the total global area over the lifetime of the project (reproduced from Hurrel and Campbell, 1992).

sensors and differences in the cloud retrieval algorithms. The modified ISCCP amount is derived using a different threshold for the infrared cloud retrieval, resulting in a smaller value of cloudiness over the oceans.

4 Radiation Budget of the Earth

Operational meteorological satellites sense, typically, in the visible and near-infrared regions of the solar spectrum and in the thermal infrared. The former senses the amount of radiation reflected by the Earth and the latter, the amount of radiation being emitted by the Earth. Careful monitoring of the character of the Earth's radiation budget is of crucial importance for climate studies. In regions where there is very little cloud, the outgoing longwave radiation (OLR) is governed by the surface temperature, whereas in cloudy regions it is the cloud top temperature which determines the amount of radiation which is emitted to space. There is, on average, more radiation absorbed in the tropics than is emitted there, while in the polar regions, a greater amount of radiation is emitted than is absorbed. This imbalance of radiation is the driving force behind much of the climate system, so the nature of the Earth's radiation budget is clearly worthy of detailed study.

The use of satellites to monitor the radiation balance of the Earth is not new. Some of the earliest satellites TIROS, ESSA and NOAA had a radiation budget monitoring capability although more recently the accuracy of systems has increased and available computer power has permitted more comprehensive analysis of the results. Kyle *et al.* (1991) presented a study of the Earth's radiation budget from Nimbus-7 although the atlas included only one year of results; further analysis is in progress (e.g. Hurrell and Campbell 1992).

Table 1 shows a non-exhaustive summary of radiation budget parameters derived from satellites over the past 20 years. It is clear from these statistics that there is no easy answer to the problem of determining the quantitative radiation budget of the Earth's atmosphere.

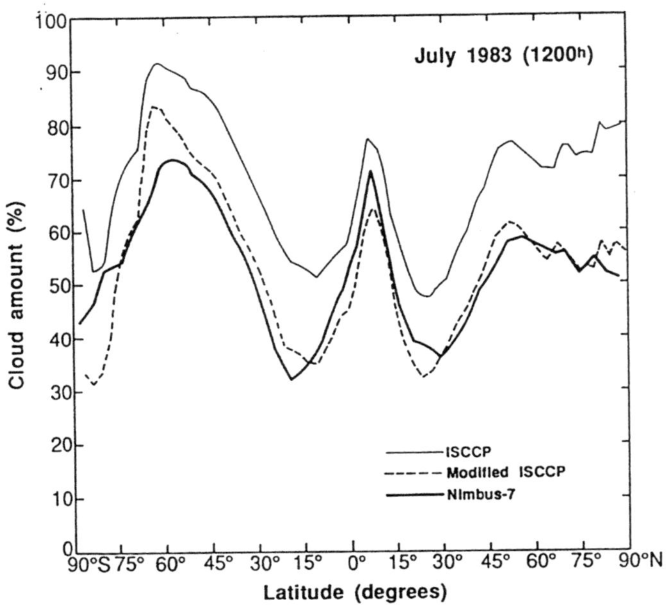

Figure 3: Comparisons of average daytime cloud amounts (zonal) at local noon during July 1983 for the Nimbus-7 preliminary ISCCP and modified ISCCP estimates (from Carleton, 1991)

The Earth Radiation budget experiment (ERBE) is a multi-satellite project which is measuring the broad band components of the Earth's radiation budget to a high degree of accuracy with diurnal resolution. Data are utilised from a number of satellites including a specialised satellite ERBS (Earth Radiation Budget Satellite). ERBS is in a 57° inclined orbit which means that each point on the Earth's surface is sampled at all times of day over a period of roughly 36 days. This kind of integration of data from many satellites is required if good observational datasets are to be compiled from satellite data, however the integration of data from different sources must be undertaken with care, since the different satellites may have temporal sampling biases which must be taken into account and often have differing spectral resolution. Radiation budget statistics can be compiled from narrow wavelength region measurements, but will obviously rely on algorithms to derive wide band statistics, resulting in further potential biases. Hurrell and Campbell (1992) have recently produced a comprehensive collection of datasets of satellite data for use in climate modelling studies and have made the datasets available for use by researchers at the US National Center for Atmospheric Research (NCAR). In their report, Hurrell and Campbell (1992) describe many of the datasets available and undertake some preliminary comparisons of the many datasets. The report is principally a discussion of the datasets archived at NCAR, but would prove valuable even to those without a means of accessing the datasets, since the raw data are available from other sources.

Source	Solar Constant	Solar Insolation	Reflected	Albedo (%)	OLR
Raschke *et al.* (1973)	1360.7	340.18	96.61	28.4	240,7
Nimbus-3 data					
Ellis & VonderHaar (1976)				30	236
Satellite composite					
Nimbus-6 WFOV	1391			31	130
Campbell & Vonder Haar (1980)					
NOAA SR				31	234
Gruber & Winston (1978)					
Nimbus-7 Original					
Jacobowitz *et al.* (1984b)					
WFOV	1371.5	342.88	104.9	30.6	228.8
NFOV			113.5	33.1	232.7
Nimbus-7 Reprocessed					
6/79-5/80 (MATRIX) WFOV					
(MLE) NFOV			104.7	30.55	236.0
2/85-1/86 (MATRIX) WFOV			103.0	20.03	235.0
ERBE (NOAA-9 ERBS)	1356	341.25	102.1	29.92	234.0
2/85-1/86 (5° scanner)					

Table 1: Radiation budget annual global values (Wm^{-2}), adapted from Kyle *et al.* (1991).

5 The Future

The large amount of data which characterises the field of satellite data retrieval is difficult to make available to a wide user community. It is principally this data problem which has prevented the use of satellite-derived data in climate studies. Many datasets are however now becoming available on compact disc since this provides an easy way to distribute large volumes of data to the user community at very little cost (the unit cost is a few dollars for each disk in addition to a cost of around $1000 for the pressing master). This method of data distribution will result in increased data availability and can only result in greater understanding of the radiation climate of the Earth. The use of satellite data for climate studies has reached an important milestone. Datasets of climatological length are just becoming available and are only now becoming readily distributable. The forthcoming decade will see a host of new data becoming available from the new Earth resources satellites which are currently being launched and planned. It will be the challenge of the next 10-20 years to make the best distillation of these data and to add significantly to the quality of our understanding of the Earth-atmosphere system.

Acknowledgments

The lectures presented drew heavily on the work of Jim Hurrell and Garrett Campbell (Hurrell and Campbell, 1992). I thank them for providing an advance copy of their report.

References

Campbell, G. G. C., and Vonder Haar, T. H., 1980, An analysis of two years of Nimbus-6 radiation budget measurements. *Atmospheric Science* Paper No 320, Colorado State University, Fort Collins, Colorado.

Carleton, A. M., 1991, *Satellite remote sensing in climatology* (Boca Raton: CRC Press).

Charlson, R. J., Lovelock, J. E., Andreae, M. O., and Warren, S. E., 1987, Oceanic phytoplankton, atmospheric sulfur, cloud albedo and climate. *Nature*, **326**, 655-661.

Ellis, J., and Vonder Haar, T. H., 1976, Zonal average Earth radiation budget measurements from satellites for climate studies. *Atmospheric Science*. Paper No 340, Colorado State University, Fort Collins, CO.

Gruber, A., and Winston, J. S., 1978, Earth-atmosphere radiative heating based on NOAA scanning radiometer measurements. *Bulletin of the American Meteorological Society*, **59**, 1570-1573.

Hurrell, J. W., and Campbell, G. G. C., 1992, Monthly mean global satellite datasets available in CCM history tape format, NCAR TN-371+STR, National Center for Atmospheric Research, Boulder, Colorado.

Jacobowitz, H., Tighe, R. J., and the Nimbus-7 ERB experiment team, 1984, The Earth radiation budget derived from the Nimbus-7 ERB experiment. *Journal of Geophysical Research*, **89**, 4997-5010.

Kyle, H. L., Hucek, R. R., and Vallette, B. J., 1991, Atlas of the Earth's radiation budget as measured by Nimbus-7: May 1979 - May 1980, NASA Reference publication 1263, NASA, Washington DC.

Ohring, G, Gallo, K., Gruber, A., Planet, W., Stowe, L., and Tarpley, J. D., 1989, Climate and global change, characteristics of NOAA satellite data, EOS (*Transactions of the American Geophysical Union*), bf 70, 889, 894, 901.

Rao, P. K., Holmes, S. J., Anderson, R. K., Winston, J.S., and Lehr, P.E. (Editors), 1990, *Weather satellites: Systems, Data and Environmental Applications* (Boston: American Meteorological Society).

Raschke, E., Vonder Haar, T. H., Bandeen, W. R., and Pasternak, M., 1973, The annual radiation balance of the Earth atmosphere system during 1969-70 from Nimbus- 3 measurements. *Journal of Atmospheric Sciences*, **30**, 341-364.

Somerville, R. C. J., and Remer, L. A., 1984, Cloud optical thickness feedbacks in the CO_2 climate problem. *Journal of Geophysical Research*, **89**, 9668-9672.

Spencer, R. W., Christie R., and Grody, N. C., 1990, Global atmospheric temperature monitoring with satellite microwave measurements: Methods and results 1979-84. *Journal of Climate*, **2**, 671-709.

Stowe, L. L., Wellemeyer, C. G., Eck, T. F., Yeh, H. Y. M., and the Nimbus-7 Cloud data processing team, 1988, Nimbus-7 global cloud climatology: Part I, algorithms and validation. *Journal of Climate*, **1**, 445-470.

Twomey, S. A., Piepgrass, M. & Wolfe, T. L., 1984, An assessment of the impact of pollution on global cloud albedo. *Tellus*, **36B**, 356-366.

The Microwave Limb Sounder and Error Reduction: Is What You See What You Get?

Iain Woodhouse

1 Introduction

There is often some debate between users and producers of atmospheric remote sensing data as to what format the data should actually take. Producers know their own instrument and are probably best suited to provide product data (in the form of maps, plots, or whatever) but what is usually the concern of users, and quite rightly so, is that it is often unclear how much of the product is actually useful. In retrieving atmospheric profiles from satellite measurements it is necessary to include *a priori* information, i.e. information that is already known about the atmosphere. If you are given a profile determined from satellite observations, it is relevant to ask just how much of the information really came from the satellite measurements and how much is just a priori information. To quantify this a term called "error reduction" is defined.

In order to illustrate this problem more clearly it is helpful to look at an example from real satellite data such as from the Microwave Limb Sounder on board the Upper Atmosphere Research Satellite (UARS). But before discussing the data and error reduction, it is worthwhile to have an overview of the instrument.

2 Upper Atmosphere Research Satellite

The $750-million Upper Atmosphere Research Satellite was launched by the space shuttle Discovery on 12 September 1991 and was put into a 600km altitude orbit with a 57° inclination. This orbit permits UARS sensors to observe up to 80 degrees in latitude, thus providing essentially global coverage of the upper atmosphere. The 7,700kg spacecraft is the most expensive and complex environmental satellite ever built and is carrying out the first systematic and comprehensive investigation of the stratosphere as well as providing important new data on the mesosphere and thermosphere.

NATO ASI Series, Vol. I 24
Remote Sensing and Global Climate Change
Edited by Robin A. Vaughan and Arthur P. Cracknell

2.1 Microwave Limb Sounder: The Instrument

One of the most significant instruments carried by UARS is the Microwave Limb Sounder (MLS) which simultaneously measures water vapour (≈15–85 km), ozone (≈15–80 km), chlorine monoxide (≈25–45 km) and pressure (≈30–60 km). ClO measurements are important since this gas is a key reactant in the chlorine chemical cycle that destroys ozone; microwave measurements are essential for observations of this species, and MLS is unique among the UARS instruments in providing microwave sensitivity. The MLS measurements will provide, for the first time, a global data set on ClO in the upper atmosphere.

The instrument has three assemblies: optics, spectrometer and power supply. The optics consists of a 1.6m scanning antenna, which receives thermal radiation from the atmospheric limb, and secondary and tertiary mirrors to redirect the radiation onto a switching mirror. The switching mirror can accept radiation either from the antenna or, for calibration, from an internal target (hot) or a space view (cold).

The spectrometer assembly divides the signal into three separate frequency bands for measurements of pressure (O_2) by the 63 GHz radiometer, O_3 and H_2O by the 183 GHz radiometer, and ClO and O_3 by the 205 GHz radiometer.

The MLS project development was coordinated by the Jet Propulsion Laboratory in the U.S. in collaboration with Heriot-Watt University and Edinburgh Universities and Rutherford Appleton Laboratory in the U.K. The 183 GHz radiometer was built at Heriot-Watt University Physics Department. Waters (1989) gives an overview of the MLS and similar instruments that are currently under consideration.

2.2 Microwave Limb Sounding: The Technique

Unlike all previous satellite-borne microwave radiometers, MLS is a limb-viewing instrument; ie, it looks sideways from the platform toward the limb or edge of the Earth's atmosphere rather than looking directly beneath the satellite (see Figure 1). This technique offers several advantages over downward-looking systems for the study of the upper atmosphere. The principle advantages are:

Cold-Sky Background: Measurements of atmospheric radiation are made with cold space as a background, thus avoiding the ambiguity that occurs when downward-looking radiometers sound the atmosphere against a background of emission and reflection from the Earth's surface.

Improved Vertical Resolution: No part of the signal received at the instrument originates below the tangent point (assuming an infinitesimally small beamwidth) and it can be shown that the path length along the line of sight is greater for the tangent layer than for any other layer. In addition, the density of the species being measured usually decreases rapidly with increasing altitude. The result of these two factors is that most of the radiation measured by the radiometer originates in a narrow vertical layer within about 3 km of the tangent point. This is equivalent to saying that limb-sounding observations are characterized by very narrow weighting functions. Figure 2 illustrates the difference between downward- and limb-looking weighting functions.

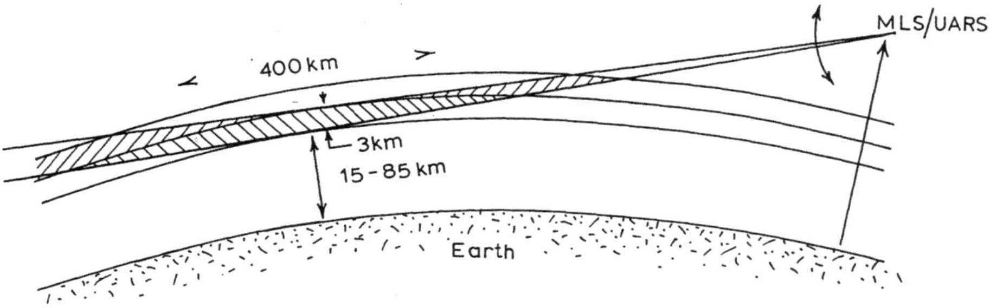

Figure 1: Geometry of the Microwave Limb Sounder. Note that the longest path length is through the tangent layer.

Improved Sensitivity: The long horizontal path length through a given tangent layer also offers the greatest sensitivity to the density of the emitting material present in the layer. This is particularly relevant for measuring high atmospheric layers where density is small and emission weak.

A more in-depth account of microwave limb sounding is given by Ulaby *et al.* (1986).

3 From Radiances to Profiles

Before discussing the final product data from the MLS and error reduction it will be instructive to some readers to have a quick (and therefore simplified) summary of the theory of atmospheric retrieval methods. (For a more complete discussion see Ulaby *et al.* 1986. Houghton *et al.* 1984).

For a given set of radiances measured by the instrument, an infinite number of solutions exist and it is therefore impossible to determine a profile from the radiances alone. What makes the problem solvable is the use of a priori information available about the character of the atmosphere for a given geographical location and given time of year.

The best way to visualise this is to think of two sets, shown in Figure 3. Set A is the set of all possible solutions (which should really be infinite, but that is not easy to visualise). Set R is all the solutions that are possible given the measured radiances. And set C is the set which we know from previous meteorological studies must contain the solution (the a priori information). Thus the actual solution lies somewhere in the intersection of R and C. The a priori information has therefore significantly reduced the range of possible solutions and if we extend this example to consider R and C as probability distributions rather than discrete sets, then there will exist an optimum estimate of the solution, marked 'x' in the figure.

However, if the optimum estimate is taken to be *the* solution, an estimate of the spread of possible solutions should also be considered, ie, a measure of the relative size

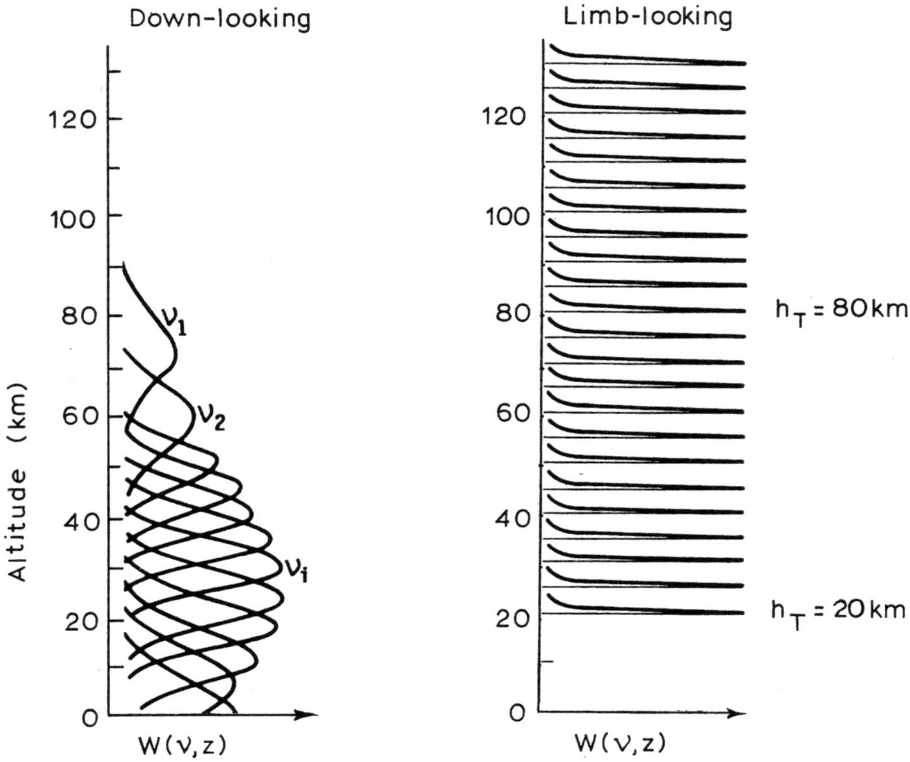

Figure 2: Downward- and limb-looking single-ray weighting functions for a species with constant volume mixing ratios.

of R ∩ C. As an extreme example to illustrate why this is important, consider the possibility that set R includes the whole of set C. In this case the reliability of the optimum estimate would be no better than the *a priori* information itself; the satellite instrument would have contributed nothing to the result.

4 Error Reduction

In order to quantify the contribution of the satellite data to the final solution we use a term called "error reduction" which is defined as the ratio of the standard deviation of the final solution to the standard deviation of the *a priori* information. If the distribution of possible solutions remains the same after using the satellite data then the instrument has given no new information and the error reduction is unity. As the contribution of information from the measured radiances increases, the error reduction approaches zero. An arbitrary value of 0.5 is used as the threshold level, above which the data is considered to be of little value.

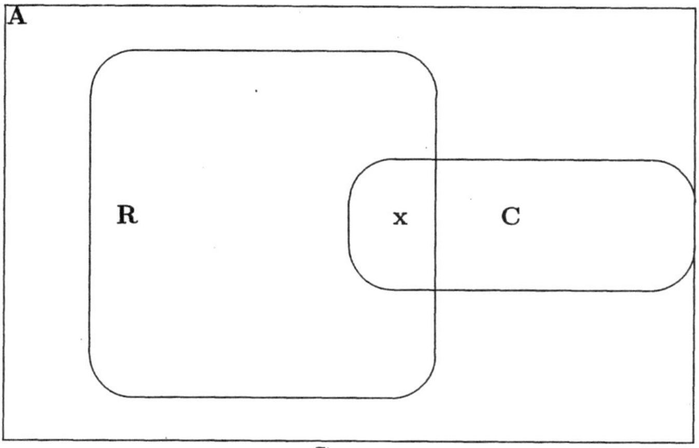

Figure 3: With the use of a priori information (C) it is possible to reduce the number of possible solutions. If we think of these sets as probability distributions rather than discrete sets, then there exists an optimum solution, marked 'x' in the figure.

The Heriot-Watt/Edinburgh MLS group is currently developing the facility to produce plots which show the retrieved concentrations alongside both the error reduction values and the original climatology (the a priori information). An example of one of these plots is shown in Figure 4. This information provides a quick assessment of the reliability of the retrieved data so that it is clear if the data is significant or not.

5 Summary

Data in the form of a ready made product is convenient, but for atmospheric remote sensing it is wise to be aware of both the limitations of the given information and the fact that this limitation can be quantified using a factor such as error reduction.

So, in future, next time you are shown atmospheric retrieval data in the form of a pretty plot remember to check that what you see is really what you get.

Acknowledgements

Thanks are due to Ewan Carr for producing the plots, Gordon Peckham for advice and help and the rest of the MLS team for all their efforts.

References

Houghton, J. T., Taylor, F. W., and Rodgers, C. D., 1984, *Remote Sounding of Atmospheres* (Cambridge: Cambridge University Press).

Ulaby, F. T., Moore, R. K., and Fung, A. K., 1986, *Microwave Remote Sensing: Active and Passive, Vol. III* (Norwood: Artech House).

Waters, J. W., 1989, Microwave Limb-Sounding of Earth's Upper Atmosphere. *Atmospheric Research*, **23**, 391–410.

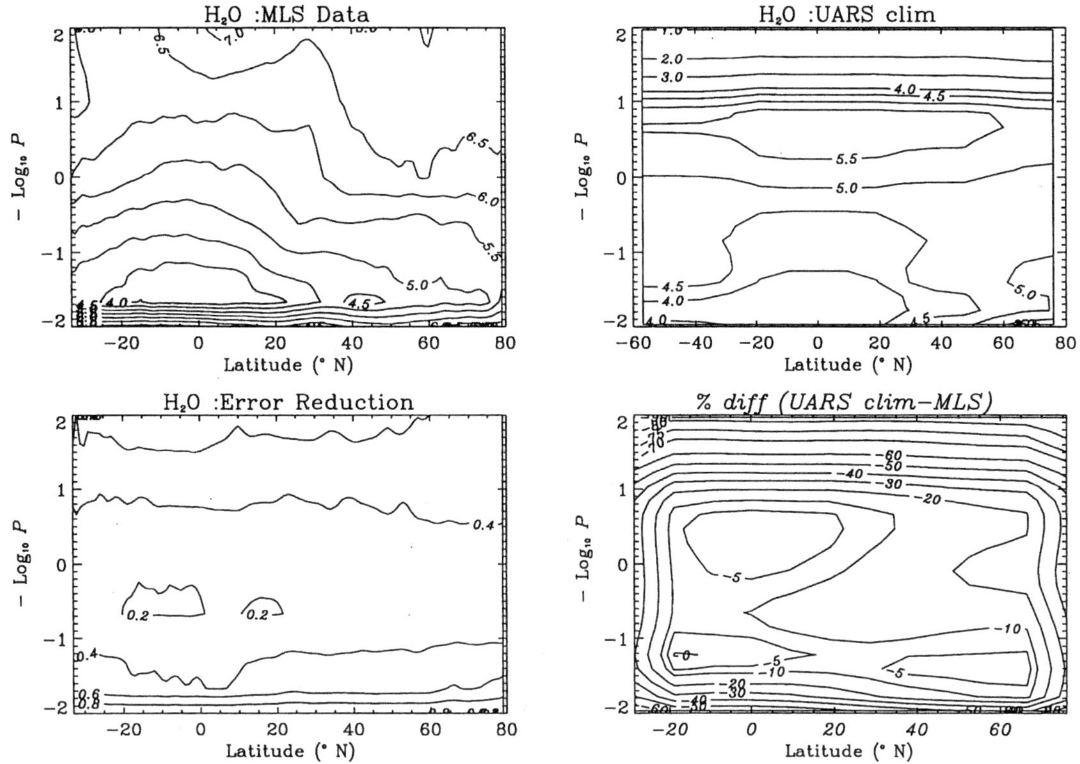

Figure 4: An example of an MLS H_2O retrieval plot showing the species concentration (ppmv), error reduction, climatology (monthly average) and the difference between the measured concentrations and the climatology. Data from 13 March 1993, UARS day 0549: North looking (Note that this is preliminary data and has yet to be verified.)

Atmospheric Trace Gases and Aerosols

S.G. Jennings

1 Atmospheric Trace Gases

1.1 Introduction

The climate of the Earth depends on the balance between the energy absorbed from solar radiation and outgoing radiation emitted in the form of long-wave radiation. Any factor that can perturb this balance and thus potentially alter the climate is called a radiative forcing agent. This balance can be affected by a number of factors such as a change in the output of solar radiation, changes in greenhouse gas concentrations and an increase in aerosol loading in the lower atmosphere.

Most emphasis, to date, has been placed on the potential effects of anthropogenic activities on the Earth's climate. The atmospheric concentrations of carbon dioxide and other radiatively active trace gases such as methane, nitrous oxide, and chlorofluorocarbons (CFCs) have been increasing as a result of an increase in energy production and other anthropogenic activities. These increases will enhance the natural greenhouse effect, which is principally caused by water vapour (66% contribution), carbon dioxide about 25% with the remainder due to methane, nitrous oxide, tropospheric ozone, and the CFCs, with an estimated contribution of about 0.2% resulting on average in an additional warming of the Earth's surface.

Measurements indicate that the rate of increase has been greatest in the more recent decades. Trace gas concentration determines its effectiveness as a radiative gas, as well as its absorptive properties in the long-wave terrrestrial radiation wavelengths.

Tropospheric water vapour is the single most important greenhouse trace gas. Water vapour is responsible for about 2/3 of the total greenhouse effect. The basic global measure of total greenhouse effect is the difference (ΔT) between the mean global surface temperature (T_s) and the equivalent blackbody emission temperature (T_e) of the Earth and the atmosphere.

NATO ASI Series, Vol. I 24
Remote Sensing and Global Climate Change
Edited by Robin A. Vaughan and Arthur P. Cracknell

Often, the distinction between total and changing greenhouse effect is not made. CO_2 contributes about 22% of the total greenhouse effect whilst chloroflurocarbons (CFCs) contribute currently about 0.2%. The concentration of tropospheric water vapour is not directly affected by anthropogenic or man-made emissions.

On assuming that water vapour behaves as an ideal gas (i.e. that equation of state holds for all pressures and temperatures), we can write

$$e = \rho_v R_v T \tag{1}$$

where e: pressure of water vapour
ρ_v: water vapour density
R_v: specific gas constant (specific to water vapour)
T: thermodynamic temperature.

One can express the amount of a trace gas present in the atmosphere in terms of its pressure and density [eq. (1)] or its mixing ratio, r. The mixing ratio, r, can be expressed in the form

$$r = \frac{e}{p-e}\frac{M}{M_d} \tag{2}$$

where p is the pressure of moist air, M is the molecular weight of the trace gas species and M_d is the molecular weight of dry air. The density of a trace gas can be determined from the equivalent (1) relation for the particular trace gas species.

1.2 The Greenhouse Effect

Let us briefly consider the mechanism by which the greenhouse gases warm the atmosphere. The greenhouse effect has been in operation for millenia. In the absence of trace gases, the heating by incoming solar radiation must be balanced by longwave terrestrial radiation emitted from the surface of the Earth.

$$Incoming\ solar\ flux = \frac{Q_\circ}{4} = \frac{1360}{4}\mathrm{Wm}^{-2}$$

If A is the fraction of solar radiation reflected (planetary short-wave albedo), then

$$incoming\ solar\ energy = (1-A)\frac{Q_\circ}{4}$$

emitted outgoing radiation $= \sigma\ T^4$
σ : Stefan's constant = 5.67 x 10^{-8} $\mathrm{Wm^{-2}\ K^{-4}}$.
Earth's mean temperature T_m given by equilibrium between the incoming (1-A) $Q_\circ/4$ and the outgoing σT_m^4, so that

$$T_m = \left[\frac{0.7 \times 340}{5.67 \times 10^{-8}}\right]^{1/4} = (41.98 \times 10^8)^{1/4} = 225 \text{ K}$$

The absorption by the atmosphere of the infrared radiation from the Earth's surface causes the Earth's surface to be some 30 °C warmer than would be if there was no atmosphere. If the concentrations of trace gases that are efficient in trapping the Earth's radiated energy are increased, then the Earth will get warmer. So one should distinguish between the greenhouse effect from non-anthropogenic trace gases and aerosols and the enhanced greenhouse effect from anthropogenic sources of trace gases and aerosols (see also section 5 of chapter 1).

1.3 Trace gas properties

First, we shall examine properties of some trace gases, which include the following:
CO_2 carbon dioxide
CH_4 methane
N_2O nitrous oxide
CCl_3 or F-11: Trichlorofluoromethane
CCl_2F_2 or F-12: Dichlorodifluoromethane
CH_3CCl_3 methyl chloroform
CCl_4 carbon tetrachloride

CO_2 constitutes some 0.035% of the Earth's atmosphere. About 0.003% of the atmosphere is composed of several trace gases, some of which are man-made; although they exist in very low concentrations, they can have a significant effect on the global environment.

Anthropogenic trace gases tend to be concentrated in the northern hemisphere where the human population and land area are concentrated. Trace gases of oceanic origin such as methylchloride (CH_3Cl) and carbonyl sulphide (COS) are more evenly distributed over the globe because of the more even distribution of ocean area in the hemispheres. Many processes - usually referred to as sinks- contribute to the removal of trace gases. Removal processes include the oceans, soils, dust storms, clouds, precipitation and photochemical and chemical reactions in the atmosphere. In the troposphere, several trace gases - particularly those with carbon-hydrogen bonds - are removed by reacting with hydroxyl radicals (OH). The chemistry of tropospheric OH is complex and concentrations reported so far are subject to considerable uncertainty (Logan *et al.* 1981). Trace gases such as carbon monoxide and methane react with the hydroxyl radicals and are thereby removed.

Trace gases that are produced in large urban centres and from more extensive agricultural lands can be displaced through transport into non-urban tropospheric regions. The concentrations of several trace gases are generally on the increase. For example the concentrtion of atmospheric methane at Point Barrow, Alaska, has increased from 1697.5 ppbv in 1983 to 1733.0 ppbv in 1985 - a 2.1% over that period (Steele *et al.* 1981).

An Atmospheric Lifetime Experiment (ALE) utilised five globally distributed coastal measurement stations to determine accurate concentrations of the four halocarbons $CFCl_3$, CF_2Cl_2, CCl_4 and CH_3CCl_3 and also N_2O: (1) Adrigole, Ireland, (2) Cape Meares, Oregon, (3) Ragged Point, Barbados, (4) Point Matatula, American Samoa, and (5) Cape Grim, Tasmania (Prinn *et al.* 1983). The measurement programme began in 1978, and is now superseded by the Global Atmospheric Gas Experiment GAGE (Prinn *et al.* 1987). Mace Head, on the west coast of Ireland has now taken the place of Adrigole since 1987. Increases of 5.7, 6.0, 8.7, 1.8 and 0.2% at the midpoint of a three year period (1978-1981) were found for $CFCl_3$, CF_2Cl_2, CH_3CCl_3, CCl_4 and N_2O (Prinn *et al.* 1983).

The best known example is the well documented increase of CO_2 measured over three decades, (Keeling *et al.* 1976; Clark, 1982; National Academy of Sciences (NAS), 1983; Keeling *et al.* (1989). The Mauna Loa record shows a 12% increase in the mean concentration in 32 years from 315.83 parts per million by volume of dry air (ppmv) in 1959 to 353.95 ppmv in 1990.

We present a brief summary of the properties of the main anthropogenic trace gases:

(i) Carbon Dioxide

The annual average concentration of CO_2 is relatively homogeneous throughout the troposphere. The ocean acts as the largest exchange pathway for CO_2. Vegetation and soils also play an important role in the exchange of CO_2: photosynthesis (takeup of CO_2 by vegetation) and respiration (production of CO_2) by plants and conversion of organic matter back into CO_2 in soils. There is approximately a balance between sources and sinks in the unperturbed habitat. CO_2 concentration in the atmosphere is increased by fossil fuel consumption increases and also through deforestation. Atmospheric CO_2 shows a seasonal cycle dominated by the seasonal uptake and release of atmospheric CO_2 by land plants and vegetation.

However even after 30 years of CO_2 measurements in the atmosphere and the oceans, the global atmospheric CO_2 budget is still uncertain. There is still not confirmation of the amount of CO_2 removed from the atmosphere by the oceans (Trabalka 1985) - one reason being that the expected increases are small compared to the natural variability. For example if half of the cumulative fossil fuel CO_2 emitted since 1850 were distributed in the thermocline (approximately the uppermost km of the ocean) the total inorganic carbon dissolved would increase by only 1%.

(ii) Methane

This is the most abundant atmospheric hydrocarbon. Methane is a radiatively active trace gas whose main sources of production are: anaerobic (oxygen deficient) decay of organic matter in rice paddies, natural wetlands, coal mining, natural gas and oil production and biomass burning. It is primarily removed by reaction with hydroxyl radicals (OH) in the troposphere and also by soils. Past concentrations of atmospheric CH_4 have been determined by analysing air bubbles trapped in ice from Greenland and Antarctica. Concentration measurements are made using gas chromatography with a flame ionization detector. Atmospheric CH_4 concentrations during the past 160,000 years are shown in Figure 1.

Atmospheric concentrations of CH_4 have been measured directly since 1978 at which time the globally averaged value was 1.51 ppmv (Rasmussen and Khalil 1981). Its value is now about 1.7 ppmv, i.e. it is increasing at a rate of about 1% per year.

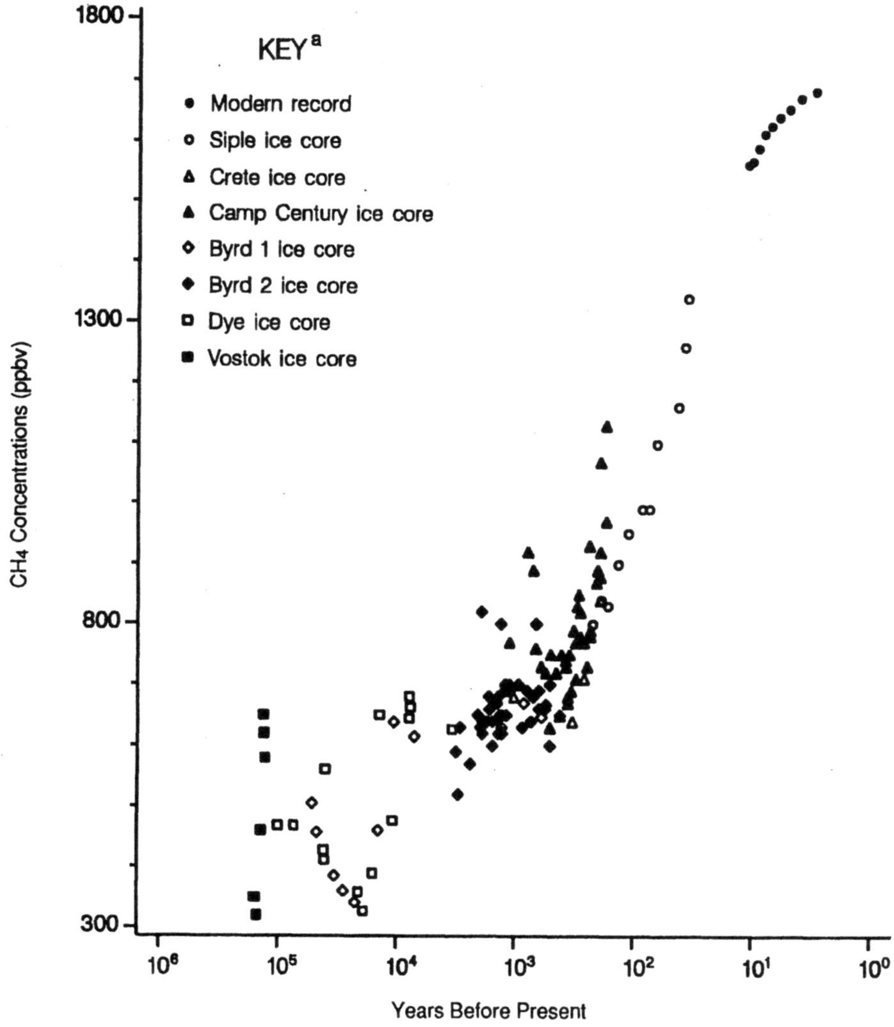

Figure 1: Atmospheric CH_4 concentrations during the past 160,000 years [derived from ice cores drilled at Vostok (Raynaud *et al.*), Bryd Station (1-Stauffer *et al.*; 2-Rasmussen and Khalil), Dye (Stauffer *et al.*), Crête (Rasmussen and Khalil), and Camp Century (Rasmussen and Khalil); and a modern record of global monthly averages derived from seven monitoring sites (Khalil and Rasmussen)].

The atmospheric lifetime for CH_4 has been estimated between 8 and 11.8 years, (Prinn *et al.* 1987). Current CH_4 concentrations are now more than double the pre-industrial (1750-1800) value of about 0.8 ppmv. Ice-core records show that CH_4 concentration were about 0.35 ppmv during glacial periods and increased in phase with temperature during glacial-interglacial periods.

Halocarbon	Concentration (pptv)	Lifetime (years)	Annual Increase rate(%)
$CFCl_2$(CFC-11)	280	65	4
CF_2Cl_2(CFC-12)	484	130	4
$CFCl_2CF_2Cl$(CFC-113)	60	90	10
CH_3CCl_3	158	7	4
CCl_4	146	50	1.5

Table 1: Halocarbon Concentrations and Trends (Watson *et al.* 1990).

(iii)Other trace gases

Some atmospheric trace gases, though less abundant than either CO_2 or CH_4 absorb terrestrial radiation and may thus contribute to global warming. Such trace gases are some man-made halocarbons including:

(a) $CFCl_3$ (CFC-11), used mainly as a propellant in aerosol sprays and as a blowing agent in foam products,

(b) CF_2Cl_2 (CFC-12), used as an aerosol propellent and as a coolant in refrigerators.

(c) $CFCl_2CF_2Cl$ (CFC-13), used as a cleaning solvent in the electronics industry.

(d) CH_3CCl_3 (methylchloroform) used as an industrial solvent.

Real time concentrations of the above gases (together with nitrous oxide, N_2O) are measured continuously at 1 hour intervals at the GAGE network sites, using gas chromatography with twin electron capture detectors. Significant correlations between the principal man-made halocarbons were observed during periods of substantial regional pollution by Simmonds and Derwent (1991) at Mace Head on the west coast of Ireland. Mean atmospheric concentrations and lifetimes of the more abundant halocarbons which are radiatively active are shown in Table 1.

Annual rates of increase are also given in Table 1. There is no major sink for the halocarbons in the troposphere and they are principally removed in the mid-upper stratosphere by photodissociation.

(iv) Nitrous oxide, N_2O

Nitrous oxide is the most important nitrogen-containing chemical from a climatic viewpoint. It is now accepted that N_2O has a very long atmospheric lifetime (> 100 years) with stratospheric photolysis the only known removal process. Sources of nitrous oxide include the ocean, due to microbial production. Other sources are attributed to nitrification under anaerobic conditions in soils and to denitrification in aerobic soils. However the quantification of N_2O sources is still open to question.

The present concentration of N_2O is 310 ppbv, representing an annual increase of between 0.2 - 0.3 % over the past decade. We note that N_2O was only discovered in the atmosphere in 1938.

1.4 Radiative effects of trace gases

The potential climatic effect of radiatively active trace gases has received most attention for CO_2. Numerous 1-, 2-, and 3- dimensional climate model calculations have estimated surface warning due to a doubled CO_2 amount (see chapter 6).

Cubasch and Cess (1990) summarise the predicted equilibrium surface temperature change ΔT_s from 22 general circulation models (GCMs) in 2 x CO_2 equilibrium experiments. Values of ΔT_s range from 1.9 to 5.2 °C.

The direct radiative effect of trace gases at terrestrial radiation wavelengths can be assessed for individual gas constituents on knowing the absorption band strength, its band width (and centre) and the absorber amount. This band treatment has been employed in a one-dimensional radiative-convective model by Ramanathan *et al.* (1985) to estimate the effect of trace gases on surface temperature. The CFCs and other trace gases are assumed to be uniformly mixed from the surface up to 12 km which is consistent with observation.

A computed surface temperature increase of 0.52 deg K is obtained from the effect of infrared CO_2 trace gas level increase for the period from 1880 up to 1980. This is enhanced by 50% (to ΔT_s = 0.79 deg K) due to the inferred changes in the other radiatively active trace gas levels.

The implication of this result is that from pre-industrial times to the present day the increase in CO_2 and the other trace gases might very likely have caused a warming of about 0.8 degK. However, the inherent variability of the climate system appears to obscure any enhanced greenhouse warning signal to date.

A sensitivity study by Ramanathan *et al.* (1985) shows a surface temperature increase of the order of 0.1 degK due to a hypothetical increase from 0 - 1 ppbv in CFC-11, CFC-12 concentration. The same study predicts an increase in surface temperature of 0.55 degK for a CFC increase from 0 to 2 ppbv. One might point out that current levels of CFC-11, CFC-12 are 0.28 and 0.48 ppbv.

The relative contribution of selected trace gases to radiative forcing expressed as a change in energy flux ΔF (Wm^{-2}) (corresponding to a change in concentration, ΔC) is examined by Shine *et al.* (1990).

Values of $\Delta F/\Delta C$ relative to CO_2 and CFC-11, are shown in Table 2 The order of uncertainty in ΔF - because of the inaccuracy (± 10 - 20%) in the spectroscopic data for many of the gases, is of the same order. It can be seen that some of the potential CFC substitutes - the hydrochlorofluorocarbons (HCFCs), and the hydroflurocarbons (HFCs) are strong absorbers of infrared radiation and will contribute as much, to global warming as the chloroflurocarbons if not more.

Trace Gas	$\Delta F/\Delta C$ per unit mass relative to CO_2	$\Delta F/\Delta C$ per unit mass relative to CFC-11
CH_4	58	–
N_2O	206	–
CFC-11	3970	1
CFC-12	5750	1.45
CFC-13	3710	0.93
CH_3CCl_3	900	0.23
CCl_4	1640	0.45
Some possible CFC substitutes		
HCFC-22	5440	1.36
HCFC-123	2860	0.72
HFC-125	4920	1.24
HFC-134a	4130	1.04

Table 2: Radiative forcing of a number of CFCs, possible CFC substitutes relative to CO_2 and CFC-11, per unit mass change in the atmosphere for present day concentrations. Source: Shine *et al.* (1990).

2 Atmospheric Aerosols

2.1 Introduction

There is now a growing awareness that, in addition to the well-publicised influence of carbon dioxide and other greenhouse gases on the warming of the Earth's atmosphere, aerosol particles may also play an important role in forcing climate change. In view of the increasing interest in global climate, the wider use of satellite systems for global measurement of some climatic parameters and the rapid development of computer capabilities to improve climatic numerical models, it is opportune to review, update, and assess aerosol-related climatic effects.

An assessment of both the direct and indirect impacts of aerosols on climate requires knowledge of the microphysical, optical properties (at solar and terrestrial middle infrared wavelengths) and chemical constituents for the major types - both natural and anthropogenic. Ideally, knowledge of these aerosol properties should be known regionally and as a function of altitude.

Atmospheric acerosol particles come from many sources such as human activity, photochemical processes, gas-to-particle conversion, wind erosion, and volcanic and marine production. Consequently, there is considerable compositional variation of aerosol particles, depending on local sources and sinks, meteorological conditions and geographical locale. Atmospheric aerosols generally consist of several modes, each mode characterised by one or more dominant constituents (and hence differing refractive indices) and their particle size ranges.

Mode	Radius Range (μm)
Nucleation (Transient)	0.001 - 0.1
Accumulation	0.1 - 1.0
Coarse	> 1.0

Table 3: Atmospheric Aerosol modes and radius ranges.

Once formed, the aerosols may undergo different changes caused by, for example, chemical reactions and processes such as diffusion, coagulation, sedimentation, evaporation and nucleation. All aerosol particle exists either as a single constituent or as a mixture of components. An internally mixed particle consists of components that are physically or chemically combined. In an external mixture, the particle exists as an aggregate of distinct particles without interacting and so the properties of the mixture are sums of the properties of the individual components.

Human activity is leading to an increase in aerosol in the lower atmosphere. The aerosol particulate matter suspended in the atmosphere is likely to play a significant role in affecting climate because it affects the system's radiative balance both directly and indirectly in several ways. The direct radiative effect is due to the fact that aerosol particles scatter, absorb, and emit electromagnetic radiation. This results in the redistribution of energy from solar and terrestrial radiation, causing atmospheric heating or cooling, which is dependent on the microphysical and optical properties, chemical constituents of the aerosol and the spatial distribution of the particles.

Aerosols are ubiquitous in the atmosphere. The atmospheric aerosol is produced by two main mechanisms : gas to particle conversion and by disintegration of solid or liquid material. Aerosol particles are removed by the combined action of Brownian diffusion, phoretic forces of thermophoresis and diffusiophoresis, inertial impaction and electrical forces. In the case of relatively large sized particles (radius > 0.5 μm), the effect of Brownian diffusion is negligible. For relatively small aerosol particles (radius < 0.5 μm), the effect of inertial impaction can be neglected.

2.2 The Natural Atmospheric Aerosol

2.2.1 Aerosol size distribution

The aerosol size distribution is considered the most prominent physical property of the aerosol. Aerosol particles possess a wide dynamic range in size from 10^{-3} μm in radius up to greater than 10^{-2} μm. The smallest particle size is defined by the condensation nucleus counter whilst the upper limit is determined by the sedimentation fall out of the particle. The aerosol size range is usually subdivided into three modes given in Table 3.

Because of the wide range in aerosol size distribution - which covers more than 5 orders of magnitude in radius and up to 12 or more orders of magnitude in number concentration, it is often more convenient to express the size distribution in logarithmic form. It has been shown that the aerosol size distribution can be represented (Davies 1974) by a combination of one or more log-normal distributions described by

$$\frac{dn}{d\ln r} = \frac{1}{2.303}\frac{dN}{d\log r} = \frac{N}{\sqrt{2\pi}\ln\sigma_g} exp\left[-\frac{1}{2}\frac{\ln(r/r_g)}{\ln\sigma_g}\right]^2 \tag{3}$$

where N is the total particle number concentration per unit volume, r_g is the geometric mean radius, $(\ln r_g = \ln r)$ and ln σ_g is the standard deviation of ln r, so that $(\ln\sigma_g)^2 = (\ln r - \ln r_g)^2$.

Such a wide range in ambient aerosol particle size can only be handled by several size measuring probes operating simultaneously. However most sizing instrumentation does not have continuous measurement capability - aerosol collection over relatively long time periods is often necessary, for subsequent analysis. Therefore the attainment of the complete natural aerosol size distribution is relatively rare.

It was first found by Junge (1953, 1963) that some aerosol size distributions conform to a power law distribution given by the expression

$$n(r) = \frac{dN}{dlogr} = cr^{-\beta} \tag{4}$$

for r > 0.1 μm. If $\beta = 3$, the corresponding logarithmic volume distribution is constant for the particular size range.

The fitting of potential log-normal distributions is achieved through the method described by Davies (1974). Alternatively, a cumulative distribution expressed in percentage format when plotted on log-probability paper will be linear if the distribution is log-normal.

The work of Whitby *et al.* (1975) demonstrated that atmospheric aerosol size distributions are almost always bimodal in surface or volume and are frequently trimodal near sources of fresh combustion aerosols. A good representation of the number size distribution, the surface size distribution and the volume size distribution is given by Jaenicke (1988) in Table 4. The natural atmospheric aerosol can be characterised in the following way (the International Global Aerosol Program (IGAP) plan): desert aerosol, remote continental aerosol, marine aerosol, polar aerosol, mid trospheric - upper tropospheric aerosol (also termed background aerosol), biogenic aerosol, cloud processed aerosol, biomass burning aerosol and stratospheric aerosol. These divisions reflect the source-derived aerosols as well as aerosol types which are characteristic of particular geographical locales.

2.2.2 Tropospheric Aerosols

The tropospheric aerosol represents the aerosols within the troposphere above the boundary layer. Most earlier measurements have been carried out at mountain stations, while more recent measurements have been taken on board aircraft or through the use of tethered balloons. The tropospheric aerosol is modified from the rural model

Size ranges Junge's types Whitby's modes		Subrange: $10^{-7}<r<10^{-5}$ cm Aitken Nucleation			Subrange: $10^{-5}<r<10^{-4}$ cm Large Accumulation			Subrange: $r>10^{-4}$ cm Giant Coarse			Complete size range: $r>10^{-7}$ cm Integrals		
Aerosol parameter	i	n_i [cm^{-3}]	R_i [cm]	$\log\sigma_i$	n_i [cm^{-3}]	R_i [cm]	$\log\sigma_i$	n_i [cm^{-3}]	R_i [cm]	$\log\sigma_i$	N [cm^{-3}]	S [cm^{-1}]	V
Integrals for subranges	Γ'	N' [cm^{-3}]	S' [cm^{-1}]	V'	N' [cm^{-3}]	S' [cm^{-1}]	V'	N' [cm^{-3}]	S' [cm^{-1}]	V'			
Polar	1	$2.17\cdot10^{1}$	$6.89\cdot10^{-6}$	0.245							$2.17\cdot10^{1}$	$2.45\cdot10^{-8}$	$1.24\cdot10^{-13}$
	2				$1.86\cdot10^{-1}$	$3.75\cdot10^{-5}$	0.300				$1.86\cdot10^{-1}$	$8.54\cdot10^{-9}$	$3.52\cdot10^{-13}$
	3							$3.04\cdot10^{-4}$	$4.29\cdot10^{-4}$	0.291	$3.04\cdot10^{-4}$	$1.73\cdot10^{-9}$	$7.56\cdot10^{-13}$
Integrals	Γ'	$1.63\cdot10^{1}$	$8.01\cdot10^{-9}$	$1.94\cdot10^{-16}$	$5.53\cdot10^{0}$	$2.13\cdot10^{-8}$	$1.98\cdot10^{-12}$	$1.53\cdot10^{-3}$	$5.92\cdot10^{-9}$	$1.02\cdot10^{-12}$	$2.19\cdot10^{1}$	$3.47\cdot10^{-8}$	$1.23\cdot10^{-12}$
Background	1	$1.29\cdot10^{2}$	$3.60\cdot10^{-7}$	0.645							$1.04\cdot10^{2}$	$1.73\cdot10^{-8}$	$5.12\cdot10^{-13}$
	2				$5.97\cdot10^{1}$	$1.27\cdot10^{-5}$	0.253				$5.97\cdot10^{1}$	$2.39\cdot10^{-7}$	$2.36\cdot10^{-11}$
	3				$6.35\cdot10^{-1}$	$2.59\cdot10^{-5}$	0.425				$6.35\cdot10^{-1}$	$3.63\cdot10^{-8}$	$3.44\cdot10^{-12}$
	Γ'	$1.03\cdot10^{2}$	$1.84\cdot10^{-8}$	$4.55\cdot10^{-14}$	$4.19\cdot10^{1}$	$2.44\cdot10^{-7}$	$2.56\cdot10^{-12}$	$7.71\cdot10^{-3}$	$3.20\cdot10^{-8}$	$3.71\cdot10^{-12}$	$1.65\cdot10^{2}$	$2.92\cdot10^{-7}$	$6.31\cdot10^{-12}$
Maritime	1	$1.33\cdot10^{2}$	$3.90\cdot10^{-7}$	0.657							$1.09\cdot10^{2}$	$2.47\cdot10^{-8}$	$9.66\cdot10^{-13}$
	2				$6.66\cdot10^{1}$	$1.33\cdot10^{-5}$	0.210				$6.66\cdot10^{1}$	$2.36\cdot10^{-7}$	$1.88\cdot10^{-12}$
	3				$3.06\cdot10^{0}$	$2.90\cdot10^{-5}$	0.396				$3.06\cdot10^{0}$	$1.71\cdot10^{-7}$	$1.32\cdot10^{-11}$
	Γ'	$1.25\cdot10^{2}$	$1.90\cdot10^{-9}$	$4.76\cdot10^{-14}$	$5.26\cdot10^{1}$	$2.90\cdot10^{-7}$	$3.12\cdot10^{-12}$	$2.84\cdot10^{-1}$	$1.22\cdot10^{-7}$	$1.29\cdot10^{-11}$	$1.78\cdot10^{2}$	$4.32\cdot10^{-7}$	$1.60\cdot10^{-11}$
Remote continental	1	$9.97\cdot10^{3}$	$1.00\cdot10^{-7}$	0.328							$5.05\cdot10^{3}$	$3.67\cdot10^{-10}$	$5.37\cdot10^{-17}$
	2	$8.42\cdot10^{3}$	$2.18\cdot10^{-6}$	0.505							$8.39\cdot10^{3}$	$7.51\cdot10^{-7}$	$1.60\cdot10^{-11}$
	3							$7.10\cdot10^{-4}$	$6.24\cdot10^{-4}$	0.277	$7.10\cdot10^{-4}$	$7.83\cdot10^{-9}$	$4.51\cdot10^{-12}$
	Γ'	$1.26\cdot10^{3}$	$1.19\cdot10^{-6}$	$2.42\cdot10^{-13}$	$8.11\cdot10^{1}$	$5.13\cdot10^{-7}$	$6.59\cdot10^{-12}$	$4.37\cdot10^{-1}$	$1.35\cdot10^{-7}$	$1.39\cdot10^{-11}$	$1.34\cdot10^{3}$	$7.60\cdot10^{-7}$	$2.05\cdot10^{-11}$
Desert dust storm	1	$7.26\cdot10^{2}$	$1.00\cdot10^{-6}$	0.247							$3.69\cdot10^{2}$	$1.53\cdot10^{-10}$	$1.25\cdot10^{-17}$
	2	$1.14\cdot10^{3}$	$1.88\cdot10^{-6}$	0.770							$1.08\cdot10^{3}$	$2.72\cdot10^{-5}$	$3.51\cdot10^{-8}$
	3							$1.78\cdot10^{-1}$	$1.08\cdot10^{-3}$	0.438	$1.78\cdot10^{-1}$	$2.00\cdot10^{-5}$	$8.49\cdot10^{-8}$
	Γ'	$1.26\cdot10^{3}$	$1.28\cdot10^{-7}$	$2.74\cdot10^{-13}$	$1.85\cdot10^{2}$	$2.52\cdot10^{-6}$	$4.68\cdot10^{-11}$	$1.47\cdot10^{1}$	$4.43\cdot10^{-5}$	$1.35\cdot10^{-7}$	$1.45\cdot10^{3}$	$4.72\cdot10^{-5}$	$1.20\cdot10^{-7}$
Rural	1	$6.65\cdot10^{3}$	$7.39\cdot10^{-7}$	0.225							$6.65\cdot10^{3}$	$7.81\cdot10^{-8}$	$3.76\cdot10^{-14}$
	2	$1.47\cdot10^{3}$	$2.69\cdot10^{-6}$	0.557							$1.46\cdot10^{3}$	$3.59\cdot10^{-7}$	$1.95\cdot10^{-11}$
	3	$1.99\cdot10^{3}$	$4.19\cdot10^{-6}$	0.266							$1.99\cdot10^{3}$	$9.30\cdot10^{-7}$	$3.32\cdot10^{-12}$
	Γ'	$8.61\cdot10^{3}$	$6.44\cdot10^{-8}$	$1.23\cdot10^{-12}$	$1.71\cdot10^{3}$	$5.80\cdot10^{-7}$	$5.14\cdot10^{-12}$	$3.44\cdot10^{-1}$	$1.42\cdot10^{-7}$	$1.65\cdot10^{-11}$	$8.79\cdot10^{3}$	$1.37\cdot10^{-6}$	$2.29\cdot10^{-11}$
Urban	1	$9.93\cdot10^{4}$	$6.51\cdot10^{-7}$	0.245							$9.93\cdot10^{4}$	$9.99\cdot10^{-7}$	$4.81\cdot10^{-13}$
	2	$1.11\cdot10^{3}$	$7.14\cdot10^{-7}$	0.666							$1.00\cdot10^{3}$	$7.84\cdot10^{-7}$	$6.43\cdot10^{-11}$
	3	$3.64\cdot10^{4}$	$2.48\cdot10^{-6}$	0.337							$3.64\cdot10^{4}$	$9.38\cdot10^{-6}$	$3.49\cdot10^{-11}$
	Γ'	$1.35\cdot10^{5}$	$6.72\cdot10^{-8}$	$1.12\cdot10^{-11}$	$1.41\cdot10^{3}$	$4.20\cdot10^{-6}$	$3.00\cdot10^{-11}$	$7.61\cdot10^{-1}$	$3.53\cdot10^{-7}$	$5.81\cdot10^{-11}$	$1.37\cdot10^{5}$	$1.12\cdot10^{-5}$	$9.87\cdot10^{-11}$
Stratosphere (20 km)	1				$4.49\cdot10^{0}$	$2.17\cdot10^{-5}$	0.248				$4.49\cdot10^{0}$	$5.10\cdot10^{-9}$	$8.34\cdot10^{-13}$
	2												
	3												
	Γ'	$4.07\cdot10^{-1}$	$3.36\cdot10^{-10}$	$9.54\cdot10^{-16}$	$4.10\cdot10^{0}$	$4.76\cdot10^{-8}$	$6.97\cdot10^{-13}$	$1.78\cdot10^{-2}$	$3.32\cdot10^{-9}$	$1.44\cdot10^{-13}$	$4.49\cdot10^{0}$	$5.10\cdot10^{-8}$	$8.34\cdot10^{-13}$

*The particle number concentration in each log-normal distribution is given by n_i; the primed integrals give the combined contribution of the three distributions to the total number N′, surface S′, and volume V′ concentration of aerosols in each particle size range; and the unprimed integrals give the concentrations for each log-normal distributions, as well as for the sum of all three distributions, given for particles larger than 10^{-7} cm. Aerosol number concentrations n_i, N′ N are given as particle number per volume of atmosphere (cm^3), surfaces S, S′ are given as area (cm^2) of particles per atmospheric volume (cm^3), and volume V, V′ are the particle volume (cm^3) per atmospheric volume (cm^3).

Table 4: The geometric standard deviation σ_i and geometric mean radius R_i of three log-normal distribution functions (i=1,2,3) which are summed to describe the size distributions of the various types of atmospheric aerosols and aerosol-particle size ranges (Jaenicke 1988) of Shettle and Fenn (1979) by eliminating the large or coarse mode component of the size distribution because of the expected fall out of the larger particles. Most tropospheric aerosol work to date has been done in the northern hemisphere and often in regions affected by sources of anthropogenic aerosols (Harris, 1981). However some measurements have been reported in regions remote from man-made sources, for example those of Patterson *et al.* (1980), Lechner *et al.* (1989) and Gras (1991).

A selection of reported tropospheric aerosol work is given in Table 5 detailing location, measurement altitudes and measurement technique used. Representative tropospheric size distribution parameters based on a survey of tropospheric aerosol work are given in Table 6.

2.2.3 Soil/desert aerosol

A good account of soil, dust and desert aerosol is given by Gillette *et al.* (1992). Their production sources, properties and influences are described. Emphasis is given to examining mechanisms of wind erosion that determine the size distribution of dust particles, the types of particles produced and their mode of transport. (The degree of soil disturbance, caused by vegetation or rocks for example, determines whether a soil surface is eroded or not). Strong surface heating and momentum transport to the surface is also required to initiate large-scale dust production. A key parameter is the minimum threshold wind velocity required to initiate soil erosion. Most of the quantitative data on dust particle characterisation is confined to mass concentration measurements. Moreover, soil dust concentration is highly variable both geographically and temporally; the major dust sources are in the more arid regions. Soil aerosols from widely different regional sources appear to be characterised by a common mode structure, with the optically important soil particles having radii between 0.5 and 10 μm. The particle size distribution is roughly bimodal and log-normally distributed.

The desert aerosol in addition to being over the deserts, extends to a considerable extent over adjacent ocean areas. Its size distribution is similar to that of the remote continental aerosol but is highly dependent on wind velocity. It is also observable after intercontinental transport , as has been confirmed by general circulation model (GCM) transport calculations.

2.2.4 Marine Aerosol

Some good review papers on marine aerosols have been published including the work of Junge (1972) who gave a survey of physico-chemical properties of the undisturbed marine aerosol. Podzimek (1980, 1982) presented research work on marine aerosol for the decade following Junge's (1972) review. Fitzgerald (1991) presented an up- to-date detailed description of the physico-chemical properties of aerosols in clean maritime air. The background aerosol in the boundary layer over the remote oceans is largely of marine origin rather than aged continental aerosol (Fitzgerald, 1991).

Fitzgerald (1991) concludes that submicrometre ($r < 0.3$ μm) - particles consist primarily (90 - 95% of number concentration) of non sea-salt sulphate which may result from gas to particle conversion of photoxidation products of sulphur containing gases - such as dimethylsulphide (DMS) emitted by the oceans. However they only constitute about 5% of total mass.

Location	Altitude	Measurement Technique
Scottsbluff Nebraska, USA	up to 10 km	single stage jet impactor
Mildura, Australia(34°S)	4 - 10 km	Light scattering counter ($0.15<r<1\mu m$)
Pacific	5 - 6 km	PMS probes: ASAS ($0.15<r<0.75\mu m$) FSSP ($0.25<r\leq3.0\ \mu m$) 2 stage impactor Electra aircraft
Laramie Wyoming USA	< 10.5 km (tropopause)	Light scattering counter $r\leq0.15\mu m$, $\leq0.25\mu m$ Volatility (heated intake) Condensation nuclei ($r>0.01\mu m$)
Western Europe	up to 6 km	Royco 220 particle counter $0.21<r<5.9\mu m$ C 130 aircraft
South-west Pacific, and North of New Zealand	4 - 5 km	PMS probes: ASASP ($0.06<r<1.6\mu m$) FSSP ($0.25<r<4\mu m$) Fokker F27 aircraft
Eastern US (42°N-85°N)	500 - 800 mb	PMS probes ASASP-X ($0.085<r<1.2\mu m$) FSSP ($0.5<r<17.6\mu m$)
Eastern Australia (4°S - 44°S)	3.6 - 6 km	CN-Diffusion battery (0.002 - 0.11 μm) ASASP-X ($0.05<r<1.5\mu m$) Climet 208 ($0.2<r<1.0\mu m$) Impactor ($1<r<20\mu m$).

Table 5: A selection of Tropospheric aerosol data sources

Location	Power Law Parameters	Log-normal Parameters					
		Nucleation Mode			Accumulation Mode		
Southern Hemisphere	β:3 $\rightarrow$ 4 (av. 3.5)	N	$r_g(\mu m)$	σ_g	N	r_g	σ_g
(3.6 - 6 km)	N: 10 $\rightarrow$ 46 r: 0.06 - 1.6 μm	100	0.02	2.0	0.03	0.3	1.9
(4 - 10 km)	β: 3.3 $\pm$ 0.2 r: 0.15 - 1.0 μm						
Continental Europe ($\sim$ 5 km)		60			0.13		1.25
Continental USA (500-800mb)	β: 3 N: 27 - 330 cm^{-3} r:0.1 - 7.5 μm	(25-327)			0.125		1.25

Table 6: Representative Tropospheric Size Distribution Parameters.

Marine aerosol particles with radii > 0.25 μm consist mainly of sea-salt and therefore the principal constituent of the coarse-particle mode. The sea-salt concentration shows a strong dependence on wind speed and varies from about 2 $\mu g\ m^{-3}$ to as much as 50 $\mu g\ m^{-3}$ for winds in excess of 15 ms^{-1}. A composite of number and volume size distributions for the marine aerosol is given by Fitzgerald (1991). The average volume distribution of sea-salt aerosol at the sea surface (6-15 m) as a function of windforce (using the Beaufort scale) at a converted relative humidity of 70% is presented by Jaenicke (1988).

2.2.5 Urban Aerosol

The work by Whitby *et al.* (1972) who measured a total of 342 atmospheric size distributions at the California Institute of Technology in Pasadena over the size range 0.003 to 6.8 μm using the Minnesota Aerosol Analysing System is one of the most comprehensive pieces of work performed on the urban aerosol. The Minnesota Aerosol Analysing system was used in collaboration with other investigators making chemical, meteorological, mass and optical measurements. It was found that the smog aerosol was universally bimodal with the saddle point in the 1 to 2 μm size range.

Size distributions of Denver aerosol are reported by Willeke *et al.* (1975). A great variety of size distributions were measured and analyzed. Size information for particles possessing radii in the range 0.00375 to 8.75 μm were presented. Whitby (1975) concludes from studies of several thousand distributions obtained at different urban sites over several years (by Clark and Whitby (1967), Whitby *et al.* (1972), Whitby *et al.* (1975), that there is practically no interaction in the aerosol state between fine particles smaller than about 2 μm and coarse particles larger than 2 μm.

The difference in the origin of the accumulation and coarse particle modes has long been recognised by the use of such terms as smoke and dust. Numerical calculations (Whitby 1978) shows that the accumulation and coarse mode particles form, transform and are removed independently. The accumulation mode is the most ubiquitous and persistent part of the size distribution. Both experimental data (Clark and Whitby 1975) and numerical calculations (Hidy and Lilly 1965) show that the nucleation mode remains unimodal and its rather narrow size range is confirmed by the measurements of Sverdrup (1977).

2.2.6 Carbonaceous Aerosol

Carbon particles form an important component of the submicrometre aerosol fraction in urban, rural and remote continental aerosols. These carbon particles are formed directly from combustion processes and can also be formed by gas to particle conversion (secondary organics). Samples of both marine and terrestrial aerosol have been found to contain elemental carbon with the largest concentrations being obtained from urban air samples (Wolff *et al.* 1982). The ratio of elemental carbon to total carbon (elemental plus organic carbon) has been found to be strongly dependant on the source. The ubiquitous presence of carbon aerosol was shown by Parkin *et al.* (1970) in one of the first scientific reports of the presence of dark carbon in collections of atmospheric dust. They observed dark carbon spherules in air sampled during a cruise between Ireland and Newfoundland in 1969. Studies by McMurry and zhang (1989) show that elemental carbon in urban aerosol normally resides in the smaller size fractions. Lindberg and Gillespie (1977) infer that sub-micron particles found in the lowest stages of Andersen sampler impactors, used to sample dust in Southern New Mexico, are most probably predominantly carbon, although direct evidence for this is not available. Ogren and Charlson (1983) discussed the factors which determine the residence time of black carbon particles in the atmosphere: initial size distribution, concentration of ambient particles, frequency and duration of precipitation episodes and the efficiency of removal mechanisms. The presence of carbon predominantly in the sub-micron size range entails that long range transport of particulate carbonaceous aerosol is likely. Residence times for particulate carbon were computed by Ogren *et al.* (1984) from field data for an elevation of 1 km. Wet removal processes accounted for a very high percentage of deposited elemental carbon. Muller (1984) determined a residence time of about 4.5 days for elemental carbon in mid-European aerosol, with about two thirds of the black carbon being removed by precipitation. Wolff *et al* (1983) attributed higher than average elemental carbon concentrations (2.6 μg m^{-3}) for air samples collected on the Louisiana Gulf Coast to the long range transport of emissions from the mid-western and north eastern United States as compared to concentrations of 1.2 μg m^{-3} obtained for samples collected from maritime tropical air masses originating in the Gulf. Studies of Cachier *et al.* (1986) indicate that the average concentration of the coarse carbonaceous aerosol of marine origins is about the same (0.071 μg m^{-3}) in both hemispheres. However the mean carbonaceous concentration for particle radii < 0.5 μm, was much lower in the southern hemisphere (0.06 μg m^{-3}) compared to the northern hemisphere (0.45 μg m^{-3}). This suggests that northern hemispheric aerosols stem more from anthropogenic sources.

The effect of carbon aerosol on degradation of visibility has led to increased attention being given to its aerosol properties. The relative chemical inertness of carbon coupled with its predominantly submicrometre size suggests that it may act as a tracer for sources of combustion type aerosols over long path distances in the atmosphere. The study of black carbon also assumes importance because of its high specific absorption of solar radiation. Thus black carbon can have important direct and indirect radiative and possible climatic effects such as warming of the atmosphere, reduction in the amount of solar radiation reaching the ground, the enhancement of cloud absorption and a reduction in the albedo of surfaces which can incorporate carbon, for example that of ice and snow. For these reasons, it is considered important to study properties of this key constituent of the atmospheric aerosol. In addition, little information is available about background values of particulate black carbon concentration in the most western regions of Europe. Some recent measurements by Jennings *et al.* (1992) at Mace Head on the west coast of Ireland, yield average carbon mass concentration levels of (38 $\pm$ 11) ng m^{-3} for clean sector sampled air masses which had traversed over the North Atlantic for several days.

2.2.7 Stratospheric Aerosol

There are many recent reviews that give excellent coverage to almost all known theoretical and experimental aspects of stratospheric aerosols (Turco *et al.* 1982, Toon and Farlow 1981, Whitten 1982, Whitten *et al.* 1980, Kondratyev 1981, Khmelevstov *et al.* 1981). These publications attest to the general and persistent interest in stratospheric aerosol. Stratospheric aerosols are not normally regarded as having a significant impact on climate. However, volcanic eruptions, coupled with longer residence times considerably enhance their potential climatic impact.

Stratospheric aerosol measurements can provide both spatial and temporal data of sufficient resolution to be of use in climate models. Relatively recent results from a wide range of instrument techniques for measuring stratospheric aerosol parameters are described by Rosen and Ivanov (1992). Such techniques include impactor sampling, lidar system sensing, filter sampling, photoelectric particle counting, satellite extinction-sensing using the Sun as a source and optical depth probing, at sites mainly removed from tropospheric aerosol sources. Some of these techniques have also had correlative and intercomparison studies.

The main methods for determining the vertical profiles of stratospheric aerosols are as follows: lidar extinction measurements from satellites, impactor measurements from balloons and aircrafts and photoelectric particle counter measurements from balloons, aircrafts, and rockets. Stratospheric aerosol mass is retrievable from a lidar backscatter signal or from satellite extinction measurements using approximate theoretically based relations but under restrictive conditions. Impactor sampling is essentially the only method in which the physical nature of the stratospheric aerosol is observed visually. Vertical profiles of stratospheric aerosol number concentration using impactor data are presented by Rosen and Ivanov (1992). Typical profiles, obtained by using a dual-size-range photoelectric dustsonde particle counter, are also given for volcanically disturbed and inactive periods.

2.2.8 Some conclusions

For some more recently recognised aerosols, such as those due to biomass burning and biological sources, there is little available data. Unconfirmed estimates by Seiler and Crutzen (1980) suggest that 15% of the burned biomass is transferred into aerosols. Cachier *et al.* (1985) attribute the main source of total particulate carbon concentration to biomass burning originating from the tropics.

The biogenic aerosol has also been neglected in the past mostly because it has been dealt with as part of the other aerosols mentioned. And indeed, it rarely constitutes an entity in itself as do the other aerosols. The aerosol consists not only of pollen, spores, fungi and bacteria as previously investigated, but broken plant parts and insects as well as abrasions belong to it. Because of the changing activity of the biosphere, its concentrations should vary during the year. A few size distributions have been produced (Jaenicke and Matthias, 1988).

Aerosol particles are reprocessed through the evaporation of non-precipitating cloud (constituting about 85% of all cloud). This has been examined by Hoppel *et al.* (1986). Cloud processed aerosol size distributions are few in number. The residue particle is larger in size than the embryonic cloud nucleus.

This transformation process requires further study and is a priority area of interest in the field of aerosol physics and chemistry.

The recent report of the Experts meeting in Aerosol Physics and Chemistry (IGAP Plan 1991) suggests that in order to overcome the problem of aerosol variability that measurement locations which are representative on as large a scale as possible, be selected. The report also suggests the need for more vertical profile measurements which also should account for particle sampling losses in the intake systems. It is also suggested that the surface size distribution is the most important moment of the distribution and should be given increased attention in future measurement and instrument development programmes.

2.3 Chemical composition

A good summary of the state of the field, the most critical scientific issues regarding chemical composition and suggested measurements is given by the report of the Experts meeting on Aerosol Physics and Chemistry (1991). Knowledge of the size distribution of the composition of atmospheric aerosols is far from definitive. A much more complete chemical characterisation of atmospheric aerosols is needed. The nature of the mixture of an atmospheric aerosol conglomerate particle is far from resolved and has a crucial effect on the optical properties of the aerosol.

Published data of the ionic composition of atmospheric aerosols is still at an incomplete stage. Warneck (1988) has compiled chemical species concentration data for 6 aerosol types. The existing data on the organic fraction is sparse, and awaits extensive measurement campaigns. For example, little work has been undertaken of the natural biogenic aerosol sources such as terpene emissions. The concentration of selected organic compounds present in a limited range of aerosol types (maritime, remote continental, rural and urban) is presented by Jaenicke (1988).

The volatile properties of the ambient aerosol have been investigated by a series of workers including Pueschel *et al.* (1973). This work has permitted the inference of key chemical constituents such as ammonium sulphate, shown to reside in the accumulation mode (Pinnick *et al.* 1987) and that of sodium chloride shown to be present in both the accumulation and coarse modes (Jennings and O'Dowd 1990) for maritime and modified maritime air masses. The technique permits the quantitative inference of the volume fraction of particular chemical species present.

The Atmospheric/Ocean Chemistry Experiment (AEROCE) is a comprehensive atmospheric and marine chemistry multi-institutional research programme with objectives (a) to develop a climatology of aerosol chemical species over the North Atlantic and to assess the impact of natural and anthropogenic emissions on atmospheric chemical processes, (b) to assess the impact of the atmospheric deposition of trace species on biogeochemical cycles in the North Atlantic and (c) to develop a climatology of aerosol physical properties over the North Atlantic, especially those related to climatic processes. These objectives are being accomplished by a co-ordinated protocol of continuous measurements in a network of four primary stations: Barbados (West Indies), Bermuda, Tenerife (Canary Islands) and Mace Head (Ireland). The research program focuses on the major aerosol chemical species in the air and in precipitation.

The knowledge of the elemental composition of soil aerosols is reasonably well established. Data on mineralogical composition of soil aerosols suggest that particles less than about 5 μm in radius are primarily clay minerals, predominantly consisting of illite and mixed-layer clays, but that quartz is also a major component far from source. In the larger size mode, feldspar and quartz particles, which characteristically are coated with clay particles, are also dominant.

Data on the composition of free tropospheric aerosol as a function of particle size and still relatively sparse. The composition of tropospheric aerosol was determined by Patterson *et al.* (1980) by a combination of chemical and microscopic techniques. They found that particles of radii $r > 0.5$ μm were generally crustal particles. Scanning electron microscopy analysis of smaller sized particles showed them to contain amorphous sulphur, indicating the presence of a sulphate species, most likely that of ammonium sulphate. Bigg (1975) found that submicrometre diameter particles were predominately ammonium sulphate near the tropopause. In the free troposphere, Gras (1991) showed that particles predominately (76%) showed sulphuric acid or partially ammoniated sulphate morphology for his 1986 tropospheric aerosol study. However in his 1983 study, 53% of the free tropospheric aerosol were identified as saltlike. So there are still uncertainties on the composition of the tropospheric aerosol and further measurements are required to confirm these findings.

A good summary of the main methods of chemical identification of aerosol particles is given by Gotz *et al.* (1991). Aerosol composition for the tropospheric background aerosol is presented as well as size distribution data for excess (non-sea salt) sulphate and sea-salt particles. The chemical composition of aerosol particles of the anthropogenic aerosol is reviewed by the work of Corn (1976) and Finlayson-Pitts and Pitts (1986).

2.4 Radiative Aerosol Properties

2.4.1 State of the Field

Aerosol particles have a direct radiative effect through scattering and absorption of electromagnetic radiation primarily over the solar wavelength spectrum. The radiative influence of aerosols through scattering at terrestrial middle infrared wavelengths is normally negligible in the absence of high concentrations of the coarse aerosol fraction. However, aerosol particles are more effective absorbers of radiation at middle infrared wavelengths. It is generally accepted that aerosols produce a net cooling effect in the heat energy balance between short wave radiation reflected back to space and that absorbed in the atmosphere.

Aerosol particles also have an indirect but very important radiative effect in their role as cloud condensation nuclei (CCN) and in the way in which they govern cloud microphysical size characteristics, which in turn influence the reflectance of short wave radiation back to space. In addition, the incorporation of sufficient (volume fractions of order 10^{-4} - 10^{-3}) absorbing aerosol into cloud droplets can significantly increase the absorptive properties of the cloud ensemble.

Another aspect of aerosol radiative properties, which is as yet little explored, is the nature of the absorbing material in aerosols. Carbon is the common absorbing aerosol constituent at solar radiation wavelengths. However, it is also known that ferric oxide material can contribute to the absorption of soil-derived and desert- derived aerosol. Little is known of the absorption by organic aerosol constituents.

2.4.2 Aerosol Optical and Radiative Measurements

(i) The complex refractive index m of individual aerosol constituents for visible, near infra-red, and middle infra-red wavelengths is generally well known. Least is known of the complex refractive index in the ultraviolet regime, where aerosol particles have the potential of backscattering some 10-15 percent of ultraviolet (UV) radiation back to space, thus contributing to the compensation for the loss of UVB radiation due to ozone depletion. Therefore, the measurement of m for aerosol constituents over the ultraviolet range is needed.

Considerable difficulty arises in the interpretation and measurement of the refractive index of the particulate atmospheric aerosol in the normal case where the aerosol consists of a range of constituents. A plethora of mixing rules exist for internally mixed aerosol constituents which predict highly variable refractive index values. Experimental validation of these so-called mixing rules is still lacking and could best be tested through microwave simulated scattering experiments using known volume fractions of the aerosol constituents.

Difficulties occur with the measurement of the refractive index of bulk aerosol particulates. Such measurement normally requires the collection of the aerosol on some filter substrate. The particle size distribution and mode of collection will render the inferred value susceptible to unquantifiable error. In situ refractive index measurements of aerosol are scarce and rely on accurate supplementary measurement or knowledge of particle size distribution, composition and scattering parameters. This is an area in need of more laboratory research work, using well characterised aerosol components, and is best done on levitated single particles. The refractive index of biogenic aerosol is largely unknown and there is considerable scope here for research.

(ii) The aerosol scattering coefficient can be routinely measured with the integrating nephelometer, taking into account angular truncation in the forward and backward directions. Measurement problems arise in the case of lightly loaded aerosol due to low signal-to-noise ratios. Therefore, further research work is needed on the careful calibration of aerosol scattering coefficient instrumentation taking into account wall effects, source and detection spectral responses and gas carrier scattering response in order to reach sufficiently low (of order 10^{-6} m^{-1}) scattering coefficient levels reliably.

(iii) Various techniques are in use to measure the aerosol absorption coefficient. All methods are indirect and require calibration. The photoacoustic spectrophone technique is regarded as the most reliable *in situ* method, but it depends on the absorption properties of a calibration gas. Other methods rely on the collection of the aerosol on a filter substrate before measuring absorption by spectrophone or diffusing plate techniques. Complications arise with the collection of aerosol: the particles become closely packed, collection efficiency as a function of particle size and flow rate needs to be accounted for and some of the more volatile particles present may be lost in the collection process. Another *in situ* technique involves the indirect measurement of absorption through the measurement of extinction and scattering. Most *in situ* techniques are presently not sufficiently sensitive to measure absorption coefficient levels of the order 10^{-5} to 10^{-6} m^{-1}, considered desirable to measure climatically significant effects (Twomey and Huffman 1982).

More research work is required to intercompare absorption measurement techniques using absorbing aerosol constituents such as carbon and haematite. Calibration of absorption instrumentation using known aerosol sources is also required.

(iv) Calculations show that the manner in which aerosol particles are mixed greatly influences the scattering parameters of the aerosol. One of the outstanding problems in the aerosol radiative research area remains how to model the aerosol constituents within an aerosol composite particle. Measurements on scattering by individual composite particles would be very useful in this regard.

(v) The primary parameters in studies of the potential effect of aerosol on climate are those of absorption and backscattering since they control the heating and cooling of the atmosphere. The study of backscatter by aerosol particles is still at an embryonic stage. Lidar systems used to detect backscatter signals rely on comparison with the assumed Rayleigh return signal at a selected altitude. Laboratory backscatter work with aerosol media is sparse over the ultraviolet - visible to middle infrared wavelengths. Backscatter measurements are needed using both absorptive and nonabsorptive aerosol for both spherical and nonspherical particles. Angular scattering measurements can best be made on individual aerosol particles.

(vi) Substantial theoretical work has been done on the scattering of radiation by irregularly shaped particles. This subject has been recently reviewed by Bohren and Singham (1991). Computational procedures now exist which are capable of predicting aerosol scattering characteristics for non-spherical particles that include the coupled dipole method (Singham and Bohren 1987) - limited to size parameter < 5 - and the extended boundary condition method (Barber and Yeh 1975, Mugnai and Wiscombe 1986) as applied to spherical particles.

Little experimental work on the scattering properties of irregularly-shaped aerosol particles has been attempted despite the theoretical work. Laboratory backscatter measurements over the ultraviolet-middle infrared wavelength range, using irregularly shaped aerosol particles of known refractive index, are required, which can then be compared with prediction. Similarly, both scattering and absorption measurements for irregularly-sized particles are lacking and could initially be carried out using carbonaceous aerosol with chainlike aggregate structures.

2.5 Aerosol effects on Climate

2.5.1 Introduction

Quantitative assessment of the climatic impact of atmospheric aerosols requires a knowledge of (a) the optical properties of the aerosol, such as total columnar optical depth, single scattering albedo, absorption and backscatter coefficient and the asymmetry factor or phase function and (b) microphysical and chemical properties of the aerosol such as particle concentration, size distribution and size-differentiated chemical speciation.

One of the most complete global climatological data sets of the atmospheric aerosol is that of d'Almeida *et al.* (1991). Over the past decade, significant quantities of data on atmospheric aerosols have been collected using space-borne sensors on satellites. The stratospheric aerosol measurement (SAM II) launched in October 1978 and stratospheric aerosol and gas experiment (SAGE II launched in October 1984), have been used (McCormick *et al.* 1979, Kent *et al.* 1991) to obtain a path average aerosol optical depth or extinction coefficient using the solar occultation technique.

Correlative measurements conducted to validate extinction data from SAGE I, SAGE II, and SAM II solar occulation data have been confined to stratospheric aserosol. These three satellites were designed for stratospheric rather than for tropospheric observations. In addition satellite tropospheric measurements are vulnerable to the increased possibility of the presence of water or ice clouds at these lower altitudes. In the absence of cloud, SAGE II data at 1.02 μm can be inverted down to the Earth's surface while the 0.525 μm channel can be inverted down to 6.5 km (Kent *et al.* 1991). Very few tropospheric aerosol optical measurements are available for comparison with SAGE II data. Long-term scattering coefficient measurements at Mauna Loa Observatory, at 3.4 km, are not at sufficiently high altitude to be truly representative of free tropospheric aerosol. Nevertheless, a comparison of the (0.525 μm/1.02 μm) wavelength extinction ratio (Kent *et al.* 1991) shows reasonably good agreement with SAGE II values (derived from a zonal average over both land and ocean). However, with the exception of the Spring data, SAGE II extinction exceeds that at Mauna Loa by up to a factor of 2.7.

Osborn *et al.* (1989) compare SAGE II aerosol extinction measurements with derived extinction values from dustsonde and lidar data over the period November 1984 to July 1986. The comparison shows that below the tropopause, SAGE extinction (at 1.02 μm) exceeds that derived from lidar profiles or from dustsonde size distributions.

Earlier work by Russel *et al.* (1981) show that below the tropopause, SAM II extinction (1.0 μm) measurements exceeds that inferred from dustsonde measurements by factors of about 2 (22 November) to at least 3 (23 November). SAGE II extinction (for stratospheric aerosol) generally exceeds by up to a factor of 2 or more, the calculated extinction based on dustsonde (Osborn *et al.* 1989, Oberbeck *et al.* 1989) impactor (Oberbeck *et al.* 1989) and lidar (Osborne *et al.* 1989, Oberbeck *et al.* 1989). There is clearly a need for further validation of the SAGE II extinction data by comparing the satellite data with more aircraft lidar and extinction measurements.

Optical depth in the atmosphere is obtained using an aerosol retrieval algorithm which is based on the observation that the diffusely reflected radiation at the top of the atmosphere is approximately linearly dependent on the total columnar aerosol burden provided the albedo of the underlying surface is very low (applicable to oceans; dark vegetation surfaces). The National Oceanic and Atmospheric Administration (NOAA)/ National meteorological Satellites, Data and Information Service (NESDIS) single channel aerosol retrieval algorithm is based on matching the visible (0.58 - 0.68 μm) radiance measured in channel 1 of the Advanced Very High Resolution Radiometer (AVHRR) aboard the NOAA polar orbiting meteorological satellite with the computed radiance at 0.63 μm for a realistic model of a turbid atmosphere. Algorithm details are found in Griggs (1975) and Rao *et al.* (1988). Limited validation of the retrived aerosol optical depth has been carried out by comparison with sunphotometer measurements at coastal and island stations with somewhat large rms deviations of 0.03-0.05 (Griggs and Stone 1984).

Aerosol optical depth retrieval methods from over land surfaces should account for the variable land surface albedo and the bidirectional nature of the land surface reflectivity. It is clear that complementary surface measurements of tropospheric aerosols, augmented by airborne observations are necessary in order to validate and correct satellite-derived optical depth measurements. Vertical profiles of aerosol extinction coefficients are required to validate the satellite and sunphotometer measurements of optical thickness. These extinction profiles cannot be measured directly but lidar systems provide a capability of measuring aerosol backscatter profiles. Then the extinction values can be inferred from the backscatter measurements using either an inversion algorithm or an assumed backscatter to extinction ratio.

Future satellite instruments (McCormick, 1992) include an advanced SAGE (SAGE III), MODIS and MISR. MODIS, the MOderate Resolution Imaging Spectrometer, is a 36-channel scanning radiometer that measures upwelling radiation. MISR is a Multi-angle Imaging Spectro-Radiometer. It measures at four spectral band images of terrestrial scenes at eight different viewing angles. SAGE III utilizes an 800- element linear array between 0.29 and 1.0 μm, plus a fixed channel at 1.55 μm to make its measurements at lunar as well as solar sunrises and sunsets, as viewed from the spacecraft. These features as well as improved electronics allow the SAGE III measurements to be made deeper into the troposphere. It is anticipated that four independent wavelength aerosol measurements will be made to 5 km, along with H_2O and temperature, allowing reasonable size distribution as well as extinction values to be retrieved. The above instruments have been accepted for flights as part of the Earth Observing System (EOS) programme beginning at the end of this decade ($\sim$ 1998).

2.5.2 Aerosol Radiative Effects on Climate

The radiative effect of aerosols is usually studied in terms of the aerosol's single scattering albedo, ω_o, namely the ratio of the scattering and extinction coefficient of the aerosol particles. Their radiative effects depends not only on their microphysical and optical properties but also on the underlying surface albedo and on solar zenith angle. Aerosols will increase the planetary albedo (and thus have a cooling effect) over weakly reflecting surfaces, such as oceans, but will have a warming effect over highly reflecting surfaces such as ice or snow. Generally speaking (unless the atmospheric aerosol contains a dominant coarse size fraction) the warming effect of the aerosol due to absorption and emission of terrestrial radiation is weak.

Aereosol models are required in order to predict average optical parameters. Such models require knowledge of the microphysical properties and chemical composition (size resolved) of the aerosol. The wide spatial and temporal variability of aerosol characteristics precludes the use of a single universal model. Thus classification of the atmospheric aerosol into different aerosol types appears to be a pre-requisite. In general, knowledge of the vertical profile of the aerosol particle size distribution is sparse. Information on the vertical variation of the aerosol refractive index is non-existent.

Aerosols act as part of the climate system through their influence on solar and infrared terrestrial radiation. Absorption by aerosols is an important mechanism from the perspective of potential aerosol climatic effects. Aerosol absorption results in the heating of atmospheric layers containing the aerosol. At solar wavelengths, the absorption is primarily determined by carbonaceous particles, and by soil particles (haematite) in the case of desert aerosols. Anthropogenic aerosols have a tendency to warm the Earth's atmospheric system, based on calculations and assumed aerosol optical properties. However, this effect is somewhat offset by the absorption and re-emission into space of infrared terrestrial radiation. Research indicates (Pueschel 1992) that anthropogenic aerosols can possibly cause local-scale warming of the atmosphere, but global-scale climatic effects remain an open question.

Stratospheric aerosol measurements can provide both spatial and temporal data of sufficient resolution to be of use in climate models. However, stratospheric aerosols are not normally regarded as a significant component of the Earth's climate system.

Nevertheless volcanic eruptions, coupled with longer residence times, considerably enhance the potential impact of stratospheric aerosols on climate. A reduction in surface temperature of about 0.1 degK due to background stratospheric aerosols has been predicted using the energy balance model. Work using various models to estimate the radiative impact of stratosphere aerosols is discussed by Rosen and Ivanov (1992).

Various climate models are used to assess the impact of aerosols on climate. A two-stream approximation to the radiation transfer equation is adequate for optically thin layers where single scattering is applicable. The approximate procedure requires, as input, the optical depth, γ, of the layer, the single scattering albedo, ρ, and the albedo, A, of the underlying system, thus permitting the calculation of either heating or cooling caused by the aerosol layer. Improved models to include aerosol terrestrial radiation effects, important feedback mechanisms and the prediction of globally and seasonally averaged surface and atmospheric temperatures are provided by the so-called radiative-convective models (RCMs). The basic structure of the RCMs, which is regarded as adequate for many aerosol climate applications, is described by Charlock *et al.* (1992). The most significant radiative impact that aerosols have on climate is probably through the role of aerosols as cloud condensation nuclei. Some RCM studies suggest that a global doubling of cloud condensation nuclei would reduce the surface temperature by about 1 degK.

The current radiative cloud and aerosol parametrisation of the three-dimensional general circulation model (GCM) model is crude. Moreover, detailed computer modelling associated with GCM climate models can often confuse the basic physics. A full-scale GCM incorporating realistic aerosol input is yet to be formulated to include regional variability of the aerosols. Investigations of a global impact of tropospheric aerosols have largely been confined to modelling efforts rather than to observational studies. A global-type aerosol then has to be assumed, which is probably unrealistic for most of the troposphere; a fairly uniform global aerosol might be approximated for the stratosphere during quiescent or so-called background periods. In addition, because of the relatively short lifetime of tropospheric aerosols and the fairly slow response of the climatic system to even a large injection of tropospheric aerosol (volcanic eruption for example), it is difficult to observe a global climatic effect. Notwithstanding those complications, independent investigations (Hansen *et al.* 1980, Charlock and Sellers 1980) found that tropospheric aerosols cool the surface by about the same amount ($\sim$ 1.3 degK) using both global and regional radiative convective models. Because volcanic aerosols injected into the stratosphere have long residence times, typically months or years, they provide a good case study of the climate response to a change in the atmospheric aerosol. However, their effect on climate remains uncertain because clear evidence of tropospheric cooling due to volcanic events is still lacking. Model predictions for more catastrophic aerosol events such as the possible occurrence of a nuclear winter or a large asteroid impact are not definitive. On the otherhand, episodic aerosol disturbances have been shown to lead to regional climatic effects. Examples of such disturbances are Arctic haze (Porch and MacCracken 1982), Saharan dust (Idso 1981, Fouquart *et al.* 1984), forest fire smoke (Robock 1988) and oil fire smoke (Bakan *et al.* 1991).

Aerosols, in their role as cloud condensation nuclei (CCN) and ice nuclei, can have a significant indirect effect on climate through their influence on the formation and microstructure of water and ice clouds. This indirect climatic effect of aerosols (CCN) as contributors to the planetary short-wave albedo is reported by Twomey (1992). The importance of the role of atmospheric aerosols and clouds in increasing the planetary short-wave albedo is illustrated by the statement that an increase of 2% in the planetary short-wave albedo alone would compensate and nullify the effects of CO_2 doubling.

Two permanent direct effects of increased pollution are discussed by Twomey (1992) (a) an increase of absorption in the visible and near infrared because of increased amounts of elemental carbon, which gives rise to a warming effect climatically, and (b) an increased optical thickness of clouds due to increasing cloud droplet number concentration caused by increasing cloud condensation nuclei number concentration, which gives rise to a cooling effect climatically. An increase in cloud albedo from 0.77 to 0.87 (corresponding for example to a change in oceanic background particle concentration from 30 to 1000 cm^{-3}, due to an assumed large increase in pollution levels) produces an appreciable climatic perturbation of cooling up to 2.5 degK at the ground, using a hemispheric general circulation model. Work by Charlson *et al.* (1992) suggests that anthropogenic (sulphate) aerosol exerts a cooling effect confined to the northern hemisphere which is comparable in magnitude but opposite in sign to that due to greenhouse gases. The effects of pollution on cloud thermal infrared absorption are negligible. The most appreciable effect of pollution appears to be the enhancement of cloud reflectance for the short-wave radiation, which constitutes a cooling influence, directly opposed to the warming influence of carbon dioxide.

There are considerable uncertainties in the prediction of climate change due to incomplete understanding of sources and sinks of greenhouse gases, spatial and temporal variability of aerosols and their consequential direct radiative effects and their indirect radiative effects on clouds, which can strongly influence the magnitude and direction of the climate change. Uncertainties also arise due to the largely unquantified natural variability of climate, in the absence of any external influences as we have seen already in chapter 1. For example, although global-mean temperature has increased by 0.3 - 0.6 °C over the past 100 years, the inherent variability of climate precludes any firm conclusions about the cause(s) of the increase. Improved prospects for detection of climate change require a long-term commitment to the measurement and determination of potential global climate forcing factors.

References

Bakan, S., Chlond, A., Cubasch, U., Feichter, J., Grassl, H., Graßl, H., Hasselmann, K., Kirchner, I., Latif, M., Roeckner, E., Sausen, R., Schlese, U., Schriever, D., Schult, I., Schumann, U., Sielmann, F., and Welke, W., 1991., Climate response to smoke from the burning oil wells in Kuwait. *Nature*, **351**, 367-371.

Barber, P., and Yeh, C., 1975, Scattering of elecromagnetic waves by arbitrarily shaped dielectric bodies. *Applied Optics*, **14**, 2864-2872.

Bigg, E. K., 1975, Stratospheric particles. *Journal of Atmospheric Science*, **32**, 910-917.

Bohren, C. F., and Singham, S. B., 1991, Backscattering by nonspherical particles: A review of methods and suggested new approaches. *Journal of Geophysical Research*, **96**, 5269-5277.

Cachier, H., Buat-Menard, P., and Fontagne, M., 1985, Source terms and source strength of the carbonaceous aerosol in the tropics. *Journal of Atmospheric Chemistry*, **3**, 469-489.

Cachier, H., Buat-Menard, P., Fontugne, M., and Chesselet, R., 1986, Long -range transport of continentally-derived particulate carbon in the marine atmosphere: evidence from stable carbon isotope studies. *Tellus*, **38B**, 161-177.

Charlock, T. P., Kondratyev, K., and Prokofyev, M., 1992, Review of recent research on the climatic effect of aerosols. In *Aerosol Effects on Climate*, edited by S. G. Jennings (Tucson: University of Arizona Press in press), pp. 233-274.

Charlock, T. P., and Sellers, W. D., 1980, Aerosol effects on climate: Calculations with time dependent and steady-state radiative-convective models. *Journal of Atmospheric Science*, **37**, 1327-1341.

Charlson, R. J., Schwartz, S. E., Hales, J. M., Cess, R. D., Coakley, J. A., Jr., Hansen, J. E., and Hofmann, D. J., 1992, Climate forcing by anthropogenic aerosols. *Science*, **255**, 423 - 430.

Clark, W. C., 1982, *Carbon Dioxide Review* (New York: Oxford University Press Inc.).

Clark, W. E., and Whitby, K. T., 1967, Concentration and size distribution measurements of atmospheric aerosols and a test of the theory of self-preserving size distribution. *Journal of Atmospheric Science*, **24**, 677-687.

Clark, W. E., and Whitby, K. T., 1975, Measurements of aerosols produced by the photochemical oxidation of SO_2 in air. *Journal of Colloid Interface Science*, **51**, 477-490.

Corn, M., 1976, Aerosols and the primary air pollutants. Noviable particles. Their sources, properties and effects. In *Air Pollution, volume 1*, edited by A. C. Steven) (New York: Academic Press), pp. 77-168.

Cubasch, U., and Cess, R. D., 1990, Processes and modelling. In *Climate Change the IPCC Scientific Assessment*, edited by J. T. Houghton, G. J. Jenkins and J. J. Ephraums (Cambridge: Cambridge University Press), pp. 69-91.

D'Almeida, G. A., Koepke, P., and Shettle, E., 1991, *Atmospheric Aerosols: Global Climatology and Radiative Characteristics* (Virginia: Deepak) .

Davies, C. N., 1974, Size distribution of atmospheric particles. *Aerosol Science*, **5**, 293-300.

Experts meeting on Aerosol Physics and Chemistry, 1991, In IGAP Plan, edited by D. A. Hegg, pp. 1-25.

Fenn, R. W., 1963, Aerosol Verteilungen und Atmospharische Streulicht, Dissertation der Naturwissenschaftlichen Facultat Munchen, Germany.

Finlayson-Pitts, B. J., and Pitts, J. N., Jr., 1986, *Atmospheric Chemistry. Fundamentals and Experimental Techniques* (New York: Jonh Wiley and Sons).

Fouquart, Y., Bonnel, B., Brogniez, G., Cerf, A., Chaoui, M., Smith, L., and Vanhouette, J. C., 1984, Size distribution and optical properties of Saharan aerosols during ECLATS. In *Aerosols and Their Climatic Effects*, edited by H.E. Gerbert and A. Deepak (Hampton: Deepak) pp. 35-62.

Gillette, D. A., Patterson, E. M. Jr., Prospero, J. M., and Jackson, M. L., 1992, Soil aerosols. In *Aerosol Effects on Climate*, edited by S. G. Jennings (Tucson: University of Arizona Press pp. 73-109.

Gotz, G., Meszaros, E., and Vali, G., 1991, *Atmospheric Particles and Nuclei* (Budapest: Akademiai Kiado .

Gras, J. L., 1991., Southern Hemisphere Tropospheric Aerosol Microphysics. *Journal of Geophysical Research*, **96**, 5345-5356.

Griggs, M., and Stone, L. L., 1984, Measurements of Aerosol optical parameters from satellites. *In IRS'84: Current Problems in Atmospheric Radiation*, edited by G. Frocco (Hampton: Deepak), pp. 42-45.

Griggs, M., 1975, Measurements of atmospheric aerosol optical thickness over water using ERTS-1 data. *Journal of Air Pollution Control*, **25**, 622-626.

Hansen, J. E., Lacis, A. A., Lee P., and Wang, W. C., 1980, Climatic effects of atmospheric aerosols. *Annal of New York Academy of Science*, **338**, 575-587.

Harris, F. S., 1981, aerosols: review of current data on physics and optical properties, Report WMO/ICSU meeting, Report 43, World Climate Research Programme, Geneva.

Hidy, G. M., and Lilly, D. K., 1965, Solutions to the equations of the kinetics of coagulation. *Journal of Colloid and Interface Science*, **20**, 867.

Hoppel, W. A., Fruk, G. M., and Larson, R. E., 1986, Effect of nonprecipitating clouds on the aerosol size distribution in the marine boundary layer. *Geophysical Research Letter*, **13**, 125-128.

Idso, S. B., 1981, Climatic change: The role of atmospheric dust, origin, characteristics, and effect on man. In *Special Paper Geological Society of America*, edited by T. L. Pewe (Washington DC: Geological Society of America), pp. 207-215.

Jaenicke, R., 1988, Aerosol physics and chemistry, In *Landolt-Bornstein Numerical data and functional relationships in Science and Technology New Series Group V : Geophysics and Space Research, 4(b)*, edited by G. Fischer (Berlin: Springer-Verlag), pp. 391-457.

Jaenicke, R., and Matthias, S., 1988, The primary biogenic fraction of the atmosphere aerosol. In *Aerosol and Climate*, edited by P. V. Hobbs, P. V., and M. P. McCormick (Hampton: Deepak), pp. 31-38.

Jennings, S. G., and O'Dowd, C. D., 1990, Volatility of Aerosol at Mace Head, on the west coast of Ireland. *Journal of Geophysical Research*, **95**, 13,937-13,948.

Jennings, S. G., McGovern, F. M., and Cooke, W. F., 1992, Carbon mass concentration measurements at Mace Head, on the west coast of Ireland. *Atmospheric Environment-Part A General Topics*, **27**, 1229-1239.

Junge, C. E., 1953, Die rolle der Aerosole und der Gasformigen Beimengungen der luft in Spurenstroff-haushalt der Troposphere. *Tellus*, **5**, 1-26.

Junge, C. E., 1963, *Air Chemistry and Radioactivity* (New York: Academic Press).

Junge, C. E., 1972, Our knowledge of the physico-chemistry of aerosols in the undisturbed marine environment. *Journal of Geophysical Research*, **77**, 5183-5200.

Keeling, C. D., Bacastow, R. B., Bainbridge, A. E., Ekdahl, C. A., Jr., Guenther, P. R., Waterman, L. S., and Chin, J. F. S., 1976, Atmospheric carbon dioxide variations at Mauna Loa Observatory, Hawaii., *Tellus*, **28(6)**, 538-51.

Keeling, C. D., Bacastow, R. B., Carter, A. F., Peper, S. C., Whorf, T. P., Heimann, M., Mook, W. G. and Roeloffzen, H., 1989, A three-dimensional model of atmospheric CO_2 transport based on observed winds: 1. Analysis of observational data. In *Aspects of Climate Variability in the Pacific and the Western Americas*, Geophysical Monograph, No. 55 edited by D. H. Peterson, 165-235.

Kent, G. S., McCormick. M. P., and Schaffner, S. K., 1991, Global Optical Climatology of the free tropospheric aerosol from 1.0 μm satellite occulation measurements. *Journal of Geophysical Research*, **96**, 5249-5267.

Khmelevtsov, S. S., Kabonov, A. S., and Kilomeyev, M. P., 1981, The effect of stratospheric aerosol on climate. *Summary Information, Meteorology Series, vol. 6* (Obninsk: VNIG-MIMID), pp. 1-42.

Kondratyev, K. Ya., 1981, *Stratosphere and climate, Itogi Nauki Techniki, Meteorologiya i Klimatologiya (The Results of Science and Technology, Meteorology and Climatology), vol. 6* (Moscow VINITI).

Lechner, I. S., Fischer, G. W., Larsen, H. R., Harvey, M. J., and Knobben, R. A., 1989, Aerosol size distributions in the southwest Pacific. *Journal of Geophysical Research*, **94**, 14893-14903.

Lindberg, J. D., and Gillespie, J. B., 1977, Relationship between particle size and imaginary index of atmospheric dust. *Applied Optics*, **16**, 2628-2630.

Logan, J. A., Prather, M. J., Wofsy, S. C., and McElroy, M. B., 1981, chemistry: A global perspective, *Journal of Geophysical Research*, **86**, 7210-7254.

McCormick, M. P., Hamill, P., Pepin, T. J., Chu, W. P., Swissler, T. J., and McMaster, L. R., 1979, Satellite studies of the stratospheric aerosol. *Bulletin of the American Meteorological Society*, **60**, 1046-1083.

McCormick, M. P., 1992, Satellite measurements of Aerosols. In *Global Atmosphere Watch, WMO report No. 79*, pp. 58.

McMurry, P. H., and Zhang, X. Q., 1989, Size distributions of ambient organic and elemental carbon. *Aerosol Science and Technology*, **10**, 430-437.

Madden, R. A., and Ramanathan, V., 1980, Detecting climate change due to increasing carbon dioxide. *Science*, **209**, 763-768.

Muller, J., 1984, Atmospheric residence time of carbonaceous particles and particulate PAH-compounds. *Science of the Total Environment*, **36**, 339-346.

Mugnai, A., and Wiscombe, W., 1986, Scattering from non-spherical chebyshev particles, 1, Cross-sections, single scattering albedo, asymmetry factor, and backscattering fractions. *Applied Optics*, **25**, 1235-1244.

National Academy of Sciences (NAS), 1983, Changing climate. (Washington, D.C.: National Academy Press)

Oberbeck, V. R., Livingston, J. M., Russell, P. B., Pueschel, R. F., Rosen, J. M., Osborn, M. T., Kritz, M. A., Snetsinger, K. G., and Ferry, G. V., 1989, SAGE II aerosol validation: selected altitude mesurements, including particle micromeasurements. *Journal of Geophysical Research*, **94**, 8367 - 8380.

Ogren, J. A., and Charlson, R. J., 1983, Elemental carbon in the atmosphere: cycle and lifetime. *Tellus*, **35B**, 241-254.

Ogren, J. A., Groblicki, P. J., and Charlson, R. J., 1984, Measurement of the removal rate of elemental carbon from the atmosphere. *Science of the Total Environment*, **36**, 329-338.

Osborn, M. T., Rosen, J. M., McCormick, M. P., Wang, P. H., Livingston, J. M., and Swissler, T. J., 1989, SAGE II aerosol validation: Profile measurements. *Journal of Geophysical Research*, **94**, 8353 - 8366.

Parkin, D. W., Phillips, D. R., and Sullivan, R. A. L., 1970, Airborne dust collections, over the North Atlantic. *Journal of Geophysical Research*, **175**, 1782-1793.

Patterson, E. M., Kiang, C. S., Delany, A. C., Wartburg, A. F., Leslie, A. C. D., and Huebert, B. J., 1980, Global measurements of aerosols in remote continental and marine regions: concentrations, size distributions, and optical properties. *Journal of Geophysical Research*, **85**, 7361-7376.

Pinnick, R. G., Jennings, S. G., and Fernandez, G., 1987, Volatility of aerosol in the arid south western United States. *Journal of Atmospheric Science*, **44**, 562-576.

Porch, W. M., and M. C. MacCracken, 1982, Parametric study of the effects of Arctic soot on solar radiation. *Atmospheric Environment*, **16**, 1365-1371.

Prahm, L. P., Torp, U., and Stern, R. M., 1976, Deposition and transformation rates of sulphur oxides during atmospheric transport over the Atlantic. *Tellus*, **28**, 355-372.

Prinn, R. G., Simmonds, P. G. Rasmussen, R. A., Rosen, R. D., Alyea, F. N., Cardelino, C. A., Cranford, A. J., Cunnold, D. M., Fraser, P. J., and Lovelock, J. E., 1983, The atmospheric lifetime experiment, 1, instrumentation and overview. *Journal of Geophysical Research*, **88**, 8353-8367.

Prinn, R. G., Cunnold, D. M., Rasmussen, R. A., Simmonds, P. G., Alyea, F. N., Crawford, A. J., Fraser, P. J., and Rosen, R. D., 1987, Atmospheric trends in methylchloroform and the global average for the hydroxyl radical. *Science*, **238**, 945-950.

Pueschel, R. F., Bodhaine, B. A., and Mendonca, B. G., 1973, The proportion of volatile aerosol on the Island of Hawaii. *Journal of Applied Meteorology*, **12**, 308-315.

Pueschel, R. F., 1992, Potential climatic effects of anthropogenic aerosols. In *Aerosol effects on climate*, edited by S. G. Jennings (Tucson: Univeristy of Arizona Press), pp. 110-132.
Ramanathan, V., Cicerone, R. J., Singh, H. B., and Kiehl, J. T., 1985, Trace gas trends and their potential role in climate change. *Journal of Geophysical Research*, **90**, 5547-5566.
Rao, C. R. N., Stone, L. L., McClain, E. P., and Sapper, J., 1988, Development and application of aerosol remote sensing with AVHRR data from the NOAA satellites. In *Aerosol and Climate*, edited by P.V. Hobbs and M.P. McCormick (Hampton: Deepak), pp. 69-80.
Rasmussen, R. A., and Khalil, M. A. K., 1981, Atmospheric methane (CH_4): Trends and seasonal cycles. *Journal of Geophysical Research*, **86**, 9826-9832.
Raynaud, D., Chappellaz, J., Barnola, J. M., Korotkevich, Y. S., and Lorius, C., 1988, Climatic and CH_4 cycle implications of glacial-interglacial CH_4 change in the Vostok ice core. *Nature*, **333**, 655-657.
Robock, A., 1988, Surface temperature effects of forest fire smoke plumes. In *Aerosols and Climate*, edited by P. V. Hobbs, and M. P. McCormick (Hampton: Deepak), pp. 435-442.
Rosen, J., and Ivanov, V. A., 1992, Stratospheric aerosols In *Aerosol Effects in climate*, edited by S. G. Jennings (Tucson: University of Arizona Press), pp. 133 - 189. (in press)
Russel, P. B., McCormick, M. P., Swissler, T. J., Chu, W. P., Livingston, J. M., Fuller, W. H., Rosen, J. M., Hofmann, D. J., McMaster, L. R., Woods, D. C., and Pepin, T. J., 1981, Satellite and correlative measurements of the stratospheric aerosol. II: Comparison of measurements made by SAM II, dustsondes and an airborne lidar. *Journal of Atmospheric Science*, **38**, 1295 - 1302.
Seiler, W., and Crutzen, P., 1980, Estimates of the gross and natural flux of carbon between the biosphere and the atmosphere for biomass burning. *Climatic Change*, **2**, 207 - 247.
Shettle, E. P., and Fenn, R. W., 1979, Model for the aerosols of the lower atmosphere and the effects of humidity variations on their optical properties, AFGL-TR-79-0214, Hanscomb, Massachusetts 01731, U.S.A.
Shine, K. P., Derwent, R. G., Wuebbles, D. J., and Morcrette, J-J., 1990, Radiative forcing of climate. In *Climate Change: The IPCC Scientific Assessment*, edited by J. T. Houghton, G. T. Jenkins, and J. J. Epharaums (Cambridge: Cambridge University Press), pp. 41-68.
Simmonds, P. G., and Derwent, R. G., 1991, Measurements of ozone and other radiatively active gases at Mace Head in the Republic of Ireland. *Atmospheric Environment*, **25A**, 1795-1808.
Singham, S. B., and C. F. Bohren, 1987, Light scattering by an arbitrary particle: A physical reformulation of the coupled dipole method, *Optical Letters*. **12**, 10-12.
Stauffer, B., Lochbronner, E., Oeschger, H., and Schwander, J., 1988, Methane concentration in the glacial atmosphere was half that of the preindustrial Holocene. *Nature*, **332**, 812-814.
Steele, L. P., Fraser, P. J., Rasmussen, R. A., Khalil, M. A. K., Conway, A. J., Crawford, A. J., Gammon, R. H., Masarie, K. A., and Thoning, K. W., 1981, *Journal of Atmospheric Chemisty*, **5**, 125-171.
Sverdrup, G. M., 1977, Parametric measurement of submicron atmospheric aerosol size distributions, Ph.D. Thesis, Particle Technology Laboratory, University of Minnesota, Minneapolis, MN 55455.
The International Global Aerosol Program (IGAP) Plan, 1991, International Commission on Clouds and Precipitation (ICCP) / International Radiation Commission (IRC) / International Commission on Atmospheric Chemistry and Global Pollution (ICACGP), Joint Working Group on the International Aerosol Climatology Project (IACP), edited by A. Deepak, and G. Gali, IGAP 1, IACP91-4.

Toon, O. B., and Farlow, N. H., 1981. Particles above the tropopause, *Annual Review of Earth and Planetary Sciences*, **9**, 19-58.

Trabalka, J. R., 1985, Atmospheric carbon dioxide and the global carbon cycle. U.S. Department of Energy, Report DOE/ER-0239, National Technical Information Service, Springfield, VA, U.S.A.

Turco, R. P., Whitten, R. C., and Toon, O. B., 1982, Stratospheric aerosols: Observation and theory. *Review of Geophysics*, **20**, 233-279.

Twomey, S., 1992, Radiative properties of clouds. In *Aerosol effects on Climate*, edited by S.G. Jennings (Tucson: University of Arizona Press), pp. 275 - 297.(in press)

Twomey, S. A., and Huffman, D. R., 1982. Workshop review. In *Light absorption by aerosol particles*, edited by H.E. Gerber and E.E. Hindman (Hampton: Spectrum Press), pp. 379-408.

Warneck, P., 1988, *Chemistry of the natural atmosphere, International Geophysics Series, vol. 41* (New York: Academic Press).

Watson, R. T., Rodhe, H., Oeschger, H., and Siegenthaler, U., 1990, Greenhouse Gases and Aerosols. In *Climate Change: the IPCC Scientific Assessment*, edited by J. T. Houghton, G. T. Jenkins and J. J. Ephraums (Cambridge: Cambridge University Press), pp. 1-40.

Whitby, K. T., 1975, Modelling of Atmospheric Aerosol Particle Size Distribution. Particle Technology Publication No. 253, Mechanical Engineering Department, University of Minnesota.

Whitby, K. T., Husar, R. B., and Liu, B. Y. H., 1972, The Aerosol Size Distribution of Los Angeles Smog. *Journal of Colloid and Interface Science*, **39**, 177 - 204.

Whitby, K. T., Clark, W. E., Marple, V. A., Sverdrup, G. M., Sem, G. M., Willeke, K. Liu, B. Y. A, and Pui, D. Y. A., 1975, Characterization of California Aerosols. I Size distribution of Freeway Aerosol. *Atmospheric Environment*, **9**, 463-482.

Whitby, K. T., 1978, The physical characteristics of sulphur aerosols, *Atmospheric Environment*, **12**, 135-159.

Whitten, R. C. (Editor), 1982, *The Stratospheric Sulfate Aerosol Layer* (Heidelberg: Springer Verlag).

Whitten, R. C., Toon, O. B., and Turco, R. P., 1980, The stratospheric sulfate aerosol layer: Processes, models, observations and simulations. *Pure and Applied Geophysics*, **118**, 86-127.

Willeke, K., Liu, B. Y. H., and Pui, D. Y. H., 1975, Characterisation of California Aerosols - 1. Size Distribution of Freeway Aerosol, *Atmospheric Environment*, **9**, 463 - 482.

Wolff, G. T., Groblicki, P. J., Cadle, S. H., and Countess, R. J., 1982, Particulate carbon at various locations in the United States. In *Particular Carbon*, edited by G. T. Wolff, and R. L. Klimisch (New York: Plenum Press), pp. 279-315.

Wolff, G. T., Kelly, W. A., Ferman, M. A., and Morissey, M. L., 1983, Rural measurements of the chemical composition of airborne particles in the Eastern United States. *Journal of Geophysical Research*, **88**, 10769-10775.

Atmospheric Ozone Concentration Measurements

C. Varotsos

1 Introductory Remarks

1.1 Basic properties of ozone

In 1840 the chemist Christian Friedrich Schoenbein observed that the odour generated during electrical discharges in air (including thunderstorms) and when decomposing water by an electric current has the same source. In his opinion this odour is attributable to a gas, which he termed ozone (from the Greek $o\zeta\omega$ meaning odour). We note some of the physical properties of ozone in Table 1. The optical absorption and emission properties of ozone are of relevance in this chapter. The absorption of energy by an ozone molecule may involve electronic transitions and changes in the vibrational and rotational energies of the molecule. Excessive electronic excitation results in ionisation while excess excitation of the vibrational energy, results in dissociation (11.09 eV). The absorption of electromagnetic radiation of wavelength λ, by ozone may be described by the Beer-Lambert-Bouguer Law

$$I_\lambda = I_{o\lambda} \times 10^{-N\alpha_\lambda x} \tag{1}$$

where N = number of molecules per unit volume and α_λ = (decadic) mulecular absoption coefficient, at wavelength λ. In practice it is the absoption of ultraviolet radiation by ozone which is of practical interest in relation to the study of the atmosphere and the global environment. It is the ozone in the atmosphere which absorbs ultraviolet radiation from the incoming solar radiation. It is the harmful effects of ultraviolet radiation which include increasing the risk of skin cancers in humans, that lie behind the present worries about the depletion of the ozone layer in the stratosphere as a result of human activities.

There are a number of absorption bands of ozone which are particularly useful in stadies of atmospheric ozone concentrations. These include:

NATO ASI Series, Vol. I 24
Remote Sensing and Global Climate Change
Edited by Robin A. Vaughan and Arthur P. Cracknell

• alotropic form of oxygen, i.e. triatomic oxygen, O_3
• molecule: an isoceles triangle (sides 1.278 Å, apical angle 116°49')
• mass of O_3^{16} molecule is 7.97 × 10^{-23} g
• atmospheric ozone contains 0.21% $O^{16}O^{18}O^{16}$, 0.41% $O^{16}O^{16}O^{18}$ molecules
• moment of inertia: I^1 = 7.87 × 10^{-40}, I^2 = 62.84 × 10^{-40}, I^3 = 70.89 × 10^{-40} (g cm^2)
• dipole moment: (-53 ± 0.02) D
• density at P=1 atm, T = 300 K is 2.144 g cm^{-3}
• specific heat C_p = 0.190 cal g^{-1} (at T = 273 K)
• it liquifies at T = 160 K
• latent heat of vaporisation is 75.6 cal g^{-1}
• it solidifies at T = 80 K
• solubility at 1 atm in water is 1.099/litre at T = 273 K and 0.869/litre at T = 283 K
• ionisation potential is 12.8 eV

Table 1: Physical properties of ozone

(i) the **Hartley band** (200-310 nm) with an absorption peak near $\lambda = 225$ nm for which $\alpha_\lambda = 1.2 \times 10^{-17}$ cm^2 $molecule^{-1}$)

(ii) the **Huggins bands** (310-355 nm) which are nearer and with a periodic structure

(iii) the **Chappuis band** (450-750 nm) which is rather weak ($\alpha_\lambda = 5 \times 10^{-21}$ cm^2 $molecule^{-1}$) and

(iv) absorption lines in the thermal infrared; absorption near to $\lambda = 9.6$ μm is frequently used for measurements. There is a temperature dependence of the α_λ for the Huggins bands.

1.2 Characteristics of atmospheric ozone

The role of atmospheric ozone includes the following:

- it is a very important trace molecular species
- it provides the shield against solar ultraviolet radiation (λ <290 nm)
- it causes the stratosphere to heat up absorbing solar ultraviolet radiation
- it is involved in the nitrogen, carbon, hydrogen, oxygen and halogen chemical cycles.

The main features of the vertical distribution of ozone are:

- approximately 10% of the total O_3 is in the troposphere (from ground level to heights of 10-18 km)
- approximately 90% of the total O_3 is in the stratosphere (10-18 km to 50 km)
- only a trace of O_3 is in the mesosphere (50 km to approximate 85 km)

The density of O_3 in the atmosphere peaks at 5×10^{12} molecules cm^{-3} at an approximate altitude of 20 km; it falls to 108 molecules cm^{-3} at an altitude of 75 km and is between 2×10^{11} and 10×10^{11} molecules cm^{-3} in the troposphere. The total O_3 amount in a vertical column in the atmosphere is usually measured in Dobson units (DU) [1 DU = milliatmosphere centimetres (matm cm) of O_3 at STP]; i.e. it is the amount of ozone that would occupy a column of height 10^{-5} m at STP. Typical values of total O_3 concentration are 300 DU in autumn and 480 DU in spring at the pole and 250 DU at the equator. O_3 is produced photochemically, generally nearer the equator than near the poles; it is then transported dynamically towards the poles.

1.3 High-flying aircraft

The general question of the effects of high-flying aircraft on the concentrations of minor greenhouse gases was considered in chapter 1 (section 5.3.4.). The main and specific question of the effect of high-flying aircraft, particularly supersonic aircraft, on the concentration of ozone is of interest here. In the USA NASA has started examining the question of developing a new generation of supersonic - passenger - carrying aircraft (HSCT - High Speed Civil Transport) that would fly faster than Concorde and carry more passengers than Concorde. Some figures for comparison are:

	Concorde	Projected HSCT
Range (km)	5,600	9,000 - 12,000
Payload	100	250 - 300
Speed	Mach 2	Mach 2.4

The contractors for the HSCT would be the Boeing Co. and the Mc Donnell Douglas Co. The question arises as to whether the aviation community could produce an environmentally acceptable HSCT jet. Consider a fleet of 500 aircraft flying at a speed of Mach 2.4, then: 1. Taking account of heterogeneous chemical reactions there is a predicted 3-6% decrease in ozone leading to 6-12% more UV radiation of mid-latitudes: a 15% reduction in the concentration of mid-latitude O_3. 2. With heterogeneous reactions HSCT traffic would cause a net increase in the ozone concentration (by 0.5%). In a model without heterogeneous chemistry O_3 is sensitive to NO_x (which are the biggest destroyers of O_3). When the model includes heterogeneous chemistry, O_3 does not react strongly with NO_x (the Cl_x compounds are the biggest O_3 destroyers). Another problem is to estimate the effects of water vapour that the aircraft will release into the normally dry stratosphere.

1.4 Volcanic eruptions

The eruption of Mt. Pinatubo in June of 1991 was the largest of the century, belching out twice as much SO_2 than the eruption of Mexico's El Chichon. As winds blow the volcanic cloud, the gas molecules (millions of tons of SO_2) react with water to form droplets or aerosols of H_2SO_4, which stay in the stratosphere for 2-3 years before falling to the ground. These provide tiny surfaces on which molecules of certain nitrogen compounds can react. These reactions alter the stratospheric chemistry causing "safe" chlorine molecules to transform into ones that can destroy ozone. For instance, the Pinatubo aerosol compound increase has been calculated to be responsible for a 15% reduction in the concentration of midlatitude ozone. By reflecting sunlight back into space, the droplets arising from the eruption exert a cooling influence (counteracting the greenhouse effect). For instance, the El Chichon eruption depressed global temperatures by 0.2-0.3 °C. Pinatubo could have an even greater effect. An increase in stratospheric turbidity due to an eruption (i) would reduce the solar ultraviolet radiation reaching the ground and (ii) would decrease the solar ultraviolet radiation attenuation from tropospheric O_3, due to the higher diffuse fraction in the radiation reaching the troposphere.

2 Direct - Absorption Measuring Instruments

Direct-absorption measuring instruments involve directly measuring the absorption features of ozone (i) with a path defined by the instrument and (ii) by using an extraterrestrial source (such as the Sun). The physical principle of the measurement is use of the Beer-Lambert-Bouguer Law, see equation (1). Assuming that α_λ is known for the wavelength in question, then the average concentration of ozone along the path can be determined. The advantage of using direct absorption is that the data-processing algorithm is generally rather simple; therefore few assumptions have to be made regarding quantities such as aerosol scattering, temperature profile etc. that cannot be measured directly. The disadvantages of using direct absorption include uncertainties in the absorption cross sections and the limited signal-to-noise ratios at the extreme ranges.

2.1 Optical rocket ozonesonde

This instrument consists of a multifilter UV radiometer launched by a rocket to a height of about 70 km and permitted to fall back to Earth with a parachute. The direct solar irradiance in several channels is measured, and the ozone concentration is determined by using the Beer-Lambert-Bouguer Law. The precision and accuracy depend on the accuracy of the ozone absorption cross sections and the knowledge of the filter transmission. The precision is given as approximately 2.5% and the accuracy is estimated at 5-7%. An example of this instrument is the ROCOZ, which uses four filters from 267 to 309 nm (see Figure 1). The shortest wavelength is used to measure

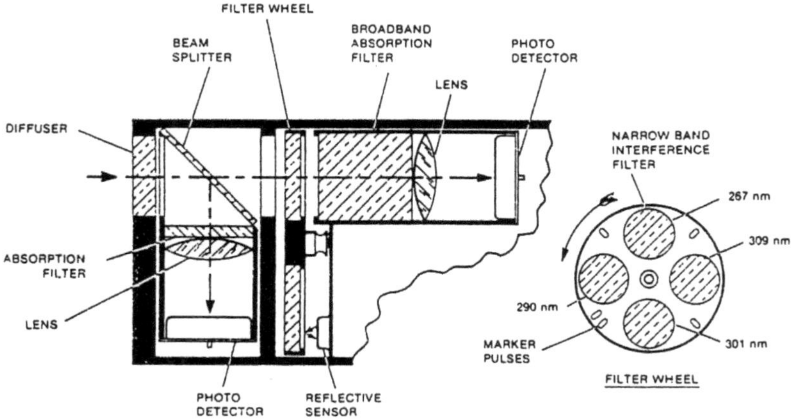

Figure 1: ROCOZ instrument

ozone from 54 to 39 km and the longer from 47 to 20 km. Samples are taken every 0.6s, giving measurements every 80 m at 55 km and every 9 m at 25 km.

2.2 Stratospheric Aerosol and Gas Experiment (SAGE)

The SAGE instrument is a satellite-borne spectometer. It measures the absorption of sunlight by ozone with four channels (SAGE I), at 0.385, 0.45, 0.60 and 1.0 μm at spacecraft sunrise and sunset (Solar occultation geometry), see Figure 2. The 0.60 μm channel is used to measure stratospheric ozone by using the Chappuis band. The 0.45 μm channel is used to measure the concentration of stratospheric NO_2 and the outer two channels are used to determine the aerosol optical depth. The vertical resolution is approximately 1 km for altitudes above 10 km.

SAGE II is a seven-channel instrument and in addition to ozone, NO_2 and aerosols, it is able to measure water vapour concentrations. It makes measurements for heights up to 65 km versus 55 km for SAGE I. The extra channels also allow for better corrections for aerosols. SAGE III is planned for the Earth Observing System (EOS programme). In SAGE III the filters will be replaced by a spectrometer, a diode array will be used for detecting radiation in the various wavelength regions and the spectral range will be extended to 1.55 μm, for the observation of aerosols with larger particles and aerosols nearer the Earth's surface. Regarding precision and accuracy, the major uncertainty in ozone concentrations as measured by either SAGE I or SAGE II is the Chappuis band ozone absorption cross sections and their temperature dependences, which are estimated to give a 6% uncertainty. The measurement precision of SAGE II is 4-7% in the 15-52 km altitude range, becoming 20% at 60 km and 50% at 10 km. The uncertainty between SAGE I and SAGE II in the ozone data is approximaely 4%. When statistical and systematic uncertanties are combined the accuracy is better than 10% for heights up to 15-52 km.

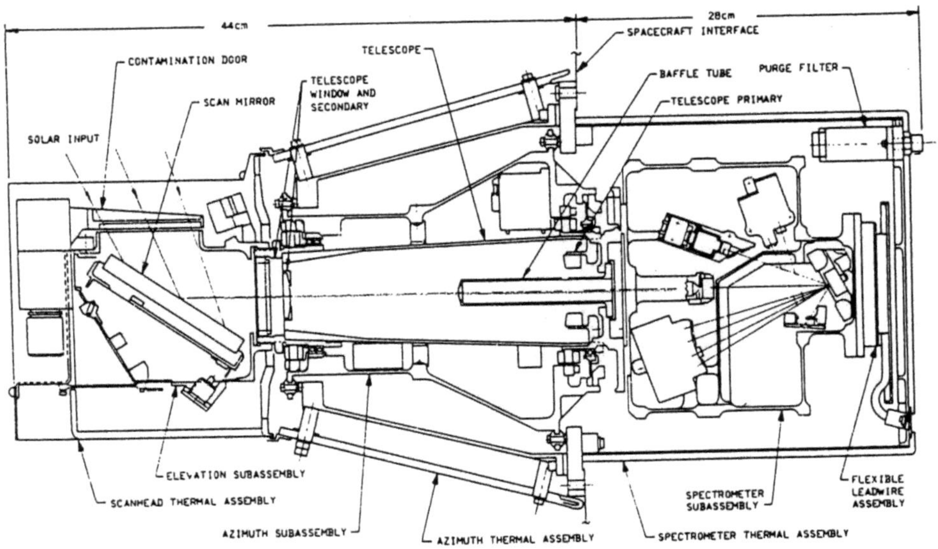

Figure 2: Diagram of the SAGE I

2.3 Laser Heterodyne Spectrometer (LHS)

This spectrometer measures ozone profiles in the 0.9 μm region in the infrared. A laser is used as a local oscillator to direct a narrow band-width beam onto a detector or photomixer. Usually a CO_2 laser is used as the local oscillator but occasionally a tunable diode laser is used. The Sun or the Moon acts as the source of radiation beyond the atmosphere. The radiation is selectively absorbed by ozone or other molecular species in the atmosphere and is detected in the spectrometer. A CO_2 and a tunable diode LHS can be used to measure the vertical profile of ozone concentration. Such measurements can be made with a vertical resolution of 4 km, which is limited by the rate of change in the spectral weighting functions with altitude or with lower resolution if an insufficient number of channels is used. The accuracy of the measurements is approximately 5-10% for column content and 10-30% for profile measurements. The precision is approximately 2%.

2.4 Halogen Occultation Experiment (HALOE)

This instrument is being flown on the Upper Atmosphere Research Satellite (UARS), see Figure 3. It has eight infrared channels and four of these are broad-band (60-120 cm^{-1}) filter radiometer channels for CO_2, H_2O, O_3 and NO_2 and four channels are gas filter correlation radiometer channels for HF, HCl, CH_4 and NO.

The measurement principle is based on the gas filter correlation technique. In this

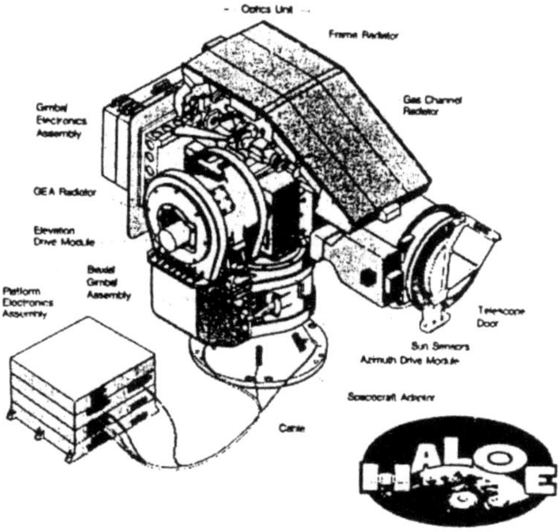

Figure 3: The HALOE instrument

technique incoming radiation is split into different channels for each species and passed through a narrow-band filter; it is then split further into two paths for each species. One path includes a cell containing the gas to be measured. As the atmosphere passes in front of the Sun, the path containing the cell will see little change in the absorption because the appropriate portion of the Sun's spectrum is already absorbed by the gas in the cell. The accuracy for ozone mixing ratios is estimated to be 10-15% in the mid-stratosphere, with a decrease in accuracy above and below this region.

2.5 The Ultraviolet Photometer (UP)

The Ultraviolet Photometer is an in situ instrument for measuring ozone concentrations. It uses a mercury lamp as the spectral source, a narrow-band filter, a silicon photodiode as the detector, and a pump to pull air through the sample volume, see Figure 4. The instrument determines ozone concentrations by measuring the absorption at the 253.65 nm mercury line (near the UV ozone peak). It uses a 40 - 1200 cm path. For half of the measurement time (one of two cells) an ozone scrubber removes the ozone to provide a reference atmosphere. The precision of this instrument is 3% and the accuracy is 5%. UV photometers provide in situ measurements for altitudes as high as 45 km. They are flown on large helium-filled ballooons and can be returned to the ground by parachute. The heavier version of this system, Dasibi ozone monitor, is often used at ground-level pollution monitoring stations.

2.6 The Differential Absorption Lidar Technique (DIAL)

The receiving instrument consists of (i) a telescope to capture the back scattered photons, (ii) a filter to reject out-of-band radiation, (iii) a photomultiplier tube and (iv) the signal-chain electronics. In this approach two laser wavelegnths are transmitted into the

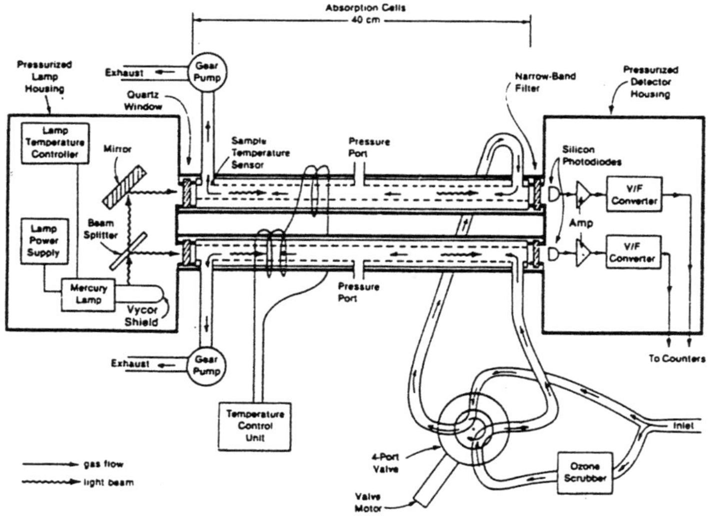

Figure 4: Diagram of the Ultraviolet Photometer

atmosphere and radiation backscattered elastically by molecules (Rayleigh scattering) and aerosols (Mie scattering) is collected by a receiver located near the transmitter. One wavelength is relatively unaffected by the species of interest and the other wavelength is tuned to an absorption feature of that species. The derivative with respect to distance of the natural logarithm of the ratio of the signals is used to determine the density profile of the species. With this technique it is possible to achieve a measurement precision better than 5% to altitudes of 25 km, decreasing to 20% at 47 km. The measurement accuracy can be limited by the presence of dense aerosols and other molecular species (e.g. SO_2). Systems of this type which have been used include XeCl excimer lasers (308 nm), an aircraft-mounted Nd: YAG-pumped dye laser DIAL system and a Balloon-borne Laser In Situ Sensor (BLISS) instrument up to a height of 38 km.

3 Indirect - Absorption - Measuring Instruments

3.1 Ground-based instruments

There are four main types of these instruments:

(i) Dobson Spectrophotometer
(ii) M-83 filter ozonometer
(iii) Brewer Spectrophotometer and
(iv) Diode Detector Array Spectrometer.

Their advantages include:

(i) They can be used to measure ozone column content with high accuracy rutinely (the GO_3OS, Ground Ozone Observing System, 140 ground stations)
(ii) They can provide vertical-profile data
(iii) They can make measurements at presunrise or postsunset ($> 90°$)

Their disadvantages include the following:

(i) They require frequency calibration
(ii) They can suffer interference from aerosols or other absorbing species
(iii) They are based on older technology; the Dobson dates from the 1920s, the M-83 and Brewer date from the 1970s and the Diode array spectrophotometer dates from the 1980s.

We shall now give some more detailed information about the first three of these various instruments.

3.1.1 Dobson Spectrophotometer

A diagram of the Dobson Spectrophotometer is shown in Figure 5. The instrument uses a double monochromator. One monochromator is used to disperse the radiation and the second is used to reject interfering scattered radiation. A pair of wavelength-isolating slits is coupled to an optical wedge that is adjusted at one slit to give a null difference reading. The principle of operation of the instrument, is based on making total ozone measurements in the Huggins bands by using direct or scattered sunlight. By using one pair of wavelengths a differential measurement is made. By using two pairs of wavelengths the scattering effects can essentially be cancelled. The vertical ozone distribution can also be determined with the Dobson spectrophotometer, using the Umkehr technique. Observations are made on the clear zenith sky as the Sun is rising or setting. Because the ozone is primarily concentrated at a height of 20-40 km, the brightness of the zenith sky at an UV wavelength that is partially absorbed by ozone will continuously decline as the Sun approaches the horizon and then, just before it sets, the brightness will increase briefly. The ozone vertical profile is obtained through an inversion technique that takes into account ozone absorption and molecular and aerosol scattering of light that passes through the atmosphere from the Sun to the instrument.

The precision of the Dobson Spectrophotometer is 0.5% and its accuracy is of the order of 2-4%. The instrument is also subject to errors of the presence of aerosols and SO_2. Dobson Spectrophotometers have been used for many years to measure atmospheric concentrations of ozone. In 1957 there were more than 30 Dobson stations in the global network and in 1963 there were 50. The Dobson network also serves a valuable function in providing ground-based calibration and validation data for satellite-based measurements that are used to measure atmospheric ozone concentrations.

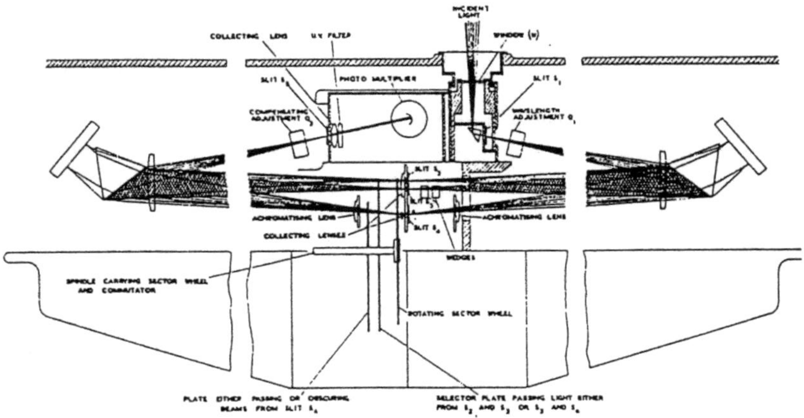

Figure 5: The Dobson Spectrophotometer

3.1.2 M-83 Filter Ozonometer

The M-83 filter ozonometer uses a 21 nm bandpass channel and a 15-nm bandpass channel. Improvements were made in 1972 by using the stated band-widths and a third filter for corrections due to aerosol effects. An improved version, the instrument M-124 (see Figure 6); it was introduced into the USSR's network of stations in 1985-6. This instrument measures ozone column content by looking directly at the Sun, using a peak spectral sensitivity at 299 nm (strong abrosption by ozone) and another peak response at 326 nm (weak absorption by ozone). The accuracy appears to be 2-4% and the precision of the measurement is approximately 5%. A network consisting of 45 stations has been operating for 25 years in the former USSR.

3.1.3 The Brewer spectrophotometer

This instrument is based on the Ebert-type spectrometer; it uses a single detector and it is generally supplied with an automatic Sun tracker. The radiances at 306.3. 310.1, 313.5. 316.8 and 320.0 nm are monitored sequentially in 1.6s. The differences between measurements of the total ozone concentration using Brewer and Dobson instruments are of the order of 3%.

3.2 Space-based instruments

These instruments measure the backscattered UV solar radiation. They operate best in the portion of the atmosphere above 15-20 km.

3.2.1 Backscattered Ultraviolet (BUV) instrument

The BUV instrument is based around a scanning double monochromator with a 1-nm bandpass, which is stepped through 12 wavelength increments every 32s through the 255-340 nm spectral region. In addition, there is a separate filter photometer with a

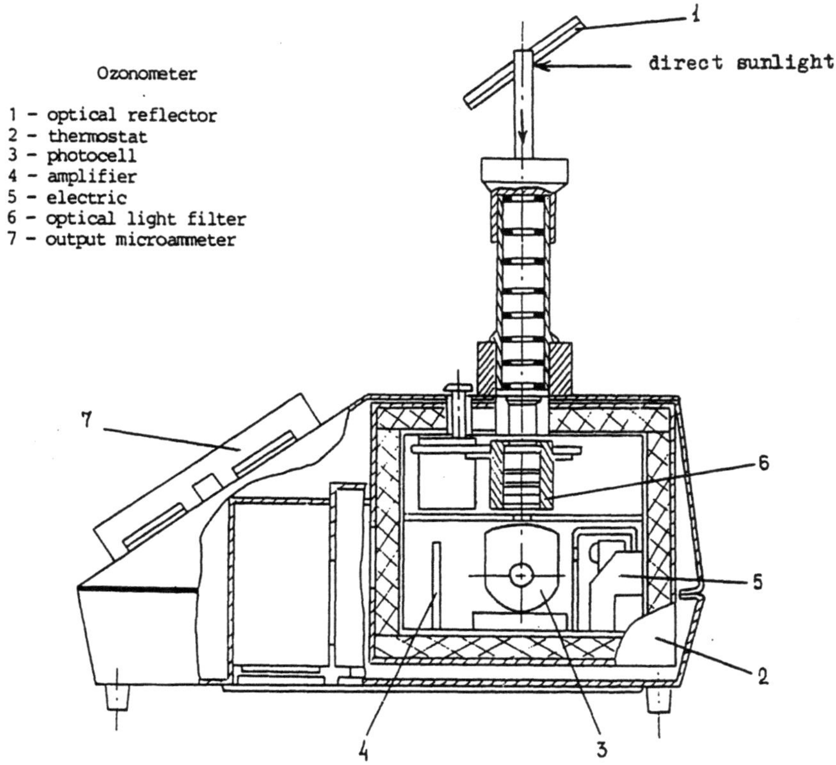

Figure 6: The M-124 ozonometer

5-nm bandpass centered at 380 nm, which is used to measure backscattered radiance. This instrument is designed to determine the vertical ozone profile while measuring backscattered solar UV radiation directed vertically upwards from the ground. Two wavelength pairs are used for measuring total ozone and eight short wavelengths are used to profile the ozone. The precision of this instrument is better than 2% for zenith angles less than about 60° and clear sky conditions. The accuracy is limited to approximately 6%. The BUV experiment was flown on the Nimbus series of satellites and was used to measure global distributions of ozone for 7 years. Complete global coverage of total ozone was obtained. An anomaly in the Earth's magnetic field in the South Atlantic, affects the distribution of the Van Allen radiation belts. These radiation levels send too much noise to the detectors; this was overcome in the Solar Backscattered UV instrument (SBUV) by adding a 50 Hz optical chopper.

3.2.2 Total Ozone Mapping Spectrometer (TOMS)

The TOMS uses a single monochromator and scans cross track over 105°. It uses an IR cloud-cover photometer to avoid problems from clouds. This instrument observes in six wavelength regions from 312.5 to 380 nm (similar to the BUV and SBUV). The

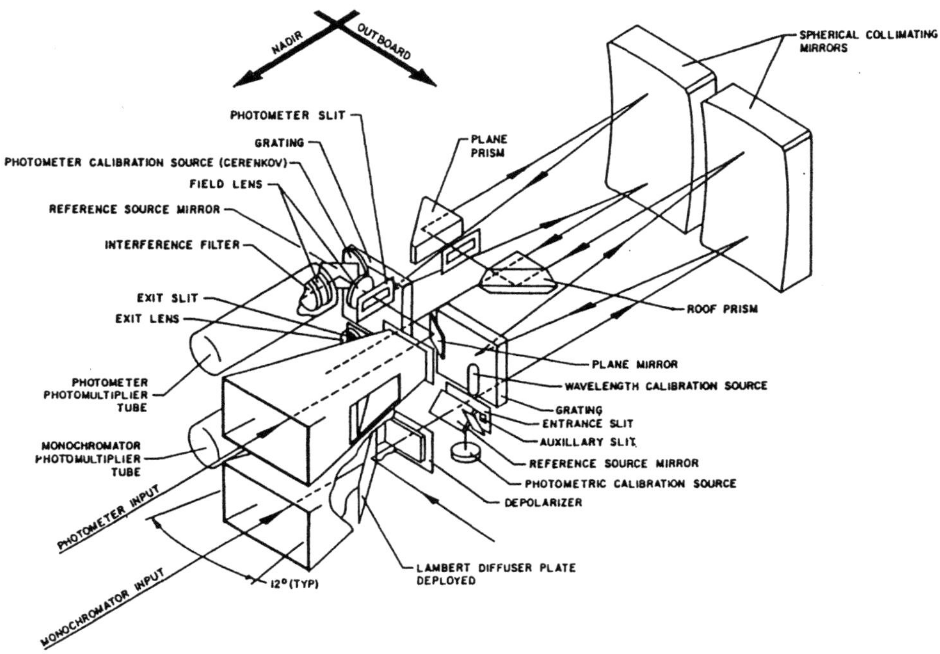

Figure 7: TOMS instrument

Temperature Humidity Infrared Radiometer (THIR) on board the satellite is used to determine cloud locations and heights.

The precision is given as 2% or better (small drift in diffuser-plate reflectance of 0.4% per year). The new absorption cross sections of ozone absorption yielded total ozone values 6% higher than the old cross sections. The TOMS instrument is flown on the Nimbus-7 satellite and has been used to measure global levels of total ozone since late 1978. Its role in relation to the study of the ozone hole over Antarctica was mentioned in chapter 1 (see section 5.3.2.).

3.2.3 Solar Mesospheric Explorer - Ultraviolet Spectrometer (SME-UVS)

This instrument has an off-axis parabolic telescope, an Ebert-Fastie grating spectrometer and a pair of photomultiplier tubes. The spectrometer is used at a pair of fixed wavelengths separated by approximately 30 nm (265.0 and 296.4 nm) and operates with a spectral resolution of 1.5 nm. The spacecraft rotates at 5 rpm and observes the Earth's limb leading and trailing the SME ground track. It measures the altitude profiles of Rayleigh-scattered UV. The random errors vary from 3.4% at a height of 48 km to 10.2% at 68 km. The temperature and pressure errors very from 4.5% at 48 km to

101% at 68 km. The SME-UVS has provided a number of ozone density data sets in the measured ozone concentration which help in the explanation of ozone photochemistry.

4 Emission - Measuring Instruments

Emission-measuring instruments involve the measurement of ozone and other species of emission, i.e. at thermal infrared wavelengths. These instruments observe the Earth's limb from a balloon or space- borne platform. Some are used in the zenith or nadir-viewing directions.

4.1 Solar Mesosphere Explorer (SME) - Near - IR Spectrometer

This instrument uses an off-axis parabolic mirror and an Ebert-Fastie spectrometer, see Figure 8. There are two channels with a resolution of 12 nm and a separation of 0.6 μm. The 1.27 μm channel is used for the emission measurements and the 1.87 μm channel is used for reference purposes, namely it is used to correct the data for radiances that are due to oxygen emission. Random error sources for the ozone density include temperature (3% at 50 km, 1% at 90 km); systematic errors include calibration errors (10% in the laboratory before launch and 20% after launch).

The SME near Infrared Spectrometer has generated a number of data sets on spatial and temporal variations of ozone concentration since its launch in October 1981. A similar instrument has been developed in Japan (Infrared Air-glow - IRA) and flown on the EXOS - C satellite.

4.2 Limb Infrared Monitor of the Stratosphere (LIMS)

This instrument is a cryogenically coolled infrared filter radiometer, see Figure 9. It has six channels which permit observation of emission by CO_2, HNO_3, O_3, H_2O and NO_2. Two channels near 15 μm determine the temperature structure. The other channels are used for molecular species measurements. Ozone, for example, is observed in the 926-1141 cm^{-1} spectral region, with a vertical field of view of 1.8 km and a horizontal field of view of 18 km. The instrument scans from above the atmosphere to below the Earth's horizon and back every 24s (144 km). The precision of this instrument was found to vary from 2-4% at 30 mb to 1% at 2 mb and 1.5% at 0.4 mb. Many experiments provided knowledge for distribution and variability of temperature and several constituents over the latitude range 84 °N to 64 °S.

4.3 TIROS Operational Vertical Sounder (TOVS)

This is a set of three instruments, the High Resoltuion Infrared Sounder (HIRS/2), the Stratospheric Sounding Unit (SSU) and the Microwave Sounding Unit (MSU). All three instruments are used to provide temperature soundings by observing emission from CO_2, H_2O, N_2O, or O_2. The surface temperature and the presence of clouds

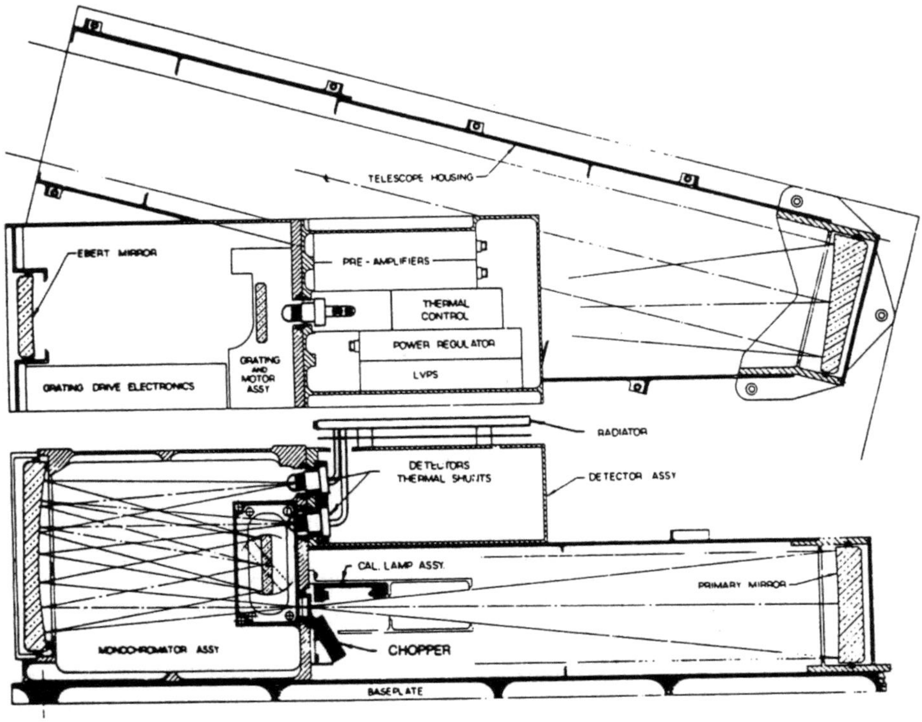

Figure 8: The SME - Near Infrared Spectrometer

can be determined. Total O_3 is determined by the use of three HIRS filter radiometer channels. The standard deviation of the measurements is approximately 7% and TOVS data have been used from 1978 to 1986.

4.4 Microwave Limb Sounder (MLS)

The MLS was originally an air-borne wavelength (1.0-1.5 mm) millimetre-wave radiometer. The present version operates in the 200-300 GHz range. Using local oscillators and filter banks (35 filters per local oscillator) the brightness temperatures at various frequencies along emission lines from various molecular species are measured. The MLS was found to have a precision of 4% and an accuracy of the order of 9%. The MLS is currently used on the Upper Atmospheric Reseach Satellite (UARS) and it is being proposed for the Earth observing system (EOS). It was also used in the Balloon Intercomparison Campaign (BOIC).

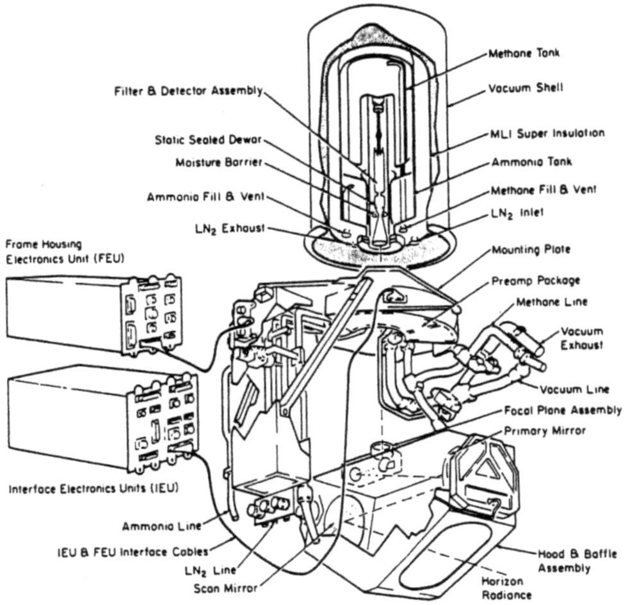

Figure 9: LIMS instrument

References

Abbas, M. M., Kunde, V. G., Mumma, M. J., Kostiuk, T., Buhl, D., and Frerking, M. A., 1979, Stratospheric sounding by infrared heterodyne spectroscopy. *Journal of Geophysical Research*, **84**, 2681-2689.

Baker, R. L., Mauldin, L. E., and Russel, J. M., 1986, Design and performance of the Halogen Occultation Experiment (HALOE) remote sensor. *Proceedings of the Society of Photo-Optical Instrumentation Engineers*, **685**, 181-191.

Barnes, R. A., and Sioneth, P. G., 1986, Design of a rocket-borne radiometer for stratospheric ozone measurements. *Review of Scientific Instrument*, **57**, 544-551.

Dobson, G. M. B., 1968, Forty Years' Research on Atmospheric Ozone at Oxford: a History. *Applied Optics*, **7**, 387-405.

Giller, J. C., and Russel, J. M., 1984, The Limb Infrared Monitor of the Stratosphere: Experiment Description, Performance and Result. *Journal of Geophysical Research*, **89**, 5125-5140.

Grant, W. B., 1989, *Ozone Measuring Instruments for the Stratosphere, Collected Works in Optics, Vol. 1* (Washington D.C: Optical Society of America)

Gushchin, G. P., Sokolenco, S. A., and Kovalyov, V. A., 1985, Total-ozone measuring instruments used at the USSR station network. In *Atmospheric Ozone* (Dordrecht: Reidel), pp. 543-546.

Health, D. F., Mateer, C. L., and Krueger, A. J., 1973, The Nimbus-4 Backscatter Ultraviolet (BUV) Atmospheric Ozone Experiment-Two Years' Operation. *Pure and Applied Geophysics*, **106-108**, 1238-1253.

Kerr, J. B., McElroy, C. T., Wardle D. I., Olafson, R. A., and Evans, W. F. J., 1985, The automated Brewer spectrometer. In *Atmospheric Ozone* (Dordrecht: Reidel), pp. 533-536.

Maudlin, L. E., Zaun, N. H., McCormick, M. P, Guy, J. H., and Vaughn, W. R., 1985, Stratospheric Aerosol and Gas Experiment II instrument: a functional description. *Optical Engineering*, **24**, 307-312.

Proffitt, M. H., and McLaughlin, R. J., 1983, Fast response dual-beam UV- absorption ozone photometer suitable for use on stratospheric balloons. *Review of Scientific Instruments*, **54**, 1719-1728.

Smith, W. L., Woolf, H. M., Hayden, C. M., Wark, D. Q., and McMillin, L. M., 1979, The TIROS-N Operational Vertical Sounder. *Bulletin of the American Meteorological Society*, **60**, 1177-1187.

Thomas, R. J., Barth, C. A., Rusch, D. W., and Sanders, R. W., 1984, Solar Mesosphere Explorer Near-Infrared Spectrometer: Measurements of 1.27 Em radiances and the inference of Mesospheric Ozone. *Journal of Geophysical Research*, **89**, 9569-9580.

Webster, C. R., and May, R. D., 1987, Simultaneous in Situ Measurements and diurnal Variations of NO, NO_2, O_3, jNO_2, CH_4, H_2O and CO_2 in the 40 to 26 Km Region Using an Open Path Tunable diode Laser-Spectrometer. *Journal of Geophysical Research*, **92**, 11931-11950.

Clouds, Cloud Observations and Cloud Feedbacks

A. Slingo

1 Introduction

The H_2O molecule is one of the fundamental building blocks of the Earth's climate system. A remarkable feature of our planet is the existence of significant amounts of water in solid, liquid and gaseous form. Water is continually recycled between these three phases of the hydrological cycle, all of which play important roles in the climate system. By far the largest mass is contained in the oceans, followed by the ice caps, ground water, lakes and rivers and finally the atmosphere.

Despite its negligible contribution to the total mass of water in the climate system, the amount of water in the atmosphere has a profound influence on the planet. The atmosphere is primarily made up of nitrogen and oxygen, which together account for about 99% of the mass. Evaporation from the oceans provides the source for atmospheric water vapour, which is very much a minor constituent. Its contribution to the total atmospheric mass averages only about 0.26% (Trenberth *et al.* 1987), although there are large geographical and temporal variations. The average residence time of water in the atmosphere is about 9 days; this is ratio of the average mass per unit area of 27 mm of water equivalent depth to the average precipitation rate of 3 mm equivalent depth per day. There is therefore plenty of time for the atmosphere to transport the water thousands of kilometres from the point of evaporation, before it is removed by condensation into clouds and subsequent precipitation. As a result, the water in the oceans is shared with the continents, making land-based life possible. Runoff through rivers which discharge into the oceans completes one of the most important elements of the hydrological cycle.

The existence of clouds and precipitation adds tremendously to the complexity of the climate system, and not just to the water budget. Evaporation cools the oceans, because it is only the most energetic water vapour molecules which can escape from the surface, so there is less energy available per molecule to be shared between those which remain. Similarly, when condensation takes place only the least energetic molecules in a gas can become bound to form a liquid, so there is more energy available per molecule for the remainder and the gas is heated by 'latent heat release'. In the atmosphere, condensation usually occurs when air is forced to rise and is thereby cooled adiabatically

NATO ASI Series, Vol. I 24
Remote Sensing and Global Climate Change
Edited by Robin A. Vaughan and Arthur P. Cracknell

until the temperature falls to the dew point. As the air continues to rise, the release of latent heat ensures that the air cools more slowly, so the temperature now follows the moist adiabatic lapse rate as opposed to the dry adiabatic. Condensation and the re-evaporation of condensate within the atmosphere are reversible processes, so the total energy content of the atmosphere is unchanged no matter how many times clouds form and dissipate. However, precipitation reaching the surface removes water from the atmosphere, which experiences a net warming because the condensation process is now not matched by re-evaporation. In the long term, the mass of water in the atmosphere is constant, so the global average precipitation rate must equal the global average evaporation rate from the surface (not only the ocean surface, but the land as well). The net warming of the atmosphere from precipitation therefore balances the net cooling of the surface from evaporation. As far as the heat balance of the system is concerned, the hydrological cycle removes energy from the surface and deposits it in the atmosphere. The distribution of the net atmospheric heating is the same as that of the precipitation; the distribution of non-precipitating cloud is irrelevant.

The cooling of the surface by evaporation and the heating of the atmosphere by precipitation are largely balanced by radiative processes. The ultimate source of energy for the climate system is the absorption of shortwave radiation from the Sun. The ultimate sink is the emission of thermal radiation to space. Over a suitably long averaging period, these must balance if the system is in equilibrium. This balance occurs only on the global scale; on a regional scale, imbalances drive the circulations of the atmosphere and oceans. For example, on average there is net radiative heating at low latitudes and net radiative cooling at high latitudes. This produces compensating circulations, the effect of which is to transport energy from the equator to the poles.

The atmosphere is relatively transparent to solar radiation. About half the radiation incident at the top of the atmosphere is absorbed at the surface. Of the rest, about 20% is absorbed in the atmosphere and the remaining 30% is reflected back to space. Clouds make the largest contribution to this reflection; the planetary albedo of 30% would be roughly halved without clouds (i.e. the planet, seen from space, would be half as bright).

The situation is more complicated for thermal (or longwave) radiation, because the absorption and emission depend not only on the distribution of absorbers but also on the temperature. As a result, both heating and cooling are possible. Water vapour, carbon dioxide, ozone and other minor and trace constituents absorb some of the upwelling radiation from the surface, replacing it by emission from the cooler atmosphere. They also radiate back downwards to the surface. The reduction of the emission to space acts as a net warming of the surface-atmosphere system, commonly known as the 'Greenhouse Effect'. The atmospheric emission acts to warm the surface, but to cool the atmosphere itself. Clouds make an important contribution to the greenhouse effect, as will be seen later.

To summarise, solar radiation is mainly absorbed at the surface, which is also warmed by longwave radiation from the atmosphere. The surface is cooled by evaporation, by the emission of longwave radiation, and to a lesser extent by direct heat transfer to the atmosphere. The atmosphere is cooled by the emission of thermal radiation and is warmed by precipitation and to a lesser extent by the absorption of solar radiation. Clouds act to cool the climate by reflecting sunlight back to space, but to warm it by contributing to the greenhouse effect. Their net effect on radiative heating is a subtle balance between these two opposing forcings.

2 Global observations of clouds from satellites

2.1 Cloud amounts and properties

The first operational weather satellite was launched in 1960 and carried television cameras to provide cloud cover information. The imaging channels on operational satellites are optimised for cloud identification, because cloud cover provides proxy information on dynamical motions and is itself an important forecast parameter. Typically, there are visible channels (e.g. 0.5-0.7 μm wavelength), infrared window channels from 10 to 12 μm and temperature and moisture sounding channels in both the infrared and microwave. The traditional qualitative uses for the images means that accurate calibration is not required. This can make them difficult to use for climate studies, although atlases of relative cloud cover have been produced (e.g. Figure 1.2 in Gill (1982)).

Quantitative information is available from the calibrated sensors on the research satellite Nimbus-7, which has been processed to create the Nimbus-7 Global Cloud Climatology (Stowe *et al.* 1989). One problem with a cloud climatology produced from a single satellite is that the diurnal cycle of cloudiness is poorly represented, which may lead to significant errors at low latitudes and over land. Sampling represents a major problem in constructing any global climatology. Proper sampling requires multiple satellites, but these are very expensive and funding is not available for multiple satellites just for research purposes. What is needed is a way of using the operational weather satellites to best advantage. This problem has been addressed in the International Satellite Cloud Climatology Project (ISCCP) by integrating data from the operational geostationary satellites with the U.S. polar orbiting satellites and by paying particular attention to inter-calibration between the various sensors (Rossow and Schiffer 1991, Rossow and Garder 1992).

ISCCP began operation in 1983 and is continuing to produce a wide range of products, details of which are given in the above references. Selected results for June 1988 are shown in the following figures. Figure 1 shows the distributions of high, medium and low cloud amount. The high cloud cover is large over the Inter-Tropical Convergence Zone (ITCZ), due to persistent deep convection. The most extensive area of high cloud is over Indonesia and the Indian Ocean. Amounts are also significant in the storm tracks in both hemispheres. In the sub-tropics there is very little high cloud and the boundary between these areas and the ITCZ is quite sharp. The distribution of medium level cloud is similar, but amounts in the ITCZ are smaller. The low cloud distribution is markedly different. For this field, the most striking features are the areas of sub-tropical stratocumulus off the coasts of California, Peru and south-west Africa. These clouds are formed at the top of the marine boundary layer, under the strong subsidence inversion produced by the stable air in the descending branch of the Hadley circulation. Significant amounts of low cloud are also evident over the oceans in the high northern (summer) latitudes.

The mean distribution of total cloud shows the result of overlapping the individual

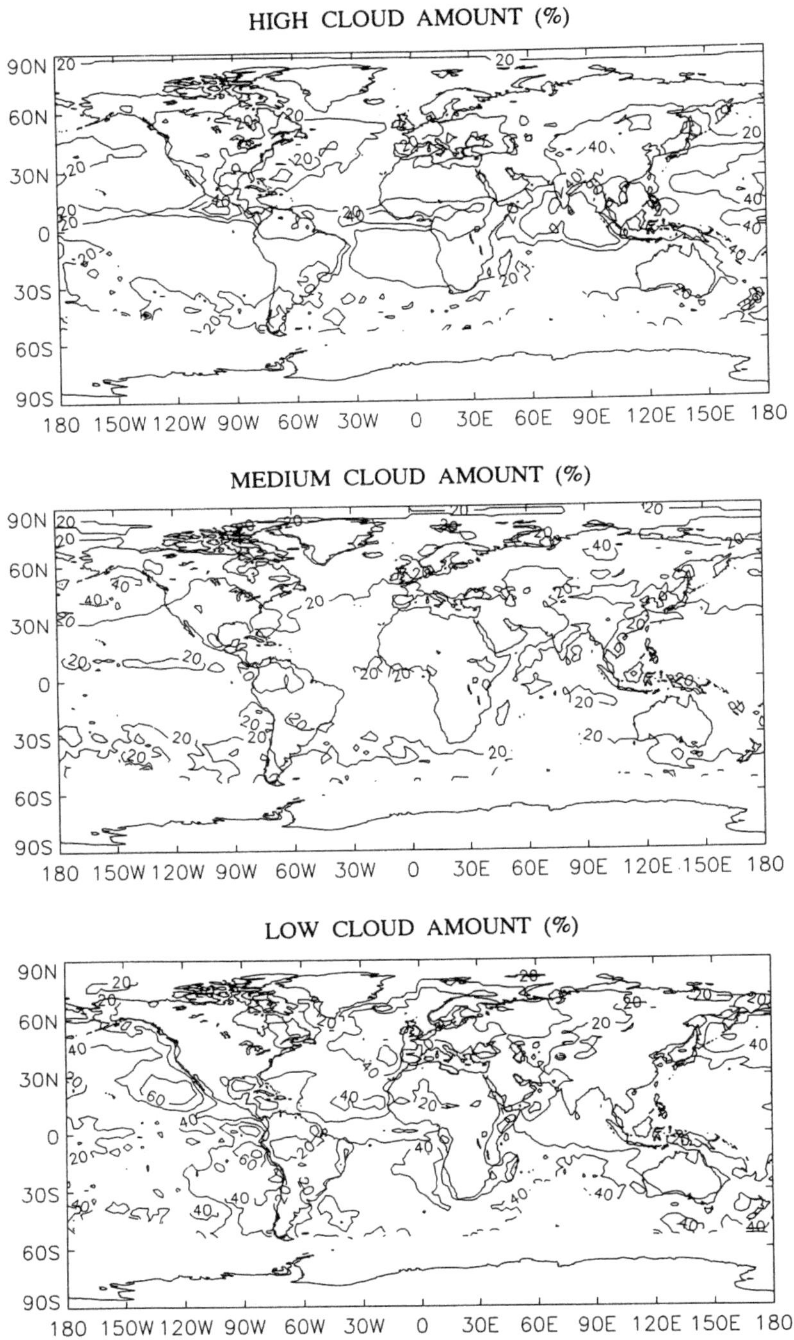

Figure 1: Monthly mean high, medium and low cloud amounts for June 1988.

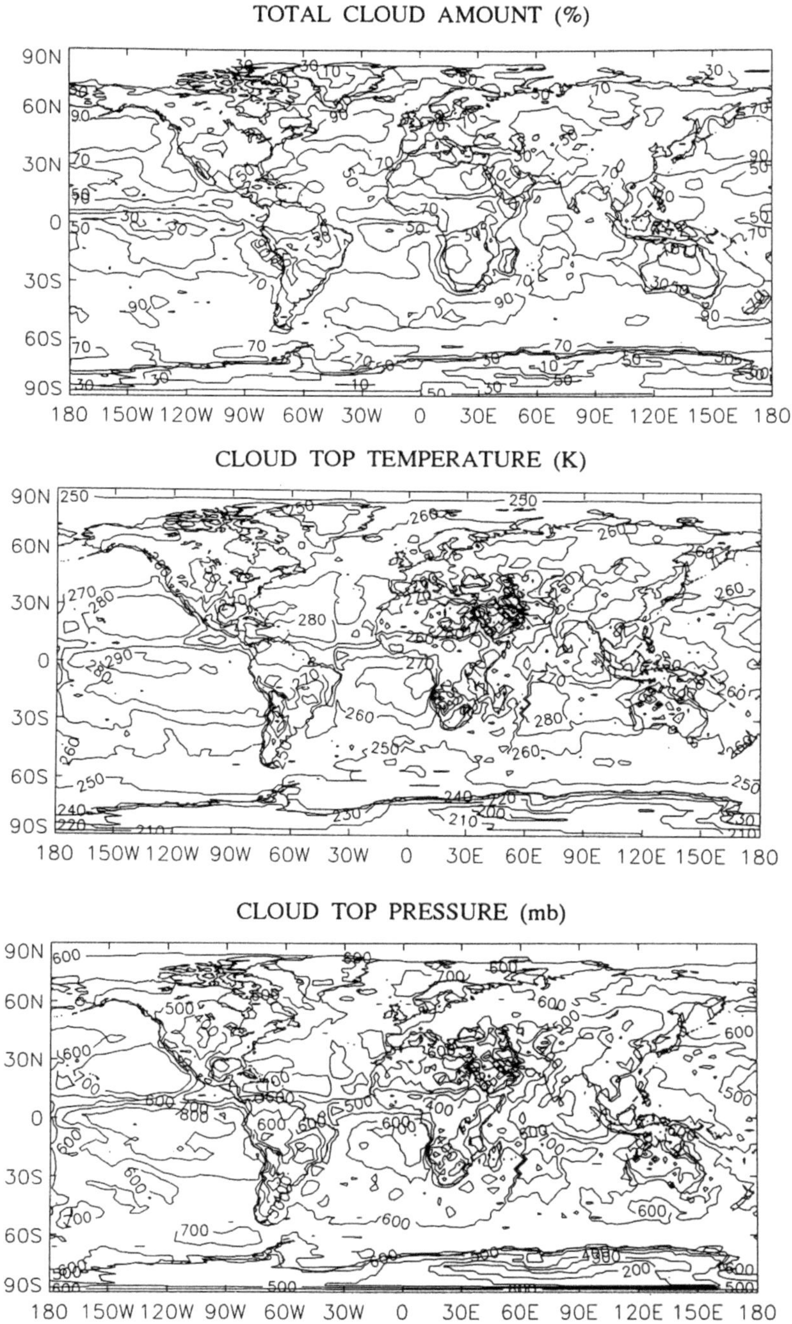

Figure 2: Monthly mean total cloud amount, cloud top temperature and cloud top pressure for June 1988.

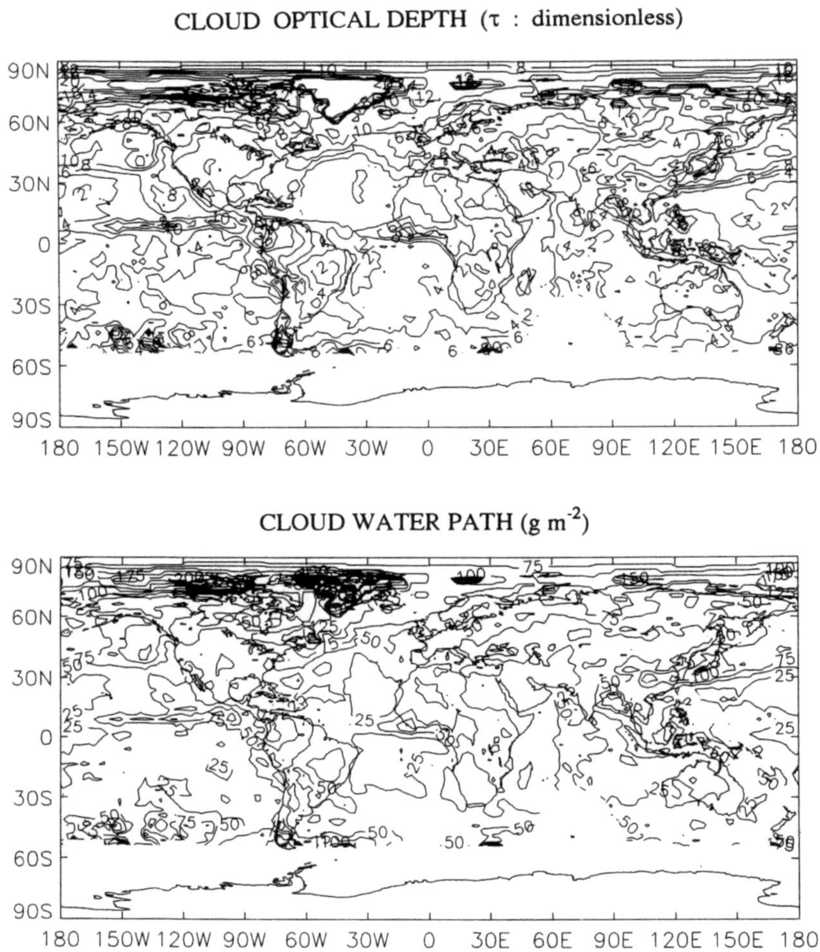

Figure 3: Monthly mean cloud optical thickness and cloud water path for June 1988, from ISCCP.

cloud layers (Figure 2). The ITCZ is still a prominent feature, but the largest amounts of cloud are over the middle and high latitude oceans, where amounts exceed 90% in several places. The mean cloud top temperature and pressure (Figure 2) are highly correlated, as expected. The highest and therefore the coldest clouds are in the ITCZ, reaching on average above 400 mb over central Africa. The clouds over the northern hemisphere land masses also reach high levels as this is northern summer and the surface is very warm, for example over north America. Over the mid-latitude oceans, the average cloud height is much lower, as a result of the large amount of low-level cloud shown earlier.

Figure 3 shows additional products derived from the visible channel data. The cloud

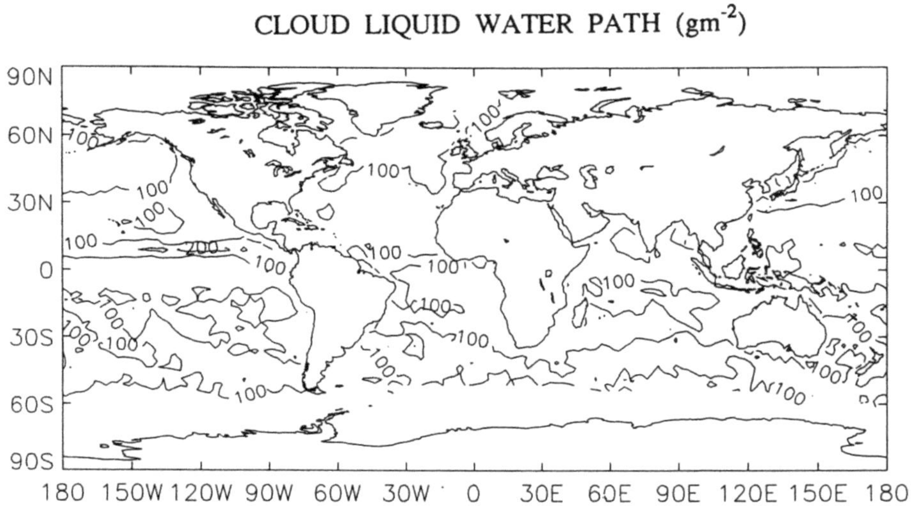

Figure 4: Monthly mean cloud liquid water path for June 1988, derived from SSM/I data (courtesy of G.L. Stephens, Colorado State University).

optical thickness shows maxima in the ITCZ, over the marine stratocumulus and over the northern oceans. At the highest northern latitudes the values are unreasonably large, due to problems with the retrieval scheme over snow-covered land and sea-ice. In general, however, the optical thicknesses are smaller than many cloud physicists would expect, but this may be due to the effect of averaging over an optical thickness distribution which is highly skewed towards low values, because of large amounts of thin and/or broken cloud. The cloud water path is also shown in Figure 3. This is derived from the optical thickness under the assumption that all clouds are composed of water drops with a mean radius of 10 μm and a particular size distribution. This is obviously an extreme assumption, particularly for high clouds which are composed of ice, but it is necessary as there is not enough information to enable an independent retrieval of particle size from the limited input radiance data. In the future, when data from more channels can be included in the retrievals, it will be possible to derive more accurate information on the cloud water paths and mean particle sizes.

Rossow and Garder (1992) present detailed comparisons of ISCCP data products with the observations performed by observers at weather stations.These observations have also been processed by Warren *et al.* (1986, 1988) to construct a parallel cloud climatology derived exclusively from surface data. The comparisons show many areas of disagreement in the details, but there is general agreement in the overall pattern and characteristics of the cloud fields. ISCCP has clearly been very successful in producing a climatology which will enable a wide range of research on clouds and climate, and in using very modest resources to derive useful products from existing operational weather satellites.

Figure 4 shows the cloud liquid water path for the same period as the ISCCP product in Figure 3, but derived by Greenwald *et al.* (1992) using data from the passive microwave sensor (SSM/I) flown by the U.S. Defense Meteorological Satellite Programme (DMSP). The values are higher than those from ISCCP, which illustrates the preliminary nature of both products as well as the need for validation with independent information, either from other satellite sensors or from *in situ* measurements. These quantities are potentially of great value in helping to develop and to check climate models, for which the cloud water path is an important predicted variable.

2.2 Effect of clouds on radiative heating

The cloud amounts and properties discussed in the previous sub-section were derived from narrow-band radiances, either from research (Nimbus-7) or operational (ISCCP) satellites. A more direct measure of the importance of clouds in climate comes from their impact on radiative heating, in particular on the reflected shortwave and emitted longwave fluxes at the top of the atmosphere, see also section 3.6 of chapter 6. Together with the incoming shortwave radiation, these define the Earth's radiation budget. The radiation budget has been derived from many satellites over the years, using both operational narrow-band data and dedicated broad-band sensors on research satellites (e.g. Kyle *et al.* 1985). The history and status of the subject a decade ago were reviewed by Stephens *et al.* (1981).

The most reliable data come from ERBE; the Earth Radiation Budget Experiment (Ramanathan *et al.* 1989, Harrison *et al.* 1990). This began operation in 1985 and produced high quality data through to 1990. ERBE used a blend of satellites to maximise the sampling but minimise the cost; three identical sets of broad-band radiometers were mounted on one dedicated research satellite (ERBS) and also on two operational weather satellites (NOAA-9 and -10). Unfortunately, all three satellites overlapped in time for only three months, but for the rest of time the sampling is still superior to that from the single satellite Nimbus-7.

An important feature of ERBE is the separate estimation of the radiation budget under clear-sky conditions. The effect of clouds is then found by differencing the clear-sky fluxes from the total values, to give the cloud radiative forcing diagnostic (see later). This has become a standard diagnostic for cloud-climate studies, not only from satellites (Ramanathan *et al.* 1989), but also from climate models. Estimates of the cloud radiative forcing have also been made by various techniques applied to data from Nimbus-7 (Ardanuy *et al.* 1989, Stephens and Greenwald 1991).

Figure 5 shows the radiation budget from ERBE for the same period as in the previous figures. The shortwave albedo (α) is the ratio of the energy reflected back to space to the incoming solar radiation. It is highest over land ice and sea-ice, for example over Greenland, and lowest where there are only small amounts of cloud overlying a relatively dark underlying surface, for example over the sub-tropical oceans. Despite small amounts of cloud, the albedo is also quite high over the bright Saharan and Arabian deserts. The ITCZ and the persistent areas of cloud at high latitudes are clearly visible. The outgoing longwave radiation (OLR) map is dominated by the minima associated with the high cold clouds in the ITCZ. The OLR is much larger in the sub-tropics, where there is little or no high cloud and so the emission comes from much

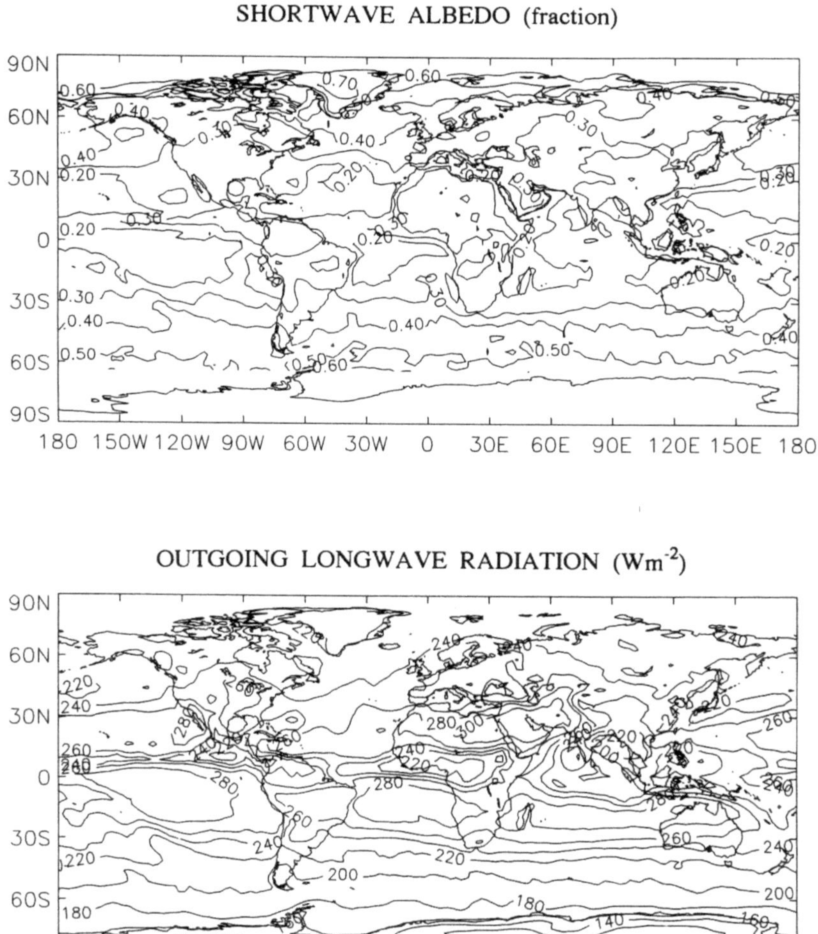

Figure 5: Monthly mean shortwave albedo and outgoing longwave radiation for June 1988, from ERBE.

lower, and hence warmer, regions. The OLR drops as one moves from the tropics to the poles, primarily because of the progressive reduction in the surface and atmospheric temperatures. Over Antarctica, for example, values less than 120 Wm^{-2} correspond to temperatures below about -60°C.

The corresponding radiation budget achieved under clear-sky conditions is shown in Figure 6. Many of the features in Figure 5 have now been removed, showing that they are due to clouds. Residual noise is due in part to the difficulty of removing the

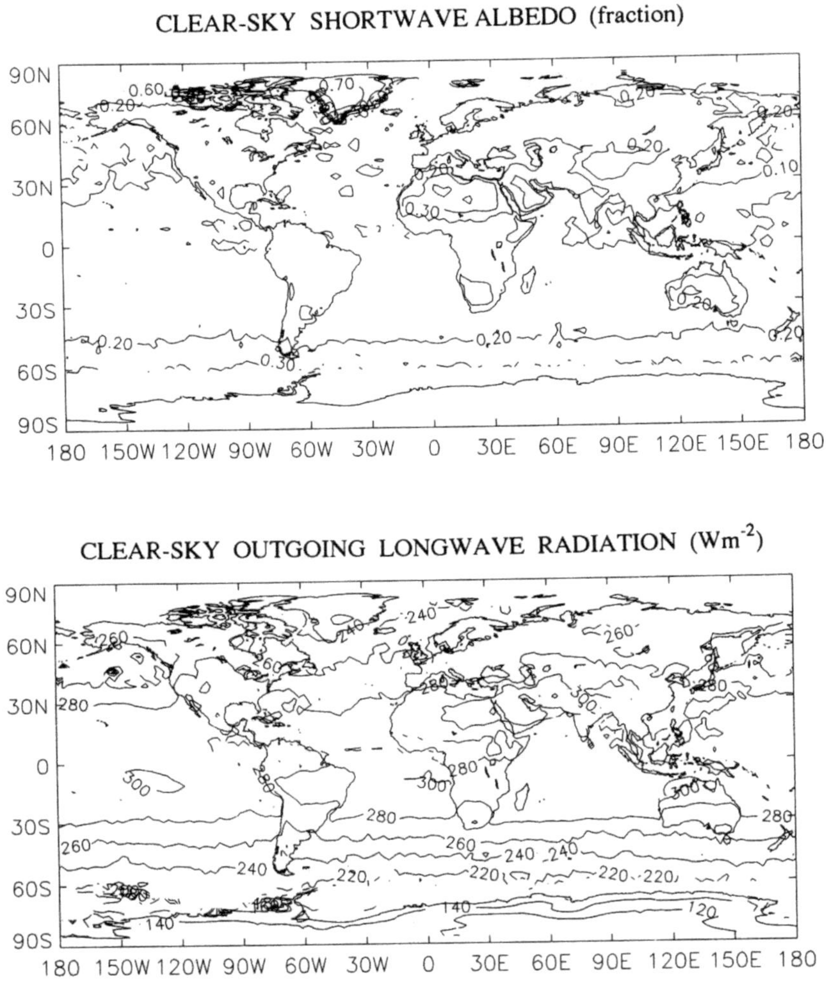

Figure 6: Monthly mean clear-sky shortwave albedo and clear-sky outgoing longwave radiation for June 1988, from ERBE.

effects of clouds completely. In the clear-sky albedo (α_C) map, Greenland and the north-African deserts are now very prominent. There is also a progressive increase in the albedo over the oceans from the equator polewards, particularly in the southern hemisphere, due to the dependence of the ocean albedo on the solar zenith angle. The clear-sky outgoing longwave radiation (OLR_C) also shows much less structure than in Figure 5. In particular, almost all of the minimum at low latitudes has disappeared. What remains is mainly due to the fact that water vapour amounts are high in the ITCZ and low in the sub-tropics, which reduces the emission in a similar way to the cloud cover.

Figure 7 shows the cloud radiative forcing diagnostic for the shortwave and longwave spectral regions and the net forcing which is the sum of the two. The cloud radiative forcing (CRF) is calculated by differencing the clear-sky (F_C) and total (F) fluxes

$$CRF = F_C - F \tag{1}$$

For the shortwave cloud radiative forcing, $SCRF$,

$$SCRF = INS(\alpha_C - \alpha) \tag{2}$$

where INS is the average value of the incoming shortwave radiation.
For the longwave cloud radiative forcing, $LCRF$,

$$LCRF = OLR_C - OLR \tag{3}$$

For the net cloud radiation forcing, $NCRF$,

$$NCRF = SCRF + LCRF \tag{4}$$

$SCRF$ is a negative quantity, because clouds are generally brighter than the underlying surface and so reflect more shortwave radiation back to space. In the shortwave, clouds thus act to cool the climate system; the average value of this cooling from the ERBE data is about -48 Wm^{-2}. Figure 7 shows that $SCRF$ is high over the ITCZ and other areas of persistent cloud cover. It is particularly large at high latitudes, both because the cloud albedo increases with solar zenith angle and because of the increase in the length of day, which increases INS in equation (2). In contrast, $LCRF$ is positive because clouds absorb the upwelling emission from the warmer underlying atmosphere and surface and replace it by the lower emission from the colder cloud tops. Clouds thus add to the greenhouse warming of the climate system, on average by about 31 Wm^{-2}. The greenhouse warming by clouds is largest over the ITCZ (Figure 7), because here there is the greatest contrast between the surface and cloud top temperatures.

The net cloud radiative forcing is negative on average (-48 + 31 = -17 Wm^{-2}), due to the large negative $SCRF$ values at high latitudes (Figure 7). The net effect of clouds on the present day climate is thus a cooling, although there are large geographical and seasonal variations in the magnitude and sign of the forcing. One intriguing feature of the $NCRF$ map is the near cancellation of the negative $SCRF$ and positive $LCRF$ values over much of the ITCZ. This is particularly striking over Indonesia. Whether this is due to some underlying physical control or simply to chance is not yet known and merits further study.

At this stage, the reader is encouraged to compare the ISCCP and ERBE maps, to see the extent to which the detailed features in the radiation budget can be explained by the cloud and surface properties. Note that the datasets come from different platforms with different time and space sampling, as well as very different retrieval and processing methodologies, so that complete consistency between the datasets cannot be expected. Much research is currently directed towards assessing the extent to which these two major climatologies can be combined to provide a more complete understanding of the system.

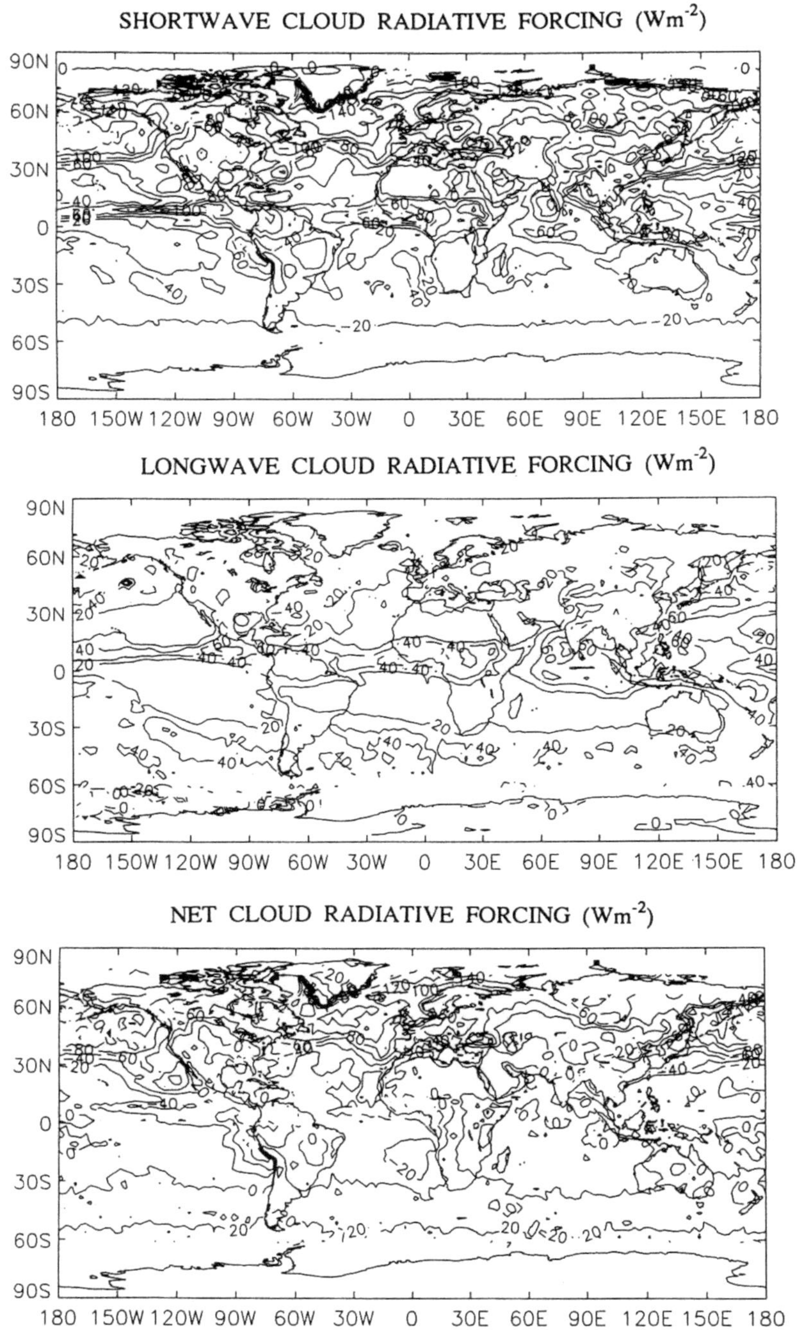

Figure 7: Monthly mean shortwave, longwave and net cloud radiative forcing for June 1988, from ERBE.

3 Clouds and the present climate

In section 1 it was shown that water is a fundamental component of the climate system and that clouds play a central role in the hydrological cycle. Many basic features of the atmospheric general circulation, such as the monsoons, the Hadley circulation and the mid-latitude storm tracks, are crucially dependent on the atmospheric heating produced by precipitation. In the previous section, it was also shown that clouds have a strong influence on the radiation balance. The ERBE data, in particular the cloud radiative forcing diagnostics, have stimulated research on the role of cloud-radiation interactions in maintaining the present climate (Hartmann *et al.* 1986). Slingo and Slingo (1988) reviewed the results of various modelling studies, carried out prior to ERBE, which demonstrated the sensitivity of the simulated climate to the radiative effects of clouds. They further investigated the role of the longwave cloud radiative forcing within the atmosphere, which is large and positive in regions of persistent deep tropical convection. In their experiments the longwave forcing interacted with the latent heat release, enhancing the precipitation in these regions and influencing both the tropical and extra-tropical circulations. The importance of the radiative forcing by these cloud systems has also been demonstrated by Randall *et al.* (1989).

A much more contentious topic has been raised more recently by Ramanathan and Collins (1991). They suggested that cirrus clouds produced by deep convection over the tropical oceans may act like a thermostat, limiting the sea surface temperatures (SSTs) in the west Pacific to below 305 K by reflecting solar radiation back to space. They used ERBE data to support this hypothesis. The data shown in Figure 8 are for the same period as in the previous figures and illustrate the argument, as well as some of the problems.

Ramanathan and Collins define the greenhouse effect, G, as the difference between the surface emission and the OLR, i,e.

$$G = \sigma {T_S}^4 - OLR \tag{5}$$

where σ is the Stefan constant and T_S is the sea surface temperature. Similarly, the clear-sky or 'atmospheric' greenhouse effect, G_A, is defined as

$$G_A = \sigma {T_S}^4 - OLR_C \tag{6}$$

The difference between G and G_A is therefore the same as the longwave cloud radiative forcing, $LCRF$.

The panels on the left of Figure 8 show the correlations between G and G_A and the sea surface temperature (SST), for 2.5 degree x 2.5 degree grid boxes over the global oceans. G_A increases with SST, mainly because of increases in atmospheric water vapour concentration as the surface temperature rises.

At the highest temperatures, the curve steepens, in part because of the non-linearity of the Clausius-Clapeyron equation and in part because of the increased importance of the water vapour continuum absorption at high temperatures. As a result, G_A increases with SST at a rate which is faster than the rate at which radiation is emitted from the surface, a situation which Ramanathan and Collins refer to as the 'super-greenhouse effect'. What balances this effect? Ramanathan and Collins argue that the

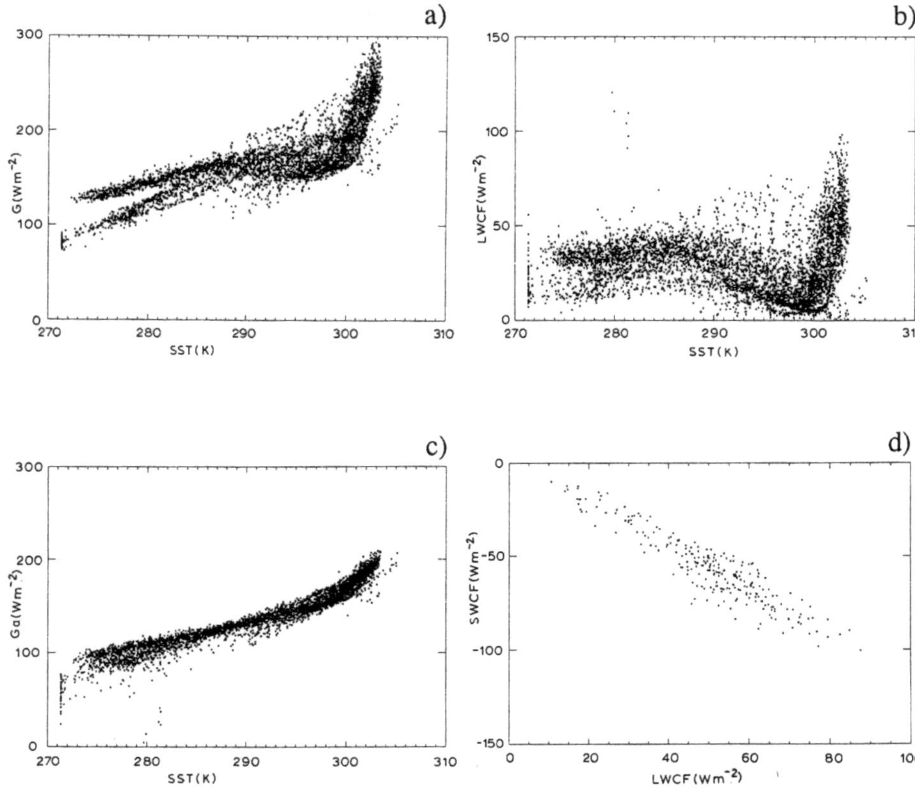

Figure 8: Scatter plots of the greenhouse effect, G, the clear-sky greenhouse effect, G_A, and the longwave cloud radiative forcing ($LCRF$ or $LWCF$) against the sea surface temperature SST (for the whole globe), and of the shortwave cloud radiative forcing ($SCRF$ or $SWCF$) against the longwave cloud radiative forcing, for the west Pacific. All the plots use monthly mean data for June 1988.

balance comes from the high clouds associated with the deep convection which forms preferentially over the warmest water. The effect of these clouds can be seen in the scatter plot of G against SST. This shows a more complicated relation than that for G_A because of atmospheric dynamics which changes the cloud cover independently of the SST. For example, the two branches at low temperatures are from the two hemispheres, with a maximum cloud contribution between 280 and 290 K from the *LCRF* provided by the storm track clouds (top right panel). In contrast, between 290 and 300 K the *LCRF* is on average much smaller, because in this SST range the descending branch of the Hadley circulation produces extensive sheets of low cloud over the subtropical oceans. These features illustrate that SST is by no means the only parameter influencing G. Nevertheless, at the highest temperatures G rises dramatically, due to the *LCRF* from the deep convective clouds in the ITCZ (top right panel). As noted in the previous section, in this region there is a remarkable near-cancellation between *LCRF* and *SCRF*, as shown in the final panel of Figure 8. Ramanathan and Collins argue that the *LCRF* is exported out of the region by the winds in the upper branch of the Hadley circulation and does not take part in the local heat budget. This leaves the large and negative *SCRF* term available to balance the super-greenhouse effect. Increases in SST thus produce extensive cloud layers which shield the surface from the solar insolation and act as a thermostat preventing further surface warming.

Many objections have been raised to this hypothesis, some of which are summarised by Stephens and Slingo (1992). The export of the *LCRF* by the Hadley circulation is not balanced by any account being taken of the effect of the convergent boundary layer circulations, which maintain and modify the cloud distribution. Evaporation is known to affect SSTs, but its importance is largely dismissed. The role of convectively-driven wind bursts, which can mix the warm surface water with the underlying cooler water, and hence limit the SSTs, is also ignored. The main theme of these and other objections is that the coupled ocean-atmosphere system in the west Pacific is believed to be much more complicated than is allowed for by Ramanathan and Collins. While their mechanism may indeed be operating at times, the importance relative to other influences is still a matter for debate. Nevertheless, their paper has helped to stimulate research on the energy budget of this most important part of the climate system.

4 Clouds and climate change

This is a large subject and it is only possible here to discuss some basic points and a few key results. For more detail, the reader is referred to Webster and Stephens (1984), to the review by Arking (1991), and to the other references encountered below.

One of the primary motivations for attempting to model climate change is the well documented increase in the concentrations of radiatively active trace gases (IPCC 1990, 1992). These increases should enhance the Earth's greenhouse effect and lead to global warming, provided that there are no compensating cooling influences. One major problem is that the radiative perturbations involved are small. Even the extreme example of an instantaneous doubling of CO_2 levels decreases the OLR by only about 2.5 Wm^{-2}, which represents a 1% change to the average OLR of about 240 Wm^{-2}. However, it is believed that the effects of such small perturbations are amplified by positive feedbacks, which magnify the sensitivity of the climate to the trace gas increase. Two positive feedbacks are well documented. First, as the climate system warms, the atmosphere is able to hold more water vapour, which is the most important greenhouse gas. This further increases the magnitude of the greenhouse effect and amplifies the importance of the trace gas warming. Secondly, as the climate warms there should be less snow cover over land and smaller amounts of sea-ice over the oceans. At present, snow and ice cool the system by reflecting solar radiation back to space, just as clouds do. A reduction in snow and ice cover would lower the surface albedo, warm the surface and hence further amplify the trace gas warming.

Cloud feedbacks are much more complicated. The reason is that there are many different types of cloud, whose properties and coverage are controlled by many different physical processes and which affect the radiation budget in many different ways. ERBE has shown that, on average, clouds cool the present climate, because the 31 Wm^{-2} enhancement of the greenhouse effect is exceeded by the 48 Wm^{-2} increase in the reflection of solar radiation to space. But Figure 7 shows that there are large geographical variations in the net cloud radiative forcing ($NCRF$), with some clouds contributing a net warming. Low cloud produces a negative $NCRF$ and so an increase would act to cool the climate and provide a negative feedback on global warming. But thin tropical cirrus cloud (which is much colder than the underlying surface and hence strongly enhances the greenhouse effect) produces a positive $NCRF$. An increase in the amount of this cloud type could therefore provide a positive feedback. The overall cloud feedback thus depends critically on the detailed changes in the three dimensional cloud distribution.

In climate model experiments, trace gas warming tends to deepen the tropical troposphere and increase the amount of cloud at the highest levels. This enhances the greenhouse effect and produces a positive feedback (e.g. Wetherald and Manabe 1988). However, the magnitude of this feedback and its importance relative to other changes in cloud amounts can vary significantly between models, or even between different versions of the same model. For example, Mitchell *et al.* (1989) performed experiments with three versions of the U.K. Meteorological Office model. Each version was used to perform a control integration, followed by one in which the CO_2 concentration was doubled, and a new warmer equilibrium climate achieved. In the first experiment with a simple cloud prediction scheme and fixed cloud radiative properties, a large positive cloud amount feedback led to a 5.2 degK increase in the global average surface temperature. In the second experiment, the cloud prediction scheme was replaced by a more physically based method, but the cloud radiative properties remained fixed; the warming dropped to only 2.7 degK, showing the strong sensitivity to details of the cloud scheme. The third experiment illustrated the effects of cloud radiative property feedbacks, which are discussed in the next section.

4.1 Cloud radiative property feedbacks

The situation is even more complicated than discussed above, because changes are also possible in the physical properties of clouds, even if the cloud distribution remains the same. The radiative properties of clouds are largely determined by the cloud optical thickness, τ, which determines the transmission, T, of a parallel beam of radiation incident normally on the cloud

$$T = exp(-\tau) \tag{7}$$

The following discussion is limited to water clouds, although much of the argument also applies to ice clouds. It is easy to show that, to a good approximation, the visible optical thickness depends on the liquid water path LWP (the vertically integrated mass of water per unit area) and on the effective radius, r_e, (a measure of the mean radius of the cloud drops)

$$\tau \approx (3LWP)/(2\rho r_e) \tag{8}$$

where ρ is the density of water. Changes in either or both LWP and r_e can therefore change τ and make clouds brighter or darker. The climate system is very sensitive to such changes. Slingo (1990) showed that the radiative perturbation from an instantaneous doubling of CO_2 concentrations (2.5 Wm^{-2}) could be balanced by any one of the following global changes in boundary layer clouds: an increase in the average cloud cover from the present value of 25% to 29-30%, an increase in LWP from 50 gm^{-2} to 60-68 gm^{-2} or a decrease in r_e from 10 μm to 8-8.5 μm (the spread comes from using different versions of the same climate model). These are modest changes, which emphasise the sensitivity of climate to these cloud parameters.

How might such changes in liquid (or ice) water paths and particle size come about? Until recently, climate models ignored changes in cloud water contents. The effects of such changes might be very important (e.g. Somerville and Remer 1984), although the dependence of water content on temperature which was derived by Somerville and Remer from limited aircraft observations does not agree with results from ISCCP (Tselioudis *et al.* 1992). The study by Mitchell *et al.* (1989) mentioned above included a third experiment in which the model calculated the cloud radiative properties from the liquid water contents given by the new cloud scheme. The surface warming on doubling CO_2 dropped further to only 1.9 degK. The drop from 2.7 degK to 1.9 degK was due to a negative cloud radiative property feedback; the clouds in the integration with doubled CO_2 had higher liquid water contents than in the control; they were therefore brighter and so reflected more solar radiation back to space and helped to reduce the warming. In their first experiment (5.2 degK warming) the overall cloud feedback was positive, but in the final experiment it was negative.

Studies of the effects of possible changes in particle size in such integrations are at an early stage. But mechanisms by which such changes might occur have been suggested and need to be examined by climate models (Figure 9). Twomey *et al.* (1984) pointed out that cloud drop sizes may be sensitive to changes in the atmospheric sulphur cycle. Cloud drops are formed by condensation on sub-micron diameter hygroscopic aerosol particles which are present throughout the troposphere. These aerosols are known as cloud condensation nuclei (CCN). Changes in the CCN concentration might alter the number of cloud drops which can grow and hence alter the mean radius for a given value of LWP.

Twomey *et al.* noted that pollution produced by burning fossil fuels consists not only of CO_2 but also SO_2, the gaseous precursor of a component of the CCN population. Increased pollution could therefore increase the number of CCN, increase the number of cloud drops and hence reduce the value of r_e (provided that LWP does not change). Equation (8) shows that this would increase the cloud optical thickness and hence the cloud shortwave albedo, leading to a cooling influence on climate and hence a negative feedback on the trace gas warming from the CO_2. It is therefore possible that the trace

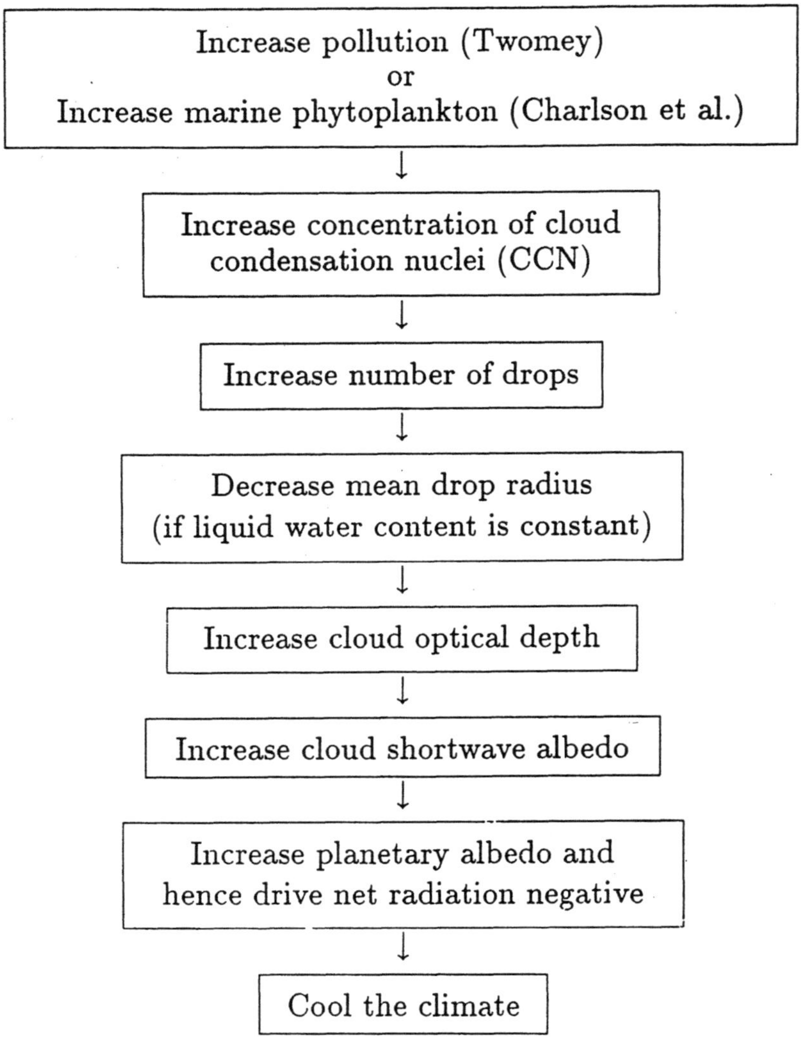

Figure 9: Simplified flow diagram of mechanisms by which changes in mean cloud particle size may influence the climate.

gas warming is being offset by the cooling produced by SO_2 emissions, so that reducing the latter might even accelerate global warming (Wigley 1991, Charlson *et al.* 1992).

A similar mechanism but involving biological organisms was suggested by Charlson *et al.* (1987). They argued that much of the sulphate aerosol over the oceans comes not from pollution but from dimethylsulphide, which excreted by marine phytoplankton. Changes in the plankton population could therefore influence cloud albedo and hence climate in much the same way as the changes in pollution discussed by Twomey.

Charlson *et al.* went further, to discuss the possibility that regulation of the climate system is therefore possible by such marine organisms, tending to keep the climate within an optimum range for life. This is an example of the 'Gaia' mechanism, proposed by Lovelock, one of the authors of the paper.

4.2 Quantitative analysis of cloud feedbacks

Understanding how cloud feedbacks operate in climate models is an important part of current research. This work has been helped by the application of simple feedback analyses and by idealised experiments. The feedback analysis described below is based on the work of Schlesinger (1989) and Cess *et al.* (1990).

In equilibrium the average net radiation per unit area at the top of the atmosphere (N) should be zero;

$$N = Q - F = 0 \tag{9}$$

where Q is the absorbed shortwave radiation and F is the outgoing longwave radiation (OLR). Climate change can occur when the system is driven out of radiative equilibrium by a perturbation to either (or both) Q and F;

$$\Delta N = \Delta Q - \Delta F \tag{10}$$

The response of the global mean surface temperature T_S may be expressed in terms of a climate sensitivity parameter, λ, which relates the initial radiative perturbation at the top of the atmosphere to the final change in T_S required to restore equilibrium;

$$\Delta T_S = \lambda \Delta N \tag{11}$$

λ can be expressed in terms of the gain of the climate system in the absence of feedbacks G_0 , plus a feedback factor f

$$\lambda = G_0/(1 - f) \tag{12}$$

If we assume that F is proportional to the surface emission ($\sigma\, T_S{}^4$), then from the above;

$$G_0 = (T_S/4F) = 0.3 \text{ K W}^{-1}\text{m}^2 \tag{13}$$

Positive values of f produce $\lambda > G_0$ and hence a positive feedback. For f=0 there is no feedback, while negative values of f produce $\lambda < G_0$ and a negative feedback.

This analysis was applied to a simple climate change experiment performed with 19 different atmospheric general circulation models by Cess *et al.* (1990). The experiment differed from the conventional approach, in which a radiative perturbation ΔN is applied to a climate model which includes a representation of the oceans. The model is then integrated to reach a new equilibrium with a different T_S, allowing λ to be found from equation (11). Such experiments take a long time to reach equilibrium, because of the large thermal inertia of the oceans. Cess *et al.* ran the conventional experiment backwards, by working solely with atmospheric models and imposing a change in T_S over the oceans and allowing the atmosphere to come into equilibrium. The difference between the top of the atmosphere radiation balance in this run and in a control provides the value of ΔN to be used in equation (11). This experiment reaches equilibrium very quickly, allowing trivial computing resources to be used. On the other hand, the ΔT_S perturbation is highly idealised, so the response will not necessarily be the same as in the conventional approach. Indeed, when the Cess *et al.* method was applied to the same three versions of the U.K. Meteorological Office model used in the Mitchell *et al.* (1989) study discussed above, the magnitudes of the cloud feedbacks changed substantially (Senior and Mitchell 1992). This demonstrates that the Cess *et al.* approach can only provide a crude first estimate of the magnitude of the cloud feedback operating during climate change; the best estimate still has to come from running conventional experiments, despite their heavy demand on computer resources.

In the Cess *et al.* (1990) study, the radiation budget was archived for both total and clear-sky conditions. This allowed the clear-sky feedbacks(primarily water vapour) to be separated from the cloud feedbacks. The upper frame in Figure 10 shows the clear-sky sensitivity parameter λ_C (open circles)and the total value λ (filled circles) for each of the 19 models. There is reasonable agreement as to the magnitude of λ_C;

$$\lambda_C \approx 0.5 \text{ K W}^{-1}\text{m}^2 \tag{14}$$

The fact that λ_C is larger than the value of G_0 given by equation (13) illustrates the positive water vapour feedback, which appears to operate in much the same way in all the models. However, the filled dots show that the overall sensitivity parameter λ varies by a factor of three between the models. The separation between the filled and open dots shows the role of the cloud feedback. For the models on the right of the frame, λ is greater than λ_C, so the cloud feedback is positive. For model numbers 1-3 and 5, λ is less than λ_C , so the cloud feedback is negative.

The disparity in the cloud feedbacks is shown in another way in the lower frame in Figure 10, where λ is now plotted against the change in the cloud radiative forcing (ΔCRF), normalised by ΔN (Cess *et al.* used G instead of ΔN). The strong correlation in this figure confirms that the spread in λ is due almost entirely to variations in the magnitude and sign of the cloud radiative feedback. The cloud feedback varies from weak negative on the left of this frame to strong positive on the right. The models thus fail to agree on either the magnitude or the sign of the feedback. Cess *et al.* examined the possibility that this disagreement was due to the fact that some of the models included the negative cloud radiative property feedback discussed earlier, but this did not explain the disparity in the results. It appears that the cloud feedback is very sensitive to details in the formulation of a model. One lesson from this comparison is therefore that more work is needed to improve the physical basis of climate models. Otherwise, uncertainties in the cloud feedback will continue to limit the ability of climate models to provide quantitative predictions of future climate change.

It is possible that the Cess *et al.* study provides a misleadingly pessimistic picture of the state of climate modelling. The study used a wide variety of models, some of which were in an early stage of development and had not been validated extensively against satellite and other data. It would be desirable to repeat the study, including only those models whose simulations are in good agreement with, for example, the ERBE data.

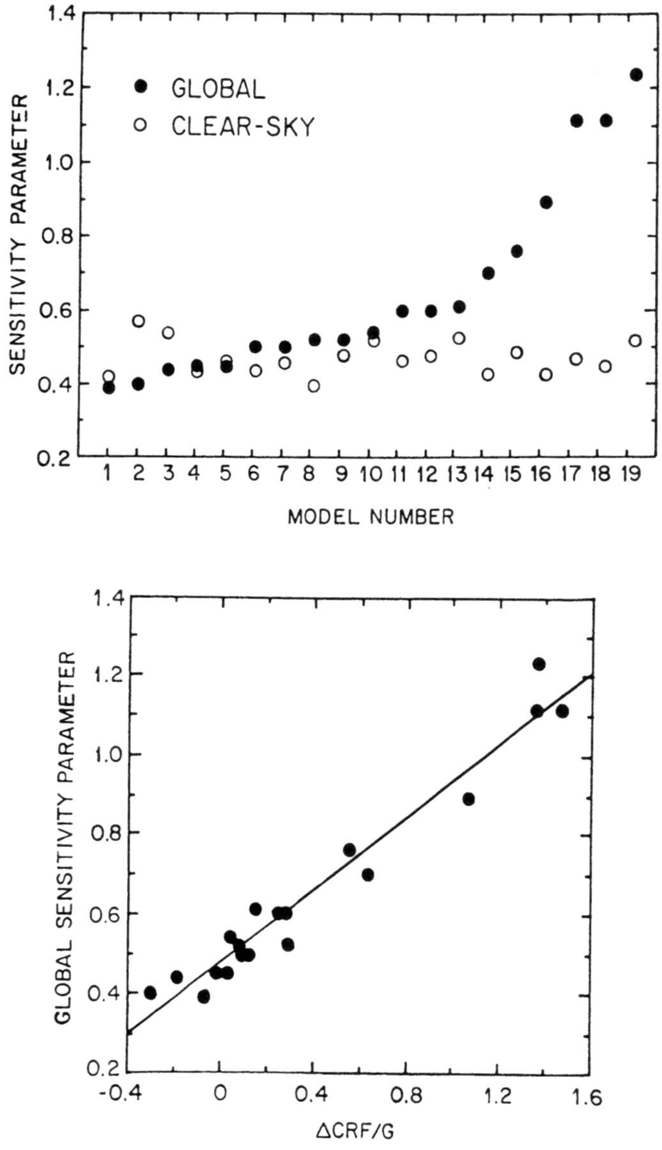

Figure 10: Global and clear-sky sensitivity parameters as functions of the model used (top) and of one measure of the cloud radiative feedback (bottom). From Cess *et al.* (1990).

5 Future satellite observations of clouds

A very brief account is given here of plans for future satellite missions which will provide information on clouds and the Earth's radiation budget.The general strategy to be followed is discussed by Schiffer *et al.* (1989) and by Kandel (1991).

The scanning radiometers on the last of the ERBE satellites ceased operation in 1990, so at present there are no broad-band measurements being made of the radiation budget. The next opportunity for such measurements will be the launch of the French SCARAB (SCAnner for RAdiation Budget) instrument on the Russian METEOR satellites from 1993 onwards. The direct descendants of the ERBE sensors are those for CERES (Clouds and the Earth's Radiant Energy System), which scan not only across the satellite track (as in ERBE) but also in other directions (Wielicki and Barkstrom 1991). The first CERES instruments should fly in about 1997 on the joint U.S./Japanese Tropical Rainfall Measuring Mission (TRMM) satellite. The main purpose of this satellite is to make measurements of tropical rainfall, using a comprehensive package of visible, infrared and microwave radiometers and the first space-borne precipitation radar (Simpson *et al.* 1988, Simpson 1991). A wide range of sensors is included in this mission, because no single sensor is sufficiently sensitive to precipitation over the very wide dynamic range which needs to be covered (Prabhakara *et al.* 1986, Arkin and Ardanny 1989).

The most detailed measurements of clouds and the radiation budget will be made from the polar platforms, to be launched by NASA as part of the Earth Observing System (EOS) and also by ESA and Japan. Several platforms should be launched from about 1998 into the next century. The full potential of the CERES system will be realised on the EOS platforms, which will also carry MODIS, the Moderate Resolution Imaging Spectrometer (King *et al.* 1992). MODIS (and similar instruments on the platforms from the other agencies) will allow retrievals to be made not only of cloud amounts but also of cloud properties, such as the water paths and particle sizes. A retrieval scheme is planned which will optimise the radiation budget and cloud retrievals simultaneously, using data from all the relevant sensors. In this way, a comprehensive dataset which combines and extends the attributes of ERBE and ISCCP should be produced.

At present, it seems that CERES may not be mounted on the ESA platforms, known as POEM (Polar Orbiting Earth Mission), but a radiation budget sensor will undoubtedly be included; possibly a derivative of SCARAB. ESA is also considering the inclusion of a radiation budget sensor on the Meteosat Second Generation satellites (MSG), due for launch around 1998. This would provide excellent temporal coverage from geostationary orbit, complementing the polar orbiters. Finally, the U.K. is examining the feasibility of building the first space-borne cloud radar, which would provide information on the three-dimensional distribution of cloud particles.

References

Ardanuy, P. E., Stowe, L. L., Gruber, A., Weiss, M., and Long, C. S., 1989, Longwave cloud radiative forcing as determined from Nimbus-7 observations. *Journal of Climatology*, **2**, 766-799.

Arkin, P. A., and Ardanuy, P. E., 1989, Estimating climatic-scale precipitation from space: a review. *Journal of Climatology*, **2**, 1229-1238.

Arking, A., 1991, The radiative effects of clouds and their impact on climate. *Bulletin of American Meteorological Society*, **71**, 795-813.

Cess, R. D., Potter, G. L., Blanchet, J. P., Boer, G. J., Delgenio, A. D., Deque, M., Dymnikov, V., Galin, V., Gates, W. L., Chan, S. J., Kiehl, J. T., Lacis, A. A., Letreut, H., Li, Z. X., Liang, X. Z., Mcaavaeny, B. J., Meleshko, V. P., Mitchell, J. F. B., Morcrette, J. J., Randall, D. A., Rikus, L., Roeckner, E., Royer, J. F., schlese, V., Sheinin, D. A., Slingo, A., Sokolov, A. P., Taylor, K. E., Washington, W. M., Wetherald, R. T., Yagai, I., and Zhang, M. H., 1990, Intercomparison and interpretation of climate feedback processes in 19 atmospheric general circulation models. *Journal of Geophysical Research*, **95**, 16601-16615.

Charlson, R. J., Lovelock, J. E., Andreae, M. O., and Warren, S. G., 1987, Oceanic phytoplankton, atmospheric sulphur, cloud albedo and climate. *Nature*, **326**, 655-661.

Charlson, R. J., Schwartz, S. E., Hales, J. M., Cess, R. D., Coakley, J. A., Hansen, J. E., and Hofmann, D. J., 1992, Climate forcing by anthropogenic aerosols. *Science*, **255**, 423-430.

Gill, A. E., 1982, *Atmosphere-Ocean dynamics.* (London: Academic Press)

Greenwald, T. J., Stephens, G. L., Vonder Haar, T. H., and Jackson, D. L., 1992, Aphysical retrieval method of liquid water over the global oceans using SSM/I observations. *Journal of Geophysical Research* (in press).

Harrison, E. F., Minnis, P., Barkstrom, B. R., Ramanathan, V., Cess, R. D., and Gibson, G. G., 1990. Seasonal variation of cloud radiative forcing derived from the Earth Radiation BudgetExperiment. *Journal of Geophysical Research*, **95**, 18687-18703.

Hartmann, D.L., Ramanathan, V., Berroir, A., and Hunt, G.E., 1986, Earth radiation budget data and climate research. *Review of Geophysics*, **24**, 439-468.

Houghton, J. T., Jenkins, G. J., and Ephramus, J. J. (Editors), 1990, *Climate Change. The IPCC Scientific Assessment* (Cambridge; New York: Cambridge University Press).

Houghton, J. T., Callander, B. A., and Varney, S. K. (Editors), 1992, *Climate Change 1992. The Supplementary Report to the IPCC Scientific Assessment.* (Cambridge; New York: Cambridge University Press).

Kandel, R., 1991, Radiation and the Energy balance. *Report of the Earth Observation User Consultation Meeting*, ESA SP-1143 (European Space Agency), pp. 100-114..

King, M. D., Kaufman, Y. J., Menzel, W. P., and Tanri, D., 1992, Remote sensing of cloud, aerosol and water vapor properties from the Moderate Resolution Imaging Spectrometer(MODIS). *IEEE Transactions on Geoscience and Remote Sensing*, **30**, 2-27.

Kyle, H. L., Ardanuy, P. E., and Hurley, E. J., 1985, The status of the Nimbus 7 Earth radiation budget data set. *Bulletin of American Meteorological Society*, **66**, 1378-1388.

Mitchell, J. F. B., Senior, C. A., and Ingram, W. J., 1989, CO_2 and climate: a missing feedback? *Nature*, **341**, 132-134

Prabhakara, C., Short, D. A., Wiscombe, W., Fraser, R. S., and Vollmer, B. E., 1986, Rainfall over oceans inferred from Nimbus 7 SMMR: Application to 1982-1983 El Niño. *Journal of Climate and Applied Meteorology*, **25**, 1464-1474.

Ramanathan, V., Cess, R. D., Harrison, E. F., Minnis, P., Barkstrom, B. R., Ahmad, E., and Hartmann, D., 1989, Cloud-radiative forcing and climate: Results from the Earth Radiation Budget Experiment. *Science*, **243**, 57-63.

Ramanathan, V., and Collins, W., 1991, Thermodynamic regulation of ocean warming by cirrus clouds deduced from observations of the 1987 El Nino. *Nature*, **351**, 27-32.

Randall, D. A., Harshvardhan, Dazlich, D. A., and Corsetti, T. G., 1989, Interactions among radiation, convection, and large-scale dynamics in a general circulation model. *Journal of Atmospheric Science*, **46**, 1943-1970.

Rossow, W. B., and Garder, L. C., 1992, Cloud detection using satellite measurements of infrared and visible radiances for ISCCP. *Journal of Climatology* (in press)

Rossow, W. B., and Schiffer, R. A., 1991, ISCCP cloud data products. *Bulletin of American Meteorological Society*, **72**, 2-20.

Schiffer, R. A., Arking, A., and Barkstrom, B., 1989, Earth radiation budget measurements for GEWEX: An assessment and proposed research strategy. In *Global Energy and Water Cycle Experiment (GEWEX). Report of the First Session of the JSC Scientific Steering Group for GEWEX.* WCRP-25 (WMO/TD-No. 321), WMO, Geneva. Appendix C

Schlesinger, M. E., 1989, Quantitative analysis of feedbacks in climate model simulations. In *Understanding Climate Change, Geophysical Monogragh, 52*, edited by A. Berger, R. E. Dickinson, and J. W. Kidson (Washington DC: American Geophysical Union), pp. 177-187.

Senior, C. A., and Mitchell, J. F. B., 1992, CO_2 and climate: The impact of cloud parametrization.*Journal of Climatology* (in press).

Simpson, J., 1991, Climate variability and tropical rain: necessity for space measurement. *Proceedings of the Second Symposium on Global Change Studies* (Boston: American Meteorological Society), pp. 5-10.

Simpson, J., Adler, R. F., and North, G. R., 1988, A proposed tropical rainfall measuring mission (TRMM) satellite. *Bulletin of American Meteorological Society.* **69**, 278- 295.

Slingo, A., 1990, Sensitivity of the Earth's radiation budget to changes in low clouds. *Nature*, **343**, 49-51.

Slingo, A., and Slingo, J. M., 1988, The response of a general circulation model to cloud longwave radiative forcing. I: Introduction and initial experiments. *Quarterly Journal of the Royal Meteorological Society*, **114**, 1027-1062.

Somerville, R. C. J., and Remer, L. A., 1984, Cloud optical thickness feedbacks inthe CO_2 climate problem. *Journal of Geophysical Research*, **89**, 9668-9672.

Stephens, G. L., Campbell, G. G., and Vonder Haar, T. H., 1981, Earth Radiation Budgets. *Journal of Geophysical Research*, **86**, 9739-9760.

Stephens, G. L., and Greenwald, T. J., 1991, The Earth's radiation budget and its relation to atmospheric hydrology: 2. Observations of cloud effects. *Journal of Geophysical Research*, **96**, 15325-15340.

Stephens, G. L., and Slingo, A., 1992, Climate feedbacks: An air-conditioned greenhouse. *Nature*, **358**, 369-370.

Stowe, L. L., Yeh, H. Y. M., Eck, T. F., Wellemeyer, C.G., Kyle, H. L., and the Nimbus-7 cloud data processing team, 1989, Nimbus-7 global cloud climatology. Part II: First Year results. *Journal of Climate*, **2**, 671-709.

Trenberth, K. E., Christy, J. R., and Olson, J. G., 1987, Global atmospheric mass, surface pressure, and water vapour variations. *Journal of Geophysical Research*, **92**, 14815-14826.

Tselioudis, G., Rossow, W., and Rind, D., 1992, Global patterns of cloud optical thickness variation with temperature. *Journal of Climate*, 5, 1484-1497.

Twomey, S. A., Piepgrass, M., and Wolfe, T. L., 1984, An assessment of the impact of pollution on global cloud albedo. *Tellus*, **36B**, 356-366.

Warren, S. G., Hahn, C. J., London, J., Chervin, R. M., and Jenne, R. L., 1986, Global distribution of total cloud cover and cloud type amounts over land. NCAR/TN-273+STR, NCAR, Boulder, Colorado, USA.

Warren, S. G., Hahn, C. J., London, J., Chervin, R. M., and Jenne, R. L., 1988, Global distribution of total cloud cover and cloud type amounts over the ocean. NCAR/TN-317+STR, NCAR, Boulder, Colorado, USA.

Webster, P. J., and Stephens G. L., 1984, Cloud-radiation interaction and the climate problem. in *The Global Climate*, edited by J. T. Houghton (Cambridge; New York: Cambridge University Press), pp. 63-78

Wetherald, R. T., and Manabe, S., 1988, Cloud feedback processes in a general circulation model. *Journal of the Atmospheric Scieces*, **45**, 1397-1415.

Wielicki, B. A., and Barkstrom, B. R., 1991, Clouds and the Earth's Radiant Energy System (CERES): an Earth Observing System Experiment. *Proceedings of the Second Symposium on Global Change Studies* (Boston: American Meteorological Society), pp. 11-16,

Wigley, T. M. L., 1991, Could reducing fossil-fuel emissions cause global warming? *Nature*, **349**, 503-506.

A Technique for Remote Sensing the Atmospheric Temperature and Pressure, Based on Fabry – Perot Interferometry of Rotational Raman Scattering Spectrum

I. D. Ivanova, L. L. Gurdev, and V. M. Mitev

1 Introduction

The methods for laser remote sensing of the atmospheric temperature are based on the temperature dependence of the interaction of the light with the atmospheric gases. A group of methods is based on the temperature dependence of some characteristics of different type elastic and nonelastic scattering, e.g. the line bandwidth of the resonance (Blamont *et al.* 1972) or Rayleigh (Fiocco *et al.* 1971) scattering, the envelope of the rotational Raman scattering (RRS) spectrum (Cooney 1972, Kobayasi *et al.* 1974, Cohen *et al.* 1976, Arshinov *et al.* 1983, 1988, Mitev and Nitsolov 1983, Kobayashi and Taira 1990) or the shape of the RRS spectral lines (Armstrong 1974, 1975), the shape of the Rayleigh-Brillouin scattering spectrum (Shimizu *et al.* 1986), etc. The temperature dependance of the light absorption in the atmosphere is also used as a basis of some DIAL methods to measure the atmospheric temperature and pressure (Mason 1975, Kalshoven *et al.* 1981, Korb *et al.* 1989).

The RRS temperature lidars have been developed most considerably, and have provided most reliable results for the lower troposphere (Cooney 1972, Kobayasi *et al.* 1974, Cohen *et al.* 1976, Arshinov *et al.* 1983, 1988, Mitev and Nitsolov 1983, Kobayashi and Taira 1990, Armstrong 1974, 1975).

The purpose of this work is the further development of one of the RRS techniques, based on the interferometric analysis of the RRS spectrum of nitrogen (Armstrong 1974). Here we discuss the possibility to use this method for simultaneous determination of the temperature and the pressure of the atmosphere.

NATO ASI Series, Vol. I 24
Remote Sensing and Global Climate Change
Edited by Robin A. Vaughan and Arthur P. Cracknell

2 Basic Relations

For monochromatic incident light with frequency ω_0, the RRS spectrum of some simple linear molecule gas component K consists of Stokes (S-branch) lines with frequencies $\omega_{J,K} = \omega^S_{J,K} = \omega_0 - 4B_{0,K}(J+3/2)$, due to transitions $J \to J+2$, and anti-Stokes (O-branch) lines with frequencies $\omega_{J,K} = \omega^O_{J,K} = \omega_0 + 4B_{0,K}(J-1/2)$, due to transitions $J \to J-2$. Here $B_{0,K}$ is the rotational constant and J is the rotational quantum number. The corresponding backscattering cross sections are $\sigma_{J,K} = (64\pi^4/45)b_{J,K}\gamma^2_K\omega^4_{J,K}$, where $b_{J,K}$ is the Placzek-Teller coefficient for the S- or O- branches and γ^2_K is the anisotropy of the molecular polarisability tensor (for more detail see Penney *et al.* 1974). The lidar equation for a single RRS line can be written in the form

$$P_{J,K} = P_i A_0 L \tau_0(z)\eta(z)A_r z^{-2} N_K \sigma_{J,K} F_{J,K}(T)\tau_{J,K}(z) \tag{1}$$

where $P_{J,K}(T)$ is the received backscattered power, P_i is the incident light (laser) power, A_0 is a lidar constant, L is the length of the scattering volume, $\eta(z)$ is the receiving efficiency of the lidar N_K is the number density of molecules. The atmospheric transmittances at frequencies ω_0 and $\omega_{J,K}$ along the propagation path to the scattering volume are denoted $\tau_0(z)$ and $\tau_{J,K}(z)$ respectively. Further we will suppose that $\tau_0(z) \approx \tau_{J,K}(z)$. The fraction $F_{J,K}(T)$ of molecules K in initial state J, at absolute temperature T is given by Penney *et al.* (1974)

$$F_{J,K} = Q^{-1} g_{J,K}(2J+1)exp(-E_J/kT) \tag{2}$$

where $Q \approx (2I_K+1)^2 kT/2hcB_{0,K}$ is the rotational partition function, I_K is the nuclear-spin quantum number, $E_J \approx J(J+1)hcB_{J,K}$ is the rotational energy, $g_{J,K}$ is the statistical weight factor, h is the Planck constant, k is the Boltzmann constant and c is the light velocity.

In the interferometric method the scanning Fabry-Perot interferometer (FPI) is used as a comb filter for processing the periodic RRS spectrum of nitrogen (Armstrong 1974, 1975, Wang *et al.* 1989). In this case, (see Armstrong 1974) the total transmitted Raman power (or energy $E_n = P_n T_i$ integrated over a period T_i) at the n th order interferogram maximum can be expressed by

$$P_n = P(\Delta_n) = \sum_{J=0}^{\infty}\left[P_{J,S}\int \Phi(x+\alpha_{J,n})T(x)dx + P_{J+2,O}\int \Phi(x+\beta_{J,n})T(x)dx\right] \tag{3}$$

where

$$\alpha_{J,n} = \Delta_0\left[\frac{(J+2)(n+\varepsilon_n)}{m_0+n} - \varepsilon_n\right],$$

$$\beta_{J,n} = -\Delta_0\left[\frac{(J+1)(n+\varepsilon_n)}{m_0+n} + \varepsilon_n\right],$$

Δ_n is the free spectral range that gives the peak Raman transmittance in the n th order of the interferogram, $\Delta_0 = 4B_0$ is the free spectral range at the matched condition, m_0 is a reference order ($m_0\Delta_0 = \omega_0 + \Delta_0/2$), ε_n is a parameter which is determined numerically. The transmittance of the FPI is

$$T(x) = (\pi\Delta\omega_F/2)(\Delta\omega_F/2\pi)/[x^2 + (\Delta\omega_F/2)^2] \tag{4}$$

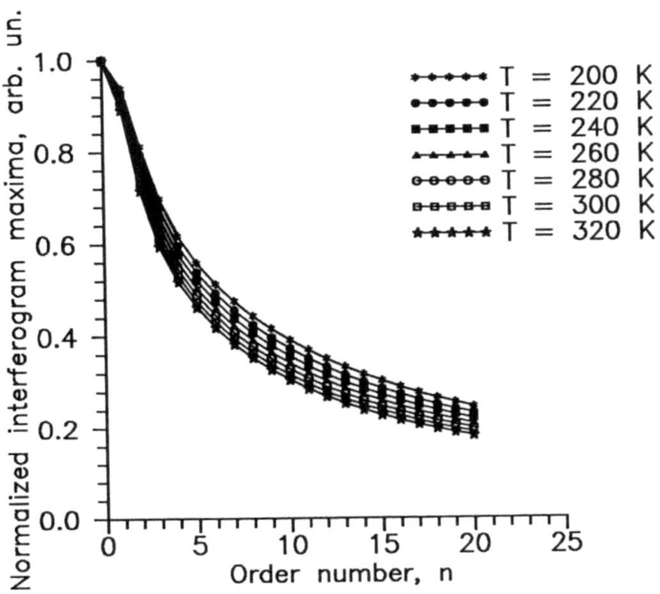

Figure 1: Normalized interferograms I_n vs the order number n for N_2 at pressure $p = 1$ atm, incident light frequency $\omega_0 = 510,6$ nm, and temperatures $T = 200$ K, 220 K, 240 K, 260 K, 300 K, 320 K.

where $\Delta\omega_F = \Delta/F$ is the instrumental function width, Δ is the free spectral range, and F is the finesse of the interferometer. In the troposphere, below 4 km at conditions of standard atmosphere, the RRS line is assumed with a good approximation to have Lorentzian shape

$$\Phi(\omega) = (\Delta\omega/2\pi)/\left[\omega^2 + (\Delta\omega/2)^2\right] \tag{5}$$

with half-maximum width (in cm^{-1}) $\Delta\omega = 0,5\sigma^2 p(MT)^{-1/2}$, where p is the pressure in atm., M is the molecular weight and σ is the optical collision diameter in Å. For nitrogen $\Delta\omega = 1,59p/\sqrt{T}$ (cm^{-1}).

Eq. (3) is obtained using the assumption that each RRS line cannot be transmitted by more than a single order of the FPI. Besides, as far as we consider here only RRS from N_2, we have replaced in (3) the subscript K by the subscripts S and O in order to denote the Stokes and anti-Stokes lines, respectively.

Due to the temperature and the pressure dependance of the RRS line width $\Delta\omega$, the Fabry-Perot interferogram maxima $P(\Delta_n)$ are also unambiguous functions of the temperature and the pressure. The normalized maxima $I_n = P(\Delta_n)/P(\Delta_0)$ vs the order number n, for different values of T and $p = 1$ atm, are represented in Figure 1.

3 Discussion of the possibility to measure simultaneously the temperature and the pressure

As mentioned above (Armstrong 1974), the parameters of the interferogram maxima $P(\Delta_n)$ (and respectively, $I_n = I(\Delta_n)$ depend both on the temperature and on the pressure. The normalized maxima $I(\Delta = \Delta_3)$ and $I(\Delta = \Delta_{10})$ are given in Figure 2, as functions of T, for different values of the pressure $p = 0.6$ atm, 0.7 atm, 0.8 atm, 0.9 atm and 1.0 atm. The sensitivity to the pressure changes seems to be strong enough to cause noticeable uncertainties in the measurement of T (Figure 3a, b). For instance, for $T = 280$ K and $p = 0.9$ atm, $dI_3/dT = 2.92 \times 10^{-4}$ K^{-1}, $dI_3/dT = 3.64 \times 10^{-4}$ K^{-1}, $dI_3/dp = 1.57 \times 10^{-1}$ atm^{-1}, $dI_{10}/dp = 5.18 \times 10^{-2}$ atm^{-1}. Consequently, it would be reasonable to look for ways for measuring T and p simultaneously. A way to do that is to use possible one-to-one correspondence between the pairs $(I(\Delta_{n1}), I(\Delta_{n2}))$ of values of two suitably chosen interferogram peaks of orders n_1 and n_2 and the pair (T, p). A fragment of diagram illustrating the one-to-one correspondence $(I_{n1}, I_{n2}) \leftrightarrow (T, p)$ is given in Figure 4 for $n_1 = 3$ (abscissa) and $n_2 = 10$ (ordinate). If we consider the dependences

$$I_{n1} = f_1(T,p) \qquad I_{n2} = f_2(T,p) \tag{6}$$

as a system of functional equations, the implicit functions $T(I_{n1}, I_{n2})$ and $p(I_{n1}, I_{n2})$ will be independent and the above one-to-one correspondence will exist when the Jacobian

$$D = [(\partial f_1/\partial p)(\partial f_2/\partial T) - (\partial f_2/\partial p)(\partial f_1/\partial T)] \neq 0. \tag{7}$$

We found here that a suitable combination of orders is $n_1 = 3$, and $n_2 = 10$.

According to the implicit function theory, from (6) we obtain

$$\begin{aligned} dp &= D^{-1}\left[(\partial f_2/\partial T)dI_{n1} - (\partial f_1/\partial T)dI_{n2}\right] \\ dT &= D^{-1}\left[(\partial f_2/\partial p)dI_{n1} - (\partial f_1/\partial p)dI_{n2}\right]. \end{aligned} \tag{8}$$

After calculating the mean-square errors $Dp = \langle (dp)^2 \rangle$ and $DT = \langle (dT)^2 \rangle$ ($\langle . \rangle$ denotes ensemble average), taking into account that $I_n = P_n/P_0$ and the mean photon counting rate $Q_n = \alpha P_n$ (α is the photodetector quantum efficiency). We have also assumed Poisson statistics of photocounts. Finally we obtain the following relation:

$$\begin{aligned} Q_{0i} &= \alpha P_0 T_i \\ &= \left[\left(\frac{\partial f_2}{\partial T, \partial p} I_{n1} - \frac{\partial f_1}{\partial T, \partial p} I_{n2}\right)^2 + \left(\frac{\partial f_2}{\partial T, \partial p}\right)^2 I_{n1} + \left(\frac{\partial f_1}{\partial T, \partial p}\right)^2 I_{n2}\right] / D^2(Dp, DT). \end{aligned} \tag{9}$$

The relation (9) is valid in the absence of noise and indicates a minimum number of photocounts which must be integrated in the 0-th order of the interferogram for the period T_i in order to achieve a given value of Dp or DT. In Eq. (9), $(\partial f/\partial T, \partial p)$ denotes the first derivative with respect to T or p, respectively; (Dp, DT) denotes that the first corresponding value is Dp and the second DT. For example, for $T = 280$ K and $p = 0.9$ atm, from (9) we obtain that for $DT = 0.04$ degK^2 and $Dp = 10^{-6}$ atm^2 the corresponding value of $Q_{0,i}$ is 1.4×10^8.

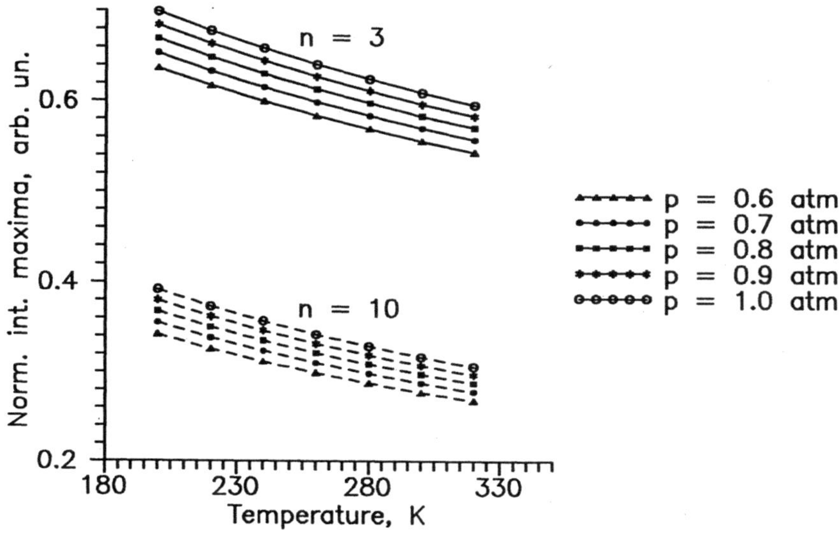

Figure 2: 10th-order and 3-th order normalized Raman interferogram maxima for N_2, vs the temperature for values of the pressure $p = 0.6$ atm, 0.7 atm, 0.8 atm, 0.9 atm, 1.0 atm.

Let us consider a laser operating at wavelength $\lambda_0 = 510.6$ nm and average power $P_i = 1$ W and assume, as well, $\tau_0(z)\tau_{J,K}(z)\eta(z)A_0 \approx 0.1$, $A_r(z) = 10^{-7}$ sr, $\alpha = 5 \times 10^{17}$ J^{-1},and the range resolution $L = 150$ m. Then using (1) and (3) we find that $P_0 \approx 3.1 \times 10^{-12}$ W, and $Q_0 = \alpha P_0 \approx 1.5 \times 10^6$ s^{-1}. Consequently, for an integration time $T_i = Q_{0,i}/Q_0 \approx 90$ s we can reach the accuracies corresponding to the given values of DT and Dp. Estimating T_i we have to take into account the possibility for technical realisations. There are two possibilities: the first is to use one interferometer, scanning in a sequence the two maxima, and the second is to divide the incoming lidar signal and analyse it in parallel employing two fixed interferometers. It is evident that any realisation will double the necessary detection time T_i.

4 Summary

The investigation of the interferometric technique shows a considerable dependence on the pressure changes which should be noted in the temperature measurements.

The interferometric method allows one, in principle, to measure the temperature and the pressure simultaneously, using the one-to-one correspondence between the values of two suitably chosen normalised interferogram maxima and the pair (T, P). The experimental tests of this possibility have to determine its further development.

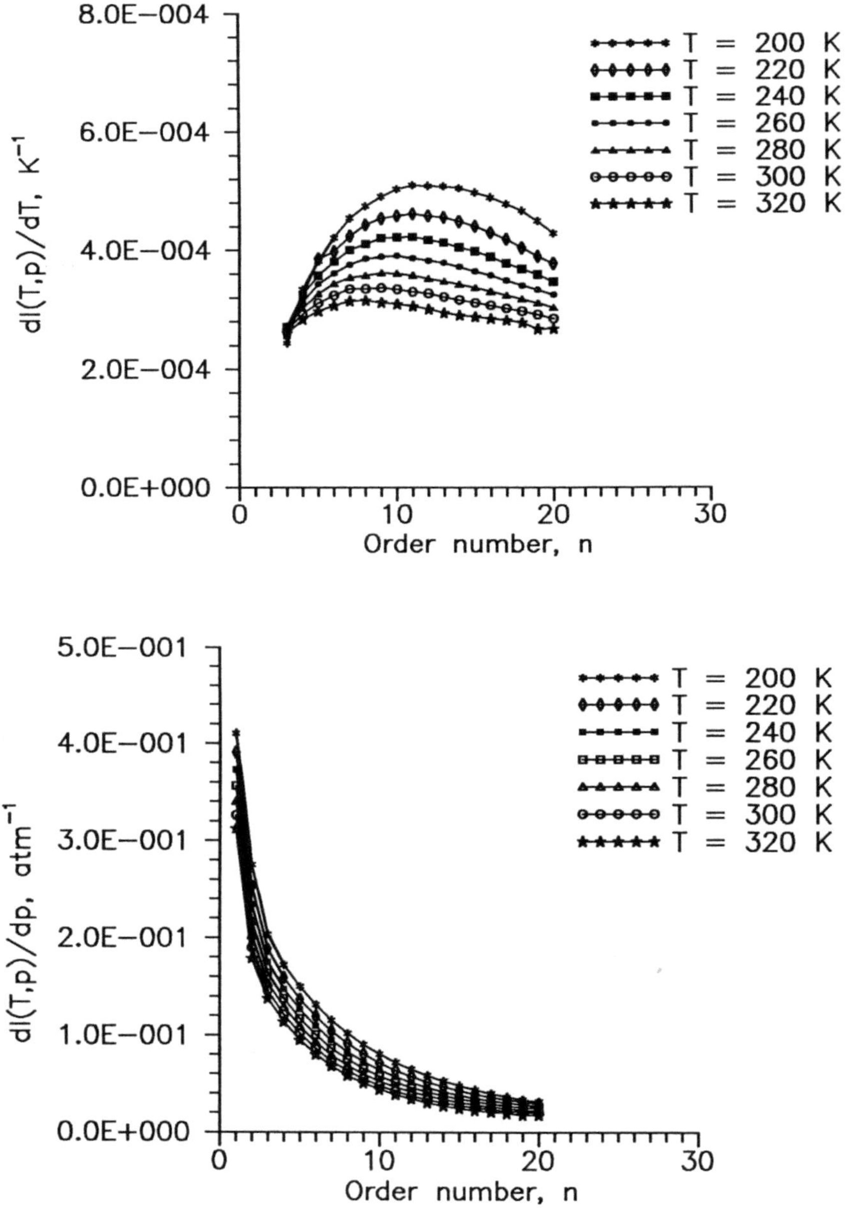

Figure 3: Derivatives of the normalized interferograms by T and p vs the order number n at pressure $p = 1$ atm, and temperatures $T = 200$ K, 220 K, 240 K, 260 K, 300 K, 320 K.

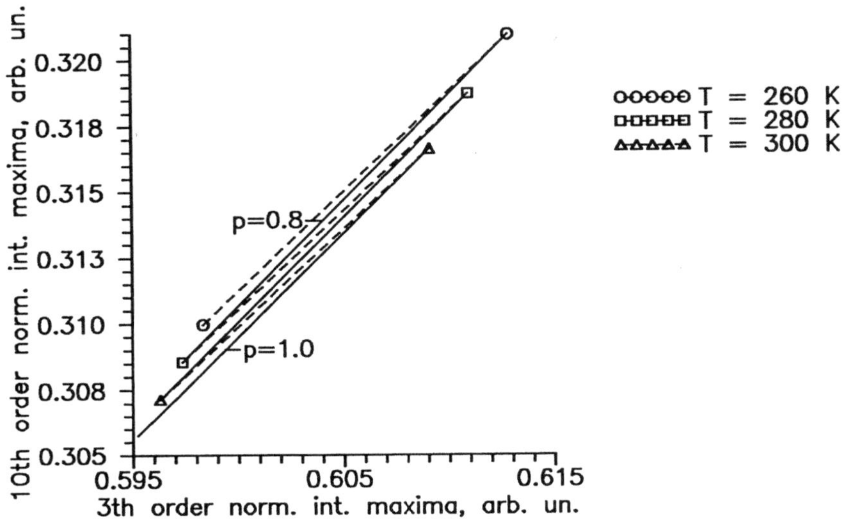

Figure 4: Diagram illustrating the one-to-one correspondence between the measured pair of values $[I(n=3), I(n=10)]$ and the temperature—pressure pair (T, p).

References

Armstrong, R. L., 1974, Analysis of Fabry-Perot interferograms of periodic spectra. *Journal of the Optical Society of America*, **64**, 871-876.

Armstrong, R. L., 1975, Rotational Raman interferometric technique to measure gas temperature. *Applied Optics*, **14**, 383-387.

Arshinov, Y. F., Bobrovnikov, S. M., Zuev, V. E., and Mitev, V. M., 1983, Atmospheric-temperature measurements using a pure rotational Raman lidar. *Applied Optics*, **22**, 2984-1990.

Arshinov, Y. F., Bobrovnikov, S. M., Volkov, Zuev, V. E., and Shumskii, V. K., 1988, *Abstracts of the 14th ILRC Italy*, pp. 63.

Blamont, J. E., Chanin, M. L., and Megie, G., 1972, Vertical distribution and temperature profile of the night time atmospheric sodium layer obtained by laser backscatter. *Annales de Geophysique*, **28**, 833-838.

Cohen, A., Cooney, J. A., and Geller, K. N. 1976, Atmospheric temperature profiles from lidar measurements of fotational Raman and elastic scattering. *Applied Optics*, **15**, 2696-2901.

Cooney, J. A., 1972, Measurement of atmospheric temperature profiles by Raman backscatter. *Journal of Applied Meteorology*, **11**, 108-112.

Fiocco, C., Benedetti-Michelangeli, G., Maischberger, K., and Madonna, E., 1971, Measurement of temperature and aerosol to molecule ratio in the troposphere by optical radar. *Nature, Physical Science*, **229**, 78-79.

Kalshoven, J. E., Jr., Korb, C. L., Schwemmer, G. K., and Dombrowski, M., Laser remote-sensing of atmospheric-temperature by observing resonant absorption of oxygen. *Applied Optics*, **20**, 1967-1971.

Kobayasi, T., and Shimisu, H., and Inaba, H., 1974, *Abstracts of the 6ht ILRC, Japan*, pp. 49.

Kobayashi, T., and Taira, T., *Abstracts of the 15th ILRC, USSR*, pp. 290.

Korb, C., Schwemmer, G. K., Dombrowski, M., and Weng, CH. Y., 1989, Airborne and ground based lidar measurements of the atmospheric-pressure profile. *Applied Optics* **28**, 3015-3020.

Mason, J. B., 1975, Ladar measurement of temperature: A new approach. *Applied Optics*, **14**, 76-78.

Mitev, V. M., and Nitsolov, S. L., 1983, Improved procedure for Raman lidar measurements of the atmospheric temperature. *Bulgarian Journal of Physics*, **10**, 86-97.

Penney, C. M., St. Peters, R. L., and Lapp, M., 1974, Absolute rotational Raman cross sections for N_2, O_2, and CO_2 *Journal of the Optics Society of America*, **64**, 712-716.

Schmidt, W., 1990, *Abstracts of the 15th ILRC, part II, USSR*, pp. 362.

Shimizu, H., Noguchi, K., and She Chiao-Yao, 1986, Atmospheric-temperature measurement by a high spectral resolution lidar. *Applied Optics*, **5**, 1460-1466.

Wang, J., Drayson, S. R., and Hays, P. B., 1989, Atmospheric temperature sensing with a multiorder Fabry-Perot interferometer. *Applied Optics* , **28**, 5038-5046.

Remote Sensing of Snow and Ice and Its Relevance to Climate Change Processes

Peter Wadhams

1 Introduction

This chapter will review those physical processes in snow and ice which are important to climate change, and will describe how remote sensing techniques are used to map the cryosphere and hence to allow the magnitude and direction of climatic change in the polar regions to be estimated. The review begins with a brief overview of the structure and morphology of ice and snow and the geophysical processes in which ice and snow are involved. We then consider the physics of interaction between electromagnetic waves and a snow or ice surface and the remote sensing techniques commonly used to study ice and snow (including acoustic as well as electromagnetic methods). The greenhouse effect is described with reference to the cryospheric processes which cause the rate of warming to be enhanced in north polar regions. We point out the importance of ice and snow processes in amplifying the rate of ground level warming, and how ice and snow parameters can be used as early indicators of climatic change in action. We go on to consider in detail how remote sensing techniques are used to monitor four key parameters. These are (1) the contribution of melting glaciers and ice sheets to the rate of global sea level rise; (2) the evidence for a retreat of sea ice in the polar regions; (3) the problem of the synoptic mapping of sea ice thickness, and the evidence for thinning of sea ice; (4) the evidence for a retreat of seasonal snowpack in the northern hemisphere.

2 The geophysics of ice and snow

2.1 Ice and snow - crystal structure

One of the most extraordinary properties of ordinary ice is that it is one of the few substances where the solid is less dense than its melt. Geophysically this is of vital importance because it implies that it floats on the surface of lakes and seas and thus makes it thermodynamically difficult or impossible for a sea to freeze completely. The

NATO ASI Series, Vol. I 24
Remote Sensing and Global Climate Change
Edited by Robin A. Vaughan and Arthur P. Cracknell

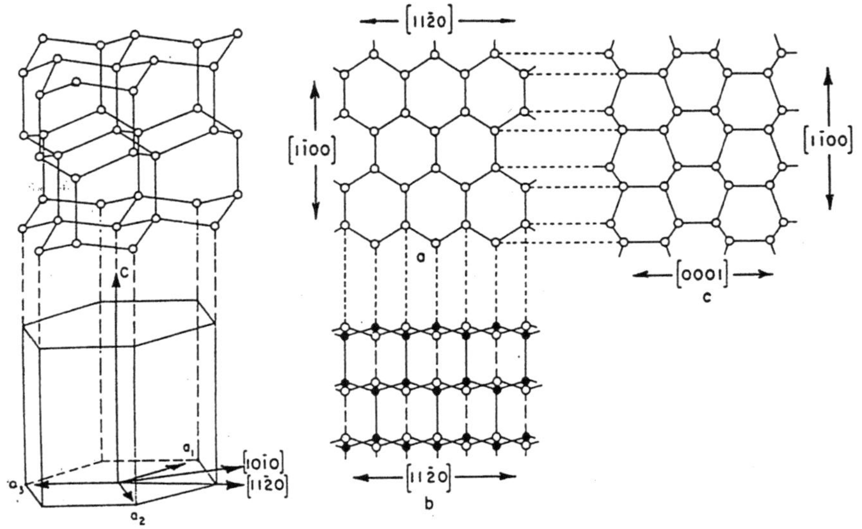

Figure 1: The structure of ice I (after Weeks and Ackley, 1986).

reason for this behaviour lies in the open structure of ice. There are at least nine polymorphs of solid H_2O under different conditions of temperature and pressure; they are discussed fully, as is the crystallography of ice, by Hobbs (1974). Under normal conditions found on the surface of the Earth, the only structure encountered is called ice I (Figure 1). It is a hexagonal structure. Each O is at the centre of a tetrahedron with four other O atoms at the apices, separated by 0.276 nm. The O atoms are concentrated close to a series of parallel planes that are known as the basal planes. The principal axis, or c-axis, of the crystal unit cell lies perpendicular to the basal plane.

This structure causes ice to be markedly anisotropic in its physical and mechanical properties. Fracture along the basal plane requires, for a unit cell, the rupture of only two bonds, while fracture along any other plane normal to this requires the breaking of at least four bonds. Thus ice glides and cleaves readily on the basal plane. When an ice crystal grows, it is energetically easier for new atoms to be added to an existing basal plane than to begin a whole new plane. Therefore ice crystals grow more readily along the a- and b-axes (axes in the basal plane) than along the c-axis. These preferred growth directions are the directions of a) the arms of snowflakes growing from vapour; b) the arms of dendritic ice crystals growing on the surface of a newly-freezing sea or lake surface; c) the preferred downward growth direction of sea ice crystals in a sheet; d) internal melt features that form inside ice crystals as a result of absorbed solar radiation.

The net of O atoms is held together by hydrogen bonds. The H atoms lie along these bonds, but in a disordered way. The position of each H atom in the bond is to be closer to one O than to the other. Each O has two H atoms near it, but there can be only one H along each bond. Subject to these two rules, the H atoms can be arranged in any way. It is the length of the hydrogen bond that creates the open structure of ice; when ice melts, some of the bonds are broken, causing a disordered structure with a higher density. But even in liquid water there is some short-range order, with a few

Typical Densities (kg m^{-3})

New snow (imediateiy after falling in calm)	50-70
Damp new snow	100-200
Settled snow	200-300
Depth hear	100-300
Wind packed snow	350-400
Firn	400-830
Very wet snow and firn	700-800
Glacier ice	830-910

Table 1: Typical Densities (kg m^{-3})

water molecules retaining the crystal-like bonded structure until this is destroyed by thermal motion; this accounts for the curious density behaviour of fresh water, where there is a maximum density at 4 °C.

2.2 Glaciers

When snow falls in a region with a sufficiently cold annual cycle of temperature, it does not melt in summer but remains to build up an ice sheet or glacier. Each successive fall of snow compresses the layers beneath, so that the underlying material gradually becomes transformed, both by pressure and by the annual cycle of partial melt and refreezing, towards ice. Typically new snow has a bulk density of only 50-70 kg m^{-3}. As it becomes compressed, and especially after it has passed through its first summer of partial melt, it becomes transformed into firn, a more granular material of higher density with larger crystal size. With further compression, the air passages which permit the circulation of air and the percolation of melt water around the crystal grains eventually become sealed off. This marks the "official" transformation of snow to ice, and it occurs at a density of 830 kg m^{-3}. Further compression takes the ice density towards the 910-920 kg m^{-3} of pure ice, with air remaining as closed bubbles of increasing pressure. Hence the familiar fizzing effect when ice from the lower part of a glacier or iceberg is placed in water (this also accounts for the underwater noise made by icebergs melting in warm water). Table 1 shows typical densities of snow and ice types.

In polar ice sheets, such as that of Antarctica, there is little or no surface melt in summer and the ice sheet is constantly accumulating new mass from precipitation. Equilibrium is maintained by ice sheet flow and the resulting calving of icebergs and melt of ice from the bottoms of floating ice shelves. Most mountain glaciers, however, occupy a range of climatic zones corresponding to the altitude range of the glacier. Figure 2 shows how this zoning affects the near-surface structure of snow and ice in the glacier at the end of summer. This is of importance to the remote sensing of glaciers. We comment on the definitions as follows:

The dry snow zone is so cold that no melt occurs even in summer. The surface snow is dry, loose, powdery and with small crystals. The dry snow line marks the lower boundary of this zone. In the percolation zone some melting occurs in summer at the

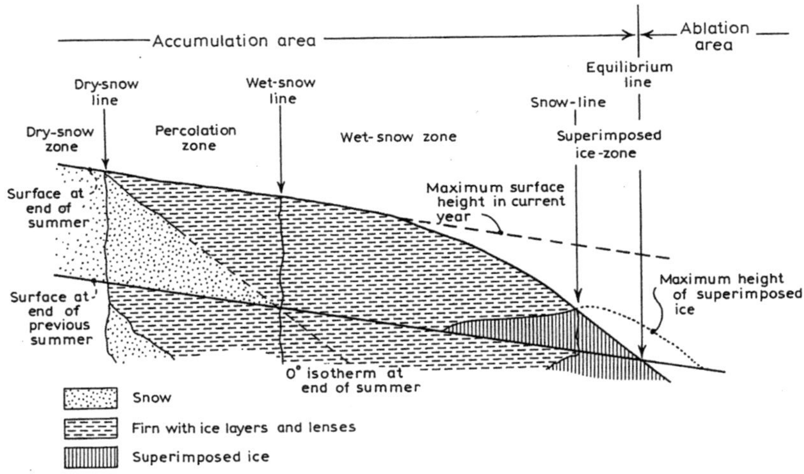

Figure 2: Zones in the accumulation area of a glacier at the end of summer (after Paterson 1981).

surface, and the water percolates down through the snow and firn to a depth which increases with decreasing altitude, where it refreezes to form ice layers or ice lenses (the vertical channels refreeze to form ice glands). This refreezing warms the snow. Eventually a point is reached at which all of the previous winter's snow is raised during the summer to the melting temperature and hence becomes wetted. This is called the wet-snow line. Below it lies the wet snow zone where all of the previous year's snow has become wetted, while the meltwater also percolates down into the previous year's snow. By now much of the meltwater is percolating out of the glacier in the form of meltwater streams, leaving the surface ablated below the level that it reaches in late winter. The snow line marks the lower limit of the region where the surface material consists of a mixture of snow, firn and ice; below this line the ice layers merge to create a continuous mass of ice, known as superimposed ice. This defines the next zone of the glacier. Below the snow line comes the point known as the equilibrium line; above it the glacier has a net gain in mass over the year, below it a net loss. The area below the equilibrium line is called the ablation zone; the area above is called the accumulation zone.

Mountain glaciers in the Arctic possess the entire sequence of zones, while the Antarctic ice sheet is entirely an accumulation zone. Temperate glaciers, however - i.e. mountain glaciers in the Alps, Rockies and other low-latitude regions - are at the pressure melting point throughout, except for a thin surface layer in winter. Such glaciers possess only the wet snow zone downwards.

Clearly these definitions are important for attempts to assess the nature of a glacier by remote sensing. For instance, even a photographic survey in late summer, so long as there have been no unusual snowfalls, will show the position of the snow line since it is here that bare wet ice begins to appear. The elevation of the snow line in relation to the shape and volume of the glacier as a whole can be used as a guide to whether the glacier is losing or gaining mass.

Depth hoar, shown in Table 1, is an unusual type of firn composed of very large crystals. It is very porous, and forms within a snowpack often in autumn from the condensation of vapour rising from a warmer ice layer beneath.

2.3 Glacier flow

As well as evolving thermodynamically, a glacier is in motion downhill. The flow may be as little as a few metres a year for small glaciers, 50-400 m y^{-1} for typical valley glaciers, and 1 km y^{-1} or more for ice streams emerging from Antarctica and Greenland. The ice flows in two ways, by internal deformation and by sliding on its bed. Internal deformation occurs because ice is a material which creeps, like blancmange, under its own weight. The creep properties are governed by the Glen flow law, that strain rate is proportional to $(\text{stress})^3$. Flow is greatest at the upper surface of the glacier, and the differential flow tends to create crevasses. Basal sliding occurs more readily in temperate glaciers where meltwater acts as a lubricant. Table 2 shows the percentage of a glacier's motion that is due to basal sliding, for a selection of glaciers, and also typical rates of motion.

Glacier surges occur when a glacier suddenly departs from its normal flow pattern and the snout advances downstream at an enormous rate, perhaps 60-80 m day^{-1}, accompanied by the massive release of meltwater. This can have disastrous effects. It is not known why glaciers surge, but one theory is that an unusual buildup of mass in the upper part of the glacier squeezes shut the ordinary pattern of meltwater channels within the glacier, forcing the meltwater to spread out thinly over the whole glacier bed, lubricating it for a rapid downstream slide.

2.4 Glacier mass balance

Over a year a glacier gains mass during the winter through snowfall and loses mass in the summer through melt. In addition, ice may be discharged from the end of the glacier, e.g. as icebergs, and mass may be lost by evaporation. Finally, the shape of the glacier may change as the snout position advances or retreats. Mass balance studies attempt to measure how the mass of a glacier changes with time by measuring all the relevant parameters which lead to gain or loss. At any point in a glacier the elevation can be measured throughout the year (Figure 3 shows the cycle and the terms used to describe gains and losses) and the results averaged over the area of the glacier. When combined with measurements of advance or retreat, an estimate can be made of mass balance. In this way it has been found that more glaciers are retreating than advancing, and it has been possible to estimate the contribution made by this loss of ice to world sea levels. Remote sensing techniques are extremely useful in such studies.

2.5 Sea ice - the ocean background

Sea ice is particularly suitable for study by remote sensing techniques. First, because of its variable brine content and surface morphology it gives a highly variable signal for both passive and active microwave sensors, enabling it to be characterised effectively

Glacier	Region	Centreline velocity (m/yr)	Comments
Lambert Glacier	East Antarctica	347	Part of largest glacier drainage system in Antarctica
Amery Ice Shelf	East Antarctica	1200	As above
Byrd Glacier	East Antarctica	760	Glacier drainage from polar plateau through the Transantarctic Mountains
Ice in West Dronning Maud Land	East Antarctica	1 to 15	Slow moving part of ice sheet where unchannellized
Jakobshavn Isbrae	NW Greenland	4700	World's fastest outlet glacier
White Glacier	Axel Heiberg Island, Canada	40	Cold, sliding valley glacier
Grosser Aletschgletscher	Swiss Alps	200	Fastest part of largest glacier in the Alps
Griesgletscher	Swiss Alps	40	Small valley glacier
Saskatchewan Glacier	Canadian Rockies	117	Valley glacier
Charles Rabots Bre	Okstindan, Norway	8	Steep, thin cirque glacier

Glacier	Region	Basal sliding %	Ice thickness (metres)
Grosser Aletschgletscher	Swiss Alps	50	137
Tuyuksu Glacier	USSR	65	52
Salmon Glacier	Canadian Rockies	45	495
Athabasca Glacier	Canadian Rockies	75	322
Saskatchewan Glacier	Canadian Rockies	20	555
Blue Glacier	Olympic Mts., Washington	90	26
Vesl Skautbreen	Norway	9	50
Meserve	Victoria Land, Antarctica	0	80

Table 2: Top. Some examples of ice velocities in different types of glacier. Bottom. Contribution of basal sliding to total glacier flow from tunnel and borehole measurements.

by remote sensing. Secondly, a very important aspect of sea ice is its motion field, which can also be measured by remote sensing techniques. Thirdly it has an enormous annual variability, rendering it of great geophysical importance; the range of variability is matched only by the seasonal snow cover on the land masses of the Northern Hemisphere.

The background to sea ice and its role lies in the physical oceanography of the Arctic and the Antarctic. The Arctic Ocean (Figure 4) consists of a three layer system.

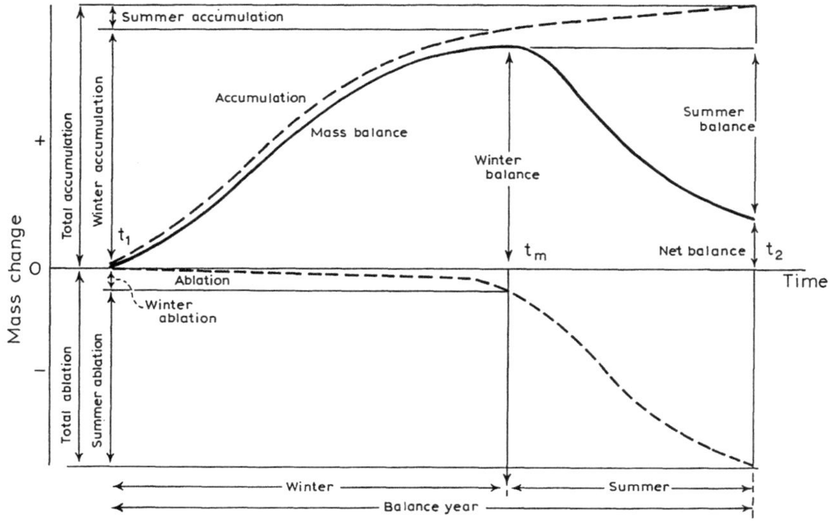

Figure 3: Definitions of mass balance terms for a glacier (after Paterson 1981).

The uppermost layer, called polar surface water, is about 200 m thick and comprises low salinity water (3.0-3.2 ‰ of salt, compared to 3.5‰ as an average for the world ocean) at or very near the freezing point. The low salinity occurs because of the very large influx of fresh water into the Arctic Ocean from large river systems such as the Ob', Lena, Yenisei and Mackenzie. Below the polar surface water lies water of Atlantic origin, which is remarkably warm (about 3° C) and saline (about 3.5‰). It has entered the Arctic through Fram Strait (Figure 5), where a warm current, the North Atlantic Current (originally the Gulf Stream) runs up the west side of Spitsbergen as the West Spitsbergen Current and then sinks as it encounters the less dense polar surface water and spreads around the Arctic at mid-depths. Deeper still lies bottom water, at or below 0°C, and again of Atlantic origin.

The surface current system in the Arctic (Figure 5) is largely wind-driven and consists of a gyre in the Canada Basin, the Beaufort Gyre, and a motion of translation, the Trans Polar Drift Stream, in the Eurasian Basin. The Drift Stream is renamed the East Greenland Current after it passes through Fram Strait between Svalbard and Greenland. This current transports most of the ice out of the Arctic and it is through Fram Strait, the only deep channel conecting the Arctic to the rest of the world ocean, that most water and heat exchange occur. The wind system which drives this flow arises because of the presence of the polar high over the centre of the Beaufort Sea, with a ridge extending over Greenland.

Antarctic water masses are more complex because of the unconfined nature of the Southern Ocean (Figure 6). Again there is a cold polar surface water mass, with a low salinity maintained by iceberg melt and ice shelf melt, but the salinity is higher than in the Arctic and so the boundary between this and the underlying water - the so-called pycnocline (density discontinuity) - is less steep and so has less effect in preventing

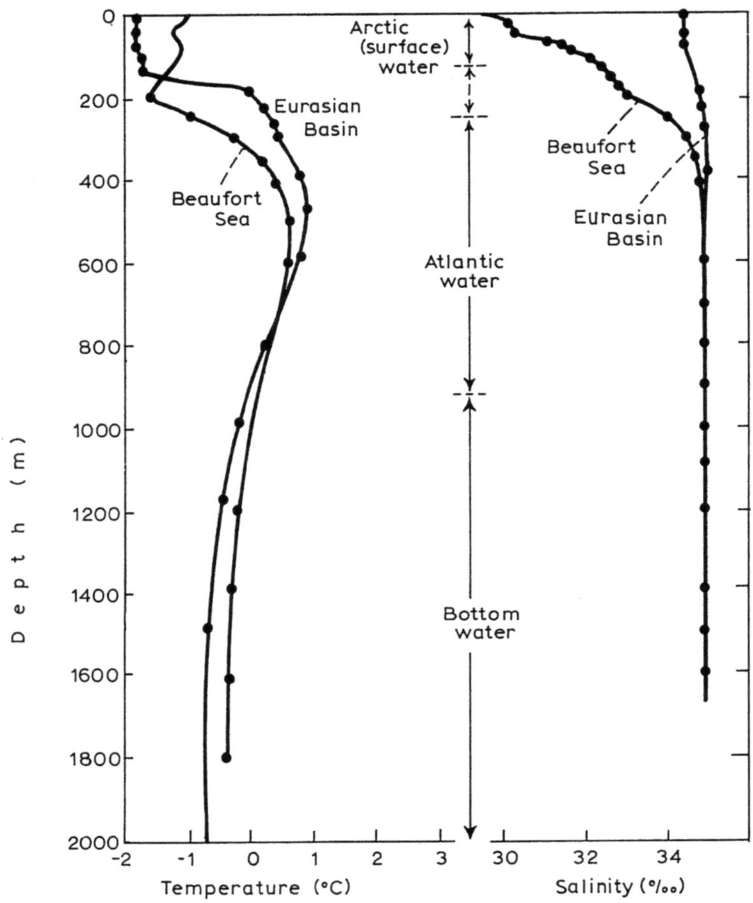

Figure 4: Temperature and salinity profiles from the Eurasian Basin and Beaufort Sea, showing division into three water masses (after Coachman and Aagaard 1974).

vertical mixing and heat transfer. Below the surface water is the Circumpolar Deep Water, again a warm water mass. There is a northern boundary in the ocean between the polar surface water and warmer water masses, known as the Antarctic Convergence or Antarctic Polar Front. Finally, there is a dense water mass near the seabed known as Bottom Water, which is produced by a complex mixing process on shelves between polar surface water and circumpolar deep water, with an additional contribution due to cooling under ice shelves and the addition of salt from sea ice formation.

The surface currents in the Antarctic (Figure 7) consist of an enormous eastward flowing current, the Antarctic Circumpolar Current, which is one of the greatest current systems in the world, transporting some 150 million m^3 s^{-1} of water (150 Sverdrups). Near the coast there is a westward flowing current, the Antarctic Coastal Current, which governs the inital drift direction of icebergs which calve from the continent.

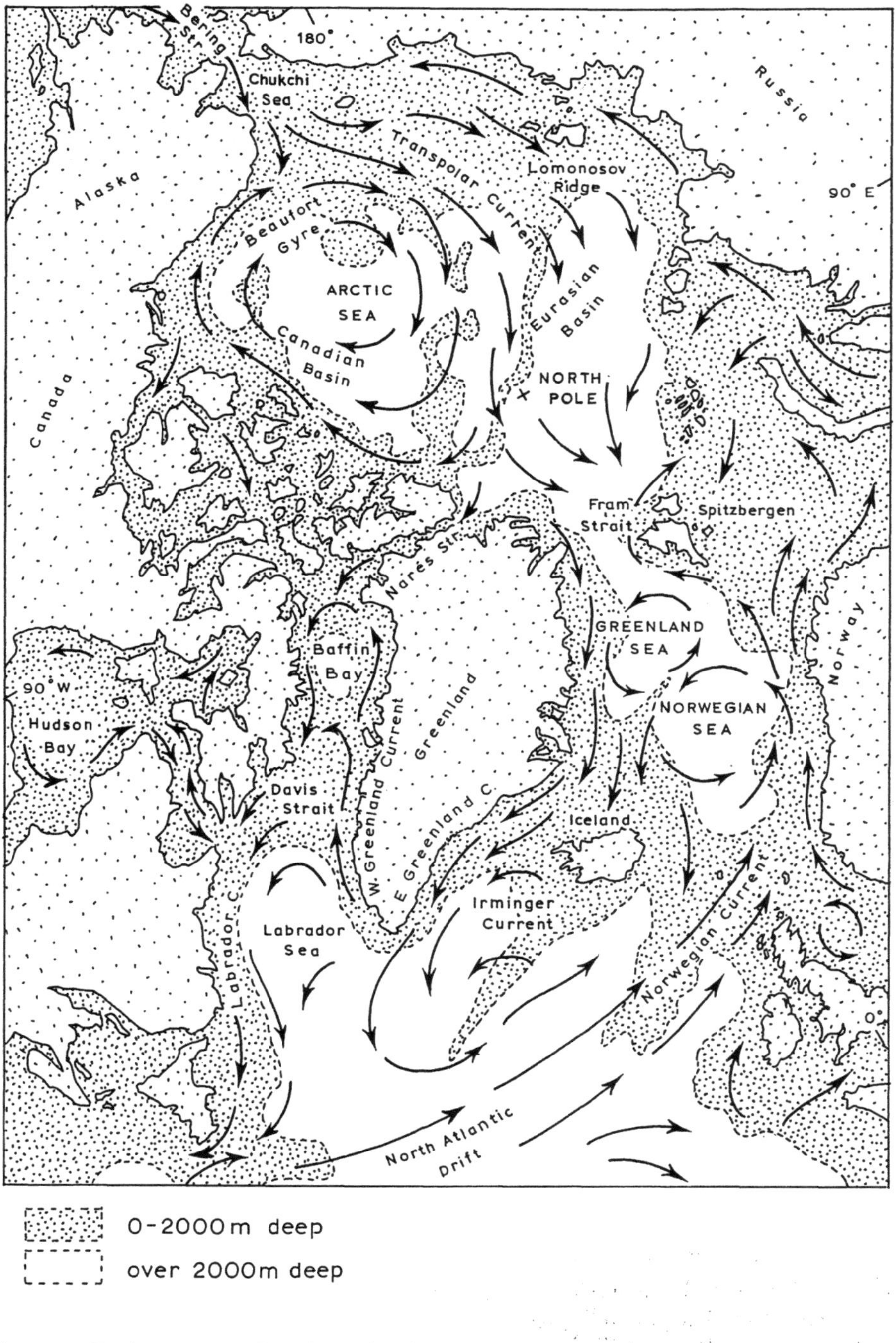

Figure 5: Bathymetry and surface circulation of the Arctic Ocean and northern North Atlantic (after Open University 1989).

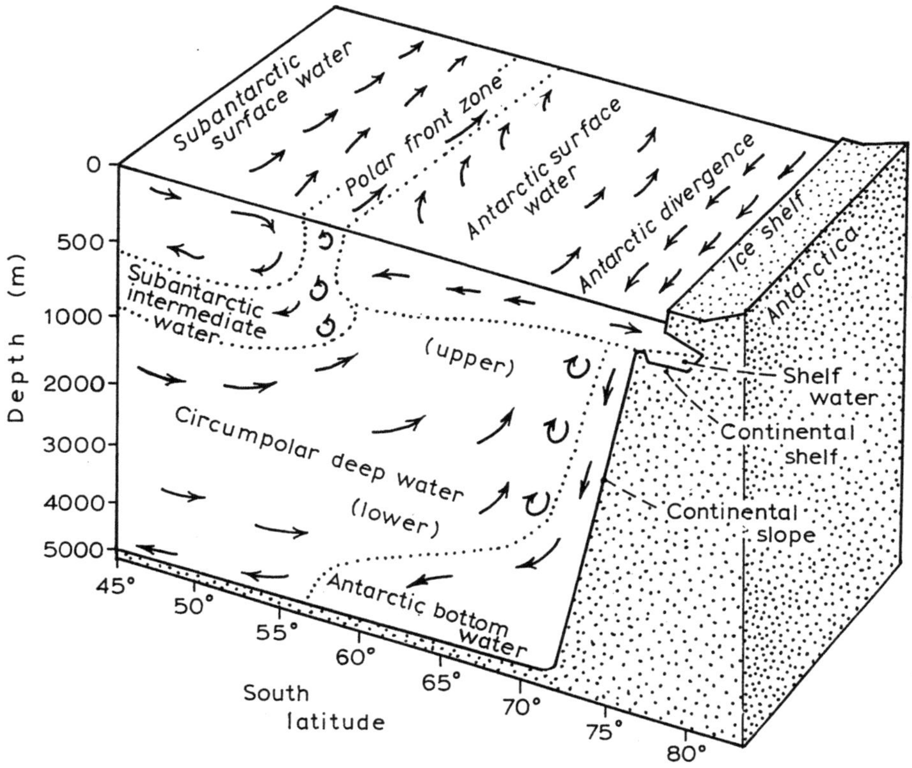

Figure 6: Schematic diagram of Antarctic water masses and motions (after Gordon and Goldberg 1970).

Finally there is evidence of gyral circulations in the Weddell and Ross Seas. Again, these motions are largely driven by the wind field.

2.6 Physics of sea ice formation

Consider a water body being cooled from above, i.e. a water mass at the end of summer, experiencing subzero air temperatures. As the water cools the density increases so the surface water sinks, to be relaced by warmer water from below, which in its turn is cooled. Thus convection sets in such that the whole water body gradually cools. When the temperature reaches 4 °C, a fresh water body reaches its maximum density. Further cooling results in the colder water staying at the surface. This thin layer can then be rapidly cooled down to the freezing point, and ice can form on the surface even though the temperature of the underlying water may still be close to 4 °C. Thus a lake or river can experience ice formation early in the autumn or winter, while considerable heat still remains in the deeper parts.

However, when the salinity exceeds 2.47‰, (Weeks and Ackley 1986), which applies to all parts of the ocean except the Baltic Sea), this temperature of maximum density

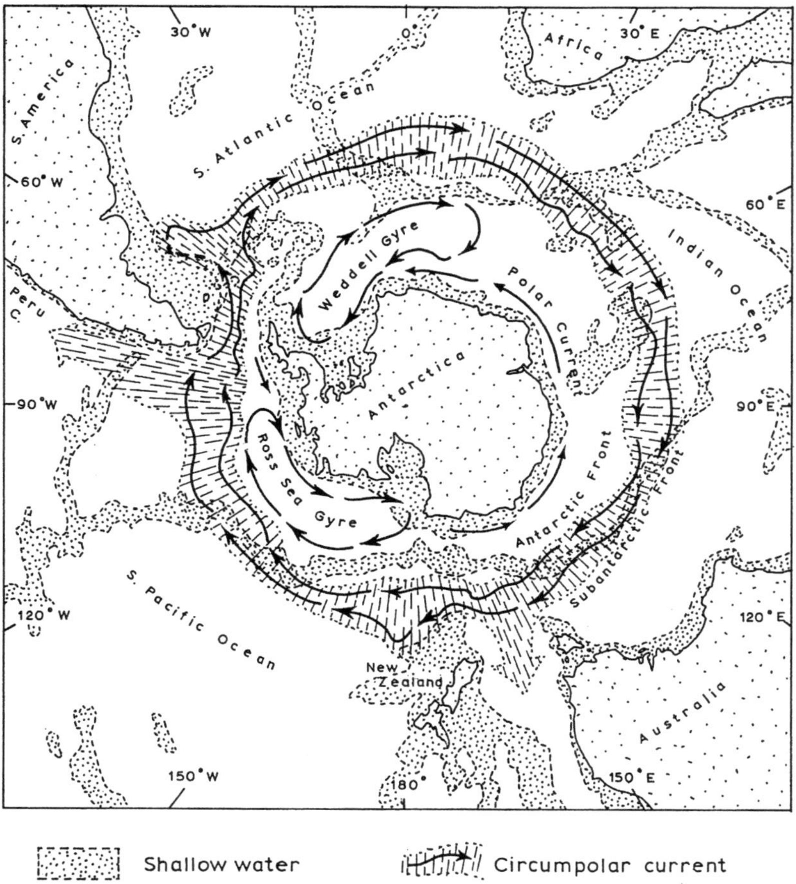

Figure 7: Schematic diagram of surface circulation in the Southern Ocean (after Open University 1989).

disappears, and further cooling continues to make the water more dense and continues to cause convection. Thus it takes much longer to cool the water mass down to the freezing point (which is itself -1.8 °C for normal sea water rather than 0 °C), and so sea ice forms later in the autumn than does lake ice in similar climatic conditions.

2.7 Sea ice growth and fabric

In quiet conditions the first sea ice to form on the surface is a skim of separate crystals which are of hexagonal, stellar form, with long fragile arms stretching out over the surface. These crystals have their c-axis vertical. The dendritic arms are very fragile, and soon break off, leaving a mixture of discs and arm fragments. With any kind of turbulence in the water, these fragments are pounded into random-shaped small crystals which form a suspension of increasing density in the surface water, an ice type called frazil or grease ice. In quiet conditions the frazil crystals soon freeze together to form a continuous thin sheet of young ice; in its early stages, when it is still transparent, it is called nilas. In turbulent conditions the frazil remains as a suspension, becoming thicker and denser until eventually the crystals freeze together in clumps or flocs to form cakes which collide with one another, thus acquiring raised edge like pancakes. This is called pancake ice. The pancakes grow by accretion from the frazil, and will eventually freeze together to form consolidated pancake ice, a very common ice type in regions where ice is trying to form in conditions of vigorous wave activity, e.g. the outermost parts of the Antarctic pack ice zone in winter. Once a continuous sheet of nilas has formed, the individual crystals which are in contact with the ice-water interface grow downwards by freezing of water molecules onto the crystal face. As we have mentioned, this freezing process is easier for crystals with horizontal c-axes (such that the growth can occur along an a- or b-direction) than for those with c-axes vertical. Thus the crystals with c-axis horizontal grow at the expense of the others, and as the ice sheet grows thicker crowd them out in a form of crystalline Darwinism. In the lower part of a first-year ice sheet the crystals all have horizontal c-axes and because of their growth history have a long vertical columnar fabric.

2.8 Brine cells

As the ice sheet grows, the salt in the sea water cannot become part of the crystal lattice and so is rejected. The ice-water interface advances in the form of parallel rows of cellular projections, with the rejected brine accumulating in the grooves between rows of cells. As these cells develop lateral connections the brine becomes trapped in the form of tiny liquid inclusions of very concentrated brine. These thus occupy vertical sheets corresponding to the former grooves.

The liquid brine retained in the growing ice sheet gives the ice a bulk salinity which is about 1.0‰ for young ice. However, the brine cells drain out of the ice as the ice sheet grows, by way of a network of brine drainage channels which they create, and as the ice sheet ages the brine concentration drops. One further mechanism for brine loss during winter is brine cell migration down the temperature gradient, whereby the upper part of a brine cell is slightly colder than the lower part, causing freezing at the top and consequent melting at the bottom. A further loss of brine occurs during the first summer when the top surface of the ice sheet melts and fresh water percolates down through the ice sheet, flushing out much of the remaining brine. This yields a stronger, fresher ice sheet to begin its second year of growth.

2.9 First-year and multi-year ice

The first summer is crucial to the development of the ice sheet. In the Arctic, it will have reached a thickness of some 1.5-2 m in one winter of thermodynamic growth (in the Antarctic only about 60 cm). In summer the overlying snow layer melts first, then the bare ice surface begins to melt, creating a network of meltwater pools over the surface. These readily absorb radiation, melting down into the ice and increasing the surface roughness. Parallel melting underneath creates an under-ice surface which also becomes rugged and bumpy instead of smooth. Further growth in subsequent years brings the ice thickness to a maximum of 3-4 m, at which point summer melt matches winter growth and the thickness oscillates through an annual cycle. This old, multi-year ice is much fresher than first-year ice; it has a lower conductivity and a rougher surface. Microwave radiation penetrates further through it than through first-year ice and an incident beam tends to be scattered within the volume of the ice sheet rather than reflected in a specular way from the surface. Its microwave properties are therefore distinct, enabling first- and multi-year ice to be well discriminated by microwave techniques.

2.10 Antarctic sea ice

Because of the high wave field in the Antarctic, new ice which forms at the edge of the advancing Antarctic ice cover in winter does so in the form of frazil and pancake ice, which cements together to form consolidated pancake ice only after a considerable belt of such ice (up to 300 km) has damped down ocean wave energy. There is a much higher ocean heat flux into the ice in the Antarctic than in the Arctic, so in its first winter ice in the circumpolar Southern Ocean region reaches a thickness of only some 60 cm. Most of it melts during the first summer - there is an enormous annual cycle of area from 4 million km^2 in summer to 20 million km^2 in winter. Ice remains only in a few locations, notably the Weddell and Ross Seas, and becomes multi-year ice which still reaches a thickness of only about 1.2-1.4 m.

While the outer ice edge advances in winter by means of the frazil-pancake cycle, the ice nearer the coast is augmented by production in coastal polynyas. These are areas of open water, kept free of a continuous ice cover by powerful katabatic winds which funnel down off the Antarctic ice sheet between mountain peaks and blow newly-forming ice out to sea. These regions are thus "ice factories" as well as being open during winter, and the best (and only) way of mapping their occurrence is by means of remote sensing.

2.11 Ice dynamics and ridging

Sea ice is constantly in motion, driven by the wind. The long term pattern of drift is identical to the surface current pattern described above. However, differential drift also occurs - the ice is pulled apart by divergent wind stresses acting over the surface, and this opens up cracks which widen to form leads. In winter leads rapidly refreeze because of the enormous temperature difference between the atmosphere (-30 °C or so) and the ocean (-1.8 °C). When a subsequent wind stress field becomes convergent, the young ice in the refrozen leads forms the weakest part of the ice cover and is the first part to be crushed, building up heaps of broken ice blocks above and below the water line. Such a linear deformation feature is called a pressure ridge, the above-water part being the sail and the below-water part (more extensive) being called the keel. Keels in the Arctic can reach down to 50 m, although most are about 10-25 m. Keels in the Antarctic are much shallower, typically only about 5 m in draft.

Ridged ice is a major component of the ice thickness distribution P(h). This is a key parameter for characterising the ice cover. In the Arctic P(h) has a long negative exponential tail containing the ridged ice, which may make up about 50% of the ice volume. Mean ice thicknesses range from 2-3 m in the seas north of Russia to 7-8 m north of Greenland and the Canadian Archipelago, where the wind drives ice towards the coast and creates enormous amounts of pressure ridging.

2.12 Sea ice and ocean convection

In the Greenland Sea in winter an overturning of the surface water occurs, allowing deep penetration of water from near the surface down to 1500-2000 m. This renews the deep water and forms part of the global thermohaline circulation pattern. It also sequesters CO_2 and hence acts as a natural brake on the rate of increase of CO_2 in the atmosphere. It is now thought that the formation of young sea ice in the Greenland Sea - especially by the processes of frazil and pancake ice production, which lead to rapid ice growth - produces enough of a salt injection into the upper ocean to trigger this deep convection process. Experiments are now being designed to test hypotheses relating local ice growth and decay in the Greenland Sea in winter to the deep convection process. The role of remote sensing in detecting this local growth process will be vital.

2.13 Lake and river ice

We have already shown how lake ice formation occurs more readily than sea ice. Lake ice has no brine inclusions, but does contain air bubbles. Its crystal fabric resembles that of first-year sea ice. River ice often forms in turbulent conditions, and frazil ice is often a dominant component. This causes severe problems in blocking inlets to power station cooling systems, and the detection of frazil ice in rivers by remote sensing is an important technical problem.

2.14 Icebergs

Icebergs partake of the crystal structure of the ice shelves and glaciers from which they calved. In the Antarctic several thousand per year calve from the ice shelves, aprons of floating ice which surround the continent and which are extruding from the continental ice sheet. Typically icebergs when they calve are tabular, some 300-400 m thick, and have a gradation of snow-firn-ice density and structure similar to the parent ice sheet. As they drift first west then north-east they gradually warm up, become eroded by wave action, and finally break up into fragments and melt.

Arctic icebergs are usually smaller and more irregular in shape, as they calve from the narrow, steeper mountain glaciers that drain the Greenland ice sheet. Since they often capsize a number of times in their early lifespans, one cannot assume that their upper surface properties are as well-defined as Antarctic icebergs. They are usually rougher, more rugged and irregular. When seen on SAR, for instance, they give a random pattern of inner surface reflections.

3 The physical basis of electromagnetic wave interaction with snow and ice

Most, but not all, remote sensing systems used for snow and ice involve either measuring the emission of electromagnetic radiation from the surface (passive techniques) or the reflection or scattering of incident electromagnetic radiation from the surface (active techniques). The physical properties of snow and ice which are important in these processes are:

- reflectivity or albedo, which is the fraction of incoming radiation which is reflected from the surface, adding together both the direct surface reflection and volume scattering from within the snow or ice

- penetration length, the distance which an electromagnetic wave will penetrate into the material before being attenuated to 1/e of its initial intensity

- backscattering coefficient, usually expressed in terms of incidence angle and polarisation of the radiation, which is a measure of the fraction of energy in an active electromagnetic beam that is reflected back along the incident path by the surface

- emissivity, the power emitted from a surface as a fraction of the power emitted at that frequency by a perfect black body of the same physical temperature.

3.1 The visible and the near infra-red

At these frequencies the Sun is the source of illumination. The penetration depth is negligible in snow and ice, and absorption and scattering are concentrated within a few mm of the surface. The scattering is therefore a function of surface roughness and of the absorption coefficient. The albedo, which expresses the joint effect of these factors, is a function of surface roughness, slope, density and wetness of the snow or ice surface, angle of illumination, snow/ice grain size, thickness (for thin sea ice floating on water) and inhomogeneities in the surface layer (e.g. air bubbles in ice). The spectral dependence of albedo for sea ice, either bare or covered with snow, has been measured by Grenfell and is shown in Figure 8. Note the decline at long wavelengths, showing that ice is "dark" in the infrared. On land, freshly fallen snow has an albedo in the range 0.75-0.90, dropping to 0.45-0.70 as the snow ages. Thus old snow on land is similar in albedo to the summer sea ice cover, with bare ice and refrozen melt ponds present (open melt ponds bring the area-averaged albedo down much lower). In thermodynamic models of sea ice growth and decay (e.g. Maykut and Untersteiner 1971) a critical role is played by the low summer albedo of a bare ice/ponded surface in the absorption of solar radiation leading to enhanced summer melt. If winter snowfall were to increase to a point where it did not melt in summer, keeping the albedo high, then the maximum thickness reached by ice in the thermodynamic growth process would become very much greater.

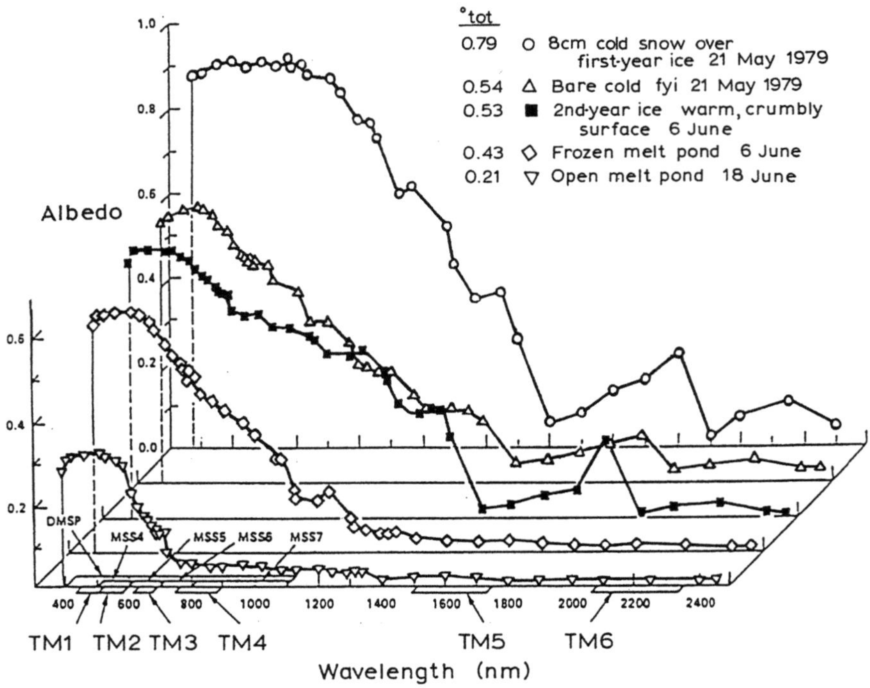

Figure 8: Sea ice albedos collected by T Grenfell off Alaska, shown in relation to some frequency bands for Landsat MSS and TM sensors (after Carsey and Zwally 1986).

3.2 Thermal infrared

In the thermal infrared (2-15 μ wavelength) we are no longer dealing with reflection of solar energy, but rather with emission of thermal radiation from the Earth. The important factors are the emissivity of the material and the penetration length, which determines the thickness of the surface layer that is effective in emitting radiation upwards. For ice, snow and firn the penetration length is only microns, and the emissivity is 0.97-1.00 (the lower figure for bare ice and melt ponds, the higher for snow). Variations in the detected signal are therefore mainly dependent on variations in the real physical temperature of the surface. The match is not perfect, because there is some dependence on viewing angle as well as on surface properties such as grain size and wetness, and there is also some atmospheric absorption, by water vapour, CO_2 and ozone. There are particular bands where this is strong, but the 8-13 μm band has high transparency and is usually used by thermal infrared sensors.

A problem with spaceborne detectors is that they have to be cooled to about 115°C to reduce thermal noise. Nevertheless, Dozier and Warren (1982) first showed that snow temperature can be mapped from space in the thermal infrared.

The advantage of thermal infrared over visible and near infrared is that it can penetrate darkness. However, like visible and near infrared it cannot penetrate clouds.

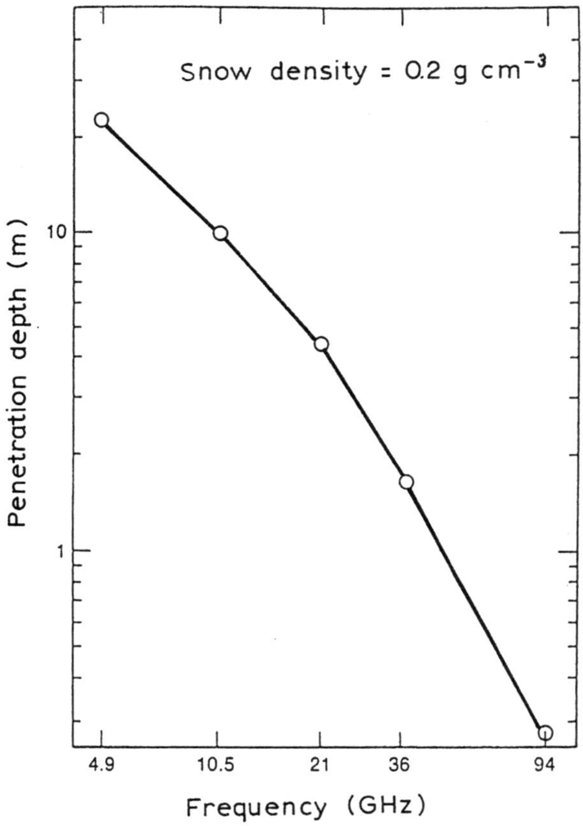

Figure 9: Penetration depth of dry snow at microwave frequencies (after Rott *et al.* 1985).

3.3 Passive microwave

Again we are dealing with the thermal emission of radiation, but at much longer wavelengths so that the energy emitted per unit area of surface is much less. Usually a sensor requires a large instantaneous field of view, or pixel size, to allow adequate radiation to be measured. This means that properties of an inhomogenous surface tend to be averaged, or smeared out, over the area of a pixel (12-25 km to a side for satellite sensors). Passive microwave radiation can, however, penetrate cloud so that it offers all-weather day-and-night coverage of the globe.

For a snow surface the microwave emissivity is found to be dependent on the rate at which the snow has accumulated, varying from 0.65 (at 10 GHz) in regions of low accumulation to 0.9 in regions of high accumulation. This is because of the varying grain size, and has been used as a way of estimating accumulation rates over the Antarctic ice sheet (see section 6.1). Emissivity also varies with snow thickness; the penetration depth in dry snow is high, especially at lower microwave frequencies (see Figure 9), and

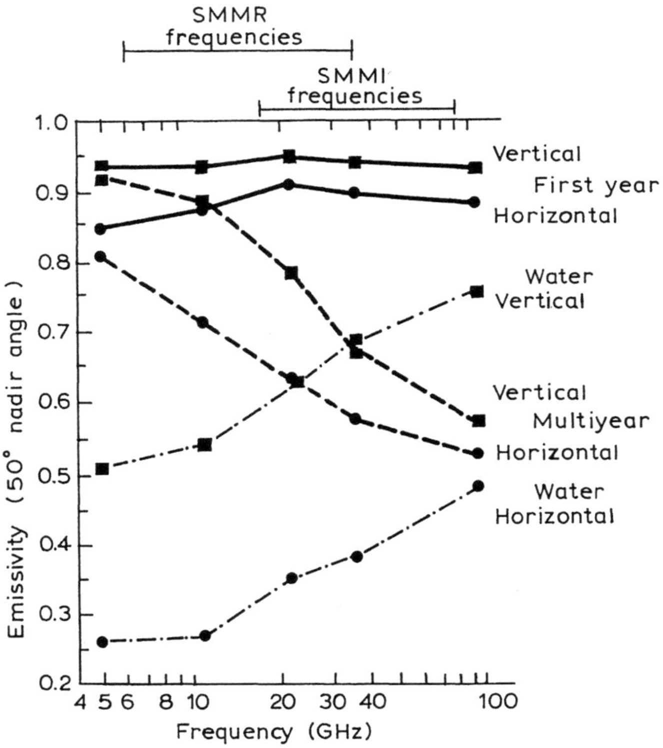

Figure 10: Frequency dependence of emissivity of open water, first- and multi-year ice in the microwave band at two polarisations (after Svendsen *et al.* 1983).

so the emission from a seasonal snowpack on land is dependent on its thickness, the basis of a technique for estimating snow thickness (see section 6.4).

For sea ice surfaces the emissivity varies greatly with ice type, and this is the basis for the use of multi-frequency passive microwave as a way of measuring the distribution of ice types in the polar oceans, one of the most valuable applications of passive microwave radiometry. Figure 10 shows some measurements of the frequency dependence of emissivity through the microwave band for first-year ice, multi-year ice and water, for both horizontal and vertical polarisations. It can be seen that the emissivity of first-year ice is high at all frequencies, while the emissivity of multi-year ice drops from a high value at low frequency to a much lower value at high frequency, in contrast to a steady rise in the emssivity of open water. A single frequency passive microwave system, such as the ESMR (electrically scanning microwave radiometer) aboard the Nimbus-5 spacecraft, can therefore be made to give good results on delineating the ice edge position and even on determining ice concentration where there is only one ice type present (e.g. Antarctic sea ice, which is mainly first-year ice, is well discriminated from open water at the 19 GHz frequency of ESMR), but it cannot give ice type discrimination. This had to await the advent of multi-frequency systems such as the SMMR (scanning multichannel microwave radiometer) aboard Nimbus-7 (1978-87) and the more recent SSM/I (special sensor microwave/imager) aboard the DMSP satellite.

To extract ice type from multi-frequency emissivity data a number of algorithms have been developed. In the case of SMMR these usually make use of the 18 GHz and 37 GHz channels at both polarisations. Two basic philosophies are used. In one various ratios are developed between emissivities at different frequencies and used for discriminating ice types, with suitable corrections for atmospheric temperature, humidity etc. The second philosophy (e.g. Comiso 1990) is cluster analysis, a self-referencing system in which all the data from a given day for the whole Arctic or Antarctic are plotted on a scatter diagram in two dimensions (typically 37H versus 37V or 18V versus 37V), whereupon a triangular shape is seen with the three vertices being "poles" representing pixels which contain purely one type of surface (first-year ice, multi-year ice or open water). Any given pixel is then assigned a weight among these three poles, depending on its position in the triangle, so that ice type discrimination is carried out without the need for further atmospheric information. This approach also permits the investigation of clusters associated with young ice types (frazil, pancake, nilas) which can show their signature in this way without being lost as in the ratio technique. A good review of passive microwave techniques for sea ice is given by Comiso (1991), while a more exhaustive review of the whole field is in preparation (Carsey 1992).

3.4 Active microwave

In this case we are measuring the power backscattered from a surface at microwave frequencies. Imaging radars (real aperture or synthetic aperture) create a scanned image of the surface, while non-imaging systems (scatterometers and altimeters) simply record the backscattered energy or the range to the surface. A large backscatter, giving a bright target, results from high surface roughness and/or from high volume scattering from the material caused by the radiation penetrating deeply into it. Materials with a high surface reflectivity tend to reflect specularly and thus give a poor return, and a dark image, for a beam not incident at nadir. In the case of sea ice, young and first year ice give a dark return, because they usually have a smooth surface with a high salinity and thus conductivity, so that reflection tends to be specular. Multi-year ice gives a bright return because much of the brine has drained out of it, leaving a low-conductivity material into which radiation can penetrate to be scattered from within the volume; the surface roughness also tends to be higher. Pressure ridges, in ice of any age, give the brightest returns of all.

Because there is no simple dependence on ice type, SAR is normally used to give high resolution delineation of the shapes of ice features and thus the velocity and deformation rates. The reviews mentioned above also survey SAR techniques. All spaceborne SARs are single-frequency (e.g. ERS-1), but, as in the case of passive microwave, it may be possible to extract more information by using multi-frequency systems. This is described by Drinkwater (1990). An analysis of the seasonal variation in SAR signature of sea ice of a well-defined type (landfast ice) has been given by Barber *et al.* (1992a).

4 Techniques and their applications

We describe each technique in turn and show examples of its applications. Since the term "remote sensing" does not apply only to electromagnetic techniques but to any method of measuring a geophysical parameter without direct contact, we shall also discuss acoustic techniques. This is a very brief summary. A valuable and complete review of current and planned satellites with a polar role, and the useful sensors carried by each, is given by Massom (1991).

4.1 Visible and infrared

Present sensors: AVHRR at visible and near I-R (1.1 km resolution); DMSP (500 m resolution, unclassified south of 60°S); Landsat MSS, TM (30 m resolution), ETM from 1992; SPOT (10 m resolution to 84° latitude)

Problems: cloud and darkness (visible) limited; cloud-snow discrimination

Uses:

(a) Low resolution imagery:- ice pilotage; generation of sea ice charts for use by mariners; cloud cover analysis for specific polar applications (e.g. passage of polar lows, cloud input to ice-ocean thermodynamic-dynamic models and albedo-feedback models); mapping of large-scale features in the polar regions such as nunataks, mountain ranges, large glaciers; mapping of large-scale time variable features such as ice shelf margins, sea ice limits, coastal and ocean polynyas; mapping of sea ice motion vectors by superimposition of successive images; mapping of seasonal snow extent on land; mapping sea surface temperature near the ice edge; extracting atmospheric aerosol content.

(b) High resolution imagery:- detection of blue ice areas as runways for wheeled aircraft in Antarctica; detailed study of iceberg sizes, floe sizes, coastline mapping, glacier snout changes including surges; ice sheet dynamics from superposition of successive images of crevasse patterns.

4.2 Passive microwave

Present sensors: SSM/I, ATSR, with others in planning stage (see Table 3)

Problems: low resolution (typically 25 km)

Uses: mapping of sea ice concentration and distribution of ice types (first year versus multi year) over Arctic and Antarctic Oceans on daily basis; detection of large anomalous features (e.g. Weddell polynya in 1970s, Odden ice tongue in Greenland Sea); census studies of overall hemispheric sea ice area and its variation; mapping of accumulation rates over Antarctica (emissivity of dry polar firn is sensitive to grain size variation with depth, which depends on accumulation rate) and other parameters of ice sheet surfaces; determination of snowpack volume (see section 6.4).

Spacecraft and year of launch	*Sensor*	*Frequency (GHz)*	*Swath width (km)*	*Spatial resolution at nadir (km)*
Mariner 2, 1962	PMW radiometer	15.8, 22.2	Planetary (Venus)	1,300
Cosmos 243, 1968	PMW radiometer	3.5, 8.8, 22.2, 37.0	—	37
Cosmos 384, 1970	PMW radiometer	3.5, 8.8, 22.2, 37.0	—	13
Nimbus-5, 1972	ESMR	19.35 (H)	3,000	25
	NEMS	22.2, 31.4, 53.6, 54.9, 58.8	185	185
Skylab, 1973	S-193	13.9	11–170	16
	S-194	1.4	280	115
Meteor, 1974	PMW radiometer	37.0	—	—
Nimbus-6, 1975	ESMR	37.0 (V & H)	1,270	20 × 43
	SCAMS	22.2, 31.6, 52.8, 53.8, 55.4	2,618	145 to 330
DMSP Block 5D, 1978	SSM/T	50.5, 53.2, 54.35, 54.9, 58.4, 58.825, 59.4	1,600	175
TIROS-N/NOAA, 1978	MSU	50.3, 53.7, 55.0, 57.9	2,300	110
Seasat, 1978	SMMR	6.63, 10.69, 18.0, 21.0, 37.0 (all H & V)	600	149 × 87 to 16 × 27
Nimbus-7, 1978	SMMR	6.63, 10.69, 18.0, 21.0, 37.0 (all H & V)	800	148 × 151 to 27 × 32
MOS-1 (1987), MOS-1b (1990), MOS-2 (1992)	MSR	23.8 (H), 31.4 (V)	317	32, 23
DMSP Block 5D-1, 1985	SSM/T-2	91.5, 150, and 3 channels near 183.0	175–200	50
DMSP Block 5D-2, 1987	SSM/I	19.35, 37.0, 85.5 (all V & H), 22.235 (V only)	1,394	16 × 14 to 70 × 45
ERS-1, 1991	ATSR-M	23.8, 36.5	500	22
NOAA 'Next', 1991 and EOS-A, 1998	AMSU-A	23.9, 31.4, 12 channels in range of 50.3 to 57.29, 89.0	2,300	50
	AMSU-B	89.0, 157.0, and 3 channels near 183.31	2,300	15
GOES 'Next', 1991	PMW radiometer	92, 118, 150, 183, 230	500	35 to 15
TOPEX/Poseidon, 1992	PMW radiometer	18.0, 21.0, 37.0	—	51, 40 and 27
EOS-A, 1998	ESTAR	1.4	1,400	10
	MIMR	6.8, 10.65, 18.7, 23.8, 36.5, 90.0 (all V & H)	1,400	60 to 4.8 (goal)

Table 3: Satellite passive microwave sensors, past, present and future.

4.3 Synthetic aperture radar

Present sensors: ERS-1, JERS-1

Problems: lack of receiving stations in Antarctica, lack of on-board storage

Uses: all-weather, day-and-night coverage at high resolution permits all the range of studies carried out by visible sensors to be done without interruption; mapping of small sea ice features (e.g. ice edge bands, leads, small polynyas); measuring sea ice motion and deformation with high accuracy and resolution using feature tracking and cross-correlation techniques; sea ice type classification, not only first-year versus multi-year, but also young ice and frazil ice; detection of wave penetration into ice and the use of wavelength changes to determine frazil ice suspension thicknesses (Wadhams and Holt 1991); development of empirical relation between SAR brightness and ice thickness (see section 6.3); iceberg detection; mapping of wet snow areas on ice sheets.

4.4 Radar altimeters

Present sensors: ERS-1, Geosat (failed 1990), TOPEX/Poseidon

Problems: interpretation of true surface location

Uses: mapping of ice sheet elevation, especially in Greenland (section 6.1); low-resolution mapping of ice edge location in Antarctic from change in pulse shape; estimation of distance of penetration of ocean waves into pack ice.

4.5 Scatterometers

Present sensors: ERS-1

Problems:

Uses: surface wind speed and direction over open sea, hence over ice edges; experimental relation between radar backscatter and ice sheet surface features (e.g. summer melt areas give large backscatter, perhaps due to subsequent formation of ice lenses).

4.6 Laser profilometer

An airborne technique which in principle offers all the mapping possibilities of radar altimeters with much better spatial and vertical resolution. The laser reflection comes from the very top surface of snow and therefore it is an accurate way of mapping ice sheet elevation and also the elevation and roughness of sea ice surfaces (including the shapes of features such as pressure ridges and the occurrence and widths of leads). Recently it has been shown that with care one can match the PDF of sea ice elevation measured from a laser with the PDF of draft measured from a submarine and hence infer ice thickness distribution from a laser alone (Wadhams and Comiso, 1992; Wadhams *et al.* 1991, Wadhams *et al.* 1992).

4.7 Electromagnetic sounding techniques

Radio echo sounding was developed to measure the volume of polar ice sheets using a pulsed radar. Its sucess depends on the low scattering by polar ice. Originally developed by SPRI, Cambridge, in 1960s, it was used to map the Antarctic ice sheet, then Greenland, recently Svalbard and soon Franz Josef Land. Problems arise from scattering loss in temperate glaciers and from beamwidth-induced effects on resolution of surface.

Eddy current sounding was developed as a way of sounding through sea ice from a helicopter flying at speed. A "bird" is towed behind the helicopter containing a coil which emits electromagnetic radiation in the frequency range 900 Hz — 33kHz. This induces eddy currents in the water under the ice, which in turn generate secondary electromagnetic fields. The secondary fields are detected by a receiver in the bird; their strength depends on the depth of the ice-water interface below the bird. The bird also carries a laser profilometer to measure the depth of the snow-air interface below the bird, and the difference gives the absolute thickness of the ice plus snow.

Impulse radar was also developed to measure sea ice thickness, and uses 100 MHz pulses to overcome high scattering by brine cells in ice. It can only be used from a slow flying helicopter very close to the ice surface, and works best over multi-year ice. It cannot sense ice thicker than 10-12 m. It is therefore of limited usefulness and is falling out of use in favour of eddy current sounding.

4.8 Acoustic sounding techniques

Upward sonar is the standard means of measuring sea ice thickness synoptically. It thus requires the use of long-range submarines, necessarily military at present. The sonar actually measures range to the underside, from which ice draft is extracted. Ice thickness distribution is then derived statistically using the known mean density of sea ice. Sonar can be mounted on moorings at key locations and hence will give a time series of ice thickness. If a line of such sonars is combined with SAR or AVHRR-based measurements of ice velocity, the result will be an ice flux estimate (at present being done for East Greenland Current). In future sonar may be mounted on autonomous underwater vehicles. In joint submarine-aircraft experiments upward sonar provides important validation possibilities for airborne remote sensors (Wadhams *et al.* 1991).

Sidescan sonar has a fan-shaped beam and gives an image of ice underside. It also must be done from a submarine or an autonomous underwater vahicle (AUV). The images show the presence and shapes of pressure ridges and leads, and texture of underside can be used to discriminate between first-year and multi-year ice (Wadhams 1988).

Interferometric sonar is a new technique which combines the virtues of upward and sidescan sonar, giving depth-contoured images of ice underside.

Acoustic tomography is a way of mapping the characteristics of a volume of ocean of typical area 300 km x 300 km and depth 1000 m by placing an array of acoustic transducers around the perimeter of the volume and measuring the travel time of acoustic pulses between every possible pair of transducers. Each acoustic pulse gives a number of arrival times based on its incidence angle at emission and hence the number of "bounces" from the water surface. Normally the technique is used to give temperature and salinity structure of the open ocean, but when a sea ice cover is present the travel times are decreased by an amount which depends on the ice thickness. Jin and Wadhams (1989) developed the theory, which is being applied to data recorded in the Greenland Sea.

4.9 Present and future needs

Having surveyed the capabilities of remote sensing systems, we end by looking at what the perceived needs are for remote sensing of polar regions. A useful summary of marine needs was recently generated by the Canadian Ice Working Group and published by Barber *et al.* 1992b.

5 The greenhouse effect and the cryosphere

We now consider the greenhouse effect and the particular snow and ice processes which cause the effect to be enhanced in polar regions, especially north polar regions. This enhancement suggests that observable effects should be seen earlier in the polar regions than elsewhere in the world and in fact may already be detectable.

Four key cryospheric indicators of climatic change in action are particularly suitable for study by remote sensing and we go on to consider in section 6 how remote sensing techniques enable us to assess whether significant changes are taking place. The topics are

- the retreat of glaciers and ice sheets and the its contribution to global sea level rise
- the retreat of sea ice
- the thinning of sea ice
- the retreat of seasonal snow cover on land.

A good summary of the physics of the greenhouse effect is given in the *Policymakers' Summary* by Houghton *et al.*(1990) (see chapter 1 of this book) and so this will not be discussed in detail in this chapter.

The most important aspect from the polar point of view is that models agree that warming at the Earth's surface will be enhanced in polar regions relative to lower latitudes. The main reason is ice-albedo feedback. A warmer Earth will lead to a retreat of the winter snow cover on land, and of winter sea ice limits, replacing snow (albedo approximately 0.8) by water, tundra or forest (albedo approximately 0.15-0.2) and hence leading to greater absorption of incoming short-wave radiation by the surface and an enhanced warming. This is a positive feedback mechanism. Since the physics are incorporated in most GCMs (general circulation models), both equilibrium GCMs (in which atmospheric CO_2 is, say, doubled, and the atmosphere allowed to come to a new equilibrium) and time-dependent GCMs (in which the CO_2 increases at a realistic rate and time-dependent effects, such as heat absorption by the ocean, are allowed for in considering how rapidly the Earth warms up), these models all predict enhanced polar warming in winter. Time-dependent GCMs, however, predict this enhancement only for the north polar regions, since the Southern Ocean has such a powerful effect in delaying warming in the Antarctic through its heat absorptive capacity. Further, since seasonal snow cover is chiefly found in the northern hemisphere, the ice-albedo feedback effect in the southern hemisphere is confined to sea ice.

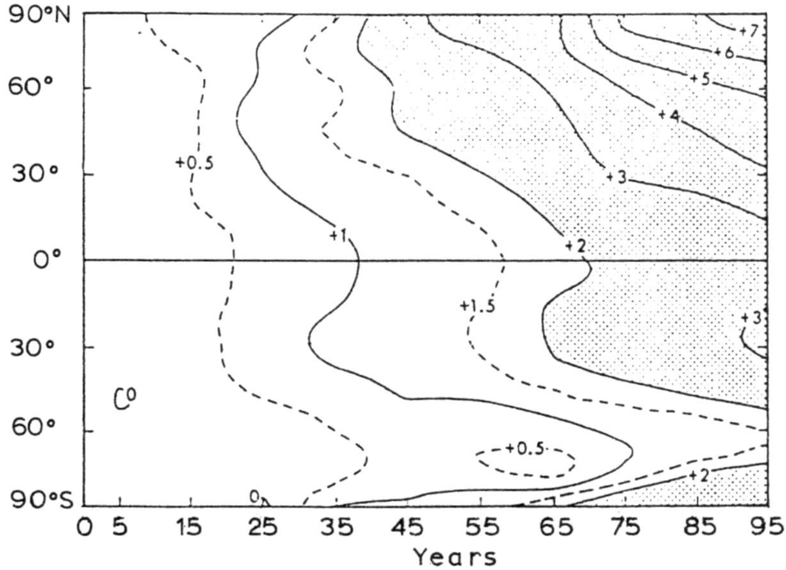

Figure 11: Latitude-time distribution of the difference in zonally-averaged, decadal-mean air temperature in °C between present and future years, from a time-dependent GCM.

Another factor in polar warming is the stability of the polar atmosphere in winter. The polar troposphere is thinner than at lower latitudes, and this again should lead to enhanced warming because a smaller atmospheric column near the surface is undergoing warming.

Offsetting factors include changes in cloudiness which may accompany warming. For instance, if sea ice retreats from the fringes of the Arctic Ocean there will be greater evaporation and cloudiness over the Arctic Ocean which may have a cooling effect.

Other completely unknown polar feedbacks are the change in methane emission from tundra which may accompany a warming of the Arctic land areas and a partial melting of the permafrost; a reduction or suppression of deep convection in the Greenland Sea, a process apparently driven by sea ice formation which also involves the downward transport of CO_2 and its sequestration in the deep ocean.

Figure 11 shows the results of the GFDL time-dependent model as a function of latitude and years into the future. It is clear that there is great enhancement in high northern latitudes and almost no warming in high southern latitudes. More recent results from four time-dependent models are shown in Figure 12 (Gates *et al.*, 1992). Here we see predictions for approximately the time when CO_2 concentration has doubled. The models agree in predicting an enhanced warming in the Arctic relative to any other part of the world, but it can be seen that the enhanced warming is least for the maritime Arctic, i.e. the Greenland Sea area, where there is much open water. This reflects the importance of the ocean's role in slowing climatic warming.

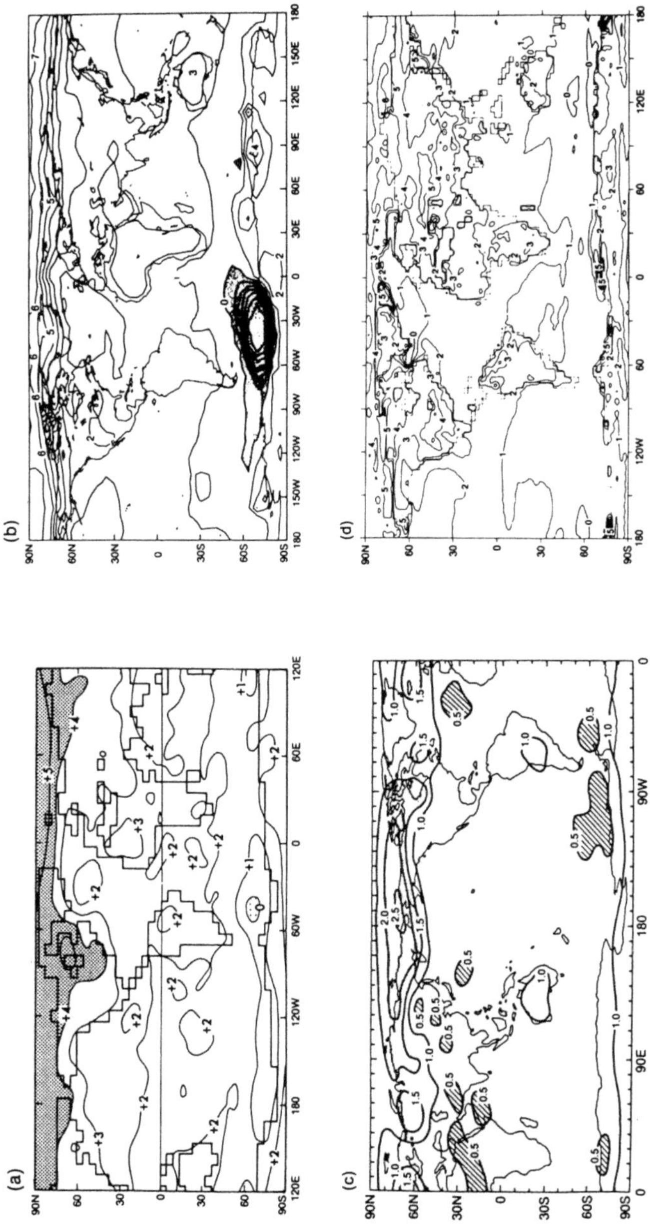

Figure 12: The distribution of change of surface air temperature (°) simulated near the time of CO_2 doubling by four coupled ocean-atmosphere GCMs in response to a transient CO_2 increase. (a) The GFDL results are averaged over years 60- 80 and referenced to the 100-year average of a control; (b) the MPI results are averaged years 56-65 and referenced to the corresponding years of a control; (C) the NCAR results are averaged over years 31-60 and referenced to the corresponding control years; (d) the UKMO results are averaged over years 65-75 and are referenced to the corresponding years of a years of a control. (Houghton *et al.* 1992).

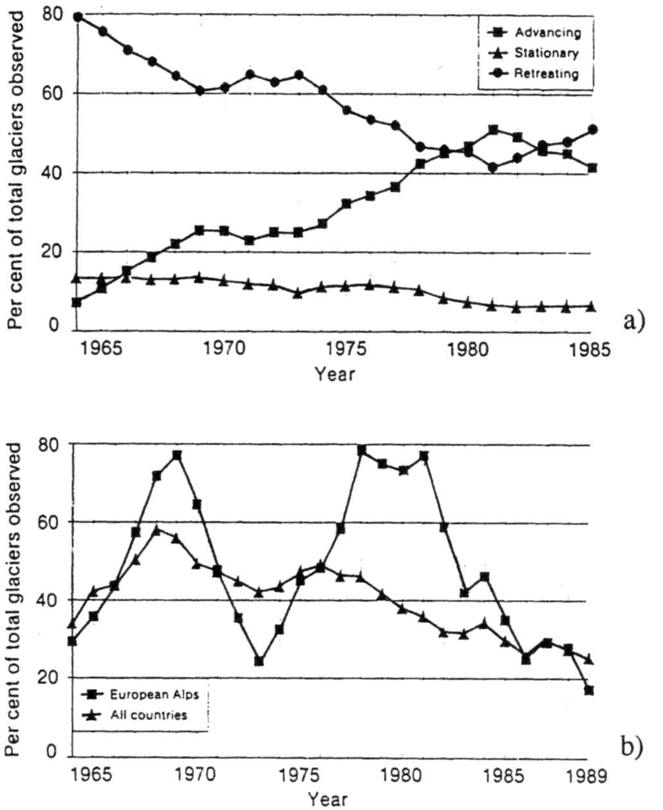

Figure 13: (a) Percentages of observed glacier fronts advancing, stationary and retreating. Data from all available countries; number of glaciers from 271 in 1959/60 to 486 in 1984/85. Values are 5-year moving averages from 1959/60-1963/64 to 1980/81- 1984/85; (b) Percentages of observed glaciers having a positive mass balance: (i) all countries (triangles); (ii) about 10 European Alpine glaciers (rectangles). Values are 5-year moving averages from 1959/60- 1963/64 (53 glaciers from all countries) to 1984/85-1988/89 (71 glaciers from all countries).

6 The cryospheric response to warming - four applications of remote sensing

6.1 Contribution of melting glaciers and ice sheets contributing to the rate of global sea level rise

It has been estimated that during the past century world sea levels have risen by 10-20 cm and that the melting of mountain glaciers is responsible for about half of this rise. The rest is due to warming of the ocean and its consequent expansion, with sea surface temperatures being estimated as having risen by 0.3-0.5 °C during the past century.

What is the present contribution of glacier melt to sea level rise, and what might it become as global warming accelerates? Remote sensing plays a large part in providing the data needed to give answers to these questions.

Table 5 shows the physical characteristics of the three regions where fresh water ice is stored on Earth (data taken from various sources). The Antarctic ice sheet excludes the floating ice shelves, and is by far the largest ice reservoir. Next comes the Greenland ice sheet. The remainder comprises small polar ice caps (e.g. on Svalbard, Baffin Island, Devon Island, Franz Josef Land, some sub-Antarctic islands) and mountain glaciers in such regions as Alaska, the Rockies, the Patagonian Andes, the Alps, the Himalayas and central Asia. The table shows that, as mentioned in section 2, Antarctica is an accumulation region with an equilibrium line at zero altitude; all loss of ice is by calving. Greenland, however, has an equilibrium line at 950 m altitude, and loses considerable amounts of ice each summer by ablation (mainly melting).

It can be seen that there is far more scope for a contribution to sea level rise from Greenland or Antarctica than from mountain glaciers - in fact a total melt would yield 7 m and 65 m respectively of water, compared to only 35 cm from small glaciers. However, we have very little information on the present mass balances of Greenland or Antarctica, whereas we do know that most mountain glaciers have been in retreat for the last century. Figure 14 shows the percentage of glaciers estimated to be in retreat, and the changes in length of some of them are shown in Figure 10 of chapter 6. In well settled areas the length of a glacier has been recorded throughout historical time, while in other accessible cases it can be deduced from a study of moraines. However, in inaccessible regions the data used to obtain estimates of the percentage of glaciers in retreat, and their likely mass balance, must be obtained by remote sensing. Typically the methods employed are visible data (e.g. Landsat) to yield changes in the glacier terminus position, and AVHRR or passive microwave to estimate mass balance by measuring the position of the snow line relative to the glacier as a whole. In the latter case the configuration of the glacier bed needs to be known, or estimated (e.g. by airborne radio echo sounding), and then the snowline position can be used to estimate the fraction of the glacier which is subject to ablation. Such estimates are not as good as direct observations on accessible glaciers, but can be used to give overall estimates of the state of glaciers in regions which are difficult of access.

The Greenland ice sheet has recently been subjected to a remote sensing study in which elevation data obtained by the Seasat radar altimeter in 1978 was compared with data from Geosat in the late 1980s, concluding that the southern part of the ice sheet has been gaining in elevation, in both the accumulation and ablation zones, at a rate of about 0.2 m y^{-1}. His analysis technique has been criticised, because of differences in the method used to interpret data derived from the two different sensors. However, it is quite possible that a warming could lead initially to a gain in mass, since warmer, moister winds will be bringing a greater amount of precipitation to Greenland. Present estimates of Greenland mass balance (Table 4) vary wildly from positive to negative, so it is clear that the contribution of Greenland to global sea level can best be assessed by carrying out a careful, systematic study of data from a single radar altimeter over a period of years. This is also true of Antarctica, where most estimates (Table 5) agree that the ice sheet will gain mass as temperature rises. This is because the temperature rise will still be inadequate to cause the ice to begin to ablate, whilst causing an influx

	Antarctica (grounded)	Greenland	Glaciers & small ice caps
Area(10^6 km^2)	11.97	1.68	0.55
Volume (10^6 km^3 ice)	29.33	2.95	0.11††
Mean thickness (m)	2,488	1,575	200††
Mean elevation (m)	2,000	2,080	-
Equivalent sea level (m)	65	7	0.35 ††
Accumulation (10^{12} kg y^{-1})	2200††	535†	-
Ablation (10^{12} kg y^{-1})	<10††	280††	-
Calving (10^{12} kg y^{-1})	2200††	255††	-
Mean equilibrium - line altitude (m)	-	950†	0 - 6,300
Mass turnover time (y^{-1})	~15,000	~5,000	50 - 1,000

(a)

Source	Accumulation	Ablation	Calving	Balance
Bader	+630	-120 to -270	-240	+270 to +120
Benson	+500	-272	-215	+13
Bauer	+500	-330	-280	-110
Weidick	+500	-295	-205	0
Reeh	+487	-169	-318	0
Ohmura & Reeh	+535			

(b)

Source	T(+1°C)	P(+5%)	C(+5%)	Remarks
Ambach & Kuhm	+0.31	-0.31		Analysis of EGIG data
Bindschadler	+0.45			EGIG data/retreating margin
Braithwaite and Olesen	+0.36 to +0.48			Energy balance calculation
Oerlemans	+0.37	-0.11	-0.06	Energy balance Model

(c)

Table 4: (a): Some physical characteristics of glacier ice on Earth.Estimated accuracy: † = 15%, †† = 30%, otherwise better than 10%.
(b): Estimates of the mass budget of the Greenland ice sheet in 10^{12} kg^{-1}.
(c): Estimates of the Sensitivity of the Greenland mass balance to climatic change. T = temperature, P = precipitation, C = cloudiness. Expressed in rate of change of global mean sea level (mm y^{-1}).

Source	Change in sea level (mm y^{-1})	Remarkes
Muszynski	-0.38	Regression on 208 data points
Fortuin and Oerlemans	-0.139 (interior) -0.061 (escarpment) -0.200 (total)	Regression on 486 data points (only grounded ice)
Proportional to water vapour mixing ratio	-0.34	20 km grid over grounded ice

Table 5: Estimates of the change in Antarctic mass balance for a 1°C warming. Δq_s represents saturation water vapour mixing ratio of air above the inversion.

	LOW	BEST ESTIMATE	HIGH
Thermal expansion	2	4	6
Glaciers/small ice caps	1.5	4	7
Greenland Ice Sheet	1	2.5	4
Antarctic Ice Sheet	-5	0	5
TOTAL	-0.5	10.5	22
OBSERVED	10	15	20

Table 6: Estimated contributions to sea-level rise over the last 100 years (in cm).

of warmer, moister winds with increased precipitation around the coastal regions of the continent.

Tables 6 gives some current estimates of the contribution of various glacial sources to sea level rise over the last 100 years. Various estimates of future global sea leval rise are given by Raper *et al.* (1990). These take no account of specially dire events that have been postulated by some (while discounted by most), such as the breakup and sliding into the sea of the West Antarctic Ice Sheet. The estimates differ greatly, and it is clear that much more systematic work needs to be done. Only satellite remote sensing can supply the data required to improve these estimates. The importance of an accurate estimate of the forthcoming rate of sea level rise cannot be exaggerated: on it will depend the measures that mankind must take to protect, or abandon, low-lying regions of the world's coastline.

6.2 Is global warming causing a retreat of sea ice?

If we adopted a simple thermodynamic view of sea ice, we might expect a retreat of sea ice limits under global warming in much the same way as a retreat of snowlines, and having much the same effect - an enhancement of warming through ice-albedo feedback. However, the dynamics of sea ice lead to complications such that the response of sea ice limits to warming is impossible to forecast (even in sign) without the application of a proper coupled model. Further, evidence has shown that despite large interannual fluctuations in energy fluxes in the Arctic, the winter and summer ice extents are very stable from year to year. Studies of ice extent using passive microwave imagery (Gloersen and Campbell 1988) have shown evidence of a small but significant retreat in winter ice extent over the Arctic Ocean as a whole since the early 1970s, with more pronounced trends (although some are positive and some negative, partially cancelling one another out) if particular sectors of the Arctic are considered individually (Parkinson 1991).

6.3 Can we map sea ice thickness synoptically, and is global warming causing a thinning?

This topic can best be discussed by referring to two papers. The first (Wadhams and Comiso 1992) deals with the question of how we can map the ice thickness distribution in the Arctic and Antarctic using remote sensing techniques. This is a very serious problem. Sea ice, as we have explained above, is a lossy material which is very difficult to sound using electromagnetic techniques. Up to now we have been dependent on upward echo sounding from submarines to map the ice thickness distribution over the Arctic, while in the Antarctic (where submarines have not operated) our knowledge, based mainly on drilling, is very poor. Arctic ice thickness mapping has not been carried out on a systematic or synoptic basis because of this dependence on submarine operations, and so it is difficult to acquire a statistically valid dataset from the same region in different years from which changes can be deduced. The paper discusses some novel remote sensing techniques which have been proposed or tried (such as impulse radar, eddy current sounding, tomography) and shows why they do not offer a solution to the synoptic mapping problem. It goes on to discuss the prospects and limitations of indirect inference of ice thickness from passive microwave and SAR. Finally, it shows how airborne laser profiling may provide an answer to the problem of synoptic mapping, so long as we can get a better idea of the seasonal variability in snow thickness over the Arctic Basin.

The second paper (Wadhams 1990), discusses some results which have actually been obtained using upward sonar from submarines in different years. A significant thinning has been observed, but there are many factors besides greenhouse warming which may have caused this, notably a change in the drift pattern of ice in the Arctic. Two surveys of the Arctic by upward sonar from a submarine were carried out in 1976 and 1987 over the same regions of the Eurasian Basin. In 1976 the conventional pattern of a build-up of pressure ridging (hence of mean ice draft) near the north coast of Greenland was observed. In 1987 this did not occur, resulting in a 15% loss of ice volume over an area of 300,000 km^2. An examination of the ice thickness distributions, combined with data from drifting Argos-tracked buoys that were present in 1987, showed that an anomalous ice drift pattern developed in 1987 whereby the Beaufort Gyre was apparently stationary during the winter while the Trans Polar Drift Stream operated normally, causing a divergence of the ice north of Greenland and the creation of much open water and young ice area. The conclusions are that substantial changes in ice thickness do occur over the Arctic, but that ice dynamics has a powerful effect which will require many years of synoptic data gathering to distinguish from any greenhouse-induced trend which may be present. This is a srong argument for the application of novel remote sensing techniques in an attempt to solve this monitoring problem, with laser profiling seeming at present to offer the best solution.

6.4 Is seasonal snow cover diminishing?

General circulation models (GCMs) agree on the enhanced warming to be expected in the Arctic mainly because they all incorporate similar physics. One of the key processes is the ice-albedo feedback effect described earlier, and the reason why this is held to be so effective in the northern hemisphere as opposed to the southern is that the large land area in the northern hemisphere carries an extensive seasonal snow cover, which responds more readily to warming than the sea ice cover which is the chief seasonally varying cryospheric parameter in the southern hemisphere. Therefore we might expect the seasonal snow extent in the northern hemisphere to be a very sensitive indicator of climate warming and to be already in measurable retreat. Is this the case?

Figure 14 (Folland *et al.* 1992) shows a time series from 1973 to 1991, with a smoothed trend generated from a 39-point binomial filter applied to monthly data. The trend shows a decline of 2 million km^2 over the 18-year period, representing about 8% of the total area. This is a significant decline. The curve also shows the smoothed average land air temperature anomalies relative to 1951-80 for the region north of 30 °N. There is a correlation of -0.41 between September-May monthly anomalies of snow cover and temperature and of -0.76 between unsmoothed annual anomalies.

How were these results obtained? They derive from weekly snow cover maps produced operationally by NOAA (National Oceanic and Atmospheric Administration, Washington) using AVHRR imagery mainly in the visible wavelengths. The dataset was generated by taking 7921 grid boxes over the hemisphere and defining them as snow or no snow. Some deficiencies of the dataset which have been noted (Folland *et al.* 1990) are that until 1975 the charts did not consistently include Himalayan snow cover (an apocryphal story is that the legend on the appropriate images masked this part of the world!), that the snow cover occasionally extended southward beyond the limits of the imagery used, that coverage was poor during mid-winter because of lack of daylight and that scattered mountain snow was omitted because of coarse grid resolution.

Problems of mapping snow extent by various means have been discussed more fully by Hall (1988), who also deals with the question of estimating snow thickness. This is very difficult with AVHRR visible and near-infrared data except in a general way whereby relative image brightness can be related to broad categories of snow depth: shallow, moderate or deep. Very shallow snow (< 5 cm) is often less reflective than deeper snow because the darker vegetation may show through a shallow snowpack,

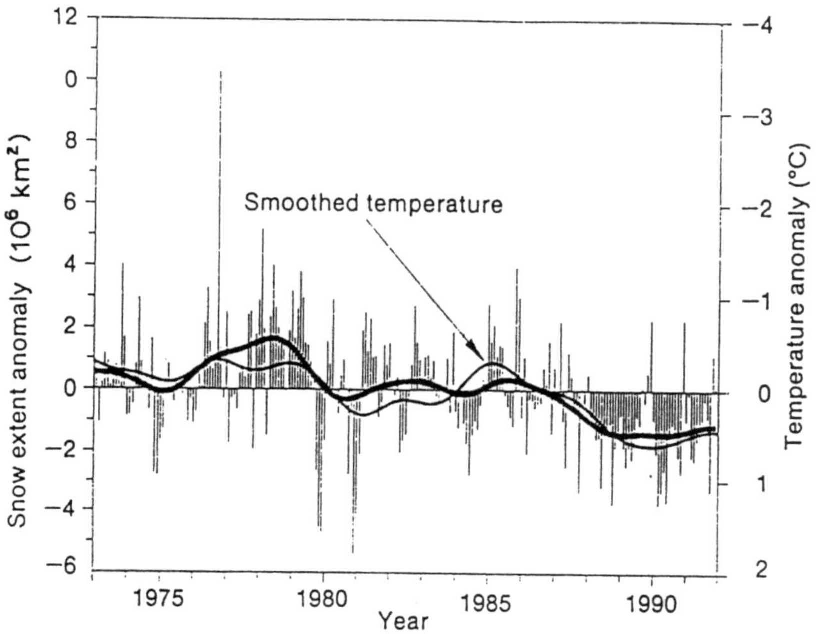

Figure 14: Northern hemisphere snow extent anomalies relative to 1973-1991 (from NOAA,USA) (vertical bar and heavy line) and Jones(1988, updated) land air temperature anomalies relative to 1951-80 north of 30 °N (thin line). Smooth lines generated from a 39-point binomial filter applied to the monthly data. Note the inverted temperature scale. (From Houghton *et al.*)

producing a mixed pixel reflectance.

Ladnsat MSS and TM sensors have been used to map snow extent, but suffer the same problems as NOAA sensors in seeing through cloud and darkness and sensing radiation from within the snowpack. However, data from Landsat-4 and -5 TM sensors can be used to discriminate between clouds and snow which is important for snow mapping. This is because in TM band 5 (1.55 - 1.75 μm) the reflectance of clouds remains high while the reflectance of snow drops to very low values.

Passive microwave data have been used both to monitor snow extent and to estimate thickness, although the resolution is crude (30 km) (e.g. Chang *et al.* 1987). Thin ($<$ 2.5 cm) dry snow is difficult to detect in this way because it does not produce enough volume scattering to alter the brightness temperature emanating from the ground. Similarly, shallow wet snow has an emissivity similar to wet bare ground and may also be difficult to detect. For this sort of snow, where the liquid water content is about 5%, X- and C-band SAR may be useful (Rott 1984); SAR also has a much better spatial resolution for detecting the position of the snowline.

Chang *et al.* (1987) and Kunzi *et al.* (1982) produced algorithms for mapping snow extent and depth from the 37 and 18 GHz channels of the Nimbus-7 SMMR. The Chang *et al.* algorithm assumes an ideal uniform snowpack of density 0.3 g cm^{-3} and

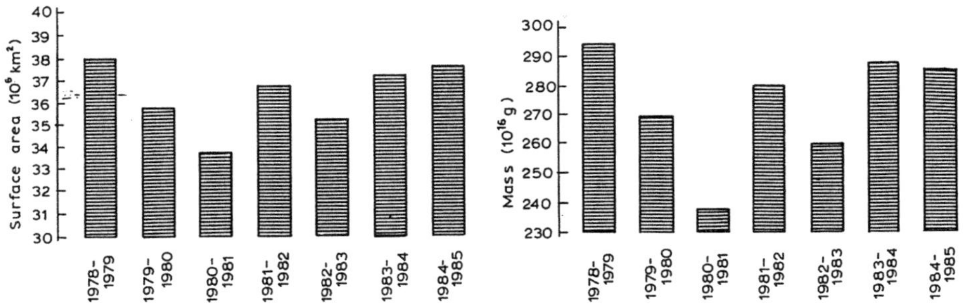

Figure 15: Bar graphs showing the extent and volume of snow in the northern hemisphere as measured using scanning multichannel microwave radiometer data

a snow grain radius of 0.35 mm, and so its validity is best for areas where the snow approximates to these properties.

As snow ages on the ground during winter, a layer of depth hoar may form at the base of the snowpack. This upsets the algorithm: depth hoar has large crystals which produce a lower microwave emissivity (Hall *et al.* 1986). If snow is lying in an area of coniferous trees, which have a high emissivity, the algorithm also fails.

Despite all these problems some brave attempts have been made to estimate northern hemisphere snow volume using SMMR data (Chao *et al.* 1987). The results, however, are not reliable enough to serve as the basis for an estimate of changes in hemispheric snow volume, unlike estimates of changes in hemispheric snow extent. Attempts to do so for 1978-1985 are shown in Figure 15. It can be seen that the snow extent charts do not match particularly well those derived from AVHRR data in Figure 16, so we can feel even less confidence in the snow volume charts. There is also a systematic offset of 5% (SMMR low) between SMMR-derived extent values and NOAA-derived values, probably because of the failure of SMMR to detect very thin snow.

It has been suggested (Barry 1984) that global warming may actually produce an increase in snow volume at high northern latitudes, due to increased atmospheric moisture flux, while giving a decrease at low to middle latitudes and a decrease in the winter snow extent.

Clearly snow volume is a parameter which it would be worthwhile to be able to derive more accurately from satellite data, since although it is snow extent which chiefly determines the albedo-feedback impact on climate change, it is snow volume and its latitude dependence which may be a quite separate element in the cryospheric climate response. We may even see a shrinking of the winter snow extent together with an increase in its hemispheric volume. As yet we cannot test these ideas, but we do see that snow extent is significantly diminishing. Remote sensing of snowpack volume on land thus remains a challenge to remote sensing.

References

Barber, D. G., LeDrew, E. F., Flett, D. G., Shokr M., and Falkingham J., 1992a, Seasonal and diurnal variations in SAR signatures of landfast sea ice. *IEEE Transactions Geoscience and Remote Sensing*, **30**, 638-642.

Barber, D. G., Manore, M. J., Agnew, T. A., Welch, H., Soulis, E. D., and LeDrew E. F., 1992b, Science issues relating to marine aspects of the cryosphere: implications for remote sensing. *Canadian Journal of Remote Sensing*, **18**, 46-55.

Barry, R. G., 1984, Possible CO_2 induced warming effects on the cryosphere. In *Climate changes on a yearly to millenial basis*, edited by N.A. Morner and W. Karlen (Hingham: Reidel), pp. 571-604.

Carsey, F. D. (Editor), 1992, *Microwave Remote Sensing of Sea Ice.* (Washington DC: American Geophysical Union. (in press)

Carsey, F. D., and Zwally, H. J., 1986, Remote sensing as a research tool. In *The Geophysics of Sea Ice*, editor by N. Untersteiner, (New York: Plenum), pp. 1021-1097.

Chang, A. T. C., Foster J. L., and Hall, D. K., 1987, Nimbus-7 SMMR derived global snow cover parameters. *Annals of Glaciology*, **9**, 39-44.

Chao, B., O'Connor, W. P., Chang, A. T. C., Hall D. K., and Foster J. L., 1987, Snow-load effect on the Earth's rotation and gravitational field, 1979-1985, *Journal of Geophysical Research*, **92**, 9415-9422.

Coachman, L. K., and Aagaard K., 1974, Physical oceanography of arctic and subarctic seas. In *Marine Geology and Oceanography of the Arctic Seas* , edited by Y. Herman (New York: Springer-Verlag), pp. 1-72.

Comiso, J. C., 1990, Arctic multi-year ice classification and summer ice cover using passive microwave satellite data. *Journal of Geophysical Research*, **95**, 13411-13422.

Comiso, J. C., 1991, Satellite remote sensing of the polar oceans. *Journal of Marine Systems*, **2**, 395-434.

Drinkwater, M. R., 1990, Multi-frequency imaging radar polarimetry of sea ice. In *Ice Technology for Polar Operations* , edited by T. K. S. Murthy, J. G. Paren, W. M. Sackinger and P. Wadhams) (Southampton: Computational Mechanics Publns), pp. 365-376.

Dozier, J., and Warren S., 1982, Effect of viewing angle on the infrared brightness temperature of snow. *Water Resources Research*, **18**, 1424.

Folland, C. K., Karl T., and Vinnikov K. Ya., 1990, Observed climate variation and change. In *Climate Change. The IPCC Scientific Assessment* , edited by J. T. Houghton, G. J. Jenkins, and J. J. Ephraums (Cambridge: Cambridge University Press), pp. 195-238.

Folland, C. K., Karl, T. R., Nicholls, N., Nyenzi, B. S., Parker D. E., and Vinnikov K. Ya., 1992, Observed climate variability and change. In *Climate Change 1992. The Supplementary Report to the IPCC Scientific Assessment* , edited by J. T. Houghton, B. A. Callander and S. K. Varney). (Cambridge: Cambridge University Press), pp. 135-170.

Gates, W. L., Mitchell, J. F. B., Boer, G. J., Cubasch U., and Meleshko, V. P., 1992, Climate modelling, climate prediction and model validation. In *Climate Change 1992*, edited by J. T. Houghton, B. A. Callander and S. K. Varney, (Cambridge: Cambridge University Press), pp. 97-134.

Gloersen, P., and Campbell, W. J., 1988, Variations in the Arctic, Antarctic and global sea ice covers during 1979-1987 as observed with the Nimbus-7 Scanning Multichannel Microwave Radiometer. *Journal of Geophysical Research*, **93**, 10666-10674.

Gordon, A. L., and Goldberg, R. D., 1970, Circumpolar characteristics of Antarctic waters. *Antarctic Map Folio Series, Folio 13*, edited by V. Bushnell (New York: American Geographical Society).

Hall, D. K., 1988, Assessment of polar climate change using satellite technology. *Review of Geophysics*, **26**, 26-39.

Hall, D. K., Chang A. T. C., and Foster J. L., 1986, Detection of the depth-hoar layer in the snow-pack of the arctic coastal plain of Alaska, USA, using satellite data. *Journal of Glaciology*, **32**, 87-94.

Hambrey, M., and Alean J., 1992, *Glaciers* (Cambridge: Cambridge University Press.

Hobbs, P. V., 1974, *Ice Physics* (Oxford: Oxford University Press.

Houghton, J. T., Jenkins, G. J. and Ephraums, J. J., 1990. *Climate Change: the IPCC scientific assessment* (Cambridge: Cambridge University Press)

Houghton, J. T., Callander, B. A., and Varney, S. K., 1992, Climate Change 1992 – The Supplementary Report to the IPCC Scientific Assessment (Cambridge: Cambridge University Press)

Jin, G., and Wadhams P., 1989, Travel time changes in a tomography array caused by a sea ice cover. *Progress in Oceanography*, **22**, 249-275.

Jones, P. D., 1988, Hemispheric surface air temperature variations: recent trends and an update to 1987. *Journal of Climatology*, **1**, 654-660.

Kunzi, K. F., Patil S., and Rott, H., 1982, Snow-cover parameters retrieval from Nimbus-7 scanning multichannel microwave radiometer (SMR) data. *IEEE Transaction on Geoscience and Remote Sensing*, **GE-20** , 452-467.

Massom, R., 1991, *Satellite Remote Sensing of Polar Regions* (London: Belhaven).

Maykut, G., and Untersteiner, N., 1971, Some results from a time-dependent thermodynamic model of sea ice. *Journal of Geophysical Research*, **76**, 1550-1576.

Open University, 1989, *Ocean Circulation* (Oxford: Pergamon Press).

Parkinson, C. L., 1991, Interannual variability of the spatial distribution of sea ice in the north polar region. *Journal of Geophysical Research*, **96**, 4791-4802.

Paterson, W. S. B., 1981, *The Physics of Glaciers. 2nd Edition.* (Oxford: Pergamon Press.

Rott, H., 1984, Synthetic aperture radar capabilities for snow and glacier monitoring. *Advances in Space Research*, **4**, 241-246.

Rott, H., Domick, G., Matzler, C., Miller H., and Lenhart, K., 1985, Study on the use and characteristics of SAR for land snow and ice applications. Institut für Meteorologie und Geophysik, Universitat Innsbruck, Mitteilung nr. 1.

Svendsen, E., Kloster, K., Farrelly, B., Johannessen, O. M., Johannessen, J. A., Campbell, W. J., Gloersen, P., Cavalieri D. J., and Matzler C., 1983, Norwegian Remote Sensing Experiment: evaluation of the Nimbus-7 scanning multichannel microwave radiometer for sea ice research. *Journal of Geophysical Research*, **88** , 2781-2791.

Wadhams, P., 1988, The underside of Arctic sea ice imaged by sidescan sonar. *Nature*, **333**, 161-164.

Wadhams, P., and Comiso J. C., 1992, The ice thickness distribution and its determination using remote sensing techniques. In *Microwave Remote Sensing of Sea Ice* , edited by F. D. Carsey (Washington DC: American Geophysical Union) (in press)

Wadhams, P., and Holt, B., 1991, Waves in frazil and pancake ice and their detection on Seasat synthetic aperture radar imagery. *Journal of Geophysical Research*, **96**, 8835-8852.

Wadhams, P., Davis, N. R., Comiso, J. C., Kutz, R., Crawford, J., Jackson, G., Frabill, W., Sear, C. B., Swift, R., and Tucker, W. B., 1991, Concurrent remote sensing of Arctic sea ice from submarine and aircraft. *International Journal of Remote Sensing*, **12**, 1829-1840.

Wadhams, P., Tucker W. B., III, Krabill, W. B., Swift, R. N., Comiso J. C., and Davis N. R., 1992, The relationship between sea ice freeboard and draft in the Arctic Basin, and implications for ice thickness monitoring. *Journal of Geophysical Research-Ocean*, **97**, 20325-20334.

Weeks, W. F., and Ackley S. F., 1986, The growth, structure and properties of sea ice. In *The Geophysics of Sea Ice*, edited by N. Untersteiner (New York: Plenum), pp. 9-164.

The Study of Antarctica Using the ERS-1 Radar Altimeter

David R. Mantripp

1 Introduction

Over the last decade, the world's attention has become increasingly focussed on the issue of climate change, and in particular change deemed due to man's activities. The impact of human activity is felt throughout the world, but is superimposed on a background of complex natural variability. In the polar regions, however, and particularly in Antarctica, some anthropogenic impacts stand out more clearly. Analysis of historical levels of atmospheric carbon dioxide from ice coring, for example, shows a clear increase correlated with industrial development over the last two hundred years.

The Antarctic is almost completely covered by a sheet of freshwater ice, which in places is known to be over 4 km thick and covers an area of about 14 million km^2. This ice, formed under the pressure of successively deposited layers of snow, slowly flows towards the continental margins and discharges into the sea. The ice cover is classified into a number of physical types: grounded continental areas are called ice sheets; areas where the continental ice floats off the bedrock are known as ice shelves; fast flowing areas of either of these two are known as ice streams. It is largely through the ice streams that mass is transferred from the ice sheets to the ice shelves. In places velocities reach up to 4 km y^{-1}, The largest ice shelves, the Ross and Filchner-Ronne, cover areas of around 500,000 km^2 (Doake 1987). The dynamic behaviour of the ice sheet and ice shelf system is of fundamental importance to the understanding of the interactions between atmosphere and cryosphere. There are many variables to be considered in the modelling of such systems, but one vital input is precise surface topography. Other than surface elevation change itself, parameters such as driving stress (for grounded ice) (Cooper *et al.* 1982) or ice thickness (for floating ice) (Zwally *et al.* 1987) can be derived from topography. Obtaining topographical data by traditional surveying methods would be both prohibitively time consuming and expensive for all but very small areas. A continent-wide dataset can only realistically be obtained by a satellite remote sensing technique and glaciologists have been eagerly awaiting data from near-polar and polar-orbiting radar altimeters. As long ago as 1966, Robin (1966) discussed the potential of obtaining Antarctic topography from satellite altimetry. Data from earlier missions have allowed encouraging glimpses of what may be possible from ERS-

NATO ASI Series, Vol. I 24
Remote Sensing and Global Climate Change
Edited by Robin A. Vaughan and Arthur P. Cracknell

1 and its successors. Now, one year into the ERS-1 mission, we can begin to evaluate whether its potential is being fulfilled.

The radar altimeter (RA) aboard ERS-1 is the latest in a series of spaceborne altimeters, which began with Apollo 19, and evolved through Skylab, GEOS-3, Seasat and Geosat. The basic principle of operation, measuring the time delay between a transmitted pulse and the first return echo, has remained constant, but increasingly sophisticated implementations have improved the precision of the instruments, to the point where a theoretical figure of 2-4 cm rms is claimed for ERS-1 1s averaged returns over ocean. The extent to which this precision can be translated into accurate measurements of surface elevation is governed by the quality of the orbit ephemerides. Despite limited polar coverage, data from Seasat and Geosat have been extensively used for polar applications. Zwally *et al.* (1983), for example, have presented maps of the topography of Greenland and Antarctica to 72 °N and 72 °S respectively; Ridley *et al.* (1989) have mapped topography and features of the Larsen Ice Shelf using both Seasat data. ERS-1 provides coverage of Antarctica to 82 °S (Figure 1), allowing full coverage of the Filchner-Ronne Ice Shelf, most of the Ross Ice Shelf and also vast areas of the Antarctic Plateau which to date have been mapped only by sparse ground and airborne surveys Drewry (1983). However, an unsurveyed hole remains, a circular area of 2.5 million km^2, including some of the most important parts of the Ross and Filchner-Ronne drainage basins. Filling this last gap can only begin with missions planned by ESA and NASA for the end of the century and beyond Massom (1991).

2 The ERS-1 Radar Altimeter

ERS-1 carries a payload of 3 instruments, one of which is a nadir-pointing, pulse-limited radar altimeter operating at 13.8 GHz (Francis 1984a). The instrument is designed to operate in one of two tracking modes, "ice" or "ocean". The novel provision of an ice mode allows the altimeter to maintain lock over surfaces too steep or undulating for previous instruments. For a full discussion of the method of operation of satellite radar altimeters the reader is referred, for example, to Rapley (1990). Briefly the ERS-1 radar altimeter transmits chirped pulses at a repetition frequency of 1020 Hz. Echoes are processed on-board to a series of 64 samples, collectively known as the return waveform. These samples are variously referred to as range gates or range bins. This waveform is analysed on board and height estimates are fed to an automatic height tracking control loop. This estimates the range of the mean surface based either on the analytic model developed by Brown (1977) when operating in ocean mode, or on a centre of gravity estimate when operating in ice mode. A second control loop, the Automatic Gain Control (AGC), tracks signal strength, allowing an estimate of surface backscatter to be derived. In the ocean mode a third loop tracks the slope of the waveform leading edge. These control loops allow the altimeter to maintain the surface return within its range window, ideally in the centre. When the surface slope is too steep, or the surface roughness is too great, and the return is lost from the window, the altimeter is said to have lost lock. It then enters an acquisition mode, where it seeks to re-acquire the surface by scanning up and down. The 64 bins of the ERS-1 radar altimeter range window correspond to a width, or vertical range, of about 30 m in ocean mode and 120 m in ice mode. One waveform, an average of 50 pulses, is

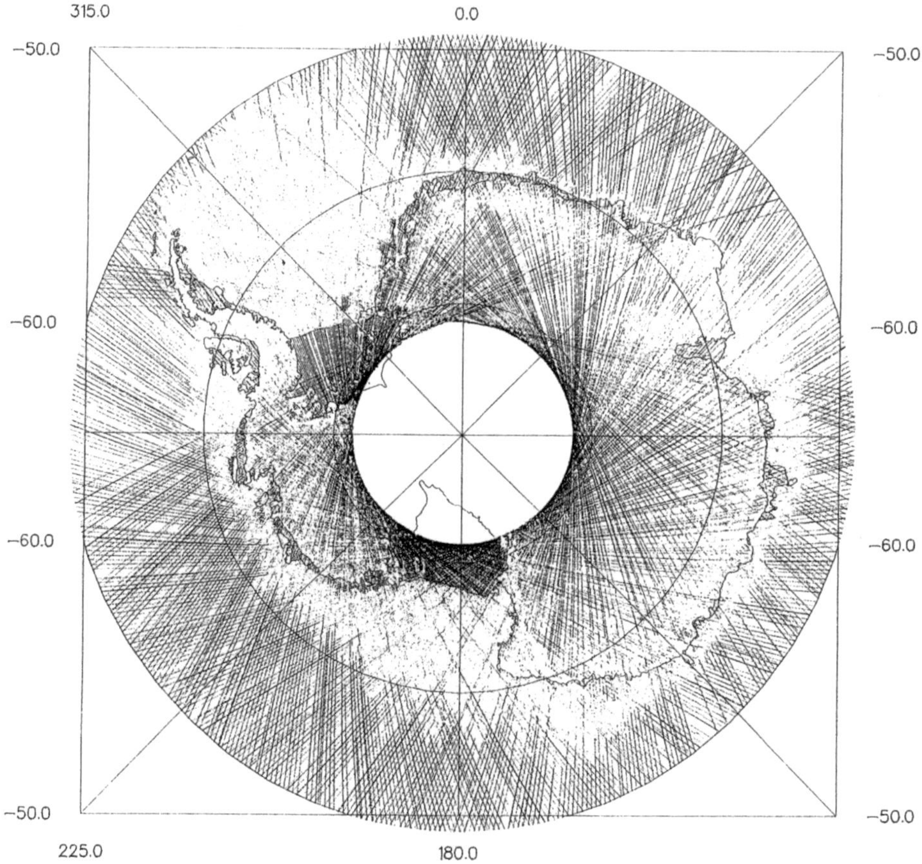

Figure 1: ERS-1 35 day repeat orbits over Antarctica, showing surfaces tracked by the radar altimeter in ocean mode. Figure courtesy of Dr J.L. Bamber.

output every 20 seconds, corresponding to an along track spacing of approximately 340 m, along with height, slope and gain estimates and supporting data. The resolution of the range measurement is dependent on the mode and the nature of the surface. In the ocean mode, for waveforms fitting the criteria of the Brown model, rms noise for a 1 second averaged height estimate is 3 cm, or below 12 cm in the ice mode (Francis *et al.* 1991). Over surfaces with a non-Gaussian height probability density function for small-scale roughness, or surfaces with significant topography, the resolution will vary due to the action of pulse-limited altimetry.

The resolutions quoted above are applicable only to ideal conditions. Radar altimeters are subject to various internal and external sources of error which, if not corrected for, can lead to greatly diminished accuracy. The sources of error are shown in Figure 2. An error budget for the ERS-1 radar altimeter is given in table 1. Error due to instrument effects is largely corrected for by an internal calibration system. This operates

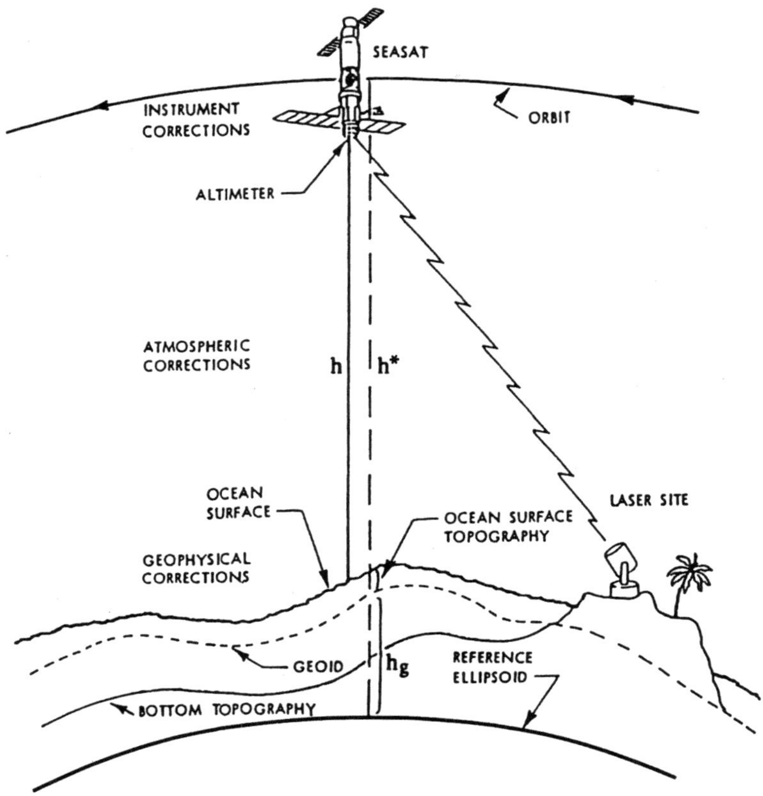

Figure 2: Errors in satellite altimeter range measurements. From Rapley (1990)

nominally every two minutes, and by coupling the transmitter directly to the receiver, allows drift affecting height and gain estimates to be measured. Ionospheric and tropospheric corrections, which compensate for atmospheric propagation delay, are based on models which tend to be less accurate in Antarctic regions due to sparse observations.

A full discussion of the interpretation of radar altimeter data is given in Rapley (1990), but here we consider two sources of error in the range measurement which must be considered specifically for polar regions. The first is slope-induced error. As the satellite passes over a surface, the altimeter ranges to the nearest sub-surface point. Over a flat surface, this point is at nadir, but over a sloping surface it is displaced in the upslope direction, resulting in errors of up to 100m in the range measurement (Brenner *et al.* 1983, Remy *et al.* 1989). The second is surface penetration. When an altimeter pulse reaches a snow surface, a proportion of the incident energy is scattered from the surface (surface scattering) and the remainder is transmitted through the interface. Part of this remainder is scattered by discrete scatterers in the sub-surface (volume scattering) but also any sub-surface layers, for example refreezing horizons or lenses. The proportion of volume to surface scattering is related to the fabric, density

Range corrections

Error source	Correction magnitude	Error in correction
Internal calibration	± 10 cm	± 2 cm
External calibration	~ 40 cm	± 5 cm
Doppler shift	± 33 cm	± 0.3 cm
Centre of Gravity offset	0 - 90 cm	± 0.1 cm
Ionospheric delay	0 - 100 cm	± 5 cm
Tropospheric delay - dry	1.7 - 2.5 m	± 1 cm
Tropospheric delay - wet	0 - 50 cm	± 5 cm
Liquid water content	0 - 3 cm	± 0.1 cm
Ocean Tide	± 50 cm	± 10 cm
Earth Tide	± 30 cm	± 1 cm
Ocean loading Tide	± 10 cm	< 1 cm
Geoid	± 100 m	± 1 m

For backscatter measurements

Internal calibration	± 0.5 dB	± 0.1 dB
Range - σ^0	0 - 0.33 dB	± 0.01 dB
Liquid water	0 - 0.12 dB	± 0.1 dB

Table 1: ERS-1 Radar Altimeter corrections

and temperature of the ice (Ridley and Partington 1988). It is in theory possible, in the absence of sub-surface structure, for pulses to penetrate down to around 20 m below the surface, but this is unlikely to occur in practice, as multiple sub-surface layering due to seasonal temperature and accumulation variations presents various additional reflecting surfaces. Finally one has to consider errors in the inversion of range to height. The range measurement, including all available corrections, is inverted to a surface elevation defined as the difference between the orbital elevation and the reference ellipsoid, which in the case of ERS-1 is GEM6. The best available calculated orbits are usually accurate to around 1-2 m, and are subject to limitations in our knowledge of high frequency variations in the Earth's gravity field. This is a particular problem for Antarctica, where errors of up to 10 m over ice shelves, and higher inland, are considered likely. These can be reduced with better knowledge of the spacecraft's orbit and of geoid to ellipsoid offsets. The satellite orbit can most precisely be determined by ground-based satellite laser rangers (SLRs), microwave rangers or transponders. A combination of these techniques can in theory provide the location of the satellite centre of gravity to less than 5cm over 2000 km arcs.

3 Antarctica from radar altimetry

Very soon after the launch of Seasat, work began on exploiting the altimeter data for Antarctic glaciology (Brooks *et al.* 1978). Various workers have exploited the datasets from Seasat and Geosat to produce results on a variety of topics, including mapping ice shelf boundaries and features (Zwally *et al.* 1987, Partington *et al.* 1987), mapping ice sheet topography (Remy *et al.* 1989), investigating specific features on ice sheets (Cudlip and McIntyre 1987) and analysing surface characteristics (Remy *et al.* 1990). An increasing body of work has also been devoted to the study of sea ice extent and type by radar altimeter, but we restrict discussion here to topics involving freshwater ice.

Radar altimeter returns from non-ocean surfaces require alternative analysis methods (Guzkowska *et al.* 1990). It is also important to consider exactly what information the altimeter is giving us. Figure 3a shows the build up of a typical altimeter return from an ocean surface, while Figure 3b describes the process for a land surface. It is important to realise that the altimeter is illuminating a wide area of the surface, the extent of which is governed by the antenna beamwidth and the surface topography. Over a smooth surface, the ERS-1 altimeter records return echoes from a circular footprint of up to 7 km in diameter. The diameter increases up to around 20 km over rougher surfaces. The height estimate calculated on board is that of a mean surface over the footprint when the altimeter is operating in ocean mode, using a simplified mean likliehood estimation (SMLE) algorithm. Over ocean surfaces, the Gaussian distribution of reflecting slopes allows an analytic description of a characteristic ocean-like waveform, but this assumption frequently breaks down over ice surfaces, although many returns from ice shelves are ocean-like (Figure 4). In the ERS-1 ice mode, the objective of the height tracking control loop is not so much to provide an accurate estimate of mean surface elevation, but more to ensure that as much of the surface as possible is tracked, so that waveforms can be recorded for post-processing. The nominal method for ice tracking is to position the centre of gravity of the return waveform at the middle of the range window. However, provision has been made for alternatively tracking at either the 1/4 or 3/4 position. The switching from mode to mode is commanded from the ground, and a number of strategies are planned, where the altimeter operates in either ice or ocean mode for full orbits, or is switched at transition points defined by participating scientists in collaboration with ESA. In order to derive a more accurate height estimate, various algorithms collectively known as retracking have been derived, which range from simple first return detection to complex fitting functions. The aim of retracking is to analyse the telemetered waveform in such a way to estimate more reliably the range of the nadir point. Retracking a waveform provides an offset which can be added to the window-centre position telemetered as the on-board range estimate.

ERS-1 was placed in orbit on 17 July 1992 and since then the radar altimeter has been operating almost continuously. The failure of the PRARE instrument, which was to be used to assist in the provision of precise orbits for ERS-1, has been a drawback, but alternative methods using satellite laser ranging and transponders (Francis 1987), and data relating to the nearby SPOT-2 should result in 1 m precision orbits. Users are supplied with data through the four ground segment Processing and Archive Facilities (PAFs). Two classes of data are supplied, from on-line (fast delivery) processing or off-line production. The fast delivery (FD) data do not contain any waveform data, have only first order corrections applied and contain only ocean mode data. They are distributed through a number of routes in a number of versions, but the most suitable source is the French PAF at Brest. The major off-line products are the 1Hz Ocean

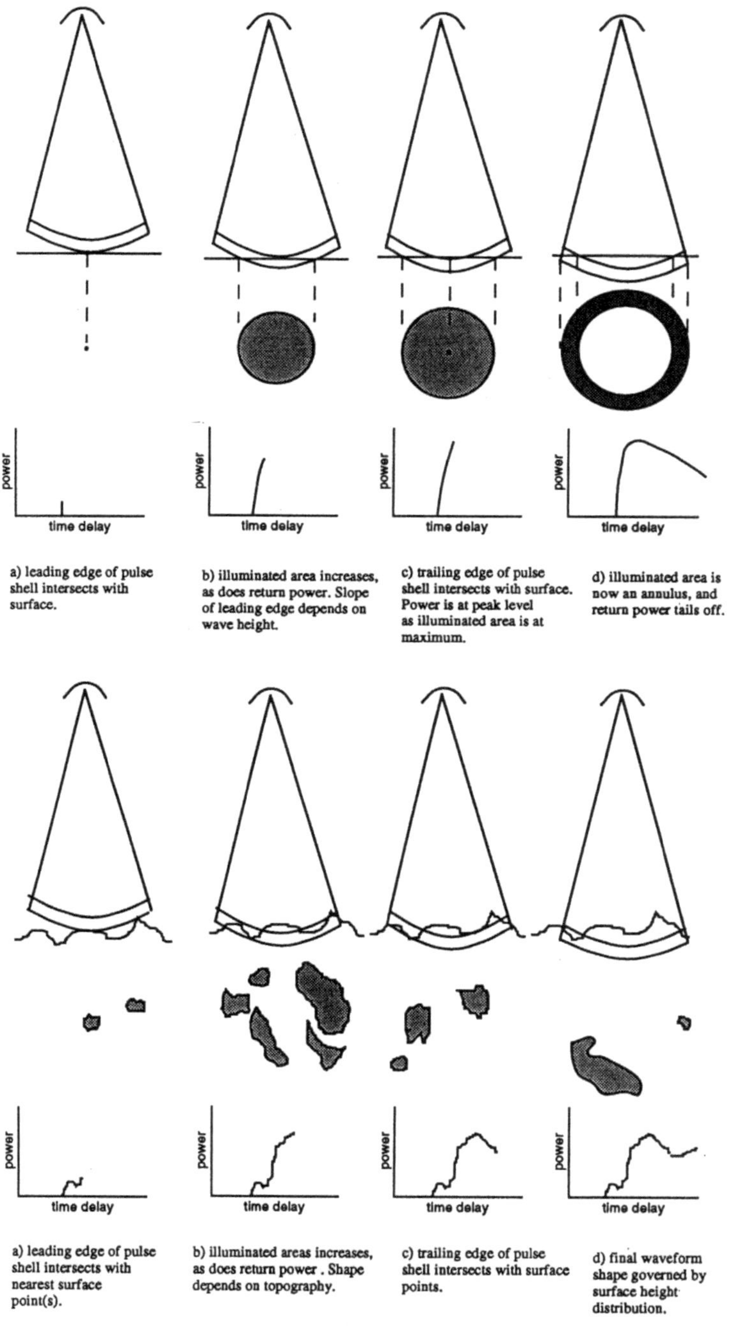

Figure 3: The generation of a satellite radar altimeter waveform over (above) ocean and ocean-like surfaces and (below) rough surfaces.

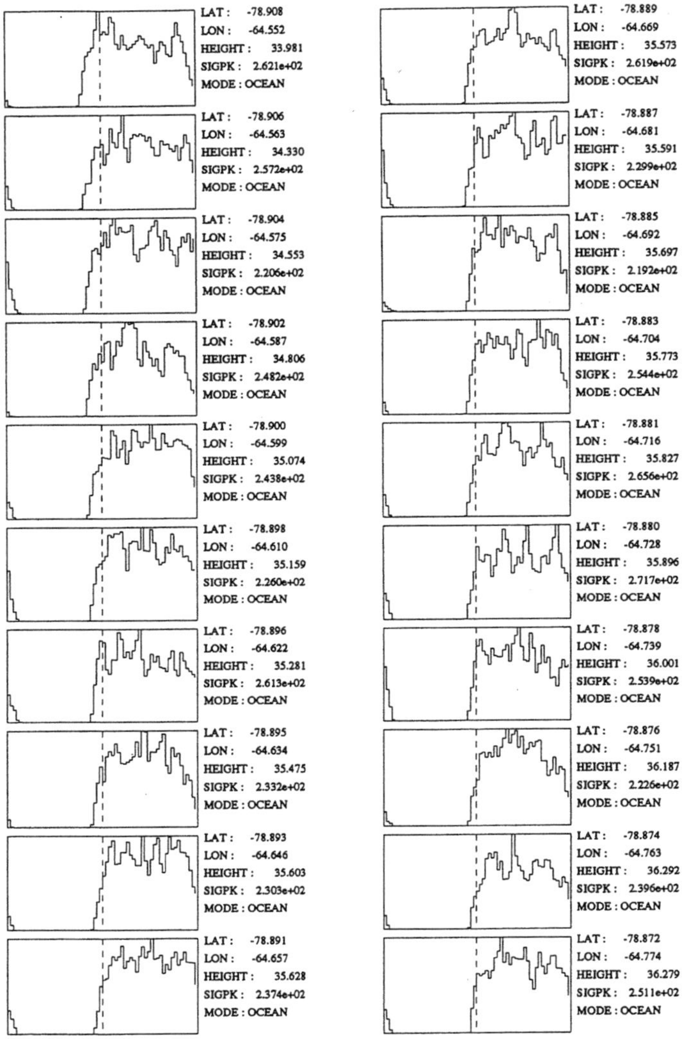

Figure 4: A sequence of ocean-like ERS-1 radar altimeter ocean-mode waveforms over Filchner Ice Shelf, Antarctica.

Product produced by the French PAF and the 20Hz Waveform Foundation Product, Waveform (Advanced) Product and Land Ice Elevation Product produced by the UK PAF at Farnborough. Advanced orbits are produced by the German PAF (D-PAF). At this stage of the mission, large amounts of data are only available in FD format, and these are the basis of the majority of the early results outlined later.

4 Glaciological Objectives

Results from radar altimetry related to glaciology can be split into two categories: applications to the field of ice dynamics, usually model inputs, and surface state analysis, usually direct observables. The study of ice sheet and ice shelf dynamics attempts to describe the evolution of these systems, both past and future, and to relate evolution to global climate. Ice shelves are more sensitive to climate warming than ice sheets, as they are generally much thinner, have mean surface elevations of only around 100 m above sea level, and are subject to interactions with ocean as well as atmospheric circulation. Furthermore, it is often suggested that the Ross and Filchner-Ronne Ice Shelves play a vital role in retaining the West Antarctic Ice Sheet, which is for the most part grounded well below sea level. The smaller ice shelves surrounding the Antarctic continent, while not playing any major dynamic role, may act as early warning signals to climatic warming. The break up of Wordie Ice Shelf over the last two decades has been described by Doake and Vaughan (1991) and the much larger Larsen Ice Shelf appears to be in decline.

The major input from radar altimetry to ice shelf dynamic modelling is the provision of dense accurate height measurements, allowing high quality topographic maps to be produced. Since ice shelves are floating, it is a simple matter to invert surface elevation to ice thickness, provided that carefully considered assumptions about ice density profiles are used. Ice thickness is important, as it is related to backpressure and longitudinal stress. Additionally, ice thickness defines grounding line position, where the ice is floated off the bedrock by hydrostatic pressure, and the extent and importance of grounded areas within the body of the ice shelf, such as ice rises or ice rumples. The retreat of the Wordie Ice Shelf is correlated with the decoupling of the ice shelf from the bed in a number of areas, thus reducing the backstress. Various results, e.g. those of MacAyeal (1987), indicate that ice streams within West Antarctica draining into the major ice shelves are sensitive to changes in ice shelf backpressure. Lingle *et al.* (1991b) present a complex time-dependent model present for Ice Stream B, a fast flowing area of the Ross Ice Shelf. This model is used to simulate response to three climatic scenarios over a 600-year time scale. The maximum change in ice thickness resulting from these simulations is of the order of -25m for the ice shelf or -50m for the ice stream above the grounding line, both over 50 years. Assuming a steady evolution, this implies that in places an annual change of 1m could exist, which would be detectable over the 3-5 year mission life of ERS-1, especially as it is expected that ERS-2, carrying a near-identical payload, will be in orbit by the time of the demise of its predecessor. Despite this, it is optimistic to expect that significant, validated change in ice thickness can be detected by radar altimetry. Lingle *et al.* (1991a) investigated changes in part of the Greenland ice sheet using Seasat and Geosat data, but could not report any statistically significant change.

The profile of the Antarctic ice sheet is only approximately known. The best available digital terrain model is quoted to be accurate to 30m (A.P.R. Cooper, personal communication), but for many parts of the continent this is derived from very sparse ground survey data. The potential for mapping the continent by altimetry has been explored by a number of workers. A particularly promising dataset comes from the Geosat Geodetic Mission, where the extremely dense coverage allows great surface de-

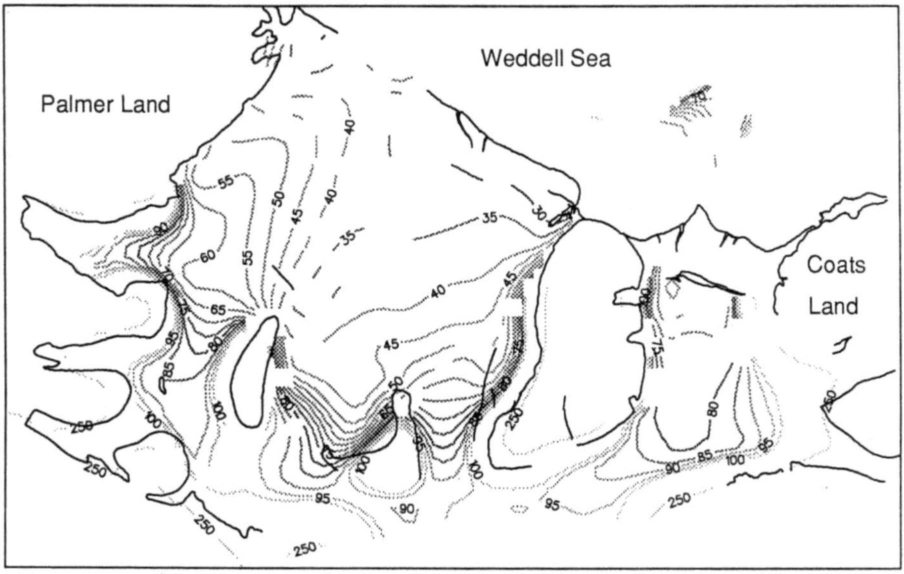

Figure 5: Topograplic map generated from ERS-1 data (contours in m).

tail to be resolved (Mantripp *et al.* 1992). Undulations in surface elevation can be related to sub-surface topography (McIntyre *et al.* 1986) and can be used to map inland ice stream boundaries. It has been suggested that volume change can be derived directly by integrating returns over large areas (Wingham in review).

All the above results are based on the exploitation of the range measurement. The intensity of the return signal, as measured by the Automatic Gain Control, provides information on the nature and structure of the scattering surface. Remy *et al.* (1990) used this information to attempt to define areas subject to important katabatic winds. Ridley and Partington (1988) discuss analysing return power together with waveform shape to derive information on the proportion of volume to surface scattering, and hence the sub-surface structure.

5 Results from the ERS-1 altimeter

The importance of ERS-1 is immediately apparent from the early topographic map produced from uncorrected FD data presented by Ridley *et al.* (1992), see Plate 1. This clearly describes the topography of the inland ice sheet and shows many features on the ice shelves. In particular, the surface topography of the Filchner-Ronne ice shelf is clearly outlined.

A part of this toporaphy is shown at a larger scale in Figure 5. The FD data must be considered to incorporate potential errors of the order 5m over the floating ice shelves and we assume that the ocean tidal signal is averaged out by the data interpolation process used before contouring. It is therefore justified to present these data contoured to 5m intervals and this map in fact presents the densest topographic dataset acquired so far for this region. Looking at relative elevations, these contours correspond extremely well to ice thickness contours presented by Robin *et al.* (1983) and appear to confirm the existence of a thinner central area downstream of Doake Ice Rumples. Although this result requires confirmation through analysis of internal and external error sources, it does demonstrate the huge advantage of satellite data over labour-intensive and time-intensive ground and air based survey methods.

An aspect of the radar altimeter which is of particular interest in the early stages of the mission is its tracking performance in ocean and ice modes. The two profiles in Figure 6 show the proportion of the surface tracked by the altimeter in either mode as it passes over the continental ice shelf. It can be seen that while there is a relatively large number of losses of lock in ocean mode, in ice mode tracking is near-continuous and lock is only lost over very rough terrain in mountainous regions. A number of profiles of the Filchner Ice Shelf and adjacent grounded regions have been examined. In Figure 6a, two profiles are shown for a track passing over the Coats Land plateau and on to the Filchner Ice Shelf. It is immediately obvious that in ice mode the sloping plateau topography is far better tracked than in ocean mode. However, where both modes track there is good agreement between the two. The important transition zone from grounded to floating ice is not tracked in ocean mode, where the tracker is observed to overshoot due to the relatively long time constant of its control loop. The more agile ice mode tracker handles the transition well and is thus well suited to provide important information on the location of the grounding line. The section over the ice shelf itself is expanded in Figure 6b. Here it can be seen that the ice mode profile exhibits considerably more noise than the ocean mode. Retracking improves matters to some extent, but ocean mode remains preferable. Towards the left of the plot, the ice mode profiles shows a large excursion. This is due to a bright target, probably the southern end of Berkner Island, entering the wider ice mode footprint, causing the altimeter to go off-range. Clearly, in circumstances where ocean mode retains lock, it is usually better to use this rather than ice mode. However, the ice mode allows the tracking of surfaces which were outside the capabilities of previous missions.

6 Conclusions

The ERS-1 Radar Altimeter is proving to be the best instrument to date for mapping the Antarctic topography from space and its output looks like becoming a major contribution to glaciology. The provision of the ice mode allows areas hitherto inaccessible to radar altimetry to be mapped and the near-polar orbit allows a far greater proportion of Antarctica to be covered than has previously been possible.

I would like to acknowledge the assistance of colleagues in the Remote Sensing Group at MSSL in the preparation of this paper, in particular Dr J.L.Bamber, for providing pre-publication material.

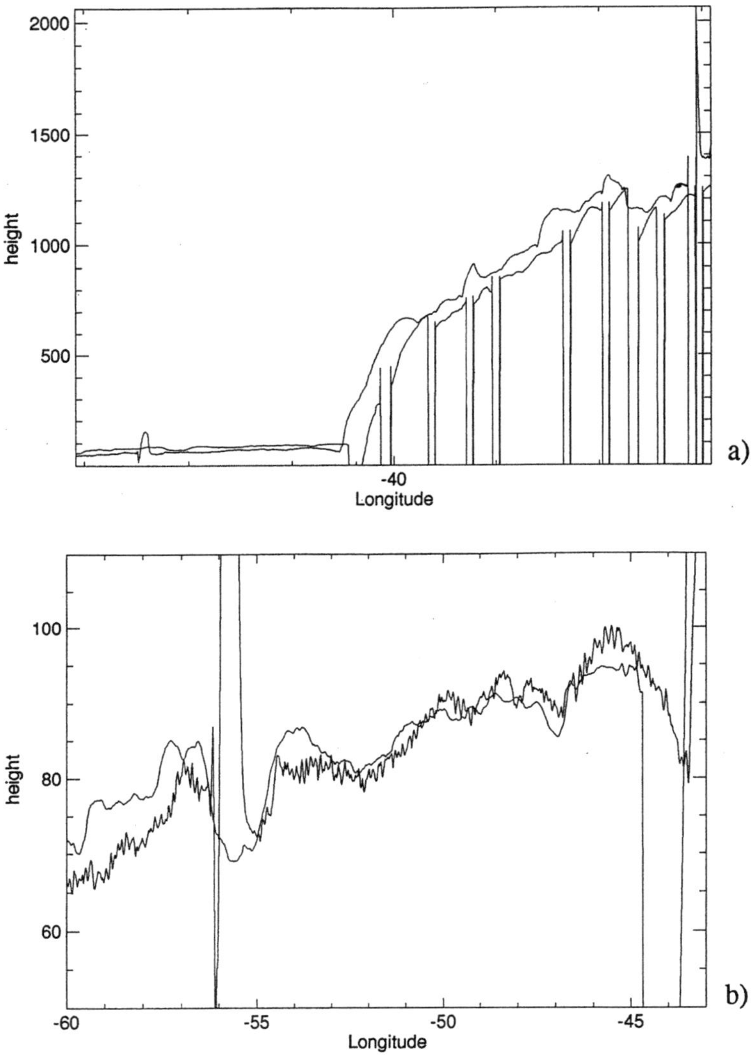

Figure 6: Surface elevation profile from two ERS-1 radar altimeter repeat tracks in alternate modes, descending from the Coats Land plateau to the Filchner Ice Shelf. Continuous profile in (a) is ice mode, offset from ocean mode for clarity. The ice shelf section is shown enlarged in (b).

References

Brenner, A. C., Bindschadler, R. A., Thomas, R. H., and Zwally, H. J., 1983, Slope-induced errors in radar altimetry over continental ice sheets. *Journal of Geophysical Research*, **88** 1617-1623.

Brooks, R. L., Campbell, W. J., Ramseier, R. D., Stanley, H. R., and Zwally, H. J., 1978, Ice Sheet Topography by satellite Altimetry. *Nature* **273**, 539-543.

Brown, G. S., 1977, The Average Impulse Response of a Rough Surface and its Applications. *IEEE Transcations on Antennas and Propagation* **AP-26**, 472-482.

Cooper, A. P. R., McIntyre, N. F., and Robin, G. de Q., 1982, Driving stresses in the Antarctic Ice Sheet. *Annals of Glaciology* **3**, 59-64.

Cudlip, W., and McIntyre N. F., 1987, Seasat altimeter observations of an Antarctic 'lake'. *Annals of Glaciology* **9**, 55-59.

Doake, C. S. M., 1987, In *Antarctic Science*. (Cambridge: Cambridge University Press), pp. 139-190.

Doake, C. S. M., and Vaughan D. G., 1991, Breakup of Wordie Ice Shelf, Antarctica. *Nature* **271**, 328-330.

Drewry, D. J., 1983, *Antarctica glaciology and geophysical folio* (Cambridge: Scott Polar Research Institute.

Elachi, C., 1987, *Introduction to the Physics and Techniques of Remote Sensing* (New York: John Wiley & Sons).

Francis, C. R., 1984a, The ERS-1 altimeter: an overview. In *ERS-1 radar altimeter data products.* (Noordwijk: Scientific and Technical Publication Branch, ESA), pp. 9-16.

Francis, C. R., 1987, ERS-1 Altimeter Calibration Plan. ERS-1 Project Report, European Space Agency, Noordwijk, Holland.

Francis, R., Graf, G., Edwards, P. G., McCraig, M., McCarthy, C., Dubock, P., Lefebvre, A., Pieper, B., Pouvreau, P.Y., Wall, R., Wechsler, F., Louet, and Zobl, R., 1991, The ERS-1 Spacecraft and Its Payload. *ESA Bulletin*, **65**, 26-48.

Guzkowska, M. A. J., Rapley, C. G., *et al.*, 1990, Developments in Inland Water and Land Altimetry.

Lingle, C. S., Brenner, A. C., Zwally, H. J, DiMarzio, J. P., 1991a, Multiyear Elevation Changes near the West Margin of the Greenland Ice Sheet from satellite radar altimetry. *Symposium Volume, International Conference on the Role of Polar Regions in Global Change.*

Lingle, C. S., Schilling, D. H., Fastook, J. L., Paterson, W. S. B., and Brown, T., 1991b, A Flow Band Model of the Ross Ice Shelf, Antarctica: Response to CO2-Induced Climatic warming. *Journal of Geophysical Research* **96**, 6849-6871.

MacAyeal, D. R., 1987, Ice Shelf Backpressure: Form drag vs. dynamic drag. In *The Dynamics of the West Antarctic Ice Sheet, proceedings of a Workshop held at Utrecht, on 6-8 May 1985*, edited by C. G. van der Veen (Hingham: D. Reidel), 141-160.

Mantripp, D. R., Ridley, J. K., and Rapley, C. G., 1992, Antarctic map from the Geosat Radar Altimeter Geodetic Mission. *Earth Observation Quaterly*, **37-38**, 6-10.

Massom, R. A., 1991, *Satellite Remote Sensing of Polar Regions.* (London: Belhaven Press).

McIntyre, N. F., 1986, Antarctic ice sheet topography and surface-bedrock relationships. *Annals of Glaciology*, **8**, 124-128.

Partington, K. C., Cudlip, W., McIntyre, N. F., and King-Hele, S., 1987, Mapping of Amery Ice Shelf, Antartica, surface features by satellite altimety. *Annals of Glaciology*, **9**, 183-188.

Rapley, C. G., 1990, Satellite Radar Altimeters. In *Microwave Remote Sensing for Oceanographic and Marine Weather-Forecast Models* (Dordrecht: Kluwer Academic Publishers), pp. 45-63.

Remy, F., Mazzega, P., Houry, G., Brossier, C., and Minster J. F., 1989, Mapping of the topography of continental ice by inversion of satellite altimeter data. *Journal of Glaciology* **35**, 98-107.

Remy, F., Brossier, C., and Minster, J. F., 1990, Intensity of satellite radar-altimeter return power over continental ice: a potential measurement of katabtic wind intensity. *Journal of Glaciology*, **36**, 133-142.

Ridley, J. K., Cudlip, W., McIntyre, N. M., and Rapley, C. G., 1989, The topography and surface characteristics of the Larsen Ice Shelf, Antarctica, using satellite altimetry. *Journal of Glaciology*, **35**, 299-310.

Ridley, J. K., Laxon, S., Rapley, C. G., and Mantripp, D. R., 1992, Topography of Antarctic Ice Sheet mapped with the ERS-1 Radar Altimeter. *Earth Observation Quaterly*, **37-38**, 14.

Ridley, J. K., and Partington K. C., 1988, A Model of Satellite Altimeter Return from Ice Sheets. *International Journal of Remote Sensing*, **9**, 601-624.

Robin, G. de Q., 1966, Mapping the Antarctic Ice Sheet by Satellite Altimetry. *Canadian Journal of Earth Sciences* **6**, 919-928.

Robin, G. de Q., Doake, C.S.M., Kohnen, H., Crabtree, R.D., Jordan, S. R., and Moller, D., 1983, Regime of the Filchner-Ronne ice shelves, Antarctica. *Nature* **302**, 582-586.

Wingham, D. J., (in review), On the relationship between a satellite altimeter echo and the volumes of large ice-sheets. *Journal of Glaciology* :

Zwally, H. J., Bindschaler, R. A., Brenner, A. C., Martin, T. V., and Thomas, R. H., 1983, Surface Elevation Contours of Greenland and Antarctic Ice Sheets. *Journal of Geophysical Research*, **88**, 1589-1596.

Zwally, H. J., and Stephenson, S., 1987, Antarctic Ice Shelf Boundaries and Elevations from Satelitte Radar Altimetry. *Annals of Glaciology*, **9**, 229-235.

The NASA Space Shuttle Earth Observations Database for Global Change Science

Kamlesh Lulla, Michael Helfert and Doug Holland

1 Abstract

The NASA Space Shuttle Earth Observations Database is a valuable source of data for research and reconstruction of Earth's recent environmental history and thus for assessment of the human impact on global Earth processes. This data source, although having the longest length-of-record of any space-derived global change database, has not been fully exploited by scientists studying global changes. With the inception of NASA's Mission to Planet Earth programme there is a need to integrate these important data into global change studies.

Global land processes such as land use, deforestation, soil and land degradation, erosion, desertification, soil salinisation, long-range dust transport, hydrological changes, and stream and reservoir sedimentation have been extensively documented during the more than 50 Space Shuttle missions flown during the past twelve years, as well as by similar baseline data acquired during the even earlier NASA manned spaceflights extending to 1961. These observations provide a unique perpective on our planetary habitat and add understanding and critically important early data points to our models of land use dynamics and their ecological implications.

This paper discusses the Space Shuttle Earth Observations database, the imaging systems utilised during Space Shuttle missions and provides a case study of an application project conducted utilising these image datasets.

2 Imaging the Planet Earth from the Space Shuttle

2.1 Overarching programme philosophy

The documentation of dynamic phenomena of Earth processes observed during manned spaceflights remains the cornerstone of the Space Shuttle Earth Observations pro-

NATO ASI Series, Vol. I 24
Remote Sensing and Global Climate Change
Edited by Robin A. Vaughan and Arthur P. Cracknell

gramme - as it was with the earlier Mercury, Gemini, Apollo and Skylab Earth observations programmes. During the Space Shuttle era, about 600 selected sites have been identified that are of interest to academic and NASA geologists, geomorphologists, ecologists, geographers, oceanographers and meteorologists. Data from these sites are acquired during missions and catalogued into an publically accessible electronic database. At this time, the database totals about 140,000 acquisitions. Figure 1 shows the distribution of Shuttle photography acquired during each of the Space Shuttle flights (1981-1992). These images are all in the public domain and the catalogues are accessible via electronic network throughout the world.

2.2 The analogue camera systems

The primary cameras flown during Space Shuttle missions are the Hasselblad and Linhof systems. NASA-modified Hasselblad 500 EL/M 70-mm cameras are extensively used with large film magazines, holding 100 to 130 exposures. Standard lenses include a Zeiss f = 50 mm CF Distagon 1:4.0, a Zeiss f = 100 mm CF Planar 1:3.5, and a Zeiss f = 250 mm CF Sonnar 1:5.6. The Aero-Technika Linhof system can be fitted with f = 90 mm 1:5.6 lenses and f = 250 mm 1:5.6 lenses. This system produces large format (100 mm x 127 mm) film.

At present, Space Shuttle window restrictions in pass-through wavelengths result in the use of two primary film bases in the Hasselblad and Aero-Technika Linhof camera systems. These camera systems are fitted with the Data Recording Module (DRM) since 1984. These two film bases are colour visible (Kodak 5017/6017 Professional Ektachrome, ASA64) and colour infrared (Kodak Aerochrome 2443) films. Additional films have been flown and will be flown, but the base films remain the two above. Despite the window restrictions, the data acquired by astronauts during their missions is of scientific value, although that value would be immeasurably increased by mounting the cameras in the Payload Bay on a pointable platform operable both by the crew and by the ground science team.

It is important to recognize the difference between the Shuttle photography and conventional aerial photography. The planners cannot direct the spacecraft where and when to fly but must use given conditions such as launch time, observational site meteorological conditions, individual Space Shuttle mission orbital characteristics and scene illumination or Sun angle. These variables are defined by the primary payload purposes of the specific mission.

Four windows in the aft part of the Space Shuttle flight deck are used for most astronaut handheld photography. Some experiments using these windows, such as comparison of colour and colour infrared films, have involved the use of a dual Hasselblad camera mount which enables simultaneous photography of observational sites.

During the 32 years length-of-record represented by this database, approximately 140,000 images have been acquired of Earth phenomena. These images are deliberate, that is, they are filtered in their meaning and utility by the intelligent sensing required in their initial acquisition. Astronaut photography of the Earth is the result of systematic Earth and environmental science training provided to each crew prior to flight. Each flight crew is tasked with specific Earth observations objectives and with acquiring data over specific sites during their mission.

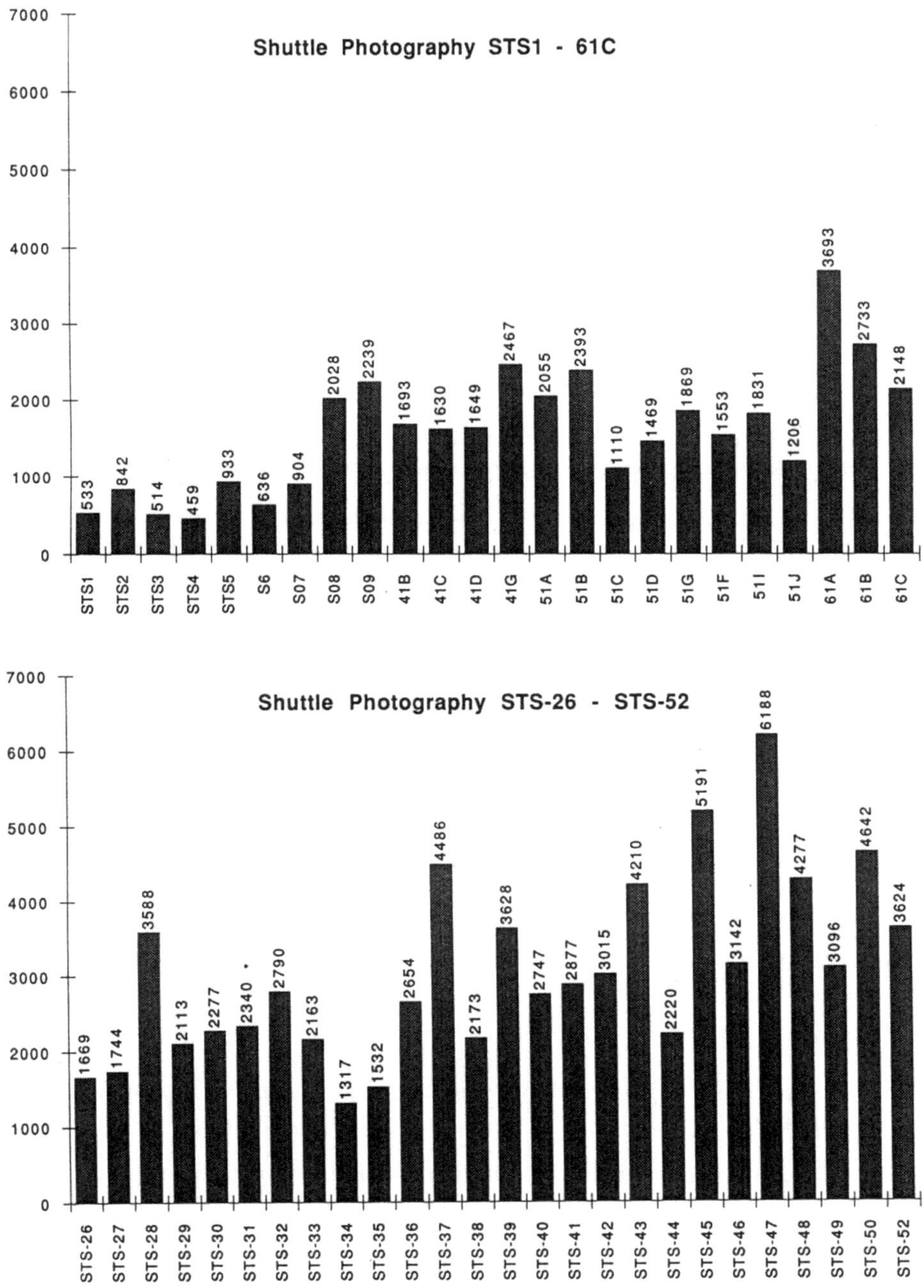

Figure 1: Space Shuttle Earth Observations Imagery acquired during the last fifty flights (1981-1992). STS refers to the Space Transportation System followed by the mission identification number. The numbers over the bars show total images acquired during the mission.

Data acquired during each mission are catalogued, archived, accessible for searching using an electronic database and available as public domain products through selected product distributors. All data are available as analogue photography and within the next year all data - including the retrospective data - will also become available in a standardised digital form.

A capsule summary of the characteristics of the camera systems and spatial resolutions is presented by Israel (1992). Among the many advantages of the astronaut photography during Space Shuttle missions, the following are worth mentioning here:

- images are obtained at a variety of Sun angles - ranging from 1° to about 80°, with the majority of pictures having Sun angles of about 30°. Very low Sun angle and some negative Sun angle photography has provided unique topographic views of remote mountainous areas otherwise poorly mapped.
- sequential photographs with different look angles provide stereo-views. Stereo strip photography has been acquired of a number of different locales.
- 75% of the photographs in the archive cover the regions between 28° N and 28° S providing coverage for many otherwise little-known tropical areas undergoing drastic environmental transformations.
- 25% of the views cover regions between 30-60 °N and S.

2.3 Digital cameras and digital imaging

In order to evolve towards state-of-the-art, multiband digital imaging by Space Shuttle and Space Station Freedom astronaut crews, NASA/Johnson Space Center has built a new digital electronic imaging camera. This camera can downlink near real-time imagery of in-cabin activities and Earth Observations. Development of this system will continue.

Thus far, this experimental Electronic Still Camera (ESC) has flown on four Space Shuttle flights. Initial results are very promising for further Earth Observation applications.

The ESC system is composed of three components:

- a hand-held battery-operated fully digital and programmable high spatial resolution camera
- a laptop-computer-based playback/downlink unit for onboard image processing
- a ground station capable of receiving Shuttle data, processing images, producing hardcopies and distributing the data and hardcopies to end users.

As flown during STS-48, the ESC was based upon a 1024x1024 charged coupled device (CCD) pixel black-and-white array. The camera is also designed to accept a 2048 x 2048 CCD pixel black-and- white array and to be upgradable to a 1024 x 1024 pixel array colour system. The camera was designed using astronomical-grade CCDs instead of video-grade CCDs. This approach has resulted in near-film-quality imaging and has far exceeded the quality of imaging that could previously be downlinked during earlier Shuttle flights, namely the Orbiter analogue video. The downlink capability and digital data are the keys to the scientific utility of the ESC. The system uses the Orbiter's K_u-band digital downlink system. At present, the system has the rather slow capability to downlink an image in 15 seconds.

2.4 Film and electronic still camera

The ESC system is an initial bridge of the gap between analogue film and digital video, bringing the advantages of both into one system. Film provides very high spatial resolution capabilities (up to 30 Mbytes of information is in each scene of fully digitised analogue film). Film is also versatile, allowing long exposures and supporting a useful variety of adaptations to scientific instruments. The ESC has demonstrated these initial desired capabilities.

Additionally, where video can be downlinked, so can be ESC imagery. In an improvement over video, ESC images can be downlinked through any transmission channel regardless of bandwidth or bit rate, the tradeoff being speed of image receipt. ESC imagery is inherently digital, producing near-error-free transmission, which is not possible with analogue signal transmission.

In an improvement over film, the ESC sensor has a high quantum efficiency (QE) which is the ratio of the number of photons converted to image signal divided by the number of incoming photons. Film generally has QEs of about 1%, whereas a CCD can have large a QE, ranging upward to near 100%.

The ESC CCD also has a spectral response exceeding that of films currently flown on the Space Shuttle. The current ESC's spectral range is 400-1100 nm. Available sensors which could later be used in the ESC system are sensitive from deep into the near infrared to soft X-ray wavelengths.

The ESC also exceeds both film and video in dynamic range. The ESC sensor is capable of an 80dB dynamic range; only 60 dB of this is employed in the current NASA ESC experimental system. The ESC is also impressive in that it does not require stowage space for film. After the images are downlinked, the removable data storage devices can be reused for later acquisitions.

A comparative study of the ESC and the Hasselblad cameras to ascertain the limits of resolution is summarised in Figure 2.

The most important aspect is downlinkability of the ESC system. Most investigators, scientists and the like, are not willing to wait until the film returns from a long-duration mission to view the results. This has particularly important implications to extending the scientific utility of both the Long Duration Space Shuttle missions and forthcoming Earth observations acquisitions during the Space Station Freedom era. By enabling near-real-time downlink of Earth observations data, the space-based crew can interact with the disciplinary expertise of ground-based science teams. This interaction optimises the yield of scientific experiments, field monitoring of ephemeral environmental situations and observations and allows for the very rapid dissemination of data throughout the scientific and educational communities.

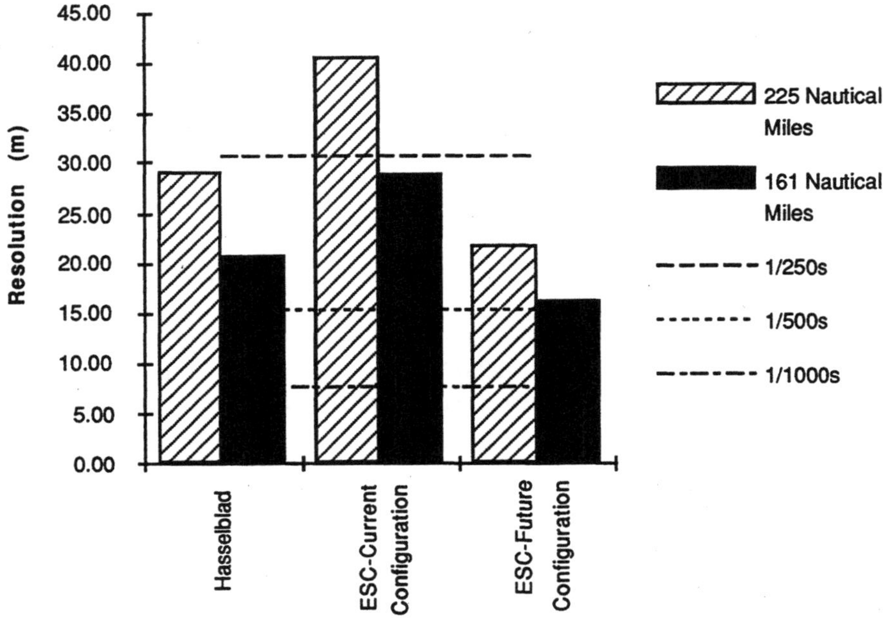

Figure 2: Limits of Resolution of Hasselblad and ESC systems in current and future configuration due to lens speed and motion blurr.

2.5 Electronic still camera and satellite data

Many Earth-observing satellites produce digital images by combining multiple linear swaths. These swaths are a certain number of kilometres wide, with the positions of different swaths determined by the position of the satellite for each individual orbital revolution.

It takes a substantial amount of sensor look-time to produce an image covering a particular area of interest. In contrast, the ESC system employs an array that can capture an area of interest in a single acquisition. There are existing satellites that have area arrays also, but the reaction time for such systems is too long to compete with what a properly equipped intelligent observer can do. For example, an astronaut observing a particular area of interest can make quick real-time decisions about the best way to document a scene. He or she can change lenses, change filters for multispectral imaging, or choose a new area of interest by quick human observations and decision-making.

The ground crew can talk to the astronaut in a higher level language than can be used to direct a satellite. This results in a higher level of intelligent and scientifically desirable results from human insight than is possible with the limited action response of other satellite systems. For instance, the lens systems for the ESC are changeable as fast as one can change the lens on a 35mm hand-held film camera. The filters screw on and off similarly. The astronaut can also screen the images before downlinking to decide which images are worthy of being downlinked and which are not. This saves precious downlinking resources and keeps the ground science team from having to go through large numbers of images to find ones of significance to them.

2.6 Case study: mapping biomass burning in the Amazon basin

Global biomass burning provides a significant input into the overall atmospheric budget of particulates and trace gases. Recent scientific findings indicate that global deforestation and biomass burning contribute perhaps 15% of the current anthropogenic emissions of greenhouse gases. There is increasing documentation from space platforms of a much larger scale of tropical biomass burning than had been previously supposed (see, for example, Helfert and Lulla 1989).

Despite our expanded view, there is not, as yet, universal agreement concerning the distribution and frequency of biomass burning in traditional savanna management, or concerning the rate and geographical distribution of the conversion of tropical moist forests to agricultural and grazing lands. These ponderables are particularly true for large areas of south and southeast Asia, Indonesia, Africa and Latin America. This paucity of data points in time series and thorough geographical coverage, especially over the global tropics (30°N-30°S), inhibits confidence in modeling atmospheric, hydrologic, and ecological changes that are connected with large-scale repetitive biomass burning. The current lack of basic geographical, temporal, chemical, and physical data on the nature and extent of biomass burning areas and the associated smoke palls will hopefully be addressed during spaceflights during the coming decade.

In order for global change models to be of real predictive value, confirmatory data on the areal extent of change processes is essential. Our objective in this study is to delineate and map the continental-scale biomass burning and smoke palls observed over the Amazon Basin using the Earth-viewing photographs taken from the Space Shuttle. The determination of the spatial extent of biomass burning and associated smoke palls is a first step towards confident quantitative determinations of their environmental impact.

2.7 An application of the database

This study focused on the geographical extent and seasonality of large smoke palls associated with tropical moist forest clearing in the Amazon Basin of South America. Data used in this study are drawn from the NASA Space Shuttle Earth Observations Project photography data base, which contains over 80,000 views taken from 1981 to 1991 and about 60,000 views acquired during the earlier spaceflights dating to 1961.

All smoke pall photographs were acquired during the Amazonian dry seasons (July-October) of 1983-1991. The crew of STS-48 observed a significant amount of biomass burning in September of 1991 and documented these smoke pall areas as had earlier crews during their flights. These new photographs have been included in this analysis below.

2.8 Techniques of mapping biomass burning and smoke palls

Interpretation of the suite of Earth-viewing photographs used in this study was based on an *a priori* knowledge of forest cover, agricultural areas, transportation networks, etc. This prior knowledge is essential in order to locate and measure with confidence the Amazonian smoke pall area. Photograph geolocation is assisted by on-orbit camera precision data recording modules and by postflight ephemeris corrections which allow us to geolocate Space Shuttle Orbiter nadir points at the moment of photograph exposure to within 6 km.

The photographs were interpreted using standard photointerpretation and image-analysis methods, and all smoke edges and smoke features on them were geolocated. A set of parameters and an interpretation key was developed for separating smoke from clouds on the basis of texture, tonal homogeneity and associated morphology. Landscape features such as mountain ridges and drainage patterns were used to interpret adjacent and overlapping photographs to ensure the matching of boundaries.

Operational Navigation Charts (ONCs) of 1:1,000,000 scale were used to derive terrain elements and measure the areal extent of the smoke palls. Lambert Conformal Conic Projection was deemed appropriate for mapping, as this projection does not pose serious distortions for large-area mapping. Scenes with smoke palls generally had unexpectedly sharp boundaries between non-smoke and smoke-covered areas. These interpretations were verified independently by three interpreters. Delineation on the ONCs was then merged. The smoke pall edges for each set of photographs for each burn year were finally transferred to a commercially available base map of the same projection for representation.

2.9 Results

Analysis of the STS-48 (September 1991) photos showed two locations of significantly large smoke palls during the mission. Figure 3 is based on the interpretation of STS-48 photos and shows smoke palls over the Amazon Basin in South America and over south and southeast Asia. Figures 4 and 5 summarise the observations from STS-48.

Trends from all of the available photographs of Amazon smoke palls during the observational period (the full observational period was actually 1961-1991) analysed show that there is gradual movement northward of the primary concentrations of smoke and of fire sources.

In 1973, the primary burning was in the large farm/landholder regions of southern Brazil, the cerrados of Goias, Bahia, and Minas Gerais States and in the forested areas of the wet-dry northeast areas (particularly the Rio Sao Francisco region) of Brazil. In the 1980s, the core of the Amazonian burning shifted to middle and northern Para, northern Mato Grosso, extreme southwest Amazonas and almost all of Rondonia State.

This 1980s north and northwestward shift is related not only to the establishment of

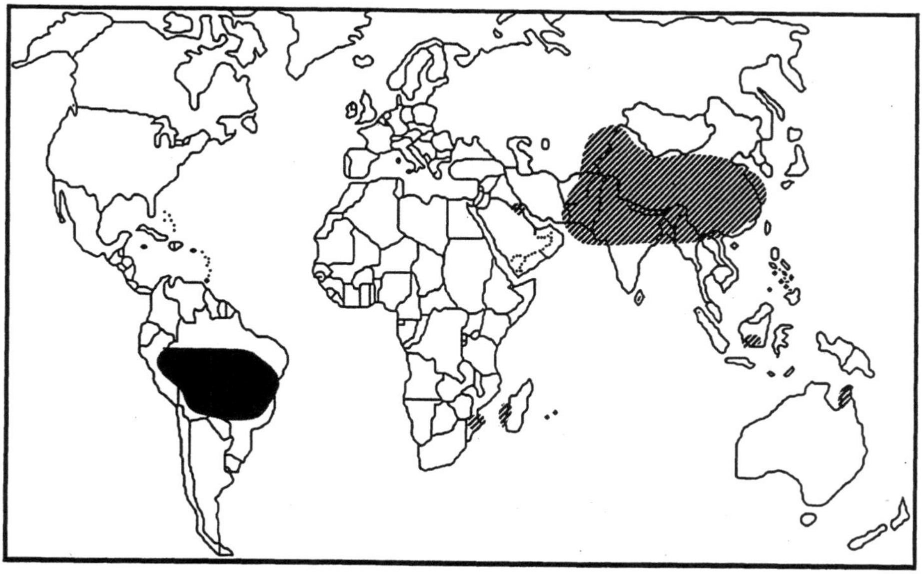

Figure 3: Map showing the location and interpreted extent of the Amazon Smoke pall and the Asian Smoke pall observed during the September 1991 flight of the Shuttle (STS 48).

agricultural areas of the Polonoroeste Project along BR-364 and its Acre State extension (BR-236) but also to agricultural/ranching developments along the Cuiaba-Tapajos and Porto Velho-Manaus highways. In contrast to the increasing smoke pall coverage in the Western Amazon, the clearing and agricultural activity along the axis of the original TransAmazon highway complex (Fortaleza-Teresina-Maraba-Porto Velho) has remained at a moderately low level throughout the entire observational period. It should be realised, however, that cloud cover and lack of nadir-looking passes over south-central Amazonas State have resulted in incomplete observations between the Rios Tocantins and Madeira between 3°-8°S (Figure 6).

3 Summary of Results

Amazonian smoke palls have shown a tenfold increase in area since 1973. No major palls were seen during the 1960s dry seasons. In 1973, a 300 000 km^2 dry season smoke pall of a discontinuous nature was documented. By 1985, dry season smoke palls had increased to become contiguous units of approximately 3 000 000 km^2, and the recent STS-48 photographs show that this increase has persisted.

The interplay of forest-clearing shifts, smoke pall areal extent, topographic influences

Figure 4: Map showing the Nadir poistions of the Shuttle during September 1991 flight (STS 48) when Fire/Smoke Pall photography was acquired.

and the seasonal persistence and long-term opacity of these smoke palls with regional meteorological phenomena cannot be determined by using photography from short-duration Space Shuttle missions alone.

It is useful to note that the Amazon dry season smoke palls have gradually moved from the southern part of the basin to the north and northwest, following the opening of the western Amazon road network. By 1985, these dry season smoke palls had increased to continental scale, almost covering the entire north and west portions of the Brazilian-Bolivian-Peruvian Amazon Basin. Coverage over Amazonian portions of Ecuador, Colombia and Venezuela during these periods did not allow us to define whether smoke palls were present in the tributaries of the Amazon (Solimoes) north of 2°S.

4 Conclusion

In our view, this study also shows that a true Earth-monitoring programme of global tropical environmental change, which includes tropical moist forest clearing, will be most sensibly done by integrating sensor suites, sensor development programmes and the availability of Earth observation data from the forthcoming unmanned polar-orbiting Earth Observations System (EOS) as well as from the human-tended platform, Space Station Freedom.

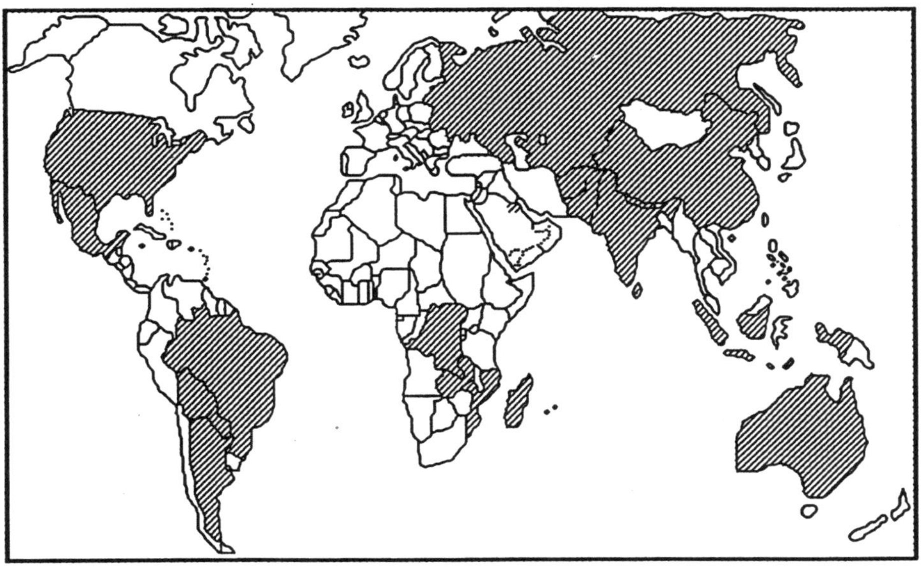

Figure 5: Map showing the global nature of biomass burning/smoke palls documented during the STS 48 flight.

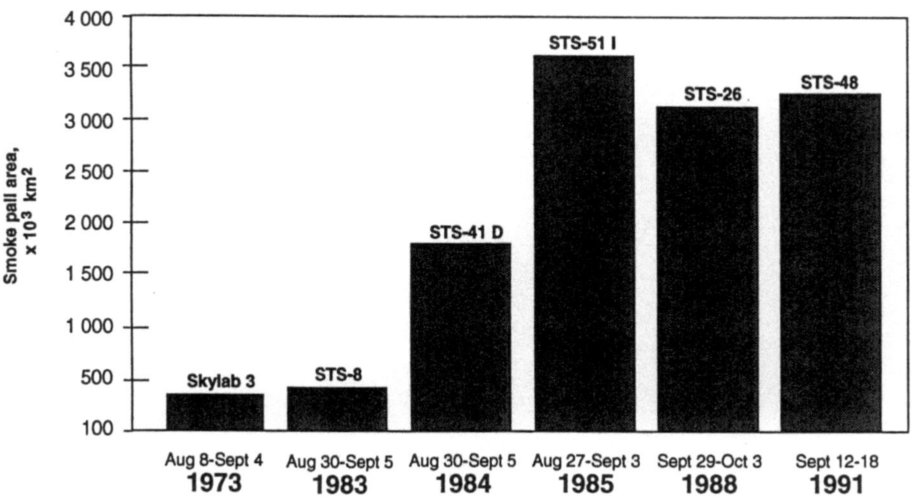

Figure 6: Growth in the Amazonian Smoke Pall areas since 1973. The data for 1973 is from the Skylab imagery. The data for 1983-1991 is from the Space Shuttle Earth Observations imagery.

Land Surface Energy Budget

H. Mannstein

1 Introduction

The motion of the Earth's atmosphere is driven by differential heating. The reservoir of the available potential energy, which describes the difference between the actual state of the atmosphere and a hypothetical atmosphere at rest, is estimated to be about 55×10^5 Jm^{-2} as a mean value around the globe. This reservoir feeds the reservoir of kinetic energy (15×10^5 Jm^{-2}), from where the energy dissipates into heat with a rate of 2.3 Wm^{-2}. As these reservoirs are fairly constant in a climatological sense, the production rate of available potential energy and the transformation rate into kinetic energy has also to be 2.3 Wm^{-2}. In traditional weather forecast models it was not necessary to put much effort into a precise description of the differential heating, the main source of available potential energy, because the time to exhaust the reservoirs mentioned above is given in months.

With the development of dynamical mesoscale models of the atmosphere and of global circulation climate models this situation has changed substantially. In the small and mesoscale dynamic models of the atmosphere, the thermally-induced wind systems with a daily period like seabreeze and mountain winds are main points of interest. The global circulation models for climatological research calculate the atmospheric motion for decades. Both rely on a realistic description of the energy input from the surface. Three different surface types are usually considered:

ocean	- which is quite homogeneous but changes its properties like temperature and roughness under the influence of the atmosphere,
snow and ice and	- which changes its areal coverage with season and climatic conditions
land surface	- which has a great spatial variability primarily due to vegetation cover and moisture content.

In connection with the aims of these atmospheric models arises the need for datasets, which make it possible to describe the transformation of energy at the surface of the Earth.

NATO ASI Series, Vol. I 24
Remote Sensing and Global Climate Change
Edited by Robin A. Vaughan and Arthur P. Cracknell

These data-sets have to cover the whole model area. The meteorological network with standard measurements is tailored for the purposes of weather forecast and climatological registration of atmospheric parameters. To obtain surface (i.e. soil and vegetation) parameters, which vary on the land surfaces in very small scales, remote sensing techniques are useful. Their advantage is the coverage of great areas, their disadvantage the restriction to only one measured quantity, the radiance, which carries different information in different spectral regions and is altered by the atmosphere if airborne or spaceborne measurements (satellite pictures) are used.

Two very different methods lead to spatially and temporally resolved information on land surface energy budget:

- Based on the idea that different surface types show different spectral signatures, a classification of the surface into classes with uniform energetic properties is possible. These properties and the spectral signatures have to be known from *in situ* measurements. The influence of the atmosphere and the illumination is reduced but not omitted by the use of ratios of different spectral channels.
- When satellite scanner data are treated as measured spectral radiances, the reflectance and the temperature of the surface can be derived after accounting for atmospheric influences. As will be shown later, the surface temperature combines the effects of all energy fluxes. With additional information on the state of the atmosphere and especially its boundary layer, multi-temporal values of surface temperature lead to estimates of the thermal properties of the uppermost layer of the ground.

2 Theoretical Aspects

2.1 Energy balance at a surface

2.1.1 Energy budget equation:

The transformation of solar radiation into other forms of energy takes place mainly at the surface of the Earth. The law of conservation of energy applied to a mathematical surface (not a surface layer) dividing the atmosphere and the ground leads to the energy budget equation

$$R + G + H + L_E = 0 \qquad (1)$$

Usually the symbols stand for flux densities of energy with the units Wm^{-2} , even if they are called more shortly fluxes. It should be noted that every flux density is a vector quantity which has a magnitude and a direction. This direction is perpendicular to the surface under consideration. R is the net radiation, G the sum of all fluxes from underneath (often reduced to the"Ground Heat Flux"), H the "Sensible Heat Flux" from or to the atmosphere and L_E the energy flux caused by the condensation and evaporation of water. All flux densities have to be counted positive when directed to the surface and negative in the other direction.

2.1.2 Net radiation:

The net radiation, which is extensively described by the difference between the spectral and hemispherical integrals of incoming and outgoing radiances $L^{\downarrow}$ and $L^{\uparrow}$

$$R = \int_{\lambda=0}^{\infty}\int_{\varphi=0}^{2\pi}\int_{\vartheta=0}^{\pi/2} L^{\downarrow}(\lambda,\vartheta,\varphi)\cos\vartheta d\vartheta d\varphi d\lambda - \int_{\lambda=0}^{\infty}\int_{\varphi=0}^{2\pi}\int_{\vartheta=0}^{\pi/2} L^{\uparrow}(\lambda,\vartheta,\varphi)\cos\vartheta d\vartheta d\varphi d\lambda \tag{2}$$

is usually simplified to only four different components:

$$R = R_S^{\downarrow} + R_S^{\uparrow} + R_L^{\downarrow} + R_L^{\uparrow}. \tag{3}$$

The components differ in their spectral range (shortwave or solar and longwave or terrestrial) and in their direction (upward and downward). This partition originates from the technique of measurement and from the different numerical treatment of these fluxes.

The downwelling solar flux is dominated (at least when no clouds obscure the Sun) by the direct solar radiation and can be parametrised as

$$R_S^{\downarrow} = \tau S_0 \cos\zeta + D \tag{4}$$

with the solar constant S_0 (= 1370 $\mathrm{Wm^{-2}}$ as given in Bolle (1982)), the transmittance of the atmosphere τ and the zenith angle of the Sun ζ. D is the downwelling scattered sunlight or sky irradiance.

The upwelling shortwave radiation is the reflected part of the downwelling flux

$$R_S^{\uparrow} = aR_S^{\downarrow} \tag{5}$$

where a is the bihemispherical reflectance or albedo, which depends on surface type and the spectral and directional distribution of the incident radiance but is usually kept constant due to the lack of more precise information.

The spectral blackbody emission from the ground $R_{BB\lambda}^{\uparrow}$ depending on the wavelength λ and the temperature T is described by Planck's function:

$$R_{BB}^{\uparrow}(\lambda) = \frac{2\pi c^2 h\lambda^{-5}}{e^{(\frac{hc}{k\lambda T})}-1} = \frac{C_1\lambda^{-5}}{e^{(\frac{C_2}{\lambda T})}-1} \tag{6}$$

with the speed of light c, Planck's constant h and Boltzmann's constant k, which are combined to $C_1 = 3.7418 \times 10^{-16}$ $\mathrm{Wm^{-2}}$ and $C_2 = 1.4388 \times 10^{-2}$ mK. The unit of$R_{BB\lambda}^{\uparrow}$ is $\mathrm{Wm^{-2}\ m^{-1}}$, λ must be given in m and the temperature in K.

An integration of Planck's function over the whole spectrum with a fixed temperature leads to the Stefan-Boltzmann-law:

$$R_{BB}^{\uparrow} = \sigma T^4 \tag{7}$$

with $\sigma = 5.6703 \times 10^{-8}$ $\mathrm{Wm^{-2}\ K^{-4}}$.

Natural surfaces do not absorb and emit like blackbodies. This is described by the emissivity $\epsilon(\lambda)$, which is a factor ≤ 1.0 depending on the wavelength and the reflectance $(1-\epsilon(\lambda))$ as the transmittance of the ground $\tau = 0$. Using an effective emittance

$$\epsilon(T) = \frac{1}{\sigma T^4} \int_{\lambda=0}^{\infty} \epsilon(\lambda) R_{BB}^{\uparrow}(\lambda, T) d\lambda \tag{8}$$

which is slowly varying with the temperature T, the upwelling longwave radiation becomes:

$$R_L^{\uparrow} = \epsilon R_{BB}^{\uparrow} + (1-\epsilon) R_L^{\downarrow} \tag{9}$$

The reflected downwelling longwave radiation $(1-\epsilon)R_L^{\downarrow}$ can be neglected in the energy budget, but has to be considered when surface temperature is derived from radiometer measurements of the upwelling longwave radiance.

The downwelling longwave radiation originates from the emission of the atmospheric constituents. Because of the line character of gaseous emission, a highly resolved radiation transfer code and the knowledge of the vertical profiles of temperature, humidity and relevant gases is necessary to compute this radiation precisely. Because most of this radiant flux originates in the lowest atmospheric layers, it can be parametrised by:

$$R_L^{\downarrow} = \epsilon_{air} \sigma T_{air}^4 (a_B + b_B \sqrt{q_{air}}) \tag{10}$$

where ϵ_{air} is the emissivity, q_{air} the humidity and T_{air} the temperature of air. a_B and b_B are the Brunt coefficients for atmospheric radiation in this semi-empirical adjustment.

2.1.3 Simple equilibria

Neglecting the atmosphere and keeping G in (1) small compared with the radiative fluxes leads to an equilibrium between incoming solar and outgoing thermal radiation at a certain surface temperature. The highest temperature a blackbody without its own heat source in the Earth-Sun distance could reach is given by

$$S_0 = R_{BB}^{\uparrow} = \sigma T^4. \tag{11}$$

With the solar constant $S_0 = 1370\ \mathrm{Wm}^{-2}$ it follows that

$$T = 394.2\mathrm{K} = 121.1^{\circ}\mathrm{C}.$$

The mean blackbody temperature of a sphere with a fast distribution of energy to the whole surface and a reflectivity a of 0.3, which is a value typical for the global albedo of the Earth, can be obtained from

$$S_0(1-a) = R_{BB}^{\uparrow} = 4\sigma T^4 \tag{12}$$

which results in

$$T = 255.0\mathrm{K} = -18.2^{\circ}\mathrm{C}.$$

	$\sqrt{\rho ck}$ (Wm^{-2} K^{-1} $s^{\frac{1}{2}}$)
Rock	1700-3200
Ice	2170
Sand (dry)	585
Sand (wet)	2510
Clay (dry)	540
Clay (wet)	2170
Peat (dry)	290
Peat (wet)	1550

2.1.4 Heat conduction

The physics of energy transfer by heat conduction is substantially different from the transfer by radiance. Usually it can be described by local gradients of temperature.

If the energy flux density E in the volume energy budget equation for a uniform thermally conducting material with variations of temperature only in the z - direction and time, which is given with the density ρ and the specific heat c as:

$$\rho c \frac{\partial T}{\partial t} = \frac{\partial E}{\partial z} \tag{13}$$

E is eliminated using the equation which defines the thermal conductivity k

$$E = -k \frac{\partial T}{\partial z} \tag{14}$$

where k is given in Wm^{-1} K^{-1} the equation of thermal diffusion results:

$$\rho c \frac{\partial T}{\partial t} = -k \frac{\partial^2 T}{\partial z^2} \tag{15}$$

An analytical solution of this equation is possible for a given periodic energy flux density $E(t) = E_0 \cos \omega t$ at $z_0 = 0$ with the angular frequency ω:

$$T(z,t) = \bar{T} + \frac{E_0}{\sqrt{\omega \rho c k}} e^{-\sqrt{\frac{\omega \rho c}{2k}} z} \cos(\omega t - \frac{\pi}{4} - \sqrt{\frac{\omega \rho c}{2k}} z) \tag{16}$$

At $z = z_0$ this leads to:

$$T(t) - \bar{T} = \frac{E_0}{\sqrt{\omega \rho c k}} \cos(\omega t - \frac{\pi}{4}) \tag{17}$$

The quantity $\sqrt{\rho ck}$, which has the unit Wm^{-2} K^{-1} $s^{\frac{1}{2}}$ quantifies the thermal behaviour of a surface and is often called thermal inertia. Price (1982) gives the following values for different surface types:

Relating these values to the period of one day by dividing by $\sqrt{24 \times 3600/2\pi}$ $s^{\frac{1}{2}}$ (= 117.3 $s^{\frac{1}{2}}$) gives the value of the"diurnal heat capacity" $\sqrt{\omega \rho c k}$ in Wm^{-2} K^{-1} $S^{\frac{1}{2}}$.

2.2 Time-dependent models

To overcome the restrictions of the analytical solution for thermal dissipation in (16), one-dimensional numerical models of the ground were developed by many authors. They all solve (15) for gridded values of temperature and include parametrisations of the fluxes above the surface. For the sake of simplicity only the treatment of the energy flux densities of (1) in a quite simple model (Price 1982) will be shown here:

2.2.1 Net radiation

The solar (shortwave) net radiation in (3) parametrised by

$$R_S^{\downarrow} + R_S^{\uparrow} = (1-a)S_0\tau\cos(\sin\vartheta\sin\varphi - \cos\vartheta\cos\varphi\cos\omega t) \tag{18}$$

with the geographical coordinates ϑ and φ(only in this equation - ϑ and φ are usually used for zenith angle and azimuth) and the local time t given in the same units as used in the angular speed ω.

The terrestrial longwave flux densities are given in(10) and (9) without the reflection of the radiance from the atmosphere at the surface.

2.2.2 Ground heat flux

The ground heat flux density is given in (14).

2.2.3 Sensible heat flux

The temperature of the air at screen height (i.e. about 2 m above the surface) is used to parametrise the sensible heat flux into the atmosphere. Neglecting the convention of the sign in (1) the sensible heat flux (density) is given by:

$$H = \rho_{air} c_p \frac{(T_s - T_{air})}{r} \tag{19}$$

with an atmospheric resistance r given by

$$r = \frac{[\ln(\frac{z_1}{z_0})]^2}{\kappa^2 u} \tag{20}$$

and the specific heat of air at constant pressure c_p. κ is the Karman constant, taken as 0.35, u the windspeed, z_1 the screen height and z_0 the roughness length for turbulent exchange.

2.2.4 Latent heat flux

$$L_E = \rho_{air} L \frac{(q_s - q_{air})}{r} \tag{21}$$

gives the latent heat flux densities with the heat of vaporisation of water L and the specific humidity at soil surface, q_s and at height z_1, q_{air}. r is the same resistance as in (19).

2.2.5 Soil hydrology

In climate models, which should be able to describe the reaction of the land surface energy budget to rainfalls and evaporation, a soil water budget is added to the ground model. Within the ECHAM, the climate model which is used at the DKRZ (German climate computing centre in Hamburg), an extension of the well known bucket model, which assumes runoff if the soil water reservoir is filled up, is used. The change of the soil water reservoir W_S given in $m^3\ m^{-2}$ by

$$\frac{\partial W_S}{\partial t} = \frac{L_E}{\rho_w c_w} - R_R - R_D + I_R. \tag{22}$$

ρ_w and c_w are the density of water and the specific heat of evaporation, R_R and R_D are the runoff rates from the surface and from drainage processes and I_R is the infiltration rate from rain and melting snow. Runoff and infiltration are modelled depending on the saturation and subscale variation of the orography.

The amount of water within the soil changes the thermal inertia substantially. An even more drastic change is caused by snow and ice.

2.2.6 Vegetation layer

The described parametrisation is sufficient for surfaces with bare ground, but most of the land surface is covered with vegetation which actively changes the energy fluxes. The most important influence is that living plants have the need and the ability to influence the transpiration of water in order to survive and to optimise their biomass production rates. Therefore they extract water from deeper soil layers and transpire even if the soil surface is dry. On the other hand many plants can vary their stomata, through which most of the water vapour is passed into the air, to avoid water stress or overheating. With this mechanism they influence the latent heat flux. Additionally, their leaves have a large surface. The necessary sensible heat flux can therefore be maintained with small temperature gradients.

A representation of a vegetation layer with a"Simple Biosphere Model" (SiB) is given by Sellers *et al.* (1986). Their parametrisation scheme models different vegetation layers as an interface between the soil and the atmospheric boundary layer. The fluxes of heat and moisture are formulated by the use of potential differences between soil, plant canopy and air and resistances, which allow one to introduce plant physiological parameters like the stomatal resistance and the leaf area index.

Within the ECHAM the model grid cells over land are divided into four variable fractions:

- snow cover - either 1 or the ratio between actual and a critical snow amount
- covered with water in a skin reservoir - the snow-free area which is wet
- covered with dry vegetation (but not with snow or water in the skin reservoir) and
- dry bare soil.

This partition into different surface types allows one to assume evaporation at the potential rate for the wet surfaces, a rate which is based on the SiB model for the dry vegetation and related to the water content of the uppermost soil layer for the bare soil fraction. It also acts like a switch if snow or rainfall occurs in the model.

2.3 The role of surface temperature

With a look at the formulation of the four distinct fluxes within the surface energy budget equation it is easy to be seen that the surface temperature occurs in every term:

- The outgoing longwave radiation (9) is dominated by the Stefan-Boltzmann law (7), where T_S enters in the fourth power.
- The ground heat flux is determined by the gradient of ground temperature, where T_S enters into (14) as the upper boundary.
- The sensible heat flux is given by the gradient of air temperature. Usually T_S gives the lower boundary of the temperature profile (nevertheless in some cases the description of surface temperature in a boundary layer model of the atmosphere does not match the skin temperature which is measured radiometrically. Traditionally meteorologists use the screen temperature measured 2 m above the surface and in model descriptions which define the surface by keeping the windspeed at zero, this surface temperature should be taken as the temperature at the level z_0, the roughness length).
- The latent heat flux depends on the gradient of specific humidity, which is defined at the surface by the saturation deficit and the surface temperature.

It is evident that the surface temperature ties together the different energy fluxes and that a measurement of surface temperature together with additional information can provide insight into the partition of incoming solar energy into the different fluxes.

3 Remote Sensing

3.1 Surface temperature

To understand the processes influencing the signal at the radiometer of a satellite or an aircraft, it is useful to follow the radiant energy from the sources to the sensor. An atmospheric radiative transfer code (LOWTRAN 6 from Kneizys *et al.* (1983)) is used here to show the spectrally resolved radiances.

3.1.1 Surface emission

The hemispherical thermal emission of a blackbody is given in Eq. (6). As a satellite sensor usually measures the radiance in a small cone, the radiant flux density given in this equation has to be converted into radiance (radiant flux density per steradian).

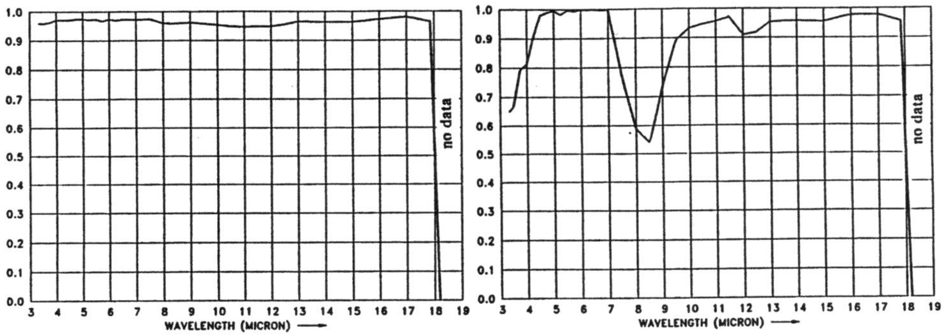

Figure 1: Spectral emissivity of an orange tree leaf (left) and coarse sand (right). Data from Sutherland (1985).

When Lambertian surfaces are considered, this is done by dividing by $\pi\Omega_0$, with $\Omega_0 = 1$ steradian as the unit solid angle. The fraction of blackbody radiance leaving a non-black surface is given by the emissivity.

3.1.2 Spectral emissivity of the surface

Two examples of the spectral behaviour of the emissivity of natural materials are given

The tree leaf data show the very smooth curve typical for vegetation. The coarse sand data should be taken as an example for bare soil, which shows the influence of minerals with pronounced reflection bands (the quartz reststrahlen bands).

The emissivity of natural surfaces varies also with the direction. At mixed surfaces, for example sparse, protruding vegetation on sand, the emissivity changes with the partition of the different surface types in the field of view of the sensor, which varies in this case from nearly only sand, looking downwards, to only plants, looking to the horizon. This should not be mixed with the effect of vertical temperature gradient in a structured vegetation layer.

3.1.3 Reflection of sky radiance

As given in Eq.(9), radiation originating from the atmosphere is reflected by the surface. This sky radiance depends on the atmospheric temperature, water content and the vertical concentration of other constituents.

This radiation is reflected by the surface according to its reflectivity (1 - ϵ). The reflected radiance has to be added to the emitted one.

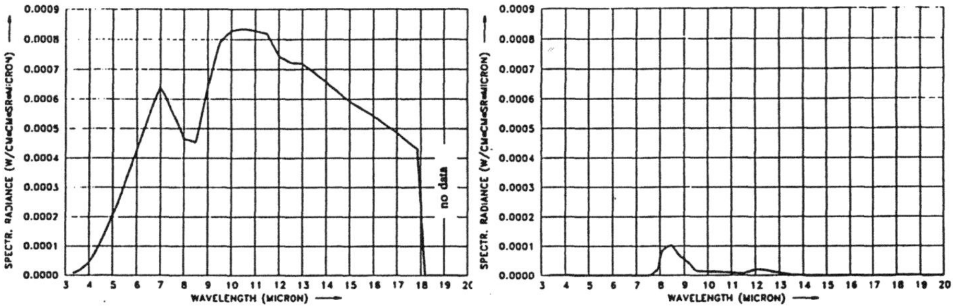

Figure 2: Radiance emitted from a coarse sand surface at 20° *C* (left) and reflected sky radiation (right), as seen by a sensor outside the atmosphere.

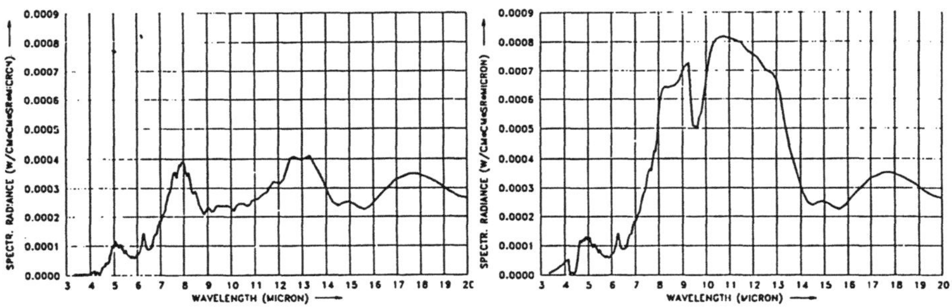

Figure 3: Radiance emitted by the atmosphere (left) and total radiance reaching the sensor (right).

3.1.4 Transmission of the atmosphere

On the way through the atmosphere a beam of radiation is changed due to absorption, scattering and emission. The radiative transfer code makes it possible to split the radiance which arrives at the sensor (in Figure 3 on the right) into radiance emitted by the atmosphere and the remaining part of the radiance, which is emitted and reflected by the surface (Figure 2) Additionally the spectrally-resolved ratio between the radiance at the sensor due to only the surface and the total radiance which is the former plus the contribution of the atmosphere is given in Figure 4. High values of the ratio show the spectral regions suitable for remote sensing of surface temperature. At lower wavelenghts ($< 5\mu$m) the scattered and reflected sunlight which is not considered in this computation disturbs the measurement at daytime.

In the window region (8-13 μm), the water vapour continuum absorption dominates the atmospheric absorption and emission. The drop at 9.6 μm is due to the O_3 (ozone) band.

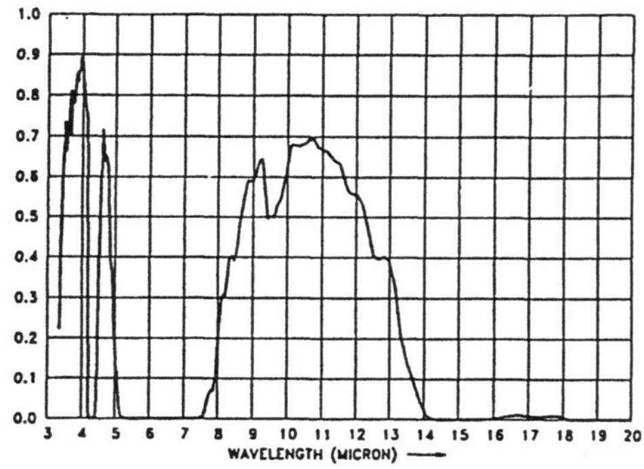

Figure 4: Part of total signal which originates from surface emission

3.1.5 Sensor and signal

Usually a sensor measures a radiant flux density. This is converted into an equivalent blackbody temperature using calibration data. For an interpretation of the blackbody temperature it is necessary to know the relative spectral sensitivity of the sensor - filter combination.

An integration of the Planck function (Eq.(6)) at different temperatures weighted with the relative spectral sensitivity gives the energy, which is measured during the calibration. Now the blackbody temperature at the sensor can be related to the incident energy flux, which does usually not follow the Planck distribution. With the radiative transfer code this energy flux can be computed depending on given surface temperatures and emissivities, when the state of the atmosphere is sufficiently well known. The difference between the surface temperature and the equivalent blackbody temperature at the sensor is the atmospheric correction, which has to be added to the measured temperature.

3.1.6 Split window method

Another way to account for the atmospheric influence is the use of the split window method for example with the channels 4 and 5 of the AVHRR on board the NOAA servies of spacecraft. The different atmospheric tranmission in these two channels allows an estimate of the combined temperature-humidity influence and therefore a correction, when the emissivity of the surface in both channels is known. Therefore this technique should be applied only to water surfaces (and perhaps ice and snow), where the surface is homogeneous with respect to spectral emissivity.

3.2 Albedo

3.2.1 Atmospheric correction

The correction of radiance measurements in the spectral region of reflected sunlight is analogous to the correction of longwave radiance described above. The main atmospheric influence is now the scattering and absorption by aerosol particles and, increasing with decreasing wavelength, air molecules (Rayleigh scattering). The state of the atmosphere has now to be described in terms of turbidity or visibility.

For the calculation of reflectances also the downwelling solar radiation and its distribution over the hemisphere above the surface has to be calculated.

3.2.2 Correction for non-Lambertian surfaces

The term albedo is used in various ways. In the context of energy budget calculations, albedo must be defined according to (5) as the ratio between upwelling and downwelling shortwave radiant flux densities. This implies, that the measurement from satellite does not give the required quantity. If no additional information is available, the most simple assumption is that of a Lambertian surface. The upwelling shortwave radiant flux density then becomes with the zenith angle ϑ and azimuth φ

$$R_S^{\uparrow} = \int_{\varphi=0}^{2\pi} \int_{\vartheta=0}^{\pi/2} L^{\uparrow}(\vartheta,\varphi) \cos\vartheta \sin\vartheta d\vartheta d\varphi = \pi\Omega_0 L^{\uparrow} \tag{23}$$

as the upwelling radiance not depending on direction.

In a more general case, the bidirectional reflection distribution function (BRDF) $\Gamma(\vartheta,\varphi,\vartheta',\varphi') = 1/a\gamma(\vartheta,\varphi,\vartheta',\varphi')$ which describes the directionality of surface reflection is used for an estimate of the hemispherical albedo

$$a = \frac{\pi\Omega_0 L^{\uparrow}(\vartheta',\varphi')}{\int_{\varphi=0}^{2\pi}\int_{\vartheta=0}^{\pi/2} L^{\downarrow}(\vartheta,\varphi)\Gamma(\vartheta,\varphi,\vartheta',\varphi') \cos\vartheta \sin\vartheta d\vartheta d\varphi} = \frac{\pi\Omega_0 L^{\uparrow}(\vartheta',\varphi')}{S_0 \cos\vartheta_0 \Gamma(\vartheta,\varphi,\vartheta',\varphi')} \tag{24}$$

with ϑ_0 as the actual Sun declination in the approximated form where the surface illumination from scattered light is neglected. BRDF's for several natural surface types are given in Kriebel (1977).

4 Applications

4.1 Development of the techniques

4.1.1 Apparent thermal inertia of dry surfaces:

With the energy budget concept (Eq.(1)), the surface temperature which is accessible by remote sensing gives the possibility to infer information of surface parameters and energy fluxes.

In geology the quantity of main interest is the thermal inertia P ($= \sqrt{\rho c k}$), because surface minerals with differences in porosity or composition can be distinguished by this parameter. P can be derived from Eq.(17) with measurements of maximum and minimum surface temperatures:

$$P = \frac{2E_0}{(T_{max} - T_{min})\sqrt{\omega}} \tag{25}$$

when E_0, the amplitude of the daily ground heat flux wave is known and the daily cycle can be assumed to be sinusoidal.

The difficulty in this approach is that the ground heat flux is usually small compared to the other fluxes. In meteorology it is often parametrised as 10% of the net radiation flux or totally neglected because of the lack of knowledge of the soil parameters.

4.1.2 Heat capacity mapping mission:

Nevertheless, NOAA gave a simple scheme for deriving an 'Apparent Thermal Inertia' (ATI) from the data of the Heat Capacity Mapping Mission (HCMM) for use in geological surveys:

$$ATI = \frac{NC(1-a)}{T_{day} - T_{night}} \tag{26}$$

with N as a scaling factor, C a correction factor depending on latitude and solar declination and a the albedo derived from measurements in the visible and near infrared part of the spectrum of the same mission.

After inspection of the lunar surface (Jaeger 1953), results of geological applications of the thermal inertia concept are reported by various authors for the eastern part of the Arabian peninsula, for the Mojave desert and for Sardegna - all rather dry regions.

Applying this technique to humid areas failed in the geological sense but led immediately to the insight that the masking effect of soil moisture and vegetation cover could be used for measurement of quantities related to evapotranspiration.

4.2 Inference of soil moisture

In a first experimental approach, Idso *et al.* (1976) normalised the remotely-sensed surface temperature difference ($T_{max} - T_{min} = \delta T$) using the ratio of air temperature amplitude to a standard value ($\delta T_{Surf,norm} = (\delta T_{Surf}/\delta T_{air}) \times 18K$). The normalized δT_{Surf} showed a good negative correlation with the soil water content in the uppermost 2cm of the test fields.

Based on experiences like this and on the fact that within a well -defined system of equations two unknowns can be computed when two values (maximum and minimum surface temperature) are given, a lot of similar but in detail different models were developed to infer soil moisture and thermal inertia from remotely sensed data - especially from the HCMM mission.

The TELL-US model by Rosema *et al.* (1978) results in a look-up table for a given set of atmospheric parameters (T_{air}, q_{air}, windspeed, roughness, albedo, ...) indicating thermal inertia and surface relative humidity for a given pair of day and night temperature. Carlson *et al.* (1981) used a much more elaborate model with four different layers spanning from the atmosphere into the ground. The model-output is converted into two second-order two-dimensional polynomial equations for moisture availability and thermal inertia depending on the surface temperatures. Other models are the Tergra model for crop surface temperatures, or a model from Axelsson (1983), which can also be applied to non-opaque surfaces such as vegetation layers or snow-cover.

4.3 Some warnings

In most of the models additional atmospheric parameters from meteorological standard measurements have to be incorporated, which do not fit to the areal coverage of satellite data. Surface roughness, which is essential for the partition of the incident energy flux into the ground heat flux and the fluxes into the atmosphere, cannot be inferred by the standard measurements and has to be estimated with a high level of uncertainty.

It is evident(and explicitly mentioned by some authors) that the derived parameters, thermal inertia and soil moisture content, are effective ones in the sense that they depend on the parametrisation scheme which is used to derive them. A reproduction of the surface temperatures and a deduction of surface energy fluxes using such values is only possible using the same parametrisation scheme.

Pratt and Ellyet (1979) showed that different composition of the soil (sand - clay) alters the relation between soil moisture content and thermal inertia, especially when different porosities are considered. They state that "*it is apparent that the mapping of absolute soil moisture content is completely impractical*".

The plant-physiological behaviour of trees was measured by Koerner (1985). The result was that different species change their hydraulic conductivity in different ways when a high demand for evaporation is given from the atmosphere, even when the soil is saturated with water. Most plants develope active strategies to avoid water stress according to their natural habitat which act like a feedforward process on evapotranspiration. A general direct relation of reduced evapotranspiration to soil water content is therefore impossible.

Price (1985) describes the errors in deriving temperatures from remotely-sensed data and the difficulties in the modelling of the surface energy budget. He comes to the conclusion that "*relative values rather than absolute measurements are a realistic goal.*"

5 The Spatial Variation of Sensible Heat Flux in the Alps

5.1 Background

The thermally driven wind systems in mountainous regions are well documented and understood theoretically. At the small scale end of these wind systems are the slope winds, blowing parallel to the gradient of the terrain and reacting directly to insolation and radiative cooling. Transferring heat to and from the whole valley atmosphere, they induce the valley winds which fill most of the valleys following the pressure gradient along the valley axis even if the bottom is nearly horizontal. In the regional (and perhaps global) scale there exists a mountain wind system, tending to follow the isobars of the heat low, which develops at daytime under favourable conditions over the mountain regions.

Differential heating, produced either by an elevated heat source or sink, e.g. a slope or a mountain plateau, or by the reduced amount of air which has to be heated in a valley, results in temperature gradients and, hence, in pressure gradients which drive these circulation systems. Dominated by orography, these concepts work without any knowledge of the actual surface sensible heat flux. It just has to be nonzero, and it should not be too different from the one in the adjacent plains. As 'sol lucet omnibus', the sensible heat flux per horizontal unit area is a constant in most of the conceptual models.

A numerical model for the quantitative simulation of real-world regional wind systems inside and around mountain regions needs information on the driving force which is the sensible heat flux, resolved on the same or a finer scale as the orography, and the resulting wind systems. The sensible heat flux could be either prescribed as a parameter varying in space and time or computed by a surface energy balance model which takes into account the thermal properties of the surface.

Surface properties in complex terrain are highly variable and a dataset with a description of the Alps which allows one to derive the thermal properties of the surface in the necessary resolution does not exist. Also most of the region is inaccessible for direct measurement of the sensible heat flux. Therefore an attempt was made to derive the sensible heat flux from satellite data.

The usual method to derive those surface parameters which influence the energy balance from satellite data is to use a set of meteorological standard data such as windspeed, air temperature and humidity, downwelling radiation and subsurface temperatures within an energy balance model for the computation of the daily course of surface temperatures for different sets of the surface parameters albedo, thermal inertia and soil moisture. The results of such computations are inverted and compared with surface temperatures measured by satellite at different times.

None of these models is designed to deal with the special effects of mountainous regions like the pronounced vertical structure of air temperature, the orographic effect on the radiation balance, the slope wind layer and so on. An extension of such a model to cope with these mountain conditions would require the use of at least three additional parameters, elevation, inclination and azimuth of the surface, and the knowledge of the variation of the meteorological parameters with them. The computation and inversion of such a model is up to now totally impractical.

Therefore another way to obtain insight into the energy balance had to be found. The theory of slope winds opens a possibility to reduce the number of parameters influencing the surface temperature at a slope at least for the daytime case with a turbulent upslope boundary layer embedded in a stable background atmosphere.

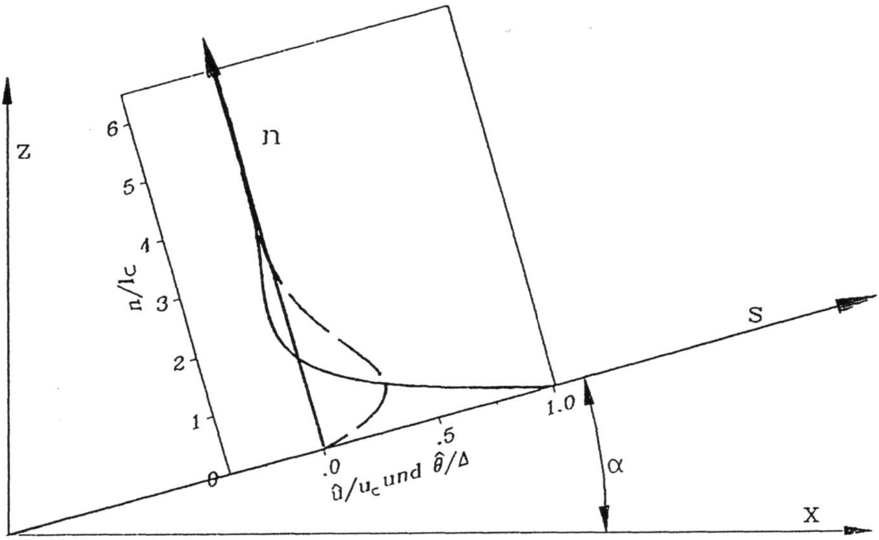

Figure 5: The slope coordinate system with profiles of non-dimensional temperature deviation $\hat{\Theta}/\Delta$ (solid line)and non-dimensional windspeed parallel to the slope $\hat{u}/u_c$ (dashed) of Prandtl's solution for a quasilaminar flow.

5.2 Sensible heat flux in the slope wind layer

5.3 Theory of upslope winds

A special form of the atmospheric boundary layer is found over heated slopes within a stable atmosphere - the upslope boundary layer. As the air is heated from the slope, a horizontal pressure gradient towards the slope evolves which organises the turbulent layer to an upslope flow. After a while, this flow is determined by the equilibrium between buoyancy forces and momentum transfer to the ground (friction). The coordinate system with a component parallel (s) and normal (n) to the slope with the inclination α is shown in Figure 5 together with the Prandtl (1942) solution for a quasi laminar flow.

5.3.1 Slope wind models

To describe the slope wind layer, Brehm (1986) chose a 1-dimensional formulation, which relates the exchange coefficient for heat K_h and for momentum K_m to the mixing lengths $l_{h,m}$ and the mean turbulent kinetic energy $\bar{E}$:

$$K_{h,m} = l_{h,m} a_1 \bar{E}^{1/2} \tag{27}$$

As l_h and l_m are determined by stability functions which reflect the properties of the temperature and wind profile, and the profile of turbulent kinetic energy is determined by a transport equation, the turbulent exchange varies with the structure of the boundary layer.

Various model runs with slopes between $2° \leq \alpha \leq 30°$ are combined to parametrisations for an analogue to the geostrophic drag c_g defined by

$$c_g = \frac{u_*}{\kappa \Delta \sqrt{\beta / \gamma}} \tag{28}$$

and ratio between the heat transfer coefficient c_H and c_g given by $\eta = c_H / c_g$, which is defined by

$$\eta = \frac{\kappa u_* \Theta_* \sqrt{\beta / \gamma}}{u_*^2} \tag{29}$$

as a function of the inclination α and a Rossby number

$$R_0 = \frac{\kappa \Delta}{\gamma \sin \alpha z_0} \tag{30}$$

with $\beta = g/\Theta_0$, $\gamma = \partial\Theta/\partial z$, $\kappa = 0.36$ (Karman constant) and z_0 as the roughness length. If we use

$$H = -\rho c_p K_H \frac{\partial \Theta}{\partial n} = -\rho c_p \kappa u_* \Theta_* \tag{31}$$

we can derive with the definition of c_g (28) and η (29) the sensible heat flux as

$$H = -\rho c_p \kappa^2 c_g^2 (R_0, \alpha) \eta (R_0, \alpha) \Delta^2 \sqrt{\frac{\beta}{\gamma}} \tag{32}$$

This equation is used for the parametrisation of the sensible heat flux H as a function of the temperature surplus at the slope Δ. c_g and η were taken from Brehm's work and interpolated to a two-dimensional grid as functions of ln (R_o) and α (Figure 6).

Large-eddy simulations (LES) of the convective up-slope boundary layer by Schumann (1990) confirmed Brehm's parametrisation in general and gave more insight into the energetics and dynamics of the slope wind later. Within a large-eddy simulation the dominant part of the turbulent motion is resolved in a dynamical, nonhydrostatical model. This includes, that the transport of kinetic energy and heat is explicitly described by advection and nearly independent of the chosen subgrid-scale description.

The results of the large-eddy simulation include steeper slopes then Brehm's results and show that the time for reaching an equilibrium boundary layer is quite long for small inclinations.

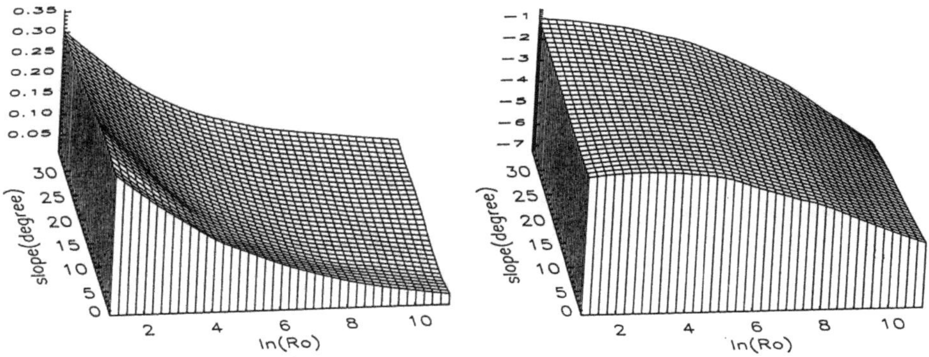

Figure 6: The analogous drag coefficient c_g (left) and the ratio η between nondimensional sensible heat flux and momentum flux (right) as a function of a Rossby - number for different slopes, taken from Brehm (1986).

stability	$\frac{\delta\theta}{\delta z}$	$= 0.005$ K m^{-1}	
slope	α	$= 2°$	(1)
	α	$= 5°$	(2)
	α	$= 10°$	(3)
	α	$= 20°$	(4)
	α	$= 30°$	(5)
roughness length	z_0	$= 0.1$ m	
pressure	p	$= 800$ hPa	
background temperature	T	$= 283$ K	

5.3.2 The difference between surface and skin temperature

The surface temperature used in numerical models of the turbulent boundary layer is usually the temperature of the lowest gridpoint. The windspeed at this level is fixed to zero. In terms of a logarithmic profile, this level has to be identified with the altitude $n = z_0$ above the surface. With a radiometrically measured surface temperature, we get the skin temperature at $n = 0$. Monin and Zilitinkevich (1986) parametrised the temperature difference between the surface and the z_0 - level Δ_G as:

$$\Delta_G = -0.13\bar{\Theta}_* \left(\frac{u_* z_0}{\nu}\right)^{0.45} \tag{33}$$

with ν as the kinematic viscosity. $\bar{\Theta}_*$ and u_* can be derived from Brehm's parametrisation depending on the temperature difference Δ between the level $n = z_0$ and the

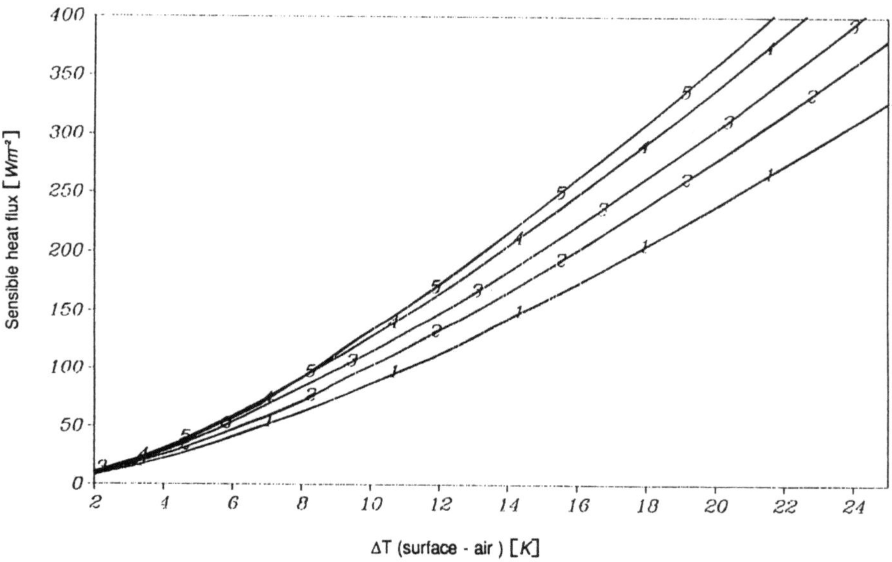

Figure 7: Sensible heat flux including the parametrisation of Monin and Zilitinkevich (1986)

undisturbed atmosphere. With equations (28) and (29) one gets:

$$\Delta_G = -0.13\kappa\eta c_g\Delta \left(\frac{c_g\kappa\Delta\sqrt{\beta/\gamma}z_0}{\nu}\right)^{0.45} \tag{34}$$

Because Δ is needed for the determination of R_0 and thus of η and c_g , a direct formulation to compute Δ from

$$\Delta_S = \Delta + \Delta_G \tag{35}$$

is impossible. Starting with a rough estimate for $\Delta(= 0.5\Delta_S)$, an iterative solution converges quickly, because for fixed values of z_0, α, γ and ρ the difference Δ_G increases steadily with Δ .

Figure 7 shows the sensible heat flux resulting from (34) and Brehm's parametrisation depending on the temperature difference Δ_S between surface and undisturbed atmosphere.

5.4 Data processing

The AVHRR data from the NOAA-9 overpass at 12:00 UTC on 11 September 1985 was chosen here as an example because the Alps were almost cloud free, the atmosphere was very clear and no snowfall had ocurred for several days. In addition, at this date and time temperature and humidity profiles were measured by aircraft in and above the basin of Klagenfurt (Austria) and 50 minutes later southwest of Munich (Germany). These profiles are used like radiosonde data together with data from the stations around the Alps.

As background information a digital terrain model (DTM) with a horizontal grid distance of 8″ × 12″, which is equivalent to about 250 m × 250 m, and a vertical resolution of 5 m was available. It was used in the region between 46 and 48 °N and between 10 and 14 °E to compute the relative direct solar radiation with the sun's position at the time of the satellite overpass, to apply an altitude-dependent atmospheric correction to the measured temperatures and to derive in combination with the air temperature field the background temperature and stability.

5.5 Surface temperature of slopes from satellite data

The information needed from the satellite data is an unbiased estimate of the thermodynamic temperature at the surface. As the satellite radiometer measures the radiance in a defined spectral band from the position of the satellite at least three different steps of correction have to be considered:

1. Count to blackbody temperature conversion

The counts transmitted from the satellite are converted to radiances expressed as equivalent blackbody temperatures by using the inflight calibration data included in the datastream (Lauritson *et al.* 1988).

2. Geometrical corrections for complex terrain

The space-oblique projection of the AVHRR data is usually mapped into a geocentred coordinate system assuming the surface to be at sea level. The variations in terrain height cause deviations up to two picture elements due to the oblique view in this overpass. Thus, the satellite data were projected onto a 3-d surface derived from the DTM. The field of relative solar irradiance computed from the DTM and mapped into the satellite projection was used for navigation of the satellite in the sub-pixel precision range.

3. Radiometric corrections

As temperature and humidity profiles of the atmosphere are known, a correction of the atmospheric effect is possible by inversion of the results of a radiative transfer model. Here the LOWTRAN-6 code from Kneizys *et al.* (1983) was used inside the SENSAT-software. Computations of the difference between the equivalent blackbody temperature and the surface temperature are done for 6 surface temperatures at 10 altitudes for two surface emissivities (0.98 for vegetation and 0.95 for rock and bare soil) each. These differences have been interpolated into a dataset with equidistant altitudes and blackbody temperatures.

5.6 Surface classification

For the emissivity correction and for an estimate of surface roughness, a knowledge of the surface type is essential. The information in Ch. 1 (visible) and Ch. 2 (near infrared) of the AVHRR was used to exclude clouds, snowfields and lakes from the data by applying thresholds. After correction for illumination effects using the computed direct solar radiation, the spectral signature of vegetation (high reflectivity in the infrared) and the difference in albedo between tall vegetation(dark) and grass (bright) was used for an estimate of the relative amount of the classes forest, meadow and rocks within one pixel.

5.7 Results

The combination of surface temperature, background air temperature and stability at the altitude of the area under consideration, inclination of the slope and surface roughness in Brehm's parametrisation results in data of sensible heat flux covering 82% of the region (colour Plate 2). The areal mean value of 108 Wm^{-2} is in the expected range. Maxima of the sensible heat flux are usually located at south facing slopes at altitudes between the snowline and the forests at lower levels. This effect can be explained by solar irradiance, the small water storage capacity of soils at higher altitudes and the variation of the vegetation canopy with height. An increase of sensible heat flux with height tends to maintain the stable stratification of the air in the valleys and therefore influences the whole circulation system.

The sensible heat flux increases towards the centre of the Alps and it is higher in the southern regions than in the northern ones. The regional distribution of precipitation could produce such a pattern either directly by its influence on the soil moisture content or indirectly by its long-term influence on the vegetation. In both cases the variations in evapotranspiration determine the sensible heat flux.

Reference

Abdellaoui, A., Becker, F., Olory-Hechinger, E., and Raffy, M., 1982, Thermal Inertia and Soil Moisture Mapping. *Final Report, Groupe de Télédétection de Strasbourg.*

Anderson, J. M., and Wilson, S. B., 1984, The Physical Basis of Current Infrared Remote-Sensing Techniques and the Interpretation of Data from Aerial Surveys. *International Journal of Remote Sensing*, **5**, 1-18

Axelsson, S. R. J., 1983, Thermal Modeling for the Interpretation of IR-Data. *Proceedings of the 17th International Symposium on Remote Sensing of Environment*, pp. 499-518.

Barton, I. J., 1983, Dual Channel Satellite Measurement of Sea Surface Temperature. *Quarterly Journal of the Royal Meterological Society*, **109**, 365-378.

Barton, I. J., 1985, Transmission Model and Ground-Truth Investigation of Satellite-Derived Sea Surface Temperatures. *Journal of Climate and applied Meteorology*, **124**, 508-516.

Becker, F., Ramanantsizehena, P., and Stoll, M. P., 1985, Angular variation of the bidirectional reflectance of bare soils in the thermal infrared band. *Applied Optics*, **24**, 365-375.

Becker, F., and Raffy, M., 1987, Problems related to the determination of land surface parameters and fluxes over heterogeneous media from satellite data. *Advances in Space Research*, **11**, 45-57.

Bolle, H.-J., 1982, Radiation and Energy Transport in the Earth Atmosphere System In The Handbook of Environmental Chemistry, Vol. 1, Part B, edited by O. Hutyinger (Berlin Heidelberg: Springer-Verlag), pp. 131-292.

Brehm, M., 1986, Experimentelle and numerische Untersuchungen der Hangwindschicht and ihrer Rolle bei der Erwaermung von Taelern. Müenchen Universitüts –Schriften, Meteorologisches Institute, Wissenschaftliche Mitteilungen Nr 54.

Büettner, K. J. K., and Kern, C. D., 1965, The Determination of Infrared Emissivities of Terrestrial Surfaces. *Journal of Geophysical Research*, **70** 1329-1337.

Byrne, G. F., Begg, J. E., Fleming, P. M., and Dunin, F.X., 1979, Remotely Sensed Land Cover Temperature and Soil Water Status - A Brief Review. *Remote Sensing of Environment*, **8**, 291-305.

Callison, R. D., and Cracknell, A. P., 1984, Atmospheric correction to AVHRR brightness temperatures for waters around Great Britain. *International Journal of Remote Sensing*, **5**, 185-198.

Cannizaro, G., Ricottilli, M., and Uliveri, C., 1985, Analysis of Different Algorithms for Sea Surface Temperature Retrieval from NOAA AVHRR Data *Proceedings of the 19th International Symposium on Remote Sensing of Environment*, held in Ann Arbor, MI.

Carlson, T. N., Dodd, J. K., Benjamin, S. G., Cooper, and J. N., 1981, Satellite Estimation of the Surface Energy Balance,Moisture Availability and Thermal Inertia *Journal of Applied Meteorology*, **20**, 67-87.

Chen, J., 1984, Uncoupled Multi-Layer Model for the Transfer of Sensible and Latent Heat Flux Densities from Vegetation. *Boundary-Layer Meteorology*, **28**, 213-225.

Colwell, R. N. (Editor), 1983, Manual of Remote Sensing, (Bethesda: American Society of Photogrammetry).

Deardorff, J. W., 1978, Efficient Prediction of Ground Surface Temperature and Moisture, with Inclusion of a Layer of Vegetation. *Journal of Geophysical Research*, **83**, 1889-1903.

Dickinson, R. E., 1983, Land surface processes and climate - surface albedos and energy balance. *Advances in Geophysics*, **25**, 305-353.

Dickinson, R. E., 1987, Evapotranspiration in global climate models. *Advances in Space Research*, **11**, 17-26.

Eagleson, P. S. (Editor), 1982, Land Surface Processes in Atmospheric General Circulation Models. (Cambridge: Cambridge University).

Flannery, B. P., 1984, Energy Balance Models Incorporating Transport of Thermal and latent Energy. *Journal of Atmospheric Science*, **41**, 414-421.

Gatlin, J. A., Sullivan, R. J., and Tucker, C. J., 1984, Considerations of and Improvements to Large-Scale Vegetation Monitoring *IEEE Transactions on Geoscience and Remote Sensing*, **GE-22**, 496-502.

Geleyn, J. F., and Preuss, H. J., 1983, A New Data Set of Satellite-derived Surface Albedo Values for Operational Use at ECMWF. *Archives of Meteorological and Geophysical Bioclinatology, Series A*, **32**, 353-359.

Hayes, L., 1985, Review Article: The current use of TIROS-N series of meteorological satellites for land-cover studies. *International Journal of Remote Sensing*, **6**, 35-45

Hechinger, E., Raffy, M., and Becker, F., 1982, Comparison Between the Accuracies of a new Discretization Method and an Improved Fourier Method to Evaluate Heat Transfers between Soil and Atmosphere. *Journal of Geophysical Research*, **87**, 7325-7339.

Idso, S. B., Jackson, R. D., and Reginato, R. J., 1976, Compensating for Environmental Variability in the Thermal Inertia Approach to Remote Sensing of Soil Moisture. *Journal of Applied Meteorology*, **15**, 811-817.

Jackson, R. D., 1982, Soil Moisture Inferences from Thermal-Infrared Measurements of Vegetation Temperatures. *IEEE Transactions on Geoscience and Remote Sensing*, **GE-20**, 282-285.

Jaeger, J. C., 1953, Conduction of heat in a solid with periodic boundary conditions, with an application to the surface of the Moon. *Proceedings of the Cambridge Philosophical Society*, **49**, 353-359.

Kahle, A. B., 1977, A Simple Thermal Model of the Earth's Surface for Geologic Mapping by Remote Sensing. *Journal of Geopherical Research*, **82**, 1673-1680.

Kahle, A. B., Schieldge, J. P., and Alley, R. E., 1984, Sensitivity of Thermal Inertia Calculations to Variations in Environmental Factors. *Remote Sensing of Environment*, **16**, 211-232.

Kidwell, K. B., 1984, NOAA Polar Orbiter Data-Users Guide. (Washington D.C: NOAA / NESDIS).

Klaassen, W., and van den Berg, W., 1984, Evapotranspiration Derived from Satellite Observed Surface Temperatures. *Journal of Climate and Applied Meteorology*, **24**, 412-424.

Kneizys, F. X., Shettle, E. P., Gallery, W. O., Chetwynd, J. H., Abreu, L. W., Selby, J. E. A., Clough, S. A., and Fenn, R. W., 1983, Atmospheric Transmittance/Radiance: Computer Code LOWTRAN 6. US Air Force Geophysics Laboratory, AFGL-TR-83-0187.

Koerner, Ch., 1985, Humidity Responses in Forest Trees: Precautions in Thermal Scanning Surveys. *Archives of Meteorological and Geophysical Bioclinatology*, **B36**, 83-98.

Kriebel, K. T., 1977, Reflection Properties of Vegetated Surfaces: Tables of Measured Spectral Biconical Reflectance Factors. Müenchner Universitäts-Schriften, Meteorologisches Institut, Wissenschaftliche Mitteilungen Nr 29.

Kriebel, K. T., Measured spectral bidirectional reflection properties of four vegetated surfaces. *Applied Optics*, **17**, 253-259.

Kriebel, K. T., 1979, Albedo of Vegetated Surfaces: Its Variability with Different Irradiances. *Remote Sensing of Environment*, **8**, 283-290.

Lauritson, L., Nelson, G. J., and Porto, F. W., 1988, Data Extraction and Calibration of TIROS-N/NOAA Radiometers. NOAA Technical Memorandum NESS 107-Rev 1.

Mahrt, L., and Pan, H., 1984, A Two-Layer Model of Soil Hydrology. *Boundary Layer Meteorology*, **29**, 1-20.

Mannstein, H., 1990, Die radiometrisch bestimmte Oberflaechentemperatur im Gebirge and die Ermittlung des Stroms fuehlbarer Waerme. DLR-FB 90-07.

Mannstein, H., 1991, The skin temperature of mountain slopes and the determination of the sensible heat flux. ESA-TT-1220.

Masuda, K., Takashima, T., and takayama, Y., 1988, Emissivity of Pure and Sea Waters for the Model Sea Surface in the Infrared Window Regions. *Remote Sensing of Environment*, **24**, 313-329.

Mccumber, M. C., and Pielke, R. A., 1981, Simulation of the Effects of Surface Fluxes of Heat and Moisture in a Numerical Model. *Journal of Geophysical Research*, **86**, 9929-9938.

Miller, D. H., 1981, Energy at the Surface of the Earth. (New York: Academic) (in Press).

Monin, A. S., and Zilitinkevich, S. S., 1986, On Description of Micro- and Mesoscale Phenomena in Numerical Models of the Atmosphere. *Proceedings of WMO-IUGG Symposium on Numerical Weather Forecasting held in Tokyo, Japan, on 26 November - 4 December 1986*, Technical Report of Japan Meteorological Agency No. 67, I.105-I.121.

Nappo, C. J., 1969, Parameterization of Surface Moisture and Evaporation Rate in a Planetary Boundary Layer Model. *Journal of Applied Meteorology*, **14**, 289-296.

Outcalt, S. J., 1972, The Development and Application of a Simple Digital Surface- Climate Simulator *Journal of Applied Meteorology*, **11**, 629-636.

Pinker, R. T., and Corio, L. A., 1984, Surface Radiation Budget from Satellites. *Monthly Weather Review*, **112**, 209-215.

Prandtl, 1942, l.: Strmungslehre. Verlag Vieweg and Sohn, Braunschweig, 369-399.

Pratt, D. A., and Ellyet, C. D., 1979, The Thermal Inertia Approach to Mapping of Soil Moisture and Geology. *Remote Sensing of Environment*, **8**, 151-168.

Price, J. C., 1977, Thermal Inertia Mapping: A New View of the Earth. *Journal of Geophysical Research*, **82**, 2582-2590.

Price, J. C., 1982, Estimation of Reginonal Scale Evapotraspiration Through Analysis of Satellite Thermal-Infrared Data. *IEEE Transactions on Geoscience and Remote Sensing*, **GE-20** 286-292.

Price, J. C., 1985, On the Analysis of Thermal Infrared Imagery: The Limited Utility of Apparent Thermal Inertia. *Remote Sensing of Environment*, **18**, 59-73.

Reginatio, R. J., Jackson, R. D., and Pinter, P. J., 1985, Evapotranspiration Calculated from Remote Multispectral and Ground Station Meteorological Data. *Remote Sensing of Environment*, **18**, 75-89.

Rosema, A., Bijeleveld, J. H., Reininger, P., Tassone, G., Blyth, K., and Gurney, R. J., 1978, "TELL-US" - A Combined Surface Temperature, Soil Miosture and Evaporation Mapping Approach. *Proceedings of the 12th International Symposium on Remote Sensing of the Environment, Environmental Research Institute of Michigan*, pp. 2267-2276.

Saltzman, B., and Pollack, J. A., 1977, Sensitivity of the Diurnal Surface Temperature Range to Changes in Physical Parameters. *Journal of Applied Meteorology*, **16**, 614-619.

Schumann, U., 1990, Large-eddy simulation of the upslope boundary layer. *Quarterly Journal of the Royal Meterological Society*, **116**, 637-670.

Schwalb, A., 1982, Modified Version of the TIROS N /NOAA A-G Satellite Series (NOAA E-J) - Advanced TIROS N (ATN). NOAA Technical Memorandum NESS 116.

Seguin, B., Lagouarde, J. P., and Kerr, Y., 1986, Estimation of Regional Evaporation using Midday Surface Temperature from Satellite Thermal IR Data. *Proceedings of the ISLSCP and Conference*, held in Rome in December 1985, ESA SP-248 Paris: European Space Agency), pp.339-344.

Sellers, P. J, Mintz, Y., Sud, Y. C., and Dalcher, A., 1986, A Simple Biosphere Model (SIB) for Use within General Circulation Models. *Journal of Atmospheric Sciences*, **43**, 505-531.

Siegel, R., and Howell, J. R., 1982, Thermal Radiation Heat Transfer (Tokyo, Kogakusha: McGraw-Hill).

Sievers, U., Forkel, R., and Zdunkowski, W., 1983, Transport Equations for Heat and Moisture in the Soil and their Application to Boundary Layer Problems. *Beitraege zur Physik and der Atmosphaere (Contributions to Atmospheric Physics)*, **56**, 58-83.

Strong, A. E., and McClain, P. E., 1984, Improved ocean surface temperature from space - comparisons with drifting buoys. *Bulletin of the American Meteorological Society*, **65**, 138-142.

Sutherland, R. A., 1985, Broadband and Spectral Emissivities (2-18 μm) of Some Natural Soils and Vegetation. *Journal of Atmospheric and Oceanic Technology*, **3**, 199-202.

Taconet, O., Bernard, R., and Vidal-Madjar, D., 1986, Évapotranspiration over an Agricultural Region Using a Surface Flux/Temperature Model Based on NOAA-AVHRR Data. *Journal of Climate and Applied Meteorology*, **23**, 284-307.

Thompson S. L., and Warren, S. G., 1982, Parameterization of Outgoing Infrared Radiation Derived from Detailed Radiative Calculations. *Journal of Atmospherical Sciences*, **39**, 2667-2680.

Townshend, J. R. G., and Tucker, C. J., 1984, Objective Assessmant of Advanced Very High Resolution Radiometer Data for Land Cover Mapping. *International Journal of Remote Sensing*, **5**, 497-504.

Watson, K., 1982, Geologic Application of Thermal Infrared Images. *Proceeding of the IEEE*, **63**, 128-137.

Wesselink, 1948, Heat Conductivity and Nature of the Lunar Surface Material. *Bulletin of the Astronomical Institutes of Netherland*, **10**, 351-363.

Wetzel, P. J., 1982, Toward Parameterization of the Stable Boundary Layer. *Journal of Applied Meteorology*, **21**, 7-13.

Wetzel, P. J., Atlas, D., and Woodward, R. H., 1983, Determining Soil Moisture from Geosynchronous Satellite Infrared Data. *Journal of Climate and Applied Meteorology*, **23**, 374-391.

Wetzel, P. J., Atlas, D., and Woodward, R. H., 1984, Determining Soil Moisture from Geosynchronous Satellite Infrared Data: A Feasibility Study. *Journal of Climate and Applied Meteorology*, **23**, 375-391,

Whiteman, C. D., and Dreiseitl, E. (Editors), 1984, Alpine Meteorology - Translations of Classic Contributions by A. Wagner, E. Ekhart and F. Defant. Pacific Northwest Lab., PNL-5141 ASCOT-84-3.

Yates, H. W., Tarpley, J. D., Schneider, S. R., Mcginnis, D. F., and Scofield, R. A., 1984, The Role of Meteorological Satellites in Agricultural Remote Sensing. *Remote Sensing of Environment*, **14**, 219-233.

Zdunkowski, W. G., Peagle, J., and Reilly, J. P., 1975, The Effekt of Soil Moisture upon Atmospheric and Soil Temperature Near the Air-Soil Interface. *Archives of Meteorological and Geophysical Bioclimatology*, **A**, **24**, 245-268.

Potential of Remote Sensing in the Thermal Band for Global Change

Marc Ph. Stoll

1 Global change scope and related activities

Global changes are the product of complex processes that occur within the Earth system and, as far as mankind is concerned, have an impact on the habitability of the Planet. Global change related activities therefore aim at understanding, modeling and monitoring the relations between Earth (as a whole), anthropogenic influence and external forcing (mainly solar input and solid Earth process). The main contribution to global change comes from the coupling between biogeochemical processes and the physical climate system : exchange of energy-mass-momentum between ocean/land and atmosphere, atmospheric chemistry and transformation of energy, for instance from radiative to non-radiative form (biosphere).

Some processes or interaction are global in nature (atmospheric circulation, atmospheric chemistry, ocean circulation, CO_2 - CH_4 - O_3 production, destruction and uptake,..); others are more regional (biomass production, water availability, etc.). The scales may be very different too (for instance surface deep ocean circulation).

Of great concern are biosphere terrestrial and marine-atmosphere interactions, biospheric aspects of the hydrological cycle, and effects of climate change on terrestrial ecosystems.

2 Documentation and study of global change

Although many kinds of observations, measurements and data records have to be used in a systematic effort to document significant global changes, remotely-sensed data, mainly from spacecraft, will constitute a major source of observations at the proper spatial/time scales. Instruments with high spatial resolution, such as SPOT or TM, are used for detailed studies and satellites of meteorological type can provide, with moderate resolution, frequent global coverage and time averages. Satellite observations are also necessary to make the spatial extension of local conventional data.

Since only a small number of physical quantities or parameters can be obtained directly from space, it is necessary to have a quality control throughout the acquisition

NATO ASI Series, Vol. I 24
Remote Sensing and Global Climate Change
Edited by Robin A. Vaughan and Arthur P. Cracknell

process to be sure that they can be used with confidence and to use pertinent indicators that trace the variation of the system under investigation. Multi-parameter analysis and time series can reveal relations between indicators and geophysical parameters that eventually define new variables based on remotely sensed data. Models should be developed in such a way that they explicitly incorporate physical quantities and variables obtainable from space. In order to test and validate remotely-sensed data and models, field measurements and experiments are still needed on various spatial scales.

3 Enhanced greenhouse effect

One central question in global change is the so-called enhanced greenhouse effect, that may, or may not, lead to an increase in the Earth's temperature, with consequences on the climate, precipitations, vegetation,..(Kasting 1989). The greenhouse effect occurs naturally and is due to the absorption of infrared radiation by atmospheric constituent gases, so that part of the radiation emitted by the Earth' surface remains in the system and is not re-radiated to space. Atmospheric gases responsible for the greenhouse effect are, in the first place, CO_2 and water vapour, but also trace gases CH_4 and N_2O.

Observations on the long term show a continuous increase in atmospheric CO_2 content (about 15% from 1960 to 1980) (Keeling 1986) and the rate of increase seems to be increasing. A similar behaviour, with an even more rapid increase, is observed for CH_4. Only part of this additional CO_2, more for CH_4, is of anthopogenic origin. More greenhouse gases mean more energy absorbed and more longwave radiation trapped in the lower atmosphere. A probable consequence of the enhanced greenhouse effect is an increase in atmospheric temperature and Earth's temperature with modifications in the atmospheric circulation precipitation regime and action on biosphere activity and biomes integrity. Due to possible feedbacks, it is not unlikely that, on the global scale, only a small temperature variation would be observed, but the expected changes would certainly, on a regional scale, depend strongly on geographical location.

4 Role of surface temperature in remote sensing

Surface temperature is a measure of the thermodynamic equilibrium, at a given time and place, of a system that undergoes energy exchange with its surroundings. Any modification in the state variables, amount of energy available or in partition between different forms of energy (radiative versus non-radiative) will result in a change in surface temperature, which is therefore an important variable in global change studies and a key indicator for the greenhouse effect.

With only a small expected change on the global scale, good accuracy in the measurement of surface temperature is required. This is not enough because surface temperature is known as a rapidly changing parameter, variable with time and place, so that one single local measurement is of little value. It is necessary to have access to its variability in order to average over a suitable area and time interval and monitor this average in the long term. It is worth saying that the very nature of this averaging is still a challenge.

As was already underlined, remote sensing is a mandatory tool for global change research, with low spatial resolution used to obtain global change indicators, monitor the greenhouse effect, space-borne or airborne instruments with higher spatial resolution being necessary to find relations between global change indicators and local physical quantities, modelling, scaling up and validation processes.

5 Remote sensing in the thermal band

The primary importance of remote sensing in the thermal band is that it gives access to temperature and other parameters of importance in radiative and thermal processes. The physical basis is Planck's law for blackbody radiance (Slater 1980)

$$B^0_\lambda(T) = C_1/\lambda^5[exp(C_2/\lambda T) - 1]^{-1} \quad \mathrm{Wcm^{-2}sr^{-1}\mu m^{-1}} \tag{1}$$

with $C_1 = 1.1904 \times 10^4$ W cm^{-2} sr$^{-1}\mu$m^4, $C_2 = 1.4388 \times 10^4\mu$m K and its corollary Wien's displacement law

$$\lambda_m T = 2898 \ \mu \text{ m K} \tag{2}$$

and the Stephan-Boltzman law for the total emittance $M = \int d\lambda \int B_\lambda(T) cos\theta d\omega$

$$M = \sigma T^4 \tag{3}$$

In (2a) λ_m is the wavelength of maximum emitted radiance at temperature T. For objects at temperatures around ordinary ambient temperature (300 K) this places emitted radiation in the 10 μm thermal infrared band (TIR).

Space remote sensing in the thermal band is possible thanks to the infrared atmospheric windows which are regions of high atmospheric transmission, the cental (8-13 μm) window and another (3-5 μm) one. Emitted radiation is negligible in the region of direct solar radiation. It is worth noting that, generally speaking, the emitted radiation is of the same order of magnitude as the reflected solar radiation (by land surfaces) in the 3-4 μm band. This will be used in a later section.

Several instruments are available for remote sensing, including in the thermal infrared (TIR), and brief details of these are given in table 1.

6 Operating remote sensing in the TIR band

The radiance signal at the satellite is built up from different processes and contributions of surface emission and atmospheric emission/absorption.

step " 0 ": spectral radiance emitted by the target

$$B_\lambda(\theta, \phi, T_s) = \varepsilon_\lambda(\theta, \phi) B^\circ_\lambda(T_s) \tag{4}$$

where $B^\circ_\lambda(T_s)$ is the Planck function at surface temperature T_s and $\varepsilon_\lambda(\theta, \phi)$ is the spectral emissivity of the target (by definition), which is material dependent.

sat / instrument	pixel size (km²)	TIR bands (μm)	repetitivity
Meteosat	5 x 5	1 10.5-12.5	48 images/day
Landsat-6 TM	0.12 x 0.12	1 10.5-12.5	16 days
NOAA AVHRR	1 x 1	3 3.5-3.9 10.5-11.3 11.5-12.5	2 to 4 images/day
ERS-1 ATSR	same as AVHRR two angles of view	high radiometric resolution/ accuracy	several days
Others planned in the future (for the EOS programme)			
ASTER	0.09 x 0.09	5 bands	
MODIS-N	0.85 x 0.85	several bands	

Table 1: Specifications of some sources of data

step " 1 " : spectral radiance at ground level

$$B_{g\lambda}(\theta,\phi,T_s) = \varepsilon_\lambda(\theta,\phi)B^0_\lambda(T_s) + (1-\varepsilon_\lambda)R_{\lambda,at\downarrow} \tag{5}$$

where the second term in the r.h.s. of (5) is an approximate, however reasonably accurate, expression for the reflection by the target of the atmospheric downward infrared radiation.

step " 2 " : spectral radiance at the satellite

$$B_{sat,\lambda}(\theta) = B_{g,\lambda}(\theta,\phi,Ts)\tau_\lambda(\theta) + R_{\lambda,at\uparrow}(\theta) \tag{6}$$

$\tau_\lambda(\theta)$ is the spectral atmospheric transmission along the path (from ground to space), and $R_{\lambda,at\uparrow}(\theta)$ is the upgoing atmospheric infrared emitted radiation in the direction of observation.

step " 3 " : radiation in channel (i)

$$R_i = R_{gi}\tau_i + R_{at\uparrow i} \tag{7}$$

X_i stands for the convolution of the spectral quantity X_λ and the spectral filter function for channel i, $F_i(\lambda)$

$$X_i = \int f_i(\lambda)X_\lambda d\lambda \tag{8}$$

It is essential to recall that in eq. (6) (i) there are two unknown physical surface parameters (ε_i,T_s); (ii) τ_i,$R_{at\uparrow i}$,$R_{at\downarrow i}$ depend on the atmospheric state, mainly temperature profile and water vapour content (and also other constituent gases CO_2, O_3)

Thus, in order to extract quantitative information from remote sensing in the TIR band, one has (i) to make atmospheric corrections in order to go from top-of-the-atmosphere data to ground-level data; (ii) to give accurate values to the channel emissivities ε_i in order to retrieve accurate values of surface temperature (better than 1% for 0.5 degK error). Emissivities are generally not known with good accuracy and need special methods to be obtained. When this is possible, information on the type of surface can be obtained from the spectral variation.

Note that equations (3) to (6) are often written in term of brightness temperature T_B instead of radiance. The definition of T_B is related to (3) :

$$\varepsilon_\lambda B^\circ_\lambda(T_s) = B^\circ_\lambda(T_B) \tag{9}$$

The brightness temperature in channel i, T_i, is thus equivalent to using R_i :

$$R_i = \int f_i(\lambda) B_\lambda d\lambda = \int f_i(\lambda) B^\circ_\lambda(T_i) d\lambda \approx \Delta\lambda_{eff} B^{irc}_{\lambda eff}(T_i) \tag{10}$$

where $\Delta\lambda_{eff}$ is the effective bandwidth of channel i and λ_{eff} the effective central wavelength of this channel.

There are several important point in TIR radiometry that need to be underlined: (i) most terrestrial materials are highly absorbing in the TIR, with the consequence that the active thickness of the material is effectively thin (often no more than a few wevelengths), and the surface temperature is but a skin temperature; (ii) land surfaces are, in most cases, heterogeneous at all scales, both in type (i.e. different emissivities) and temperature. There is no unique way to define an equivalent homogeneous medium with equivalent emissivity and equivalent temperature T_{eq}. One way to do this is to use Kirchoff's law for non-transparent media, ε = 1 - reflectivity, to define an equivalent emissivity from the measurement of the reflectivity of the heterogeneous surface. (iii) for vegetation-covered surfaces, the very definition of surface temperature is difficult. As far as remote sensing is concerned, an attractive solution is to consider radiative temperature (i.e. the temperature associated with emission of radiation measured by means of a radiometer) in a given direction and a given field of view and to use the Stephan-Boltzmann law to define surface temperature as $\int d\lambda \int B_\lambda(\theta) cos\theta d\omega = \varepsilon_{veg} \sigma T s^4$. Clearly, one needs a model to relate the surface temperature to temperature of the elements that constitute the vegetation cover.

7 Methods for surface temperature retrieval from satellites

7.1 Sea surface temperature (SST)

In this case, the problem is somewhat simplified because the emissivity of water is very constant and close to one in the TIR band (Masuda *et al.* 1988), and the temperature of a water surface is rather homogeneous and not very diffecrent from the air temperature in the lower atmospheric layer. Hence, a good value for SST is likely to be obtained, provided that the instrument is well calibrated and the atmospheric corrections are made accurately.

There are two types of situation, regarding the corrections of atmospheric effects: a) the actual state of the atmosphere is known (i.e. profiles of temperature, pressure, humidity), either from conventional radiosounding or from real-time satellite soundings using, for instance, the TOVS (TIROS Operational Vertical Sounder) system on board the NOAA platforms. Using atmospheric codes, such as LOWTRAN (Kneizys *et al.* 1983) or 3I (Chedin *et al.* 1985) or 4A (Scott and Chedin 1981), one then calculates the atmospheric quantities of interest , $R_{at\uparrow}$ and τ in a given channel, that are introduced in the radiative transfer equation (6) to give the surface brightness temperature practically equal to the SST (because the emissivity is taken equal to one). b) in case the actual atmospheric state is not known, only approximate methods can be used. A rather accurate, and very popular one, is the split-window method, usually applied with the channels 4 and 5 of the AVHRR (the instrument was designed for applying the split-window method). The split-window method is based on the difference in absorption coefficients (mostly water vapour) in two spectrally adjacent thermal infrared bands, so that the difference in brightness temperatures at the satellite, T_4-T_5, is effectively a measure of the atmospheric total water content. It can be shown theoretically (Anding and Kauth 1970, Prabhakara *et al.* 1974, Deschamps and Phulpin 1980) that, within the assumption that the different temperatures coming into play are not very different, the radiative transfer equation can be linearised and the sea surface temperature given as a linear combination of the two satellite brightness temperatures:

$$SST = T_4 + a(T_4 - T_5) + b \tag{11}$$

a and b (an offset term) can be given theoretical values (Price 1984), although they are better obtained as best fit coefficients from simulations (a large number of atmospheric situations determined by means of a data base of radiosoundings and an atmospheric code) or *in situ* data, if they are available in quantity and quality (McMillin 1975, Sidran 1980, Singh 1984, McClain *et al.* 1985). The new instrument ATSR (Along Track Scanning Radiometer) on board ERS-1 was specifically designed for accurate sea surface temperature measurement by the split window technique with the addition of two-angle viewing, nadir and forward, to improve the atmospheric corrections, and high radiometric accuracy (Delderfield *et al.* 1986). The rms error on sea surface temperature retrieval should meet the 0.3 degK requirement for ocean studies.

7.2 Land surface temperature (LST)

The situation is more complex, as opposed to SST, because land emissivities are variable with wavelength, sometimes significantly less than one (0.9 to 0.98) (Hovis *et al.* 1968, Nerry *et al.* 1988, 1990), LST is inhomogeneous, with high dynamical range (by daytime ΔT_s 30 degK), and LST can be very different from the near surface air temperature (by daytime). Regarding the correction of atmospheric effects, the situation is the same as for SST, whether the actual real time atmospheric profiles are known or not. However, from the physical point of view, the appropriate approach depends on whether the channel emissivities are known or not (Becker 1987, Wan and Dozier 1989).

a) ε_i are unknown: the channel emissivities can be set to one (in a particular channel) or given a *priori* values from the knowlege of the type of surface and field measurements data base. Then the radiative transfer equation is solved with real-time profiles in a particular channel, or an approximate method is applied. Although the conditions described above are not in favour of the split window approach, it can be shown, both theoretically and from simulations, that a split window method is still applicable for LST retrieval, provided that the coefficients a and b of Eq. (10) are now written as functions of average emissivity, $\varepsilon_{av} = (\varepsilon_4 + \varepsilon_5)/2$, emissivity difference, $\Delta\varepsilon = \varepsilon_4 - \varepsilon_5$, and integrated water content IWC. The split window is now a local one (Becker and Li 1990a) since the coefficients are pixel dependent.

b) physical decorrelaton ε/Ts : as was emphasised in sections 5 and 6, the radiometric equation in the TIR has no exact solution with only measurements of emitted radiation, because of the coupling between temperature and emissivity. There are methods for ε/T decorrelation that rely on multi-channel data and additional iterative (Susskind *et al.* 1984) or statistical methods. However, to decorrelate the emissivity and temperature physically one needs an independent measurement related to the surface emissivity. This is possible, from satellites, thanks to a method developed by Li and Becker (1992) for use with the AVHRR/2 instrument. The method is based on the radiance in channel 3 being, by daytime, the sum of the emitted radiance and the reflected solar radiance, and the TISI (Temperature Independent Spectral Indices) in the TIR (Becker and Li 1990b, Li and Becker 1990b) concept. (The TISI are in fact combinations of channel emissivities and can be obtained from satellite brightness temperatures). Using consecutive daytime and night-time images of the same area, and with the assuptions that the channel emissivities are unchanged between day and night and that the surface behaves approximately as a Lambertian surface (no anisotropy), the $TISI_{345}$ (combination of $\varepsilon_3, \varepsilon_4$, and ε_5) are obtained from the night-time image and the emissivity in channel 3 from the reflected solar radiation (Kirchoff's law) by daytime. The emissivities in channels 4 and 5 are then extracted from the TISI. The importance of the method is that it gives access to the channels' emissivities directly from satellite data at the scale of the pixel. LST can then be obtained with the local split window where the previously retrieved spectral parameters have been introduced. The LST obtained in this way should be the true thermodynamic surface temperature. The counterpart of the method is that it requires two consecutive images, pixel by pixel superposition of images, and relies on assumptions that are difficult to evaluate.

7.3 Conclusion on surface temperature retrieval

Remote sensing in the TIR band can provide access to surface temperature. For SST determination, the split window method is quite efficient and, although the rms eror is of the order of 0.5 degK; this could be improved to meet the 0.3 degK requirement for global change studies. For LST determination, the situation is much more difficult and uncertain, due to the physics of radiation by land surfcaces. No reasonably significant estimate of the uncertainty on LST can be given (perhaps it is a few degrees) and there is still much to do to meet the 1 degK error required for land surface studies. In association with LST determination, spectral information (at least relative information) on the surface can be extracted. This information is most important and reliable with high spatial resolution studies.

8 Role of remotely sensed surface temperature in global change studies

8.1 Monitoring the biosphere

SST is most important in oceanography and in global climate studies. Over continental areas LST is most important for monitoring the biosphere. As was already mentioned, the enhanced greenhouse effect induces climate modifications that, in turn, result in more or less Sun radiation (notably photosynthetically active radiation, PAR), precipitation (i.e. moisture available) and air temperature. All this has an action on the vegetation or biomes. To study and monitor the greenhouse effect, it is thus mandatory to monitor the vegetation state and behaviour (Townshend and Justice 1986). Two major parameters are surface temperature and the vegetation index (for instance NDVI from AVHRR channels 1 and 2) (Choudhury 1987, Becker and Choudhury 1988). Surface temperature is a good indicator of the energy exchange between vegetation (soil) and atmosphere and of moisture availabilty (both in soil/vegetation and in lower atmospheric layer), whereas vegetation index is an indicator of the vegetation state and ability to transform energy from radiative to non-radiative form (PAR). A vegetation cover that functions well should develop a lower surface temperature due to evapotranspiration, inasmuch as the atmospheric moisture allows enough evaporation.

8.2 What can be done with surface temperature ?

a) study homogeneity and type of surface.

Most continental surfaces are heterogeneous. For studying land surface processes it is important to consider separately the different components and thus to identify the types of surface cover. Surface temperature can be used for this purpose through its value and areal distribution. A homogeneous vegetation cover has a low surface temperature (slightly below air temperature) and narrow histogram, whereas a bare soil area exhibits a majority of higher surface temperature (by daytime), often large dynamical range and strong time variation. Mixed areas are revealed as widely spread histograms.

b) obtain surface energy fluxes.

Surface temperature is a measure of the thermodynamical equilibrium at the surface. Its knowlege is needed for flux determination, although surface temperature alone is not enough; meteorological and surface data are required at the same time. The net radiative flux at the surface, RN, is given by

$$RN = (1-a)RS + Ra_{\downarrow} - R_{T\uparrow} \tag{12}$$

where a is the surface Sun albedo, RS the global solar radiation at ground level, $Ra_{\downarrow} \cong \varepsilon_{air}\sigma T_a^4$ and $RT \cong \varepsilon\sigma T_s^4$. The balance equation is

$$RN = H + LE + G \tag{13}$$

where H is the convective flux, LE is the latent heat or evaporation flux, and G is the ground geophysical flux. It is enough to recall here that $H \sim (T_a - Ts)$, $LE \sim (e_a - e_s)$, the difference in specific humidity of air and surface cover, where $e_s(T_s)$, and G is proportional to the ground temperature gradient at the surface. A simplified relation has been developed (Seguin and Itier 1983) to estimate the latent heat flux (daily average)

$$LE = RN - \alpha(T_s - T_a) + \beta \tag{14}$$

where T_s is the maximum (daytime) surface temperature of the vegetation cover, T_a is the meteorological air temperature and α and β are coefficients that depend on various biometeorological parameters.

c) study correlations between remotely sensed parameters

Although, generally speaking, correlations between physical parameters should not be confused with modelling, this type of investigation is important in remote sensing. The reason is that the transition from local small scale physical variables or parameters to large scale parameters is still poorly understood, whereas remote sensing, particularly with meteorological satellites, produces large scale integrated parameters.

The analysis of multi-parameter space-derived quantities is one way to improve the understanding or characterisation of biosphere behaviour, provided that the correlation study is performed with physically independent parameters, most frequently derived in different spectral domains. One type of correlation that is now currently beeing investigated is T_s versus NDVI, introducing a coupling between short wave (visible-near infrared) remote sensing and TIR remote sensing . Ideally, a plot of T_s versus NDVI is linear with a negative slope. This is because low surface temperature values are associated with dense, healthy, strongly evaporating vegetation cover, whereas high surface temperatures are associated with stressed vegetation or soils. Extrapolating this linear dependence to small NDVI values (appropriate to bare soil) can serve to estimate the soil surface temperature, and, on the other side, NDVI values appropriate to full cover determine an estimate of the canopy-only surface temperature. In turn, this parameter can be used to estimate $e_s(T_s)$ and then the evaporation flux LE. It is worth noting that the ratio of radiative energy RN to evaporation flux is a key indicator of a system's development. It has been suggested (Nemani and Running 1989) that the slope of the T_s versus NDVI diagram is related to the antecedent moisture available. This is an illustration of a new macroscopic variable measured directly from space without reference to detailed local quantities.

9 Conclusion

The objective of this chapter was to underline the role of remote esensing in global change studies, with particular reference to remote sensing in the TIR band, giving access to the Earth's surface temperature and specific surface material quantities (emissivity and reflectivity). For ocean and global climate studies, SST determination is relatively easy and accurate, although improvements are still needed, but seem at hand, to meet less than 0.3 degK error in a production environment. For continental surfaces, LST is much more difficult to obtain and much more uncertain. It will be necessary to document and study the role of spectral surface quantities, anisotropy effects and try to estimate error bars for LST retrieval from space. In both cases, SST and LST, it is important to develop and improve real time access to atmospheric profiles.

Remote sensing in the TIR band, through surface temperature, was shown to be a tool to monitor and study the enhanced greenhouse effect and the effect of climate change on the biosphere. In the framework of global change research that implies systematic large scale investigations over long periods of time, the following objectives should be adressed:
- search for an operational algorithm for LST retrieval from space (ultimate accuracy is not the primary requirement);
- study systematically and interpret relations between independent space-derived parameters, such as T_s versus NDVI;
- use the complementarity between the three spectral domains of remote sensing: Visible/near infrared (solar energy input and uptake), TIR (thermodynamics-energy exchanges), microwave (moisture-roughness);
- search for large-scale, space-derived, new global change variables to be introduced in soil-biosphere-atmosphere models.

Acknowledgements

The author is indebt to Professor F. Becker and Dr. Z.L. Li for frequent and fruitful discussions.

References

Anding, D., and Kauth. R., 1970, Estimation of sea surface temperature from space. *Remote Sensing of Environment*, **14**, 217-220.

Becker, F., 1987, The impact of spectral emissivity on the measurement of land surface temperature from a satellite. *International Journal of Remote Sensing*, **10**, 1509-1522.

Becker, F., and Raffy, M., 1987, Problems related to the determination of land surface parameters and fluxes over heterogeneous media from satellite data. *Advances in Space Research*, **7**, 45-57.

Becker, F., and Choudhury, B. J., 1988, Relative sensitive of Normalized Difference Vegetation Index (NDVI) and Microwave Polarization Difference Index (MPDI) for vegetation and desertification monitoring. *Remote Sensing of Environment*, **24**, 297-311.

Becker, F., and Li, Z. L., 1990a, Towards a local split window method over land surface. *International Journal of Remote Sensing*, **3**, 369-393.

Becker, F., and Li, Z. L., 1990b, Temperature Independent Indices in Thermal Infrare. *Remote Sensing of Environment*, **32**, 17-33.

Chedin, A., Scott, N. A., Wahiche, C., and Moulinier, P., 1985, The improved initialisation inversion method: a high resolution physical method for temperature retrivals from the Tiros-N series. *Journal of Climate and Applied Meteorology*, **24**, 124-143.

Choudhury, B. J., 1987, Relationships between vegetation indices, radiation absorption and net photosynthesis evaluated by a sensitivity analysis. *Remote Sensing of Environment*, **22**, 209-233.

Delderfield, J., Llewllyn-Jones, D. T., Bernard, R., de Javel, Y., Williamson, E. J., Mason, I., Pick, D. R., and Barton, I. J., 1986, The along track scanning radiometer (ATSR) for ERS-1. In *Instrumentation for optical remote sensing from space*, edited by J. S. Seeley, J. W. Lear, A. Monfils, and S. L. Russak, Proceedings Vol 589, pp. 114-120.

Deschamps, P. Y., and Phulpin, T., 1980, Atmospheric correction of infrared measurements of sea surface temperature using channels at 3.7, 11, and 12mm. *Boundary - Layer Meteorology*, **18**, 131-143.

Hovis, W. A., Blaine, L. R. and Callaghan, W. R., 1968, Infrared aircraft spectra over desert terrain 8;5-16 μm. *Applied Optics*, **7**, 1137-1140.

Kasting, J. F., 1989, Long term stability of the Earth's climate. *Global and planetary change*, **1**, 83-95.

Keeling, C. D., 1986, Atmospheric CO_2 concentrations-Mauna Loa Observatory, Hawaii 1958-1986. NDP-001/R1. Carbon Dioxide Information Analysis Center. Oak Ridge National Laboratory. Oak Ridge, Tennessee.

Kneizys, F. X., Shettle, E. P., Gallery, W. O., Chetwynds, J. H. Jr., Abreu, L. W., Selby, J. E. A., Clough, S. A., and Fenn, R. W., 1983, Atmospheric Transmittance Radiance: Computer Code Lowtran 6. Optical physics Division, U.S. Air Force Geophysics Laboratory, Hanscom, AFB, MA.

Li, Z. L., and Becker, F., 1990, Properties and Comparison of Temperature-Independent Thermal Infrared Spectral Indices with NDVI for HAPEX data. *Remote Sensing Environment*, **33**, 165-182.

Li, Z. L., and Becker F., 1992, Feasibility of land surface temperature and emissivity determination from AVHRR data. *Remote Sensing of Environment*, **33**, 165-182.

McClain, E. P., Pichel, W. G., and Walton, C. C., 1985, Comparative performance of Avhrr-Based multichannel sea surface temperatures. *Journal of Geophysical Research*, **20**, 11587-11601.

Masuda, K., Takashima, T., and Takayama, Y., 1988, emissivity of pure and sea waters for the model sea surface in the infrared window regions. *Remote Sensing of Environment*, **24**, 313-329.

Mcmillin, L. M., 1975, Estimation of sea surface temperature from two infrared window measurements with different absorption. *Journal of Geophysical Research*, **80**, 5113-5117.

Nemani, R. R., and Running, S. W., 1989, Estimation of surface resistance to evapotranspiration from NDVI and thermal-IR AVHRR data. *Journal of Climate and Applied Meteorology*, **28**, 276-294.

Nerry, F., Labed, J., and Stoll, M. P., 1988, Essai de dätermination absolue des signatures spectrales de sols nus dans l'infrarouge thermique, en ämission and en räflexion. (see original page). *Proceedings of the 4th international colloquium on spectral signatures of objects in remote sensing held at Aussois, France, on 18-22 January 1988*, ESA SP-287, (Noordwijk: European Space Agency), pp. 185-188.

Nerry, F., Labed, J., and Stoll, M. P., 1990, Spectral properties of land surfaces in the thermal infrared band. Part I : Laboratory measurements of absolute spectral emissivity and reflectivity signatures. *Journal of Geophysical Research*, **95**, 7027-7044.

Prabhakara, C., Dalu, G., and Kunde, V., G., 1974, Estimation of sea surface temperature from remote sensing in the 11 and 13mm window region. *Journal of Geophysical Research*, **79**, 5039-5044.

Price, J. C., 1984, Land Surface Temperature measurements from the Split Window Channels of the NOAA 7 Advanced Very High Resolution Radiometer. *Journal of Geophysical Research*, **89**, 7231-7237.

Scott, N. A., and Chedin, A., 1981, A fast line by line method for atmospheric absorption computations: The Automatized Atmospheric Absorption Atlas. *Journal of Climate and Applied Meteorology*, **20**, 801-812.

Seguin, B., and Itier, B., 1983, Using midday surface temperature to estimate daily evapotraspiration from satellite thermal IR data. *International Journal of Remote Sensing*, **4**, 371-383.

Sidran, M., 1980, Infrared sensing of sea surface temperature from space. *Remote Sensing of Environment*, **10**, 101-114.

Singh, S. M., 1984, Removal of atmospheric effects on a pixel by pixel basis from the thermal infrared data from instruments on satellites AVHRR. *International Journal of remote Sensing*, **5**, 161-183.

Slater, P. N., 1980, *Remote sensing, optics and optical system* (Reading, Addison-Wesley).

Susskind, J., Rosenfield, J., Renter, D., and Chahine, M. T., 1984, Remote sensing of weather and climate parameters from HIRS2/MSU on TIROS-N. *Journal of Geophysical Research*, **89**, 4677-4697.

Townshend, J. R. G., and Justice, C. J., 1986, Analysis of the dynamics of African vegetation using the normalized difference vegetation index. *International Journal of Remote Sensing*, **7**, 1435-1445.

Wan, Z., and Dozier, J., 1989, Land surface temperature measurement from space: physical principles and inverse modeling. *IEEE Transaction on Geoscience and Remote Sensing*, **27**, 268-278.

Whitehead, V. S., Johnson, W. R., and Boatright, J. A., (1986), Vegetation Assessment Using a Combination of Visible, Near-IR, and Thermal-IR AVHRR data. *IEEE Transactions on Geoscience and Remote Sensing*, **GE-24**, 107-112.

Comparison Between Albedo Derived from Satellite with Different Calibration Coefficients

S. Zientek

International programs like IGBP (the International Geosphere-Biosphere Program) use NOAA satellite data to calculate the (normalised difference vegetation index) to monitor the growth of arid or semi-arid regions of planet Earth. Therefore well calibrated AVHRR data of channel- 1 and -2 are needed with errors less than 5%. The aim of the work is to show the impact of different calibration coefficients on the calculation of Albedo from channel-1 and -2 NOAA-11 AVHRR data. In addition, the sensor degradation was monitored and a radiative transfer model was applied to remove atmospheric effects.

With the launch of NOAA-11 in late 1988 pre-flight calibration coefficients for channel-1 and -2 NOAA-11 AVHRR were supplied. In July 1991 NOAA NESDIS published the report "NOAA Polar Orbiter Data User Guide" (Kidwell 1991) where pre-flight calibration coefficients for NOAA-11 AVHRR channels 1 and 2 were published which are considerably different from the 1988 released calibration coefficients (Table 1).

$$CA_j = G_j \cdot COUNT_j + I_j$$

where CA_i = calibrated data, G_j = gain, $COUNT = AVHRR$ data, I_j = intercept and j = channel.

	Gain 'G'	Intercept 'I'	
Channel 1	0.0906	-3.73	1988 published Pre-flight Calibration Coeff.
Channel 2	0.0827	-3.39	
Channel 1	0.095	-3.8	1991 published Pre-flight Calibration Coeff.
Channel 2	0.1061	-3.6	

Table 1: Comparison of Calibration Coefficients and Calibration Equation

NATO ASI Series, Vol. I 24
Remote Sensing and Global Climate Change
Edited by Robin A. Vaughan and Arthur P. Cracknell

Orbit Number	COS(SZA)	Technical Albedo 'TA' CHANNEL 1_{88}	CHANNEL 1_{91}	CHANNEL 2_{88}	CHANNEL 2_{91}
TEST SITE DESERT					
4359	0.928	0.330	0.344	0.315	0.394
8351	0.891	0.327	0.350	0.301	0.412
14419	0.871	0.333	0.350	0.315	0.412
TEST SITE SEA					
4359	0.904	0.050	0.054	0.032	0.049
8351	0.87	0.044	0.048	0.026	0.042
14419	0.864	0.032	0.035	0.015	0.027

Table 2: Comparison of calibrated and normalised data. Subscript 88: Calibration with 1988 published coefficients Subscript 91: Calibration with 1991 published coefficients

and j = channel.

$$TA_j = \frac{CA_j}{cos(SZA)}$$

where TA_j = technical albedo, CA_i = calibrated data, and cos(SZA) = cosine of Sun zenith angle.

To compare the different calibrations two homogenous test sites were selected: desert (Algerian Erg) and sea (Mediterranean Sea). Each orbit was carefully examined on cloud-free conditions and the test sites were close or within the nadir track of NOAA-11. 10 pixels × 10 pixels were extracted for each test site for channel 1 and channel 2 data, calibrated with the two sets of pre-flight calibration coefficients and normalised with the cosine of the Sun zenith angle. The comparison of the two calibrated and normalised data sets show that channel 2 is approximately 25 % underestimated by the 1988 released calibration coefficients and channel 1 by about 4% (Table 2).

Furthermore the reflectance of desert surfaces for spectral ranges similar to those of channel 1 and 2 show the characteristic that channel 1 < channel 2 (Henderson-Sellers and Wilson 1983) this is reversed by applying the 1988 pre-flight calibration coefficients. However the 1991 published pre-flight calibration coefficients reproduce the known reflectance of desert surfaces and sea surfaces. Therefore the 1991 published calibration coefficients should be applied to calibrate channel-1 and -2 NOAA-11 AVHRR data. In a next step the sensor degradation was monitored. 24 orbits between 1 January 1989 and 31 December 1991 were chosen. Each orbit was carefully examined on cloud-free conditions and the test sites were close or within the nadir track of NOAA-11. 10 pixls × 10 pixels were extracted for each test site, calibrated and normalised. The resulting data were plotted against time. The result shows no sensor degradation for the given time (Figure 1).

The final step was to apply a radiation transfer model to remove atmospheric effects and to upset the relation between the surface albedo and the albedo at the top of the atmosphere for each orbit respectively (Figure 2).

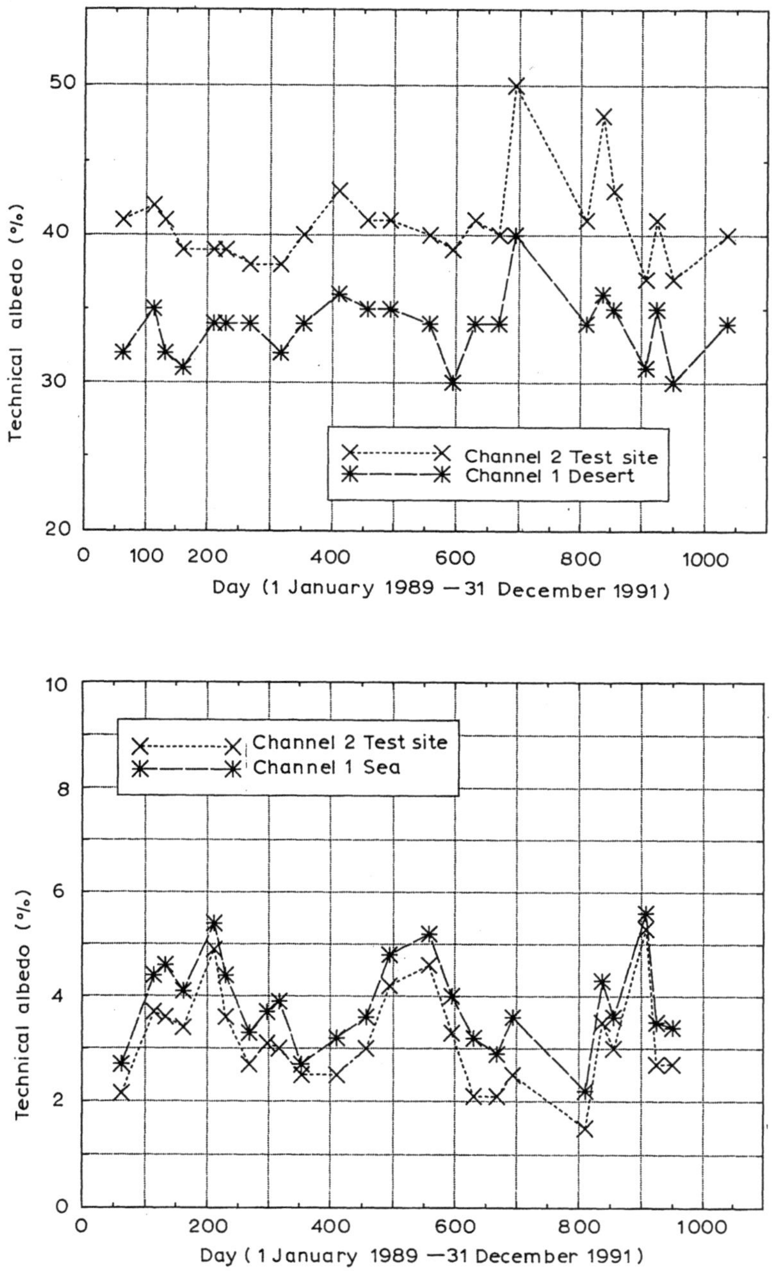

Figure 1: Monitoring sensor degradation. Calibrated and normalized data for test site: desert (upper) and sea (lower)

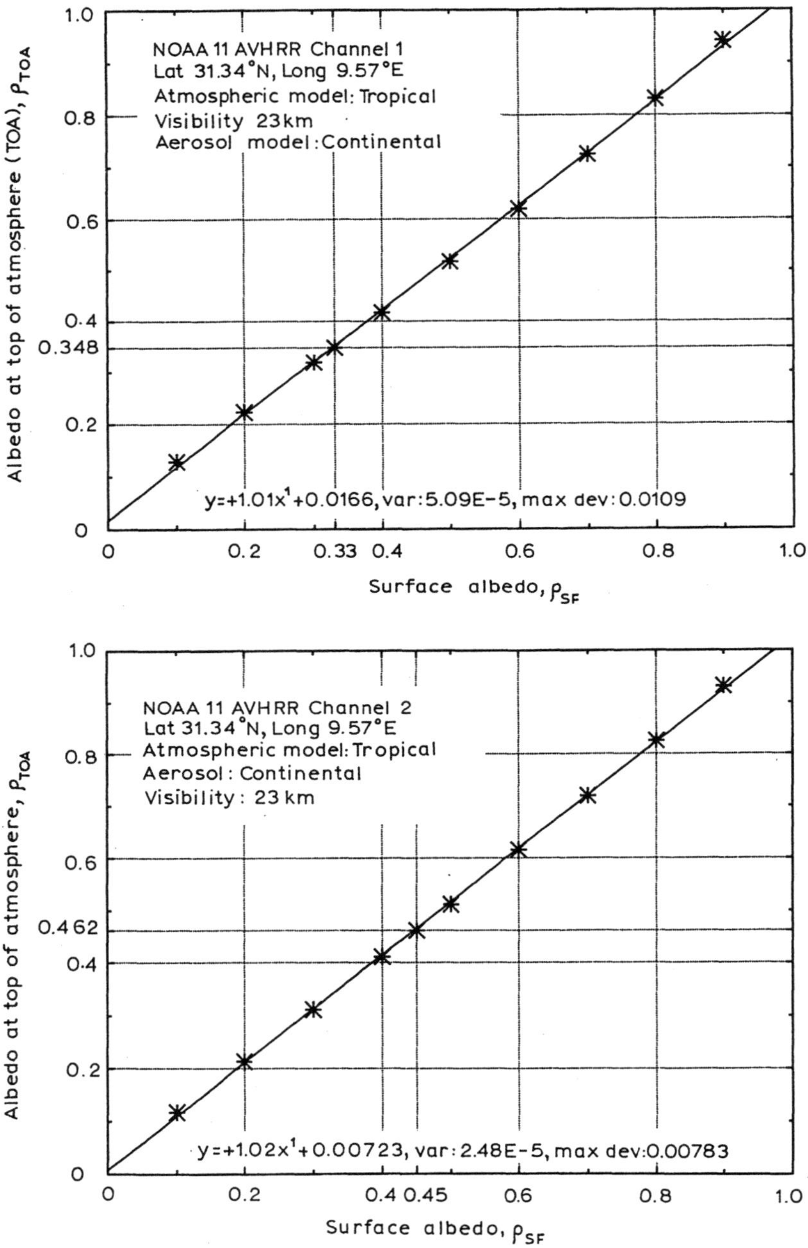

Figure 2: Applying radiation transfer model to remove atmospheric effects.
The model S5 was applied. First, the surface albedo was simulated and secondly the relation between the albedo at the top of the atmosphere and the surface albedo was established. Upper, channel 1 and lower, channel 2.

References

Henderson-Sellers, A., and Wilson, M. F., 1983, Surface Albedo Data for Climate Modeling. *Review of Geophysics and Space Physics*, **21**, 1743-1778.

Holben, B. N., Kanfman, Y. J., and Kendall, J. D., 1990, NOAA-11 AVHRR visible and near-IR inflight calibration. *International Journal Remote Sensing*, **11**, 1511-1519.

Kidwell, K., 1991, NOAA Polar Orbiter Data User Guide, NOAA / NESDIS.

Tanré D., Deroo, C., Buhant, P., Herman, M., Morcrette, J. J., Perbos, J., and Deschamps, P. Y., 1986, *Simulation of the Satellite Signal in the Solar Spectrum* (Villeneuve d'Asq: Laboratoire d'Optique Atmosphérique, Université des Sciences et Techiques de Lille).

Some Notes on the Oceanic Aspect of the Remote Sensing and Global Climate Change Issue

Serge V. Victorov

1 Introduction

Various previous chapters have been concerned with several aspects of the oceanic factor in global climate change and it is not intended to repeat that material in this chapter. This chapter, therefore, will concentrate on practical applications of satellite data to marine science in the general context of global climate change.

2 World Ocean and Climate Change

2.1 International conference on ocean monitoring for global climate change (Tokyo, 24 February - 2 March 1992)

Recently, the climate issue has become one of the vital concerns for all human beings. Comprehensive scientific efforts for better understanding and reliable prediction of the climate system have been urged by the world's scientific communities. The ocean system plays one of the essential roles in climate change. In this regard, relevant international organisations have already initiated operational and research programmes for ocean monitoring, such as:

- Tropical Ocean and Global Atmosphere (TOGA)
- World Ocean Circulation Experiment (WOCE)
- Integrated Global Ocean Services System (IGOSS)

as well as individual national activities.

The conference on global ocean monitoring was organised to provide participants with an informal but comprehensive forum to exchange information on:

NATO ASI Series, Vol. I 24
Remote Sensing and Global Climate Change
Edited by Robin A. Vaughan and Arthur P. Cracknell

- the present status and future plans of global ocean monitoring
- the role of the oceans in global climate change
- the development of the Global Ocean Observing System (GOOS).

The Global Ocean Observing System (GOOS) will be built on present systems. It is not a new programme but rather an integrating process of using data obtained in other projects relevant to world ocean research. Other programmes to be used are:

- Global Sea Level Observing System (GLOSS)
- International Oceanographic Data and Information Exchange (IODIE)
- Marine Pollution Monitoring System (MARPOL)

and others.

Two interesting features of the GOOS programme are worth mentioning. Using US Navy Geosat Altimeter data, sea level variability can be retrieved with an error of about 5 cm nowadays, while on 10 August 1992 the TOPEX/POSEIDON satellite was launched (which promises better accuracy in future).

The passive Special Sensor Microwave/Imager (SSM/I) on board the US Defense Meteorological Satellites is the most important source of data on surface wind speed and water vapour content (and precipitation). Three SSM/Is have been launched and the four DMSP will be launched before the end of the 1990s. Five SSM/Is are planned. Therefore SSM/I observations will be continued for over 20 years. This is an impressive example of a climate change-oriented satellite sensor.

2.2 Ocean variability.

One should consider the variability of oceanographic parameters on diurnal, synoptic, seasonal, annual scales (Morel 1991). (World-wide sea surface temperature records are only available since the beginning of the century.)

The physical quantities which characterise the interaction between the atmosphere, the global ocean and ice, the land surface hydrology and vegetation, are interface fluxes of momentum, radiation, heat and water.

The diurnal cycling of solar radiation reaches 1000 Wm^{-2} in the tropics.

The hemispherical-mean seasonal variations of the net energy flux across the air-sea interface is 50 Wm^{-2}.

The flux anomaly associated with a large El Niño event may be as large as 100 Wm^{-2} over an oceanic domain of several million square kilometres during several months.

In contrast, the radiative energy flux increment associated with the change in the concentration of greenhouse gases is so far only about 2 Wm^{-2}.

The most important oceanographic observations from space are

- sea surface topography to study upper-surface currents

- wind stress (friction)
- ocean surface temperature with a meaningful precision of 0.2-0.3 deg K
- ocean surface colour
- extent of sea-ice.

2.3 Some notes on modelling

T Nagai at the conference in Tokyo in 1992 (see section 2.1) reported on a coupled model to study atmospheric and oceanic variability in the Pacific for a long term (30 years) time integration experiment. It consists of 2 components:

- the atmospheric component is a global circulation model (GCM) with 5 levels in the vertical and $4°lat \times 5°long$ grids in the horizontal.
- the oceanic component is a GCM for the Pacific with 19 levels in the vertical and $1° \times 2.5°$ grids in the horizontal.

The model includes seasonal varying solar radiation as forcing. E1 Niño events can be seen. It takes 700 hours of super-computer processing for the 30-year run.

Now let us ask ourselves: are there many institutions which can afford running numerical modelling activities of this kind? Are there many countries which can afford participating in full-scale global change data collecting, processing and analysis? The answers to these two questions lead us to the necessity of thinking over the plausible ways of participating in the global change research for the bulk of the countries.

Figure 1 helps to illustrate the complex interference between global climate change, local and regional environment, species (including human beings) and local (regional) ecosystems. We could also apply change to all the constituent parts, thus emphasising the supercompexity of what we dare to study. Speaking about change we should also talk into account the natural variability of climate, environment, species (and thus ecosystems)... we cannot say that we know much about it on a long term scale.

What are we actually monitoring? The state of ecosystems? The state of the local environment? Or globe climate change? Is it possible to separate them?

It seem to me that, bearing in mind the limited amount of manpower and other resources, the plausible way out for everyone is monitoring of the regional environment (in the broad meaning of the terms "monitoring" and "environment"). I mean that collecting, storing and analysing the time series of information on the state of the regional environment is a natural task of each country or geographical cluster of countries.

The data collected may also be used for global climate research, but for each country the highest priority will always be using these data for regional planning, development of industry and agriculture.

Some aspects of the regional and global approaches are given below for satellite oceanography, with the emphasis on regional satellite oceanography.

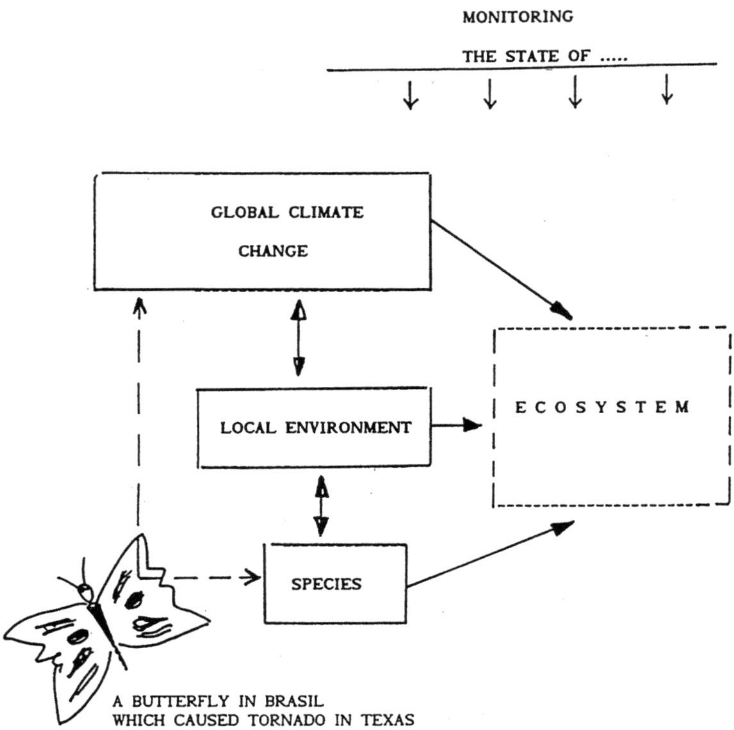

Figure 1: Interrelation between global climate change and the local and regional environment.

3 Global and Regional Satellite Oceanography

There are many differencies between regional satellite oceanography and global satellite oceanography. The two differ in:

- objects and objectives of studies
- requirements for satellite data (spatial, temporal) which lead to different modes of payload operation
- *insitu* availability which leads to much better accuracy of information retrieval in regional satellite oceanography
- algorithms of data processing
- users etc...

(for details see Victorov 1988)

Parameter	problems	
	General	Local
Dimensions of studied area (km)	50 – 1,000	1 – 50
Ground resolution (m)	100 – 1,000	20 – 100
Temporal resolution (h)	3 – 24	3 – 6

Table 1: Some requirements for satellite data in problems of regional satellite oceanography

Let us note some points in regard to users. Satellite oceanography being just a tool, one can often judge the real skill of conventional oceanographers when they are given satellite information. The inability to use the satellite data, the lack of skill to assimilate them with conventional knowledge, the lack of numerical models with fine enough spatial resolution to fit the existing satellite data, the fear of the large amount of data in digital satellite imagery (as compared to the routine amount of data collected by research vessels), all these factors often lead to either ignoring the satellite information by conventional oceanographers or to another phenomenon. Namely, they put forward such requirements for satellite data that cannot be met at all or during the forthcoming decades! Hence mutual understanding and close cooperation between the remote sensing community and conventional oceanographers is a necessity.

3.1 Sensors for regional satellite oceanography

The instruments available are listed in Table 2 along with some of their parameters.

3.2 General strategy of regional satellite oceanography

1. The preliminary stage consists of

- Analyses of general problems of a certain sea (or part of the sea)
- Selection of the most important problems
- Priority nomination
- Development of a strategy of studies based on satellite data
- Selection of current and future studies.

2. The second stage consists of collecting information for current studies and includes

- Regular satellite data managing
- Data storage in data banks
- Managing of complex subsatellite experiments, ysynchronous and quasi-synchronous *insitu* data.

sensor	Type of satellite	Band (nm)	Ground solution (m)	Swath (km)
MSU–S	Meteor–30	500–700 700–1,100	240	1,400
MSU–SK	Kosmos–1939	500–1,100 (4 channels)	175 × 200	600
MSU–E	Kosmos–1939	500–600 600–700 800– 900	30 × 45	43
SAR	Kosmos–1870	10 cm	20–30	20
AVHRR	NOAA	580–680 725–1,100 3,500–3,900 10,300-11,300 11,500–12,500	1,100	3,300
TM	Landsat	450–2,350 10,400–12,500	30 120	185
HRV	SPOT		20 (10)	
VTIR	MOS	500–750 6,000–7,000 10,500– 11,500 11,500– 12,500	1,000	1,500
OSM	ERS-1	415–11,500	800 × 800	900
SAR		5.3 GHz	30(100)	80
CZCS	Nimbus–7	430–800 (20 resol.) 10,500–12,500 (total 6 chan.)	800 × 800	900

Table 2: State-of-the-art satellite sensors of interest for regional satellite oceanography and marine environment studies

3. The third stage consists of data processing and analysis including

- Using routine technology, including algorithms, hardware and software
- Analyses of the restrictions of routine technology
- Account of regional peculiarities
- Development of new technology
- Synergetic approach.

4. The fourth stage involves the use of information in various ways including

- Direct use in marine research

- Direct use in marine technology
- Assimilation in traditional marine sciences
- Operational use in cruise researches.

4 Monitoring the seas from space: case study of the Baltic Sea

Oceanographic phenomena in the Baltic Sea recorded in satellite imagery (during the period of experimental monitoring in 1985-1989) include:

- upwellings
- discharge plumes
- eddy structures
- mushroom-like structures
- discharge lenses
- water exchange through the straits (inflow/outflow)
- still zones
- coastal fronts.

Figure 2 shows the number of cloud-free days and satellite passes in the Baltic Sea for the summer period of 1986. For example, in May 25 days out of the total of 31 were cloud free, 73 passes of 153 recorded were cloud-free at least for one of the 5 parts of the Baltic Sea.

Some oceanographic phenomena recorded in satellite imagery are shown in Figure 3 (upwelling areas), Figure 4 (eddy structures) and Figure 5 (mushroom-like structures).

5 Monitoring the local environment from space: case study of the River Neva mouth.

In the River Neva delta there is the city of St. Petersburg, with its 5 million citizens and a large number of industrial enterprises using old technologies and inadequate filtering systems. Thus the problem of the spread and spatial distribution of the River Neva pollutants in the Gulf of Finland and in the Baltic Sea is obvious.

This is not a simple scientific problem and it has not yet been solved. The situation became still more critical when the construction of the flood barrier - the dam protection of the city of St. Pertersburg against the floods from the west, - was started in the early 1980s (see figures 6 and 7).

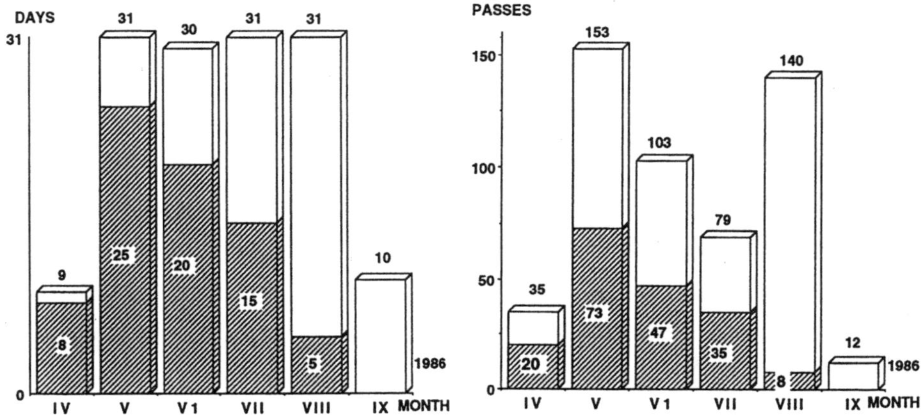

Figure 2: Cloud free days and satellite passes in Baltic Sea region

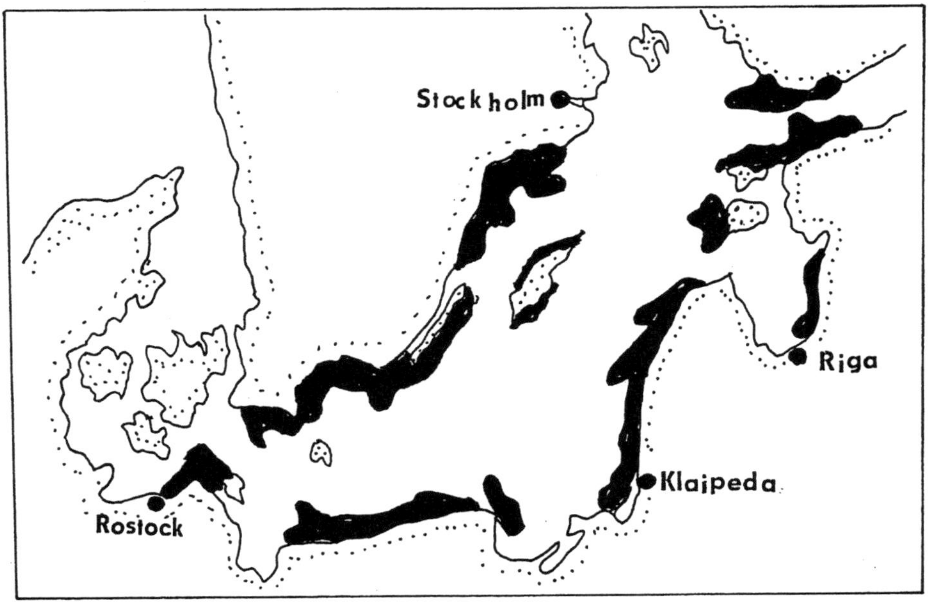

Figure 3: Areas of upwelling in the Baltic Sea.

	Phenomenon	Tracer	Period of observation, (month)	Typical dimensions (km)
1.	Upwellings	SST	4–10	10–20 (1) 200–300 (2)
2.	Discharge plumes	SST, suspended matter	4–11	10–100 (2) 10–30 (1)
3.	Eddy structures	SST, suspended matter, ice, algae	round the year	1–50 (3)
4.	"Mushroom"-like structures			20–50 (1) 2–8 (4)
5.	discharge lenses	SST, suspended matter	4–11	10–30 (3)
6.	Water charge through the straits (inflow/outflow)	SST	round the year	10–20 (1)
7.	Still zones	SST	4–8	10–100 (3)
8.	Coastal fronts	SST, suspended matter	4–8	10–20 (1)
Notes: (1)-length	(2)-width	(3)-diameter of cap	(4)-thickness of foot	(5)-depends on wind situation and river flow.

Table 3: Oceanographic phenomena in the Baltic Sea recorded in satellite imagery at the period of experimental monitoring in 1985-1989

When the barrier was almost built the construction was nearly stopped under the pressure of public opinion. What has happened?

There are two principal questions which are under discussion:

- Has the construction of the barrier changed the aquatic environment in the River Neva mouth and the neighbouring water?
- Has the barrier influenced the state of the environment in the St. Petersburg region?

In this chapter only one aspect of this problem is considered.

Remote sensing satellite and airborne data and *in situ* information have been used

	Temperature difference (degK)	Appearance frequency	Structure features in SST (or brightness) field	Life-time (days)
1.	2–10	up to 30% per month	Jets up to 100 km (2) Cores about 10 km (3)	0.5–14
2.	2–4	up to 100% per month	Right-hand asymmetry	Seasonal phenomenon
3.	0.5–1	accidentally	Core and peripheral parts. Mainly cyclonic rotation. Eddy cascades	1–7
4.	0.5–1	accidentally	Pair of eddies	1–3
5.	About 1	accidentally	Area of lens increases with distance from source	1
6.	1–2	continuously	Mushroom-like structures may occur	Connected with synoptic processes
7.	3–4	accidentally	Warm core	1
8.	1–4	continuously	Meanders on the front	Seasonal phenomenon

Table 3 continued

River	Drainage area (km^2)	Discharge (m^3s^{-1})
Kymi	37,000	517
Neva	281,000	2,463
Luga	13,002	124
Narva	56,200	437

Table 4: The Gulf of Finland rivers

to monitor the spatial distribution of suspended matter in this region for a number of years.

The Sources of information used in this work included satellite data from:

- Meteor and Kosmos satellites with sensors MSU-M (1000 m ground resolution), MSU-S (30 m) since the mid 1970s.
- Landsat.
- Kosmos-1870/Almaz SAR (synthetic aperture radar) (15-25 m).

as well as airborne survey data.

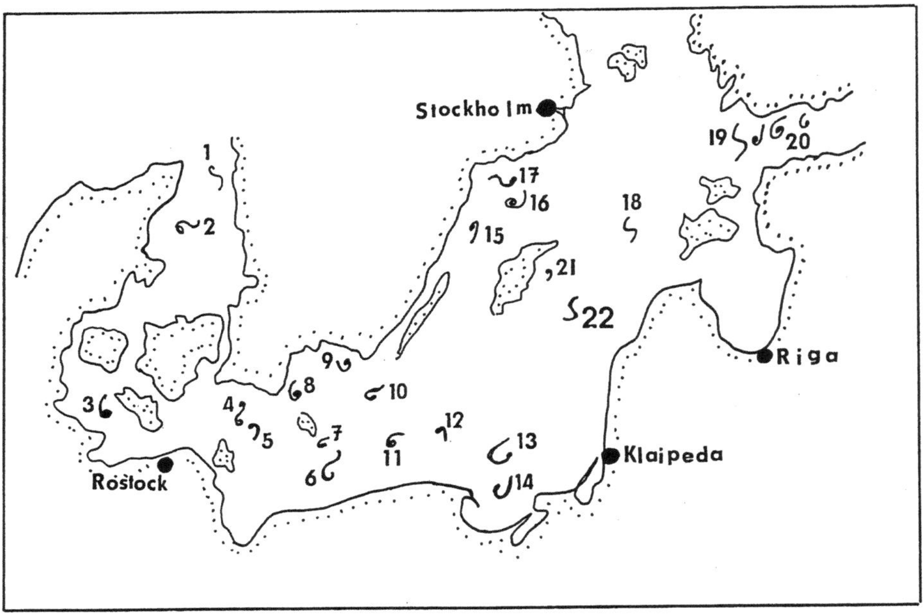

Figure 4: Eddy structures in the Baltic Sea recorded in satellite imagery on many different dates.

Additional information from databases and knowledge base included:

- navigational map
- mean velocities, general structure and schemes of currents
- characteristics of suspended matter including the size distribution of suspended particles and sediments
- data on seston concentration and regression characteristics "transparency - concentration"...

Additional information *in situ* data was available from

- data on the sources of suspended substances
- historical data on the spatial distribution of suspended matter under various hydrometeorological conditions
- meteorological situation before, after and at the moment of the satellite overpass (wind characteristics, atmospheric pressure, clouds, horizontal visibility)
- water lever

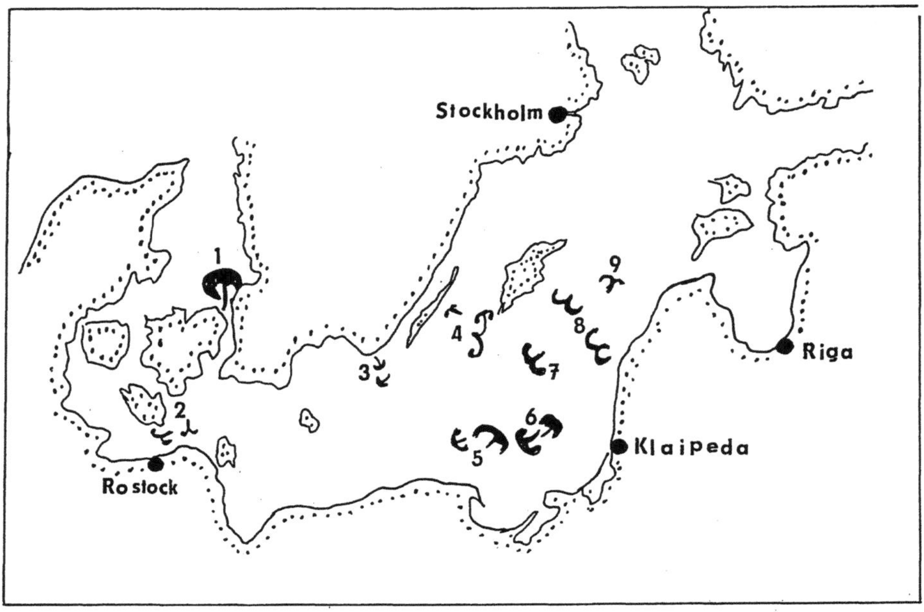

Figure 5: Mushroom-like structures observed in the Baltic Sea in satellite imagery on many different dates.

- *in situ* measurements of water transparency, temperature and salinity
- airborne measurements of sea surface temperature and optical characteristics of the water upper layer.

Additional subsatellite information helps to process the satellite imagery and makes the output user-oriented products more reliable.

In terms of before and after it is not so simple to answer the two questions about the effects of the construction of the barrier. There are at least 3 types of temporal variability:

- there is a very high variability of feature characteristics of suspended matter patterns on the **synoptic** scale
- as for the **seasonal** variability, there is a tendency that the total area of zones of suspended matter within the scene increases from spring to the end of autumn. One of the reasons may be the increasing activity of dredging and other operations in the summer period plus seasonal growth of wind wave activity
- it is difficult to show any tendency in **annual** variability.

No reliable remote sensing data exist which could show changes in the general patterns of suspended matter fields before and after the construction of the flood barrier.

Some lessons in the field of scientific research can be learned from the flood barrier:

- adequate observation of the water body requires regular satellite monitoring providing high-resolution images once every 3-4 days and medium-resolution images daily,
- specific spectrometric remote sensing measurements should be performed along with the existing broad-band satellite images
- location of regular *in situ* observation stations must be specified in accordance with the current system and spatial distribution of suspended matter
- prior to the construction of major engineering structures and also during the progrss of them, the storage of homogeneous time series of remote sensing and relevant *in situ* data is necessary
- those data should be accumulated in integrated geographical information systems (GIS) of the type suggested by Victorov (1990)
- further analysis of satellite imagery demands the efforts of specialists in hydrodynamical modelling who could use satellite data to determine the initial and boundary conditions and some other parameters in numerical models. These models for the River Neva mouth should account for two-dimensional sources of pollution
- the zero-pollution approach suggested for the Great Belt Link is a good example to follow in future major engineering structure projects.

There are some lessons to be learned in the field of public relations:

- for a major engineering structure that could possibly affect a great number of citizens or a large water body, or cause trans-generational effects, international examination of the project is recommended
- scientists should not ignore public discussions. Sometimes scientists simply ignore non-specialists' opinions. Very soon a large amount of non-professional opinions expressed in letters, published in newspapers or appearing in TV-programmes becomes so large that it is almost impossible to ignore the public
- try to avoid manipulating with terms: pollution, mud, clean, dirty; all these terms should be defined strictly before one starts discussions with laymen
- it should be clarified that two environmental events occurring at the same time or one after another might not necessarily be causally related to one another. one by another. Even this very simple thought has to be explained to many people over and over again

- experts in public relations in the field of the environment should have a good knowledge of modern sociology, psychology, mass-media communication skills, etc.
- the first ideas of ecology and environmental concern must come to people at a very early age, maybe in their childhood.

6 Geographical Information System (GIS) "Baltic Europe".

"Baltic Europe" is the Baltic Sea itself and its catchment area. The modern tendency is to create integrated GISs comprising information on the state of the environment and the situation with the population health which makes it possible to process these data together.

The data to be collected (measurable and descriptive) include:

- meteorological
- hydrological
- water quality control
- air pollution control
- seismic
- electric field
- magnetic field
- radiation
- land-use
- geological
- soil contamination
- noise
- malodour
- population health.

In connection with global climate change one should pay attention to the problem of chemical time bombs and GIS could help to monitor this issue.

7 Conclusion

Nowadays for many countries direct participation in global climate change studies is not of the highest priority in terms of costs and benefits.

While monitoring the local and regional environment, analysing the climatic issues on the regional scale, these countries could contribute directly or indirectly to the world climate investigations. Through what mechanisms?

The plausible and convenient mechanisms could be discussed and some regional pilot projects could be launched. Perhaps, the WMO Programme "Regional Seas" could be used for the oceanographic segment of these activities.

But, at the same time, I understand that the involvement in such a large number of different international projects in Earth sciences is far beyond the real capabilities and resources of mankind (and not all of these projects are being managed very well). Thus it is necessary to provide an overview and to concentrate our joint efforts on the key issues.

One should understand that to choose the key issues, the key problems or the key points is again a very difficult scientific problem. It would be very attractive to find some sensitive points, (areas) for detailed monitoring which could be considered as representative of some regions (or of some natural phenomena and thus to save money and manpower. In this situation remote sensing data could be used to harmonise the measurements in local points (areas) in order to create comprehensive pseudo-field of variables. But even in the absence of those wonderful key points the harmonising effect of remote sensing data – as applied to point measurements – is hard to overestimate.

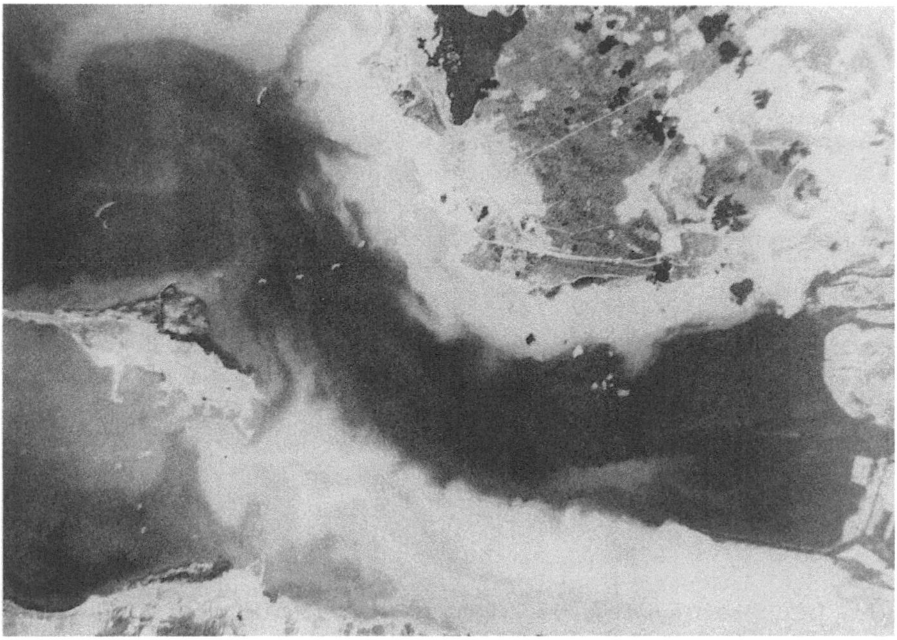

Figure 6: Satellite image of the River Neva Mouth 9 July 1981 (before construction of the flood barrier)

Acknowledgements

The author is grateful to Prof. A. P. Cracknell and Dr. R. A. Vaughan who invited me to give lectures at the Summer School and made possible my attendance. I also thank all the participants in the experimental satellite monitoring of the Baltic Sea carried out in my Laboratory for Satellite Oceanography in 1985-1989.

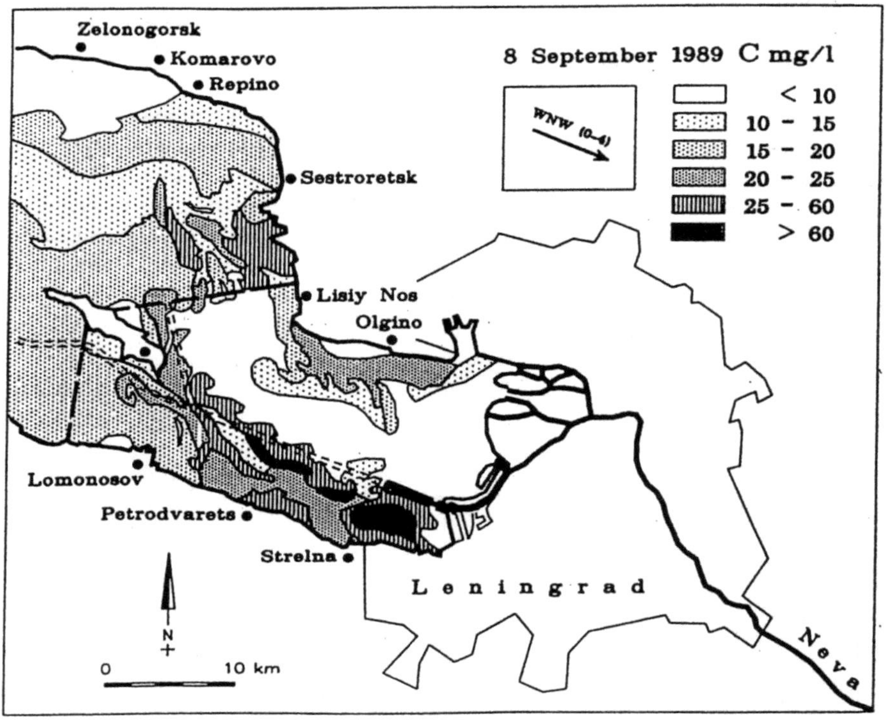

Figure 7: Suspended matter distribution in the River Neva Mouth. Map based on satellite image of 8 September 1989 (construction of the flood barrier is nearly completed)

References

Morel P., 1991, Space Observation Requirements for Global Climate Science and Prediction. *Proceeding of the EURISY Symposium on the Earth's Environment - An Assessment from Space held in Venice, Italy, on 10-11 April 1991, ESA SP-337* (Noordwijk: European Space Agency), pp. 57-59.

Victorov S. V., 1988, *Regional Satellite Oceanography* (Moscow: Gidrometeoizdat) (In Russian).

Victorov S. V., 1990, Environmental Monitoring of the Gulf of Finland: an Approach Based on Remote Sensing Data. *Statistical Journal of the United Nations Economic Commission for Europe*, **7**, 149-162.

Victorov S. V., 1991, Water dynamics and Pollution nearby Leningrad Dam./In: The Interaction between Major Engineering Structures and the Marine Environment. *Proceeding of the International Association for Bridge and Structural Engineering Colloquium, held in Nyborg, Denmark, in May 1991, IABSE Reports* **63**, pp. 215-223.

Victorov S. V., and Sukhacheva L. L., 1992, Satellite Monitoring of the Gulf of Finland: State-of-the-Art and Needs for Development. Paper submitted to the ERIM First Thematic Conference on Remote Sensing for Coastal Waters Monitoring, held in New Orleans, USA, on 15-17 June, 1992).

Remote Sensing Applications to Environmental Monitoring of the Coastal Zone

Mohd. Ibrahim Hj. Mohamed

1 Introduction

The rapid development of satellite oceanography since the launch of satellites designed to carry marine observation radiometers has provided a valuable tool to study the oceans and its processes in a new light. Sensors operating in the visible, infrared or microwave part of the electromagnetic spectrum represent the only means of providing synoptic observations of four basic sea surface properties i.e. 1) colour, 2) temperature, 3) height and 4) roughness (Allan 1991). The interpretation of these four basic properties has formed the basis for the determination of phytoplankton, turbidity, algal blooms, sea surface temperature, ocean fronts, geostrophic currents, wind stress and wave fields of the oceans.

Environmental applications of remotely-sensed data, however, have not developed as rapidly as in oceanography. Yet the potential benefits of near-real-time data (when available), synoptic coverage of the data, archived database and the visible impact of glossy colour photographs from satellites are obvious dividends that can be achieved from the use of remotely-sensed data.

Recognising the important potential for environmental applications of remotely-sensed data, the ASEAN (Association of South East Asian Nations) Australia Programme in Marine Science incorporated a remote sensing component into its Coastal Living Resources Programme. This programme, which started in 1984, covers an ambitious attempt to determine the distribution, status and ecology of three major ecosystems found in the ASEAN and Australian geographical area, namely the mangrove forests, coral reefs and what is in between, i.e. the soft bottom communities.

Unique in its approach, the studies involve a network of scientists from five ASEAN countries of Indonesia, Malaysia, Philippines, Singapore and Thailand and Australia employing similar data collection methods, data management, analysis routines and constant calibration and standardisation of methodologies to ensure the comparability of results over the whole region. The project is now into its second 5-year phase.

NATO ASI Series, Vol. I 24
Remote Sensing and Global Climate Change
Edited by Robin A. Vaughan and Arthur P. Cracknell

2 The objective

The primary objective of the remote sensing component of the programme was generally two pronged i.e:-

i. To develop a network of remote sensing specialists in the region competent in the application of remote sensing of the marine and coastal areas;

ii. To explore and develop the applications of remote sensing to environmental problems and management of the coastal zone and marine areas.

3 The remote sensing programmes

The remote sensing component of the coastal living resources programme initially sought to explore the various possibilities of promoting the understanding of the coastal zone particularly to take advantage of the large amount of ground and sea truth data presently generated by the manual sampling undertaken in the project.

3.1 Coral reefs

Some major considerations in coral reef studies are the danger associated with underwater observations, the limited amount of expertise available to do large-scale projects and the time and cost requirements of mounting the projects. The application of remotely-sensed data is therefore directed towards the solution of the present limitations as briefly described below:-

i. Shallow water mapping. The blue-green wavelength of the electromagnetic spectrum have a depth of penetration in clear tropical waters of up to 30 m. This ability to capture the spectral signatures of bottom features can thus be used to detect different irradiance values of features in the water and map shallow water features such as coral reefs and sandy and rocky substrates.

ii. Classification of coral reefs. As an extension of mapping coral reef areas, the occurrence of coral reefs is normally associated with species zonation which are depth related. If the depth occurence of corals can be determined, then broad classification of species can be determined in an overal effort to determine the distribution and status of the coral reef.

iii. Water Quality. Clear blue ocean waters, as opposed to those containing a high sediment load, terrigenous runoff or chlorophyll/phytoplankton can be differentiated from the reflected radiance captured by satellite-borne radiometers. Thus the occurrence of high sediment outflows such as those from rivers can clearly be distinguished from surrounding waters. Present knowledge, however, is inadequate actually to determine the total amount of suspended solids in the water without simultaneous ground truthing of the area with the satellite overpass.

iv. Water circulation studies. The process of ocean circulation, whether wind, temperature or geostrophically driven, will affect the sea surface temperature and/or plankton distribution in the water surface or water column of the area. These features can be used as markers to detect eddies or fronts in the water and shows the shape of the water circulation pattern in the coastal zone.

v. Recruitment Studies. Eddies and sheltered areas of islands where the speeds of water movements are low allow a better chance for the planktonic juveniles of coral polyp to settle to on the substrate. Knowledge of these areas determined from remotely-sensed data will allow reef scientists to pinpoint areas where more detailed surveys can be mounted.

3.2 Mangrove Studies

Remote sensing applications were also used for mapping, determination and delineation of mangrove areas, and ground cover mapping of areas felled to make way for development, aquaculture ponds and other uses. Together with a geographical information systems (GISs), this application is regularly used to determine the extent of coastal resources for coastal zone management and planning. The project is also attempting to use Landsat TM data to classify mangrove species and zonation, distribution and status and to age the stands of mangrove in a managed mangrove forests.

4 Monitoring the environmental impact of development; a case study.

In the first phase of the Coastal Living Resources project, the coral reef studies part of the project focussed on the determination of baseline information of 10 island groups from the Malacca Straits to the Sulu Sea. In the second five-year phase of the project, the emphasis of the project was shifted towards a more detailed study of the ecology of coral reefs and its connections with coral reef fish. Redang Island archipelago, designated a Marine Park in 1988, was selected as the study site because it gave the added impetus of studying the effects of marine park establishment on the coral reef fauna.

In 1991, a private developer started development of the first phase of the Pulau Redang Island Resort comprising an 18-hole golf course, a clubhouse, a 250-room hotel, a horse ranch, 45 holiday villas, a sports complex, farms, a tourist centre, staff quarters and all its supporting infrastructure on 157.9 ha of land surrounding the Kuala Sungei Redang.

Soon after the development started, a big controversy ensued when it was apparent that the development works were causing a large amount of sediment runoff seriously threatening the sensitive coral and other marine flora and fauna of the marine park. Scientists, environmentalists and the general public are concerned that the on-going development of the four phased project is not congruent with the professed goals and objectives of the establishment of the Marine Park. Furthermore, massive earthworks associated with the development will cause erosion and sedimentation on the sensitive mangrove ecosystem of the island and destroy them together with the lucrative fishing industry which depended on the mangroves as its breeding grounds.

The total and cumulative impact of development of the Pulau Redang Island resort on the marine flora and fauna of the area is almost impossible to predict. What is certain is the increasing trend of sediment loads, total suspended solids in the water, leaching of fertilizers into the water from the estimated total of 125 t y^{-1} to be applied to the golf course and the leaching of pesticides from the golf course development, vegetable and other farms. Increased effluent flow into the environment as a result of increased numbers of visitors to the island will all result in a matrix of intricate and detrimental impacts individually and cumulatively on the delicately balanced ecosystem of Pulau Redang Marine Park.

The development of the Island Resort in the middle of a marine park provided a golden opportunity to extend the use of remote sensing to the monitoring of development activities and its impact on the fauna and flora of the marine park and coastal zone. A sampling programme of the mangrove areas for biological data, sedimentation, accretion rates, transect data, water quality, total suspended sediment and sedimentation rates, ground truthing and aerial photography was launched. Simultaneously, a succession of Landsat TM data were obtained for analysis of ground cover, ocean colour, bathymetry, turbidity, and coral mapping. This sampling programme which started in November 1991, is over and above the second phase of the Coastal Living Resources programme specifically to address the impacts of development on the fauna and flora both near to and far away from the development sites and to interface with the longer-term objectives of the second phase Coastal Living Resources project which started in July 1990.

4.1 Results of the study

4.1.1 Ambient water conditions

The water temperature and salinity profile of the area was within the normal range. Surface, middle and bottom water temperatures of the study area ranged from 27 °C to 30 °C, with higher temperatures found in surface waters in the early afternoon. The salinity values were also found to be within the normal range from 27‰ to 29‰. Lower values were observed at the mouth of the Redang River while higher values of 27.5‰ to 28.5‰ were observed in the eastern part of Pulau Pinang, the site of the Marine Park office.

4.1.2 Total Suspended Solids

The total suspended solids (TSS in mg l^{-1}) in the water at all stations in February 1992 had increased tremendously compared to values obtained in December 1991 (see Table 1). The percentage increase in TSS over the two-month period ranged from 33.4% to 599.5%. What is most significant here is that the increase is evident in all stations. Massive earthworks which included hill cutting, terracing and landfill together with depletion of the vegetation cover coupled with heavy rains during the northeast monsoon resulted in a greater outflow of terrigenous water from the land and the development area into the waters of the marine park.

Station	Dacember 1991	February 1992	% change
ss01	10.90	24.45	+133.50
ss02	10.90	14.55	+33.37
ss03	9.09	19.99	+119.91
ss04	12.73	16.36	+28.51
ss05	9.09	12.73	+40.40
ss06	1.82	12.73	+599.50
ss07	10.90	10.90	0
ss08	7.27	16.36	+125.0
ss09	5.46	23.64	+333.80
ss10	7.27	21.82	+200.13

Table 1: Total suspended sediment (mg l^{-1}) in Redang Island waters at the vicinity of phase 1 development.

4.1.3 Sedimentation Rate

The results of the sedimentation studies carried out over a year around the island from July 1990 to September 1991 showed sedimentation rates ranging from 0.01 mg cm^{-2} day^{-1} to 1.84 mg cm^{-2} day^{-1}. These values are consistently low and as the sampling was done over a one-year period can be taken as ambient sedimentation values for the area.

Since the commencement of the north east monsoon, sedimentation rates had risen to between 2.67 mg cm^{-2} day^{-1} and 25.21 mg cm^{-2} day^{-1} depending on their distance from the development area. Stations near to the development area exhibited large increases of sedimentation while those farther away from the development area exhibited only slight or no increases in sedimentation rates. Table 2 is representative of the trend.

The study further revealed that the impact of high sedimentation rates was found to be localised at the present time and the causes of the high total suspended solids in the waters of the marine park can be attributed totally to the massive earthworks related to the development of the island resort.

4.1.4 Coral reef distribution

From the results of the transect studies, no discernible trend can be derived from the results of the transect data on the distribution or state of the coral reefs. The data shows no trends and bears no specific relation between coral reef cover and sedimentation rate. Similarly, analysis of the data from the other stations shows no specific relation at the present moment.

General observation of the area, however, showed that the increase in total suspended sediment and sedimentation rates are clearly visible in the waters of Kuala

Station	26/04/91	17/06/91	01/09/91	19/21/91	12/02/92
ss03 (near)	0.875	0.82	1.84	9.10	10.71
ss09 (away)	0.525	0.73	0.57	-	0.68

Table 2: Sedimentation rate in mg cm^{-2} day^{-1} at stations near to and away from the first phase development.

Sungei Redang. The sediment trap placed in the area is covered by sediment after only one month placement. Turbidity of the area has also increased causing reduction of visibility. Typical to the area, coral abundance and diversity is very low and the low recruitment levels of a few *Pocillopora spp.* and *Acropora spp.* indicate a stressed environment.

Further symptoms of a stressed environment such as tissue death, as shown by the appearance of white spot disease, were reported for the first time for the area. Small coral heads of the same area were unable to shed large sediment particles from their surfaces. This phenomenon will be followed by the natural process of algal encroachment, further sediment deposition and the eventual death of the coral. Coral reef distribution and abundance in sites away from the development areas have not shown any visible signs of degradation.

4.1.5 Mangrove forest

Large tracts of mangroves were found to have perished due to sedimentation burying of their root system. On the western side of the Redang River, the mortality rate was determined to be as high as 60% in some areas. The impact of landfill and embankment to accomodate the development have effectively isolated some of the mangrove areas from intertidal flow of water resulting in death to the mangroves. With the onset of the monsoon, the inadequacies of the various mitigating measures to control soil erosion were clearly highlighted. Sediment ponds were quickly filled up, drainage rendered ineffective and sediment traps became inoperative, effectively turning the mangrove forests into sediment sinks rather than areas to be safeguarded from the sediment outflow.

4.1.6 Ground cover

Two Landsat TM images of the area were obtained for the overpass period of 7 September 1990 and 9 August 1991 (attempts to obtain a later image have so far been unsuccessful) was obtained. The images were rectified to the latest 1:50,000 topographic map of the area. The images were then separated by spectral digitization for land-only and water-only images, while for the mangrove area they were manually digitised. Cloud cover was also digitised out of the water images to provide good image enhancement. Classification for ground cover pattern on the land-only and mangrove-only images were performed using the classification methods available with the MicroBRIAN image analysis system. Channels 1, 2, and 3 of the TM image were used for ground cover classes of the land-only images. Analysis of the water-only images was performed by density slicing to show suspended sediment concentration (SSC) in channels 1, 2, and 3 seperately. The mangrove areas were classified for changes in ground cover and calculation of the amount of mangrove areas felled for development were performed. Finally the land, water and mangrove-only images were pasted together to form one composite image (see for example plates 3 and 4). These were than compared with the results of ground truthing, aerial photography, coral reef studies and the topographic maps to assist in the analysis of the classified image.

In general little change in the ground cover pattern can be determined from the images except some the earthworks. From the first image (7 September 1990), the presence of land clearing and roadwork joining the the present fishing village to the resettlement site can be detected. This development performed in 1987 and 1988 is clearly revealed. The second image (9 August 1991) of the same area shows the extent of the development that had taken place. On both sides of the Redang River the ground cover had been removed and cut and landfill of part of the mangrove forest had been performed for the development of the golf course. Although after only 7 months of development, the earthworks have been rapid and and the removal of the original lowland forest and mangrove is clearly illustrated.

4.1.7 Mangrove resources

Delineation of the mangrove areas was performed using channel 4 of the Landsat TM image. Classification of the mangrove area using semi-supervised methods was then reduced using minimum spanning tree algorithms to reveal 4 superclasses. Calculation to determine the area of mangroves felled was performed by delineating the classified image and the conversion of TM 30m pixels to hectares. The result obtained showed that the original area of mangroves in Redang Island was 40 ha. In August 1991 the mangrove forest have been reduced to 29 hectares.

4.1.8 Water quality

Density sliced images of the water-only image analysed for channels 1,2 and 3 of the TM image do not show any discernible pattern sediment outflow from the Redang River. Channels 1 and 2 gave the best resolution and differentiation of the water areas while the resolution from channel 3, i.e. the spectral irradiance of the surface waters, showed the least features of the marine waters, and thus differences of water colour of Redang Island.

5 Discussion

The application of remote sensing for the monitoring of development impacts on the flora and fauna of Redang Island Marine Park showed a lot of potential as an important tool to provide an effective, inexpensive and timely means of monitoring the development impact on the ecosystem of the Pulau Redang Marine Park. The analysis of land images representing a one-year period and about 7 months since the development started on Redang Island Resort traces closely the activities of ground clearing and earthworks derived from ground truthing, manual sampling of the terrestrial area and aerial photographs. The data also provided a quick estimation of areas felled for development and can be used as a check on the activities of private developers developing unalienated lands.

Since the start of development, 6 ha. of mangrove forest have been felled. In itself this may not be significant except that the conditions of approval of the project stipulated that no more than 2.9 ha of mangrove in Pulau Redang are to be developed and the rest should be left as forests. This finding clearly show that one of the major conditions for the approval of the project has been breached. Since then, the development of the island resort is proceeding rapidly. The 9 holes of the golf course that adjoin the mangrove forest have been laid out and more cutting of the hill is being done for the site of the second 9-holes, clubhouse and chalet sites.

The density slice for suspended sediment concentration performed for the waters of Redang Island showed little outflow of sediments from the Redang River in August, 1991. Attempts obtain further images during the time of ground data acquisition have so far been unsuccessful for reasons of cloud cover, as well as difficulties in obtaining quicklooks to ascertain whether the 50% cloud cover actually covers Redang Island. As the results of the density slice analysis from channel 1 of the data showed more resolution than channel 3, it can be suggested that the features detected in the water are perhaps more a reflection of bottom features rather than water colour denoting suspended sediment concentrations. The high sediment load and localised effect of the sediment outflow revealed by the field sampling have not been shown in the analysis of the satellite data. This is expected as the field data was obtained in December 1991, and February 1992 and the overpass data was in August 1991. August is a dry period for the area with little terrigenous runoff from the land areas of the island (Mohamed and Bujang 1992).

The use of remote sensing on clear tropical waters requires more detailed studies on light attenuation, depth of penetration, water colour and the atmospheric corrections to be applied. The area studied, considered as classical class-2 waters as defined by Sathyendranath and Morel (1983), exhibited different physical characteristics from those found in temperate areas and may require different classification based on visibility criteria.

The findings of this study have prompted the relevant authorities to impose more rigorous conditions for the continuation of the development in the marine park area. Among these are a more rigorous monitoring of the terrestrial and marine areas and mandatory reporting of the results on a quarterly basis, development to be limited to designated areas only, immediate rehabilitation and stabilisation of cut hill slopes and terraces simultaneous with opening of new areas and improved design and construction of drainage and sediment control measures.

Several problems were highlighted during the study. They included the timely acquisition and availability of cloud-free data, especially during the north east monsoon months of November to March. Besides the use of visible bands of Landsat MSS, TM and SPOT, and AVHRR, synthetic aperture radar (SAR) should be used to overcome cloud the cover problem so prevalent in tropical areas. A more rigorous ground sampling regime concomitant with satellite overpass or other remote data acquisition methods is required to produce standards and models for codifying satellite data for application to environmental monitoring in the tropics.

6 Future challenge

The present study on the impact of development on the flora and fauna of Redang Island Marine Park has expanded in its scope to cover monitoring the development impacts of a resort in a sensitive environment, but is still restricted to the ecology and the related fishery of the area. The opportunities presented by the development of the island resort, on what is effectively a controlled environment away from the immediate effects of land, presents a natural laboratory to study comprehensively the impact of resort development on a marine park, before, during and after development has taken place. Predictive modelling of the development impact on an island environment will naturally follow in the future.

Besides the rapid detection, identification of a potential environmental problem and immediate implementation of mitigating measures, experiences learned from Redang Island can be beneficial to any such development planned in the future.

Remote sensing studies and applications will play a pivotal role in there comprehensive studies to obtain a complete picture in this natural experiment. The advantage of synoptic observations of terrestrial and ocean features and processes on a timely basis is essential to ensure that the coastal areas will be managed on a rational and sustainable basis for future generations to come.

Acknowledgments

The authors acknowledge the support, contribution and encouragement of the Universiti Pertanian Malaysia, ASEAN-Australia Living Coastal Resources Programme, WWF (Malaysia) and the University of Dundee.

References and Bibliography

Allan, T. D., 1992, The marine environment. *International Journal of Remote Sensing*. **13**, 1261-1276.

Ch'ng Kim Looi, 1990, National Marine Parks Malaysia. Policy and concepts. Deptartment of Fisheries, Ministry of Agriculture, Kuala Lumpur, Malaysia.

Dartnall, A. J., and Jones, M., 1986, A Manual of survey methods: Living resources in coastal areas, AIMS Townsville, Australia.

Dodge, R. E., and Vaisnys, J. R., 1977, Coral population and growth patterns, responses to sedimentation and turbidity associated with dredging. *Journal of Marine Research*, **35**, 715-730.

Gower, J. F. R., and Apel, J. R., (editors), 1986, Opportunities and problems in satellite measurements of the sea. *Unesco technical paper in Marine Science*, **46**, pp. 70.

Harrison, B. A., and Jupp, D. L. B., 1989, *Introduction to remotely sensed data* (Victoria: CSIRO).

Harrison, B. A., and Jupp, D. L. B., 1990, *Introduction to image processing* (Victoria: CSIRO)

Jupp, D. L. B., Mayo, K. K., Kuchler, D. A., Van, R., Classen, D., Kenchington, R. A., and Guerin, P. R., 1985, Remote sensing for planning and managing the great Barrier Reef of Australia. *Photogrammetria*, **40**, 21-42.

Malayan Nature Society, 1990, Pulau Redang State Park: Terrestrial resources , management-problems and proposed management guidelines. A report submitted to the Depertment of Fisheries, Ministry of Agriculture, Malaysia.

Mohd. Ibrahim Hj. Mohamed, and Japar Sidek Bujang, 1992, Development Impact on coral reefs and mangroves of Pulau Redang. WWF Malaysia Project MYS 223/91. 62.

Mohd. Ibrahim Hj. Mohamed, Hilal Abdul Aziz, and Shamsul Bahar Ahmad, 1992, Impact of resort development on the coral reefs of Pulau Redang Marine Park. Paper submitted to the 3rd ASEAN Science and Technology Week Singapore. on 23-25 September 1992.

Robinson, I. S., 1985, Satellite Oceanography. (Chichester: Ellis Horwood).

Rodgers C. S., 1990, Responses of coral reefs and reef organisms to sedimentation. *Marine Ecology Progress Series*, **62**, 185-202.

Salm, R. V., and Clark, J. R., 1984, Marine and Coastal Protected areas. *A Guide for Planners and Managers* (Switzerland: IUCN, Gland).

Sathyendranath, S., and Morel, A., 1983, Light emerging from the sea:- Interpretation and uses in remote sensing. In *Remote sensing applications in marine science and technology*, edited by A. P. Cracknell (Dordrecht: D. Reidel), pp. 325-358.

Marine Surveillance and Information System

P Singhasaneh

1 Introduction

SEAWATCH THAILAND is a complete marine environmental monitoring and forecasting system which integrates data collection, data analysis, environmental modelling and forecasting with an advanced computerised system for the distribution of marine information and forecasts to interested operators and/or authorities. This three-year project is being implemented under close co-operation between the National Research Council of Thailand (NRCT), OCEANOR, the Oceanographic Company of Norway, and various other parties, including the Harbour Department, the Meteorological Department, the Port Authority of Thailand, the Naval Hydrographic Department, the Department of Fisheries, the Petroleum Authority of Thailand, the Marine Police Division, Chulalongkorn University, Kasetsart University, Prince of Songkla University, Burapha University, etc.

2 Data acquisition module

A real-time data coverage is provided by a network of moored data buoys (called TOBIS buoys), which includes meteorological parameters (wind speed and direction, air temperature and pressure) and oceanographic parameters (oxygen/algae/nutrient contents, waves, currents, temperature/salinity profile, radioactivity). The buoys have also their own data logging equipment, on-board processing (for data analysis and quality control), and a transmission system. The collected data are then transmitted to a shore station through the ARGOS satellite communication system on NOAA.

The total of seven buoys are planned to be deployed in the Gulf of Thailand under the programme.

NATO ASI Series, Vol. I 24
Remote Sensing and Global Climate Change
Edited by Robin A. Vaughan and Arthur P. Cracknell

3 Data storage, analysis and presentation module

ORKAN is a complete flexible system for processing, presentation and storing of environmental data collected by the buoys. The system core is an efficient database for measured time series and output from numerical models. It is supplied with a wide range of modules for data import and export, time and frequency analysis and presentation tools for graphics and tables.

4 Environmental modelling and forecasting module

A number of pieces of numerical modelling and forecasting software will be available under this programme, including:

NOMAD. This is a model for the simulation of spreading and dilution of aqueous effluents and particulate discharges. It simulates diffusion and advection in a prescribed current field, by using the 'particle-in-fluid' concept. The discharge is represented as a cloud of particles released continuously during the discharge period. The particles are either passively buoyant or assigned a prescribed sinking/rising velocity determined by the size distribution and density of the discharged material. The ocean currents are prescribed from local current measurements (vertical profiles) or imported from hydrodynamic simulation models, like HYBOS. The concentration field is derived from the particles' distribution by weighting the individual particles into a grid with resolution chosen according to the length scale appropriate for the specific case.

HYBOS. This is a three-dimensional ocean circulation model for the simulation of currents, sea elevation, salinity, temperature and density. The model is used to simulate ocean parameters in offshore and coastal regions, according to the user's selected data input. The HYBOS ocean current simulation model is therefore specially useful in offshore design and operations, storm surge/current forecasting, estimation of mean water level and water circulation/quality studies. The simulated currents fit into discharge diffusion transport models like NOMAD which are used in water quality studies. The horizontal grid spacing is varied using boundary fitted co-ordinates, thus making it easier for the user to control computation. Sigma transformation is used in the vertical dimension.

OILSPILL. This is a PC forecast model for the prediction of drift and fate of an accidental oil spill at sea from offshore platforms or ships. The simulations are based on wind observations and forecasts and ocean current data (background currents and tides). The latter data are permanently stored in the OILSPILL database. Simulations are run on the basis of user-supplied wind forecasts. The OILSPILL model also accounts for loss of the type of oil involved, due to evaporation and entrainment in the water column. In order to make the program applicable to any specific region, relevant data must be made available on ocean currents and the specific oil types involved.

OILSTAT. This is a PC system for the graphical presentation of statistical data on the drift of oil from potential oil spills in the offshore or coastal region of concern. OILSTAT is specially designed to be used in feasibility studies and contingency planning. The oil drift data stored in the system are computed in advance with a statistical trajectory model (DRIFTMAP). This model covers a large number of potential spill sites distributed in a regular grid. The output from the trajectory model for each spill site comprises minimum drift time to shore, probability for stranding and expected stranding of oil. The model also enables the user to extract oil drift data for specific offshore locations or coastline sections for graphical presentation.

5 Data / forecasts / users-relevant information distribution module

An information system or THAINET will be developed to allow users to have easy access to the various data collected and other derived information including the forecasts on a near real-time basis via PC terminals. The system will consist of different modules to perform the following tasks.

- communication with the data collection buoys for systematic data identification and recording
- input of external data sources including news, abstracts of reports and laws/regulations related to marine activities
- transfer of data to various users as well as those in-house in accordance with specified requests
- system operation for the THAINET management
- information extraction with generic and specialised models
- presentation of all measured and analysed data.

6 Advantages of SEAWATCH THAILAND

SEAWATCH forecasts and data stored in the central database can be accessed through a personal computer via a modem. The user whose livelihood depends on such information can use this information to support their decision making, such as;

1. Planning managing and monitoring aqua-cultures, plants and commercial fishing

Results from the models can be combined with information from other sources to generate suitability maps of fish-farm locations and fishing spots which, in turn, help fishermen save money and man-power spent on random search.

THAINET provides information concerning not only weather conditions but also impending changes in the sea and/or advance warning of pollution threats such as oil slicks, poisonous algae, jellyfish and substandard water quality and advises on countermeasures.

Further research can be done to set up strategies and inspection routines to reduce cage wear and tear, corrosion and marine growth in order to prevent cage damage and fish disease.

2. Environmental preservation for tourist industry and future generations

   Some of Thailand's coastal valuables, namely beaches, coral reefs, and mangrove forests, etc., are being threatened by tourism, construction and environmental pollutants. Using SEAWATCH information, environmental monitoring and protection plans can be set up in order to conserve and maintain these national treasures.

3. Providing design parameters for onshore and offshore structures

   Environmental data is usually necessary for establishing design parameters for feasibility studies and designs of projected locations for production platforms, pipeline routes, harbours, oil terminals, breakwaters, etc. These data includes data on wind, waves, currents, rate of marine growth and rate of corrosion.

4. Analysis of marine pollution from sewage outfalls, industry and offshore installations

   The SEAWATCH system can be used to monitor the water quality and organisms in the study area and aid in the assessment of the long-term effects of environmental pollutants. Environmental investigation, such as how a given pollutant will dilute and spread, how it will affect organisms in the area, and how much of the pollutant the environment can absorb without harm, can be carried out.

5. National security

   The temperature profiles from the buoys are highly valuable for the detection of illegal penetration by foreign submarines.

6. Weather forecast

   Since each buoy is carrying its own meteorological station, the measured parameters can be used with data from other stations and/or other sources in the forecasting process.

7 Progress of the programme

The National Research Council of Thailand has successfully deployed four TOBIS buoys at Sichang Island, Rayong Bay, Chang Island and Plathong Oil Platform. The other three buoys will be in place within the next six months. The buoy-transmitted data is transferred from the ARGOS office in France to the National Research Council of Thailand on a daily basis. The data is then processed and stored in the central database.

The users can access the data in two ways:

1. Via modem communication. The communication between the user's remote PC and the central database is managed through a modem connection. The communication is asynchronous with a speed of at least 2400 baud, preferably higher. The desirable speed is 9600 baud.

2. By requesting hard copies directly from the National Research Council of Thailand. The followings are the report's formats which can be generated by the database:

- Table of parameters
- Joint occurrence table
- Time series plot
- Rose vector diagram
- Progressive vector diagram
- Extreme parameter analysis.

For further details of the programme, please contact:

SEAWATCH THAILAND Programme,
National Research Council of Thailand,
196 Paholyothin Road, Chatuchak,
Bangkok 10900, Thailand.
Telephone : 579-0733 , 579-8940
Telefax : 579-3035 , 579-8940

8 Conclusion

The SEAWATCH THAILAND programme is being established through networking of data collection buoys. The observed data can be integrated with data from other sources or used as an input to various numerical models. Results from the models can be further combined with the information from the buoy network to produce user-oriented forecasts. The distribution of marine environmental data and forecasts is a PC-based system that would allow users to log on and retrieve the information directly.

The State of the Problem of Ecological Monitoring in the City of St Petersburg and Its Region

Kirill Ya Kondratyev and Leonid P Bobyliov

1 Characteristic features of the complex of ecological problems of the St. Petersburg region

The most important ecological problem of the St. Petersburg Region is that of Lake Ladoga. Its basic aspect is the continuously deteriorating water quality in Lake Ladoga and the entire system of the Lake Ladoga, the Neva River, the Neva Mouth and the eastern part of the Gulf of Finland. This deterioration is a consequence of human economic activity in the lake's basin (Voropayeva and Rumyantsev 1991).

Lake Ladoga is the largest lake in Europe and is a unique fresh water body. It is of strategic importance for the development of the St Petersburg urban agglomeration and the entire north-eastern part of Russia. Lake Ladoga is the principal source of drinking water supply for the 5 million inhabitants of the city of St. Petersburg.

The basin of Lake Ladoga is characteristic of intense and multi-factor anthropogenic impact on the environment. In the long run, this impact affects not only Lake Ladoga and the Neva River, but also the Gulf of Finland and the Baltic Sea, the preservation of which is of international importance.

Another problem is that of the influence of the set of constructions intended for the protection of the city of St Petersburg from inundations (the "Dam" problem) on the ecological state of the Neva Mouth and the eastern part of the Gulf of Finland.

2 The Programme "Ecology of St. Petersburg" and its section - "Monitoring"

The Programme under consideration was outlined at the St. Petersburg Scientific Centre of the Russian Academy of Sciences (RAS) in 1990.

NATO ASI Series, Vol. I 24
Remote Sensing and Global Climate Change
Edited by Robin A. Vaughan and Arthur P. Cracknell

The Programme embraces interdisciplinary fundamental and applied studies aimed at the development of a theory, as well as practical recommendations, concerning the stabilisation and improvement of the current ecological situation in the St. Petersburg region.

The Programme consists of 5 Sections:

1. Forecasting of the ecological situation in the region.

2. Development of the scientific basis, methods and means of regional monitoring of the natural environment ("Monitoring").

3. Ecology, human health and the state of the animal and plant life in the region.

4. Development of concepts of environmental management in the region.

5. Development of recommendations concerning the improvement of the ecological situation in the region on the basis of the implementation of contemporary technologies.

We shall dwell in more detail on Section 2 ("Monitoring").

Four main problems are considered here:

• Development of a concept of ecological monitoring in the region of the city of St. Petersburg

• Working out of methods for ecological monitoring

• Development and creation of a sub-system of aerospace monitoring of the environment in the region

• Development of technical means for the implementation of the monitoring including data processing.

During the initial stage of the realisation of Section 2 of the Programme in 1991, an analytical survey of the state of the problem of monitoring the development of the basis for the concept of creating a system of natural environment monitoring in the region of St Petersburg was performed. The research was carried out by an implementation team, which included the leading experts in the field of ecology representing various relevant institutions of the city of St. Petersburg.

3 A brief characterisation of the state of the environment and ecological situation in the St. Petersburg region

The St. Petersburg region is characterised by an intense technogenical impact on the environment causing a high level of pollution of all media: the air basin, waterbodies, soils and underground waters. The St. Petersburg region has been defined as the St. Petersburg province proper, together with the basin area formed by the water system of Lake Ladoga, the Neva river, the Neva mouth and the eastern part of the Gulf of Finland. It should be emphasised here that the Lake Ladoga basin is of special importance for the above-mentioned region. Its schematic map is given in Figure 1 (Ivanova and Kirillova 1966).

The level of pollution of the air basin of the city of St. Petersburg and its region is

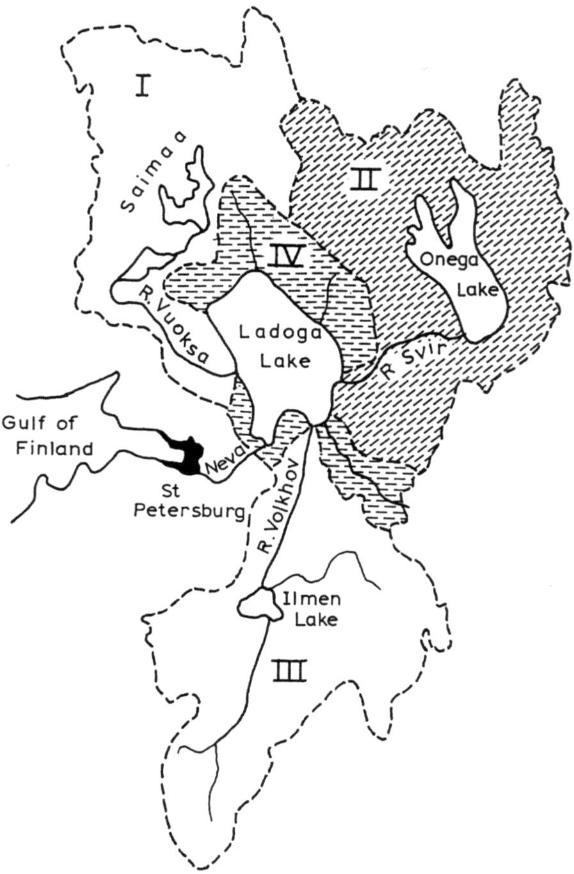

Figure 1: Schematic map of lake Ladoga basin and St. Petersburg region.

fairly high, although the city proper is probably somewhat less polluted in comparison with other cities with approximately the same population in Russia and in the world. The polluted area around St. Petersburg has a radius of about 100 km. The most polluted districts in the St Petersburg province are the Kirishi, Kirovsk, Slantzy and Kingisepp districts (The Annual Report of Atmospheric Air Pollution ..., 1990).

The most dangerous environmental situation is observed in the inland waterbodies of the region, especially in the system of Lake Ladoga, the Neva River, the Neva Mouth and the eastern part of the Gulf of Finland.

The Neva Mouth waters can be classified in the category of moderately polluted - polluted waters of the III - IVth class, and the bottom sediments - in that of moderately polluted-polluted ones of the III-Vth class. The most polluted part of the Neva Mouth is that nearest to the city of St. Petersburg. If during the period 1977-1989 the Neva Mouth was a mesotrophic waterbody, in 1990 its water area was characterised as eutrophic. Before 1989, the Eastern part of the Gulf of Finland was a mesotrophic object on the whole; since 1989, it is characterised as a eutrophic one (The Annual Report on Sea Water Quality ..., 1990).

The situation with Lake Ladoga is rather severe. Its catchment basin embraces a system of three lakes, Saimaa, Onega and Il'men', and occupies an area of 258 600 km^2 . The intensive development of economic and industrial activity over the water catchment territory of Ladoga has had a very negative effect on the water quality and the trophic status of the lake.

Beginning with the 1960s, the lake was becoming more and more subject to growing anthropogenic pressure which started the process of anthropogenic eutrophication. By the mid-1980s, the concentration of phosphate in Lake Ladoga increased by a factor of 3. In less than 20 years, the lake was transformed from an oligotrophic waterbody to a mesotrophic one. The input of phosphate there is approaching a critical level when the lake can become eutrophic (Voropayeva and Rumyantsev 1991).

The heavy metal pollution of the lake has reached a considerable amount. Particularly high are the concentrations of lead, copper, zinc, manganese and iron. A considerable bacterial polution is also observed which at certain points exceeds the admissible level by 3-4 orders of magnitude (Voropayeva and Rumyantsev 1991).

Soils in the St. Petersburg region have also been subject to marked changes. The arable land is characteristic of phosphate overloading. The soils are contaminated with heavy metals. The average concentration of copper in some areas exceeds the background level by a factor of 32, that of zinc by a factor of 19 and that of lead by a factor of 2.5. High pesticide loadings also exist. The level of soil pollution in the St. Petersburg region has been defined as a critical one (Voropayeva and Rumyantsev 1991).

Continuous deterioration of forests is also a characteristic feature of the St Petersburg region. The situation has been caused mainly by the contemporary technologies of forest cutting, the considerable recreation impact on forests and the effect of atmospheric pollution. Apparently forests, especially coniferous woods, are damaged due to atmospheric contamination, although the extent of this damage should only be determined in the course of monitoring the state of forests (Mironova and Slepyan 1983).

4 Assessment of anthropogenic impact on the regional environment

The anthropogenic impact on the atmosphere is to a great extent determined by the amount and composition of pollutant emissions into the atmosphere. The total number of industrial enterprises in St. Petersburg which contribute to the emissions of harmful matter is about 400, with over 22,000 sources of release.

Apart from stationary sources of emissions, automobile transport contributes very considerably to atmospheric pollution. In such towns as Vyborg, Gatchina, Kingisepp, Luga, Priozersk, Tikhvin and Tosno, the automobile transport contributes over a half of all emissions to the atmosphere.

The emissions of harmful substances to the atmosphere in 1990 reached 436.4 thou-

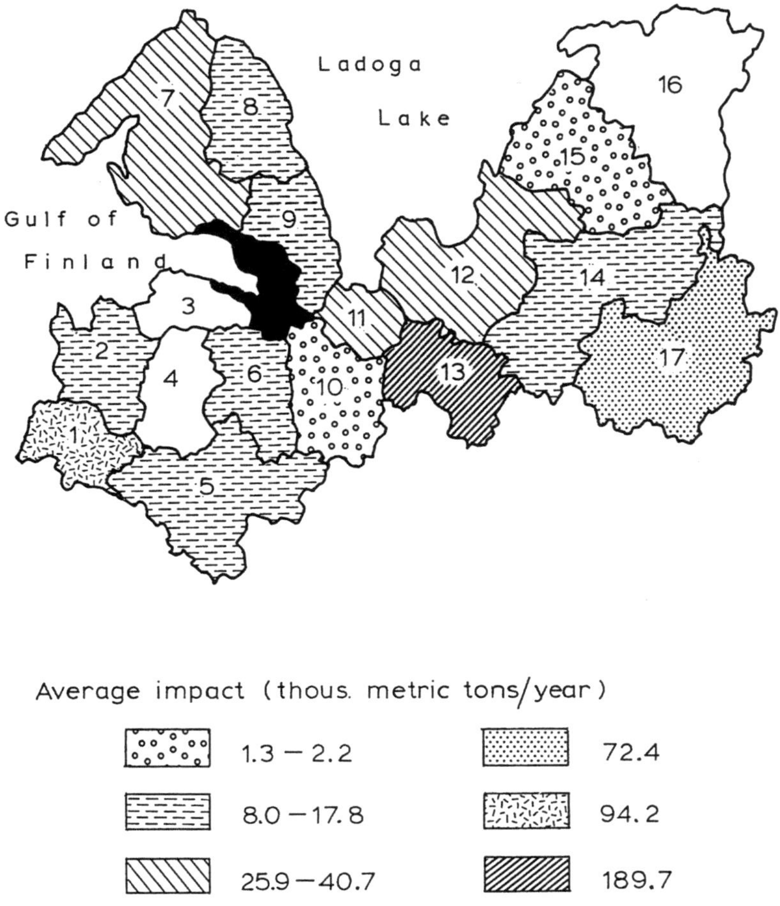

Figure 2: Anthropogenic impact on the atmosphere in the St. Petersburg region.

sand tons in St. Petersburg, and 740.0 thousand tons in the St. Petersburg province. A map of anthropogenic impact on the atmosphere (emissions to the atmosphere) for the St. Petersburg province, compiled at the Institute of Hygiene and Professional Pathology, is presented in Figure 2. It shows that the Kirishi and Slantzy districts are subject to the strongest anthropogenic impact.

Substantial pollution of the atmosphere, waterbodies, rivers, soils and forests is caused by the transboundary transport.

Lake Ladoga is subject to a strong anthropogenic loading. 13 big paper mills and timber industrial enterprises are located in its basin. Their industrial waste contains over 300 toxic substances. The development of large-scale animal farms in the Lake Ladoga catchment is responsible for the input of biogenes to the lake waters. 81 out of 218 big farms are located within the borders of water protection zones.

Other waterbodies and the inland water network of the region are also subject to

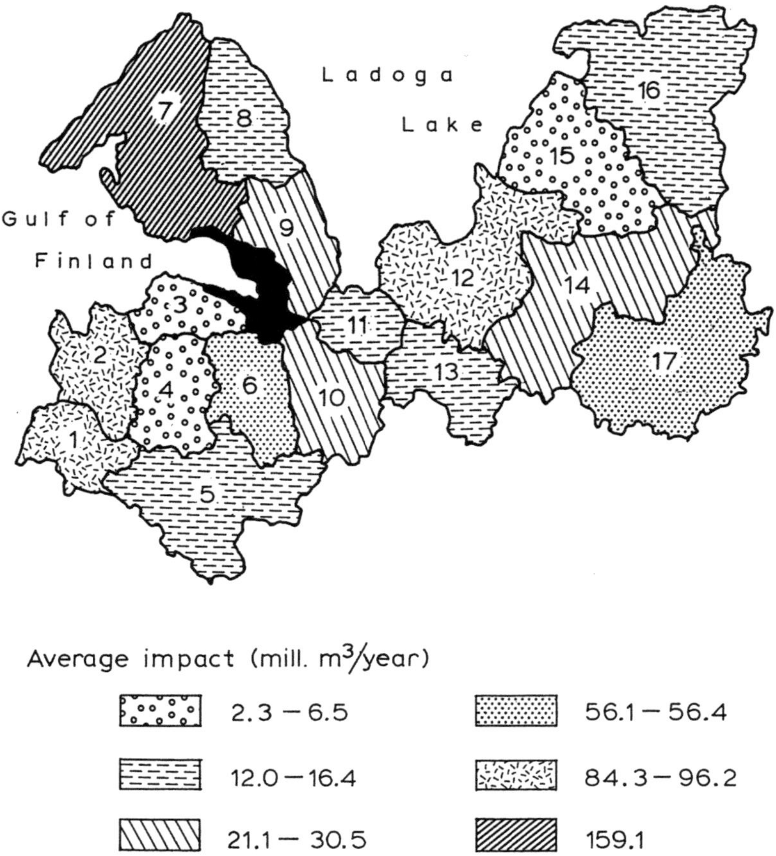

Figure 3: Anthropogenic impact on the waters in the St. Petersburg region.

a strong anthropogenic impact. The total number of industrial enterprises producing emissions into water objects of the St. Petersburg province is 635. A map of anthropogenic impact on the provincial water objects is presented in Figure 3.

Apart from the emissions of industrial enterprises, water objects are also polluted by runoff from the fields, where pesticides are employed, and by precipitation from the atmosphere. For instance, according to the data available, the level of pollution of Lake Ladoga through the atmosphere (wet and dry deposition) is presently quite comparable with its contamination from other sources.

On the basis of the data on total industrial emissions and on the level of pollution of the atmosphere and water objects for the recent 5 years, a complex hygienic assessment of the entire province has been performed. As a result, four districts have been singled out as the most polluted and having the strongest anthropogenic loading in the St. Petersburg province; they are the Vyborg, Volkhov, Kingisepp, and Kirishi districts. The most serious ecological situation has been found in the Kirishi district. The least polluted districts are the Podporozhye, Luga, Tosno and Lodeynoye Pole districts.

5 The state and efficiency of the functioning observation systems used for control of the state of the environment

The system of monitoring which is being created should be based first of all on the existing observation networks, such as:

- the networks of the ROSKOMGIDROMET (Russian Committee on Hydrometeorology)
- the system of sanitary control stations of various administrative hierarchy levels
- the observational system of the Geological Service
- the information service for plant protection.

Here is a brief characterisation of the observational networks functioning in the city of St. Petersburg and its province;

(1) Network of air pollution observations (ROSKOMGIDROMET):

- 18 stationary posts in 5 towns in the St. Petersburg province (Kirishi - 3 posts; Vyborg, Kingisepp, Luga - 1 post in each town); 12 posts (out of 18) are located in St. Petersburg
- 2 laboratories for atmospheric air analysis (St. Petersburg and Kirishi). In Vyborg, Luga and Kingisepp pollution control is carried out on an occasional basis. The number of pollutants observed is 32
- 29 "en-route" points of air control in St. Petersburg
- 3 stations in the St. Petersburg province for the observation of the chemical composition and pH of precipitation
- a number of mobile laboratories are available.

(2) Network of water pollution observations (ROSKOMGIDROMET):

- analysis of surface waters (52 ingredients)
- 2 laboratories (St. Petersburg, Kirishi)
- selection and analysis of samples in the St. Petersburg province - at 32 rivers (52 points) and 2 lakes (15 points)
- 13 points for sample selection in the city of St. Petersburg

- selection of water samples at the Kronstadt Hydrometeorological Laboratory, at hydrometeostations in Schlisselburg, Vyborg, Tikhvin, Liuban', Tolmachovo, Kingisepp and Kirishi.

(3) The observation network of VODOKANAL:

- drinking water quality control at water-supply station laboratories
- purification station observations (emission control laboratories)
- the control-analysis department - 3 laboratories
- tap water analysis
- effluent analysis.

(4) Network for stationary observation of the regime of underground waters (the Geological Service):

- 512 points (wells) in the St. Petersburg province
- 42 (out of 512) specialised points for routine observations of underground waters.

(5) Provisional information service in the plant protection system:

- laboratory for diagnostics and forecasting of the plant protection station
- 15 points of agricultural consulting and forecasting
- a network of inter-district quarantine inspectorates.

The problem is to assess the adequacy of the existing observational means from the viewpoint of their sufficiency, as well as their spatial and temporal representativity. The complementary use of remote sensing techniques is very important in this respect.

6 The place and role of aerospace remote sensing techniques in the system of regional monitoring of the environment

The following problems may be solved using aerospace techniques of remote sensing aimed at the monitoring of the environment:

(1) Atmospheric pollution control:

- concentrations of $S0_2$ and aerosols
- the ozone concentrations

- concentrations of CO, hydrocarbons, CO_2 , nitrogen oxides, etc.

(2) Observations of oceans, seas and inland waterbodies:

- spatial-temporal concentration fields of phytoplankton chlorophyll
- organic and mineral suspended matter concentrations in water
- areas and sources of pollution of water areas
- water pollution by hydrocarbons
- the state of bottom, shallow water and coastline
- water surface temperature variations.

(3) Land surface observations:

- the soil mineral composition (the component concentration, salinisation, littering)
- soil moisture
- erosion processes.

(4) Observation over urbanised territories:

- changes in areas of such territories and forecasting of their growth dynamics
- the ecological state of urbanised territories.

(5) Control of dynamic exogenic geological processes:

- formation and development of territories of exogenic processes
- processes of fertile soil layer washout, bogging, desertification, soil erosion, vegetation damage
- water use.

(6) Global lithospheric processes.

(7) Technological accidents, natural disasters and extreme ecological situations.

Data required for solving problems of ecological monitoring of the St. Petersburg region are available and are obtained from the following satellites:

- Cosmos - 1939 - digital data (surface images) of multi-spectral scanners of medium (MSU-SK) and high (MSU-E) resolution;
- NOAA – digital data (images) of the AVHRR scanner;

- ERS-1 – the Earth's surface images obtained using a synthetic aperture radar (SAR);
- Almaz-1 – the Earth's surface images obtained using SAR.

A very important element of an aerospace monitoring sub-system should be special aircraft laboratories. Regrettably, such flying laboratories equipped by a complex of remote sensing instruments, are available in the city of St.Petersburg only on an occasional basis.

7 Basic elements of the of regional environmental monitoring

7.1 Introductory Notes

The eventual aim is to create, in the long run, a system of complete ecological monitoring. Such a system should embrace geophysical and biological monitoring in their close interaction.

It should be noted that the current world-wide tendency is to create at local and regional levels integrated systems of monitoring for natural environment plus human health.

7.2 Regional monitoring of the environment and human health: objectives and tasks

The system of regional monitoring is intended for observation of the state of the environment and the health of the population of the region, as well as for purposes of diagnostics and forecasting. The principal objective of regional monitoring is detection of the main harmful consequences of the impact on the environment and the population, as well as working out of recommendations for necessary and adequate management decisions to undertake relevant measures for the protection of the environment and the population.

Concrete tasks of regional monitoring are as follows:

- control of sources of pollution (at a local level) of natural and anthropogenic origin (industrial emissions, the effect of agricultural farms and fields after fertilisers and pesticides are introduced, the influence of transport and communal sources, etc.)
- gathering of data on the cycles of pollutants, their transformation and accumulation. The determination of the main routes of migration of pollutants
- control of levels of environmental pollution by harmful chemical compounds, of thermal pollution, levels of noise and various emissions, detection of areas of ecological risk and dangerous conditions

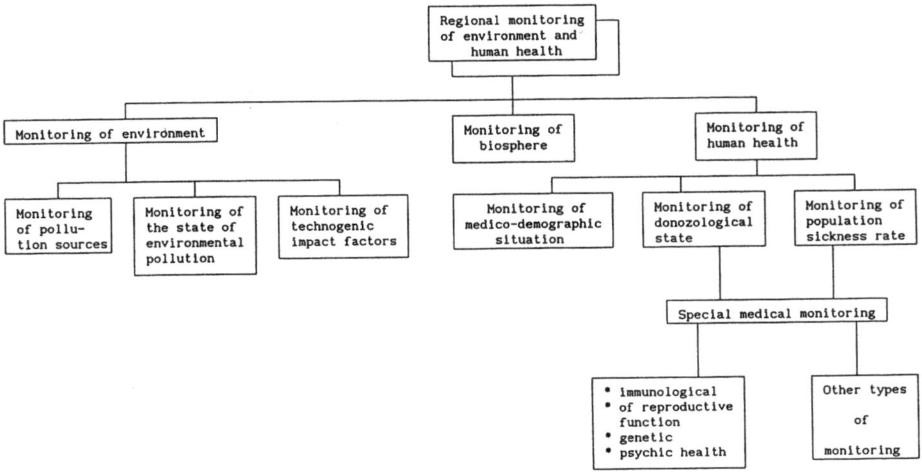

Figure 4: Regional monitoring of the environment and human health.

- control of disturbances in the vegetation cover of the region, the use of renewable and non-renewable resources, etc.
- observation of the response of biota to various impact factors
- observations of the response of the human organism to various impacts.

7.3 A system proposed for regional monitoring of the environment and human health

The principal scheme of regional monitoring of the environment and human health is based on the consideration of three main components (Figure 4):

- monitoring of the environment (the abiotic component)
- monitoring of the biosphere (the biotic component)
- monitoring of human health.

With regard to human health monitoring, the St. Petersburg Institute of Hygiene and Professional Pathology has been developing for a number of years a methodology for the early detection of harmful factors of regional impact and changes in human health caused by these factors.

The developed concept of creating an integrated national information system of ecological control implies that it is a composite part of a similar international system, and as such, it should involve regional information-processing systems (IPS) which must integrate the IPS of individual cities and provinces. The latter, in their turn, must include the IPS of districts linked by communication lines with local centres for information collection (LCIC). The local centres are intended for the collection and transformation of signals from primary sources into a digital form for their transmission and processing on a computer. Primary information sources are linked with the collection channels and are at a distance of about 1 km from the centre.

The present network of the Russian Academy of Sciences can be used as a basis for an information-processing system of environmental pollution control. The network involves the information-processing centres of major industrial and agricultural centres over the territory of the north-western region of Russia, and it is fairly well developed in the city of St. Petersburg.

Taking account of the already existing possibilities both in St Petersburg, and its region, the elements of a future system of regional monitoring can be created, and in fact are being created now.

The structural scheme of such a system, developed at the St Petersburg Institute of Informatics and Automatisation of RAS, is presented in Figure 5. In perspective, it can be transformed into a system of environmental monitoring of the entire north-western region of the country.

7.4 Regional geographical information system (GIS) for ecological purposes

An important component of the system of regional environmental monitoring is a regional geographical information system (GIS) for ecological purposes. The fundamentals of such a GIS have been worked out at the St. Petersburg Division of the State Oceanographic Institute. The concept of the GIS "Baltic Europe" may be considered as an initial approximation for determining the structure of a regional GIS intended for ecological purposes.

A regional GIS must involve the following components:

- a database "Ecology" (environment)
- a database "Health" (human health)
- a unit for adjustment of the databases
- a unit of models of the "Priroda" (Nature) class (diagnostic and prognostic models for operation with data on the environment and human health)
- a unit of models of the "Decision" class (elaboration of versions of management decisions, etc.)
- a unit of output information.

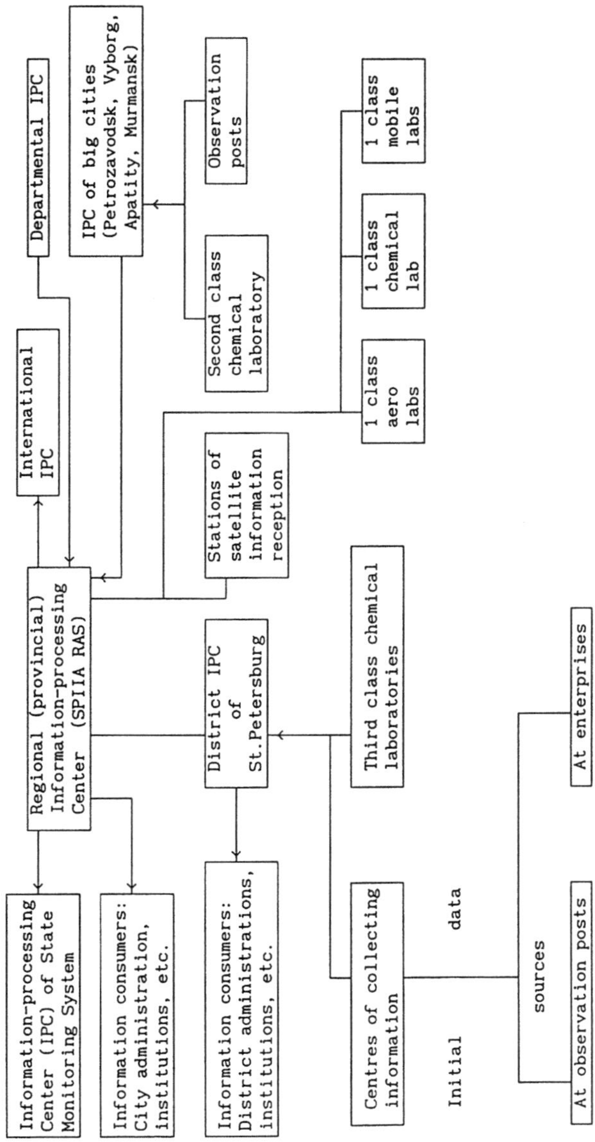

Figure 5: Structure scheme of environment and human health monitoring system in the St. Peterburg region.

8 Proposals concerning international cooperation

The St. Petersburg Region, or in other words, mainly the basin of the water system "Lake Ladoga and the Neva Mouth" may be regarded as a suitable site for international cooperation in studying regional ecological problems. Naturally, when global changes are investigated, special attention should be given to regional projects which must examine, under well-controlled conditions, the mechanisms of the interaction of various media with certain anthropogenic impacts (Kondratyev, 1990; Marchuk and Kondratyev, 1992). In this case, it becomes possible to provide representative in situ data for the investigations, as well as remote sensing information.

From this view point, the selection of the water system Lake Ladoga, the Neva River, the Neva Mouth and the eastern part of the Gulf of Finland as a regional object seems to be a good choice.

The choice is also due to the fact that this water system is subject to a technogenic pressure and can therefore also be a natural site for the validation of techniques of remote sensing for monitoring the state of the environment. Here it should be noted that this system has been thoroughly investigated.

Apart from this, the natural system under examination is of considerable interest, both in the scientific and practical sense, not only for Russia, but also for other states in the Baltic region. It is therefore most suitable as an object for international cooperation.

Besides, the problem of the ecological condition of the eastern part of the Gulf of Finland is urgent in connection with the building of protective constructions against floods in the city of St. Petersburg.

Finally, the programme "Ecology of St. Petersburg" has been developed and is being realised in St. Petersburg. It is within the framework of this programme that the regional system of ecological monitoring is now being created.

References

Ivanova, Z. M., and Kirillova, V. A., 1966, A calculation of the river inflow to the Ladoga Lake and the Neva-River discharge, Trans. Limnol. Lab. In *Hydrological Regime and Water Balance of Ladoga Lake, vol. XX* (Leningrad: Leningrad University Press), pp. 119-132.

Kondratyev, K. Ya., 1990, *Key Problems of Global Ecology, Itogi Nauki i Tekhniki, Theoretical and General Aspects of Geography, vol. 9* (Moscow: VINITI).

Marchuk, G. I., and Kondratyev, K. Ya., 1992, *Priorities of Global Ecology* (Moscow: Nauka).

Mironova, T. I., and Slepyan E. I., 1983, *Leningrad Region Environment and Its Protection* (Leningrad: Lenizdat).

The Annual Report on Atmospheric Air Pollution of Urban Territories in the area controlled by North-Western Branch of "Roskomgidromet", 1990, (Leningrad: Gidrometeoizdat).

The Annual Report on Sea Water Quality in the Eastern Part of the Gulf of Finland from hydrobiological data, 1990, (Leningrad: Gidrometeoizdat).

Voropayeva, G. M., and Rumyantsev, V. A., 1991, *A Concept of Improvement of Environmental Management in the Ladoga Lake Basin, 1991* (Leningrad: Gidrometeoizdat).

A Second-Order Two-Scale Model for Rough Surface Scattering with Applications to Scatterometry

N. I. Nickolaev and M. A. Michalev

1 Introduction

Scatterometry is a powerful method for studying the dynamics of the ocean-atmosphere interactions. It is based on the fact that the normalised backscattering radar cross section (NBRCS) σ^o is sensitive to the surface friction velocity (Jones and Schroeder 1978). The algorithms used to retrieve the surface wind speed are based on empirical models of the microwave scattering signatures derived from experimental data acquired under different atmospheric conditions. The physical mechanisms of the interaction between the microwaves and the sea surface have been the subject of active study for many years but no a theory which explains completely the microwave scattering properties of the sea-atmosphere interface has yet been proposed. The so-called two-scale model (TSM) is still the most widely accepted theoretical treatment of radar backscattering from the sea surface measured at small and moderate incidence angles (Ulaby *et al.* 1982).

The basic idea of the two-scale model is to divide the sea waves spectrum into two parts corresponding to large-scale and small-scale components of the wave system. The dominating scattering mechanism at moderate angles of incidence (the case which will be considered in this chapter) is the resonant Bragg scattering from the small-scale waves which are modulated by the large-scale waves. Thus the radar cross section given by the two-scale model is simply the radar cross section calculated from the small perturbation theory (SPT) (Rice 1951) and averaged over the local slopes of the large-scale waves. The usual practice is to use first-order perturbation expressions only (with respect to the small parameter $(k\eta)$, k being the radar wavenumber and η being the rms height of the small-scale waves) assuming that the second-order terms do not contribute appreciably to the backscattered signal. However, it has recently been demonstrated (by comparison with numerical experiments in the cases of scattering of scalar waves from one-dimensional Gaussian perfectly sound-soft surfaces) that for certain scattering circumstances the second-order terms may produce up to 10 dB contribution to the scattering cross section (Thorsos and Jackson 1989, Thorsos 1990).

In this chapter we consider the contributions of the second-order scattered fields to

NATO ASI Series, Vol. I 24
Remote Sensing and Global Climate Change
Edited by Robin A. Vaughan and Arthur P. Cracknell

the normalised backscattering radar cross section values calculated within the two-scale model for the realistic case of radar backscattering by the ocean surface. To this end, we have extended the approach used by us earlier (Nickolaev *et al.* in print) to account for the second-order fields. The results are compared with both experimental results and the predictions of the first-order two-scale model.

2 Formulation of the second-order two-scale model

Following an approach described by Ulaly *et al.* (1982) we write down an expression for the normalised backscattering radar cross section σ^o_{HH} for the case of horizontally polarised transmitted as well as received radiation which is valid up to the second-order of the scattered fields:

$$\begin{aligned}\sigma^o_{HH} = &\left\langle \left(\hat{h}\hat{h}'\right)^4 \left(\sigma^{(11)}_{h'h',h'h'} + \sigma^{(22)}_{h'h',h'h'}\right)\right\rangle + \left\langle \left(\hat{h}\hat{v}'\right)^4 \left(\sigma^{(11)}_{v'v',v'v'} + \sigma^{(22)}_{v'v',v'v'}\right)\right\rangle \\ &+ \left\langle 2\left(\hat{h}\hat{h}'\right)^2 \left(\hat{h}\hat{v}'\right)^2 \left(\sigma^{(11)}_{v'v',h'h'} + \sigma^{(22)}_{v'v',h'h'} + 2\sigma^{(22)}_{h'v',h'v'}\right)\right\rangle \\ &+ \left\langle 4\left(\hat{h}\hat{h}'\right)\left(\hat{h}\hat{v}'\right)^3 \sigma^{(22)}_{v'v',h'v'}\right\rangle + \left\langle 4\left(\hat{h}\hat{h}'\right)^3\left(\hat{h}\hat{v}'\right) \sigma^{(22)}_{h'h',h'v'}\right\rangle\end{aligned} \tag{1}$$

where the angular brackets denote averaging with respect to the large-scale wave slopes; $\hat{h}$ and $\hat{v}$ are the polarisation vectors in the main coordinate system system; $\hat{h}'$ and $\hat{v}'$ are vectors determining the polarisation with respect to a coordinate system associated with the local tangent plane. The backscattered cross sections computed in the local coordinate system using small peturbation theory are denoted by $\sigma^{(nn)}_{r'_1t'_1,r'_2t'_2}$ where the superscripts denote the orders of the backscattered fields contributing to this term. The subscripts r'_1 and r'_2 denote the polarisation of the received radiation and the subscripts t'_1 and t'_2 the polarisation of the transmitted one.

Similarly, for the case of cross-polarisation, the expression for the normalised backscattering radar cross section σ^o_{VH} reads:

$$\begin{aligned}\sigma^o_{VH} = &\left\langle \left(\hat{v}\hat{v}'\right)^2 \left(\hat{h}\hat{v}'\right)^2 \left(\sigma^{(11)}_{v'v',v'v'} + \sigma^{(22)}_{v'v',v'v'}\right)\right\rangle \\ &+ \left\langle \left(\hat{h}\hat{h}'\right)^2 \left(\hat{v}\hat{h}'\right) \left(\sigma^{(11)}_{h'h',h'h'} + \sigma^{(22)}_{h'h',h'h'}\right)\right\rangle \\ &+ \left\langle \left(\hat{v}\hat{v}'\hat{h}\hat{h}' + \hat{h}\hat{v}'\hat{v}\hat{h}'\right)^2 \sigma^{(22)}_{h'v',h'v'}\right\rangle \\ &+ \left\langle 2\left(\hat{v}\hat{v}'\right)\left(\hat{h}\hat{h}'\right)\left(\hat{h}\hat{v}'\right)\left(\hat{v}\hat{h}'\right) \left(\sigma^{(11)}_{v'v',h'h'} + \sigma^{(22)}_{v'v',h'h'}\right)\right\rangle \\ &+ \left\langle 4\left[\left(\hat{v}\hat{v}'\right)^2\left(\hat{h}\hat{v}'\right)\left(\hat{h}\hat{h}'\right) + \left(\hat{v}\hat{v}'\right)\left(\hat{h}\hat{v}'\right)^2\left(\hat{v}\hat{h}'\right)\right] \sigma^{(22)}_{v'v',h'v'}\right\rangle \\ &+ \left\langle 4\left[\left(\hat{h}\hat{h}'\right)^2\left(\hat{v}\hat{h}'\right)\left(\hat{v}\hat{v}'\right) + \left(\hat{h}\hat{h}'\right)\left(\hat{v}\hat{h}'\right)^2\left(\hat{h}\hat{v}'\right)\right] \sigma^{(22)}_{h'h',h'v'}\right\rangle\end{aligned} \tag{2}$$

It should be noted that in both (1) and (2) it is assumed that, due to reciprocity, backscattered hv and vh fields are equal. Therefore, the expression for σ^o_{HV} (vertical transmitted and horizontal received radiation) is identical to (2) whereas the expression for σ^o_{VV} can be obtained from (1) simply by replacing the symbols h with v and v with

h (Ulaby *et al.* 1982). Thus, the total number of second-order terms that should be accounted for in the second-order two-scale model is six.

The normalised backscattering radar cross section $\sigma^{(11)}_{r_1't_1',r_2't_2'}$ given by the first-order small peturbation theory can be found elsewhere (see for example Ulaby *et al.* 1982). For the second-order small peturbation theory contributions $\sigma^{(22)}_{r_1't_1',r_2't_2'}$ to the total normalised backscattering radar cross section we can write the following general expression:

$$\sigma^{(22)}_{r_1't_1',r_2't_2'}(\theta') = 2\pi k^4 \cos^4\theta' \times \int_{-\infty}^{+\infty}\int_{-\infty}^{+\infty} \mathcal{K}_{r_1't_1',r_2't_2'}(u,v,k,\theta')\, W(u+k\sin\theta',v)\, W(u-k\sin\theta',v)\, du\, dv \quad (3)$$

where θ' is the local angle of incidence, k is the radar wavenumber and W is the two-dimensional spectrum of the small-scale waves. Explicit expressions for the six kernels $\mathcal{K}_{r_1't_1',r_2't_2'}$ are readily found in a manner analogous to that described by Valenzuela (1967). As in Nickolaev *et al.* 1992, we shall use a model for the sea surface spectrum which consists of the Pierson and Stacy radial part and a new angular factor which pertains to both large and small scale waves. This spectrum depends on only one external parameter — the wind speed at height 19.5 m, $u_{19.5}$.

3 Results and discussion

The evaluation of the second-order contributions in (1) and (2) requires quadruple integration — the inner integral is given by (3) and the outer double integral is over the principal slopes of the large-scale waves. Since the integrands in (3) possess very sharp peaks, a straightforward multi-dimensional numerical integration is hardly efficient (if realisable at all). Moreover, as has been shown by Nickolaev *et al.* 1992, a realistic model for the radar sea backscatter should account for the non-gaussian nature of the large-scale sea slopes. This excludes simplifications that can be made using symmetry arguments. To avoid these difficulties, we calculate for the given radar frequency (radar wave number k) and the given sea surface spectrum W (i.e. given wind) all six second-order terms given by the small peturbation theory only for a limited number of angles of incidence θ'. These results serve for the construction of spline functions which are then used to interpolate the actual continuous dependences of the second-order terms on θ'. Since all dependences $\sigma^{(22)}_{r_1't_1',r_2't_2'}(\theta')$ are smooth functions, the interpolation errors are small.

The values of σ^o are obtained following a specific multi-step procedure described in detail by Nickolaev *et al.* 1992. The only free parameters in this approach are the cutoff wavenumber values that separate the spectral intervals of the large-scale (K_L) and small-scale (K_S) wave components and a parameter ε that accounts for the modulation of the small ripples by the large sea waves. The criteria used to determine these free parameters are also discussed by Nickolaev *et al.* 1992.

Figure 1a presents the calculated K_u band ($f = 14.6$ GHz) azimuthal dependences of the second-order normalised backscattering radar cross section (solid lines) for VV polarisation at a wind speed of 11.14 m s^{-1} and incidence angles of 40°, 50° and 60° (top to bottom, respectively). The small circles correspond to the empirical Wentz model

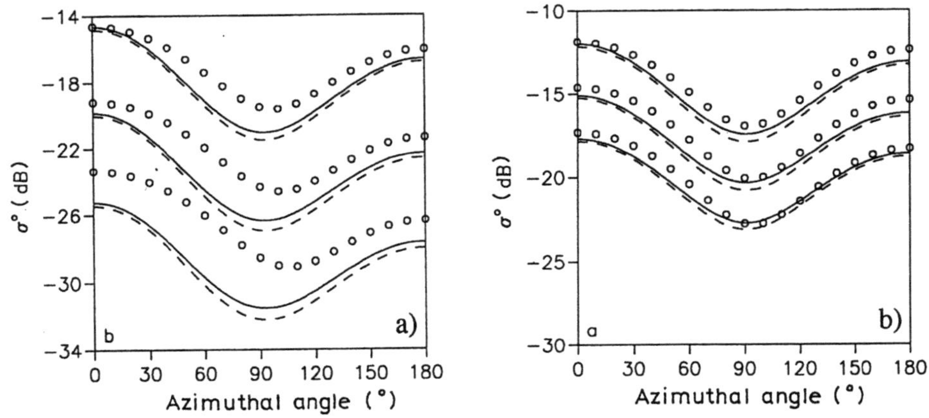

Figure 1: Second-order two-scale model predictions for the azimuthal dependence of the backscattering cross section σ^o for VV polarization (solid lines) for three different angles of incidence: 40°, 50°, 60° (respectively, top to bottom); $U_{19.5} = 11.14$ m s^{-1}, $f = 14.6$ GHz, $K_L = 1.4908$ rad cm^{-1}, $K_S = 1.262$ rad cm^{-1} and $\varepsilon = 0.5$. The predictions of the first-order tow-scale model are illustrated by dashed lines. The circles represent the Wentz empirical model function. The azimuthal angle of 0° corresponds to the upwind direction. (b) the same as (a) except for HH polarization.

function and the dashed lines represent the results of the first-order two-scale model (see Figure 5a Nickolaev *et al.* 1992). It is seen that the second-order terms lead to better agreement with the experimental results; the contribution of the second-order terms is more appreciable in the crosswind direction. The maximum difference between the first-order curves and the second-order curves is 0.5 dB and the maximum difference between the experiment and the second-order theory is less than 1 dB (about 0.8 dB).

The curves presented in Figure 1b correspond to the same sea and radar conditions as in Figure 1a except that the polarisation is HH. Although the contribution of the second-order terms is greater than in the case of VV polarisation (up to 0.7 dB), the difference between the theory and the experiment is still significant (up to 3.0 dB). This discrepancy is expected , since it is believed that for this polarisation it is not only the Bragg scattering mechanism which plays an important role but additional effects are involved especially at higher incidence angles (see for example Donelan and Pierson 1987).

We compared our model also with X-band experimental data (Long 1983, Masuko *et al.* 1986). The agreement between the theory and the experiment for the like-polarised cases is very good with approximately the same contributions of the second-order terms as those for the case of K_u-band. For the cross-polarised states, the main part of the backscattered power is accounted for by the first two terms of equation (2), while the contribution of the other four terms to σ^o is of order tenths of dB.

The fact that contributions of the second-order terms to the normalised backscattering radar cross section in all the cases considered are much smaller than the first-order ones is an expected result for the scattering from surfaces with wide roughness spectra such as those of the sea surface (see the comments by Nicklaev *et al*). 1992. It has recently been confirmed by the results of Thorsos (1990). He reports a somewhat higher contribution of about 2 dB, but for scattering from a rather rougher surface ($k\eta = 1.79$). For the case of moderate small-scale roughness ($k\eta = 0.45$) that is comparable with what we considered here ($k\eta = 0.32$) Thorsos found that the contribution of the second-order small peturbation theory terms is less than 1 dB. Thus, the second-order two-scale model may prove useful for the interpretation of relatively fine effects which require a precision within 1 dB.

4 Conclusions

We have developed a numerical algorithm which evaluates the contribution of the second-order scattered fields to the normalised backscattering cross section in the framework of the two-scale model. It allows more precise exploration of the features of the azimuthal dependencies of σ^o and in particular the upwind-downwind and the upwind-crosswind ratios that are often as large as tenths of a dB. On this basis, a variety of inverse problems related to the two-dimensional spectral properties of the waves on the sea surface can be addressed.

References

Donelan M. A., and Pierson, W.J., 1987, Radar scattering and equilibrium ranges in wind generated waves with application to scatterometry. *Journal of Geophysical Research*, **82**, 4971-5029.

Jones W. L., Schroeder, L.C., 1978, Radar backscatter from the ocean: Dependence on surface friction velocity. *Boundary-Layer Meteorol*, **13**, 133-149.

Long M. W., 1983, *Radar reflectivity of land and sea* (Dedham: Artech House).

Masuko H., Okamoto, K., Shimada, M., and Niwa, S., 1986, Measurements of microwave backscattering signatures of the ocean surface using X-band and K_a-band airborne scatterometer. *Journal of Geophysical Research*, **91**, 13065-13083.

Nickolaev N. I., Yordanov O.I., and Michalev, M.A., 1992, Non-Gaussian Effects in the Two-Scale Model for Rough Surface Scattering. *Journal of Geophysical Research – Oceans*, **97**, 15617-15624.

Rice S. O., 1951, Reflection of Electromagnetic Waves from Slightly Rough Surfaces. *Communications on Pure and Applied Mathematics*, **4**, 351-378.

Thorsos E. I., and Jackson, D. R., 1989, The validity of the perturbation approximation for rough surface scattering using a Gaussian roughness spectrum. *Journal of the Acoustical Society of America*, **86**, 261-277.

Thorsos E., 1990, Acoustic scattering from a "Pierson-Moskowitz" surface. *Journal of the Acoustical Society of America*, **88**, 335-349.

Ulaby F. T., Moore, R. K., and Fung, A. K., 1982, *Microwave Remote Sensing. Active and Passive* (Reading: Addison-Wesley). **2**, 12.

Valenzuela G. R., 1967, Depolarization of EM Waves by Slightly Rough Surfaces. *IEEE Transaction on Antennas and Propagation*, **AP-15**, 552-557.

Politics and Climate Change

K. Ya. Kondratyev, A. P. Cracknell and
R. A. Vaughan

1 Introduction

During the course of the summer school, Professor Kondratyev suggested that we might hold a panel session or round table discussion on politics and global climate change. This chapter comprises a summary of the discussion on this subject.

Participants were invited to submit questions in advance of that session to form the basis of the discussion. The following questions were submitted and are reproduced verbatim and in no particularly special order:

> Why is the USA not positive in reducing emissions of greenhouse gases, but the western European countries are?
>
> Which is more important - developing the economy or reducing greenhouse gas emissions in the developing countries?
>
> Considering the energy supply in the next decades, should climatologists support the nuclear option?
>
> Why are thousands of pounds spent on research in different institutes, universities etc., carrying out research in the same fields which most of the time come up with different results? Is this a point of uncertainty for the politicians?
>
> Are there differing perceptions of tropical forest between national and global interest?
>
> How can we balance potential positive effects of climate change against potential negative effects?

NATO ASI Series, Vol. I 24
Remote Sensing and Global Climate Change
Edited by Robin A. Vaughan and Arthur P. Cracknell

What agricultural policies are there or should there be to reduce the losses by farmers due to climate change?

Should scientists get much more involved in politics?

How do we overcome the problem of impending global warming without having to take a cut in our standard of living?

Why has the second United Nations Conference on Environment and Development not been a success?

What are the implications of technology transfer from north to south or from the developed to the developing world?

How can the developing countries develop without environmental destruction and degradation?

Can the magnitude of the global climate change be reduced by strict economic awareness without the threat of north - south political confrontation?

What are the technical and political means to reduce emissions of greenhouse gases, in particular CO_2?

A general session was set aside to consider how to address these questions, and at this, after some general discussion, it was agreed to form four working groups and that each Working Group would consider one of the following four themes, each of which would embrace a number of questions from the above list:

1. Politics and science.
2. Energy production.
3. A possible environmental tax.
4. North-south relations.

Each working group spent some time on its deliberations and the conclusions of each group's discussions were presented by the rapporteur for that group at a later session. These reports form the basis of this chapter and have been only very lightly edited.

2 Working Group 1 - Politics and Science

The discussion started off with the general question of how scientific expertise could be integrated into policy decisions at all levels of decision making. This led to the more specific, and urgent, question of how to raise the awareness of politicians etc. about global climate change and other environmental issues. It was accepted that there is a limit to what scientists can do, but that they do have an important role to play (a larger role than at present).

The following simple education model was proposed:

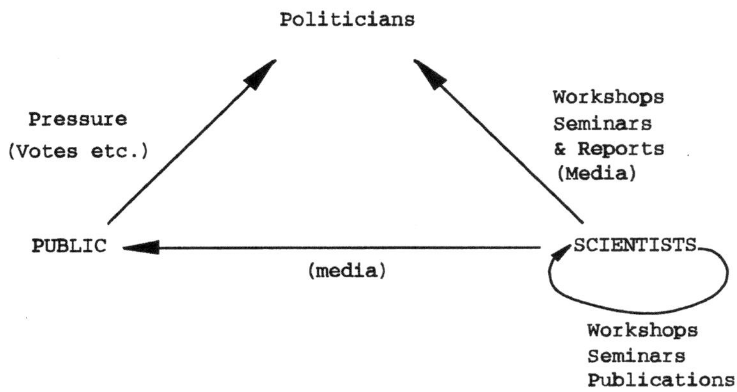

A number of difficulties were considered. The first was that politicians and decision makers, as a rule, do not understand science or how it evolves (e.g. learning from mistakes). It was concluded that much better education in scientific matters was needed both for the general public but in particular for politician, managers, economists etc. There was need for more input by scientists from, for example, an environmental advisory committee and that much stronger guidance was required from the Royal Society, the National Academy of Sciences etc.

Another problem considered was that politicians operate on a different (very short) time scale as compared to scientists. There did not seem to be an easy solution to this. The suggestion was made that politicians should be pressurised by scientific groups for making "illogical and destructive decisions", on the basis that scientists understand science best and should therefore have a much greater input to education and policy decisions. Then, if the worlds leaders and hence the general public decide to follow the road to Armageddon, at least they will do it on an informed basis.

3 Working Group 2 - Energy Production

This group first looked qualitative at those considerations relating to sustainable global energy production and consumption from a climate point of view. It was suggested that the present status could be summarised as follow:

> Huge differences exist between the energy consumption (EC) per capita in so-called more developed countries (MDC) and the lesser developed countries (LDCs).
>
> A global increase of EC is inevitable in LDCs and can be justified. The saving potential cannot counter-balance this completely, hence a net increase

of global pollution has to be expected.

Present Energy Production (EP) is highly dominated by burning fossil fuels.

Present (research on) energy production systems is mainly concentrated on large centralised systems based on "high tech".

Present energy prices are oriented towards (national) economics, not at ecological or international interests.

Most people (in MDCs) are even not aware of their energy consumption.

It was concluded that present EP/EC is not sustainable because of physical, social and ethical reasons, but that a "back to nature" scenario seems **NOT** to be an adequate solution.

Before considering possible answers to the questions posed, the group considered one or two other aspects related to sustainability. For example, we need a system of environmental protection based on cost-benefit analysis and monetary evaluation of environmental asset. This requires a balance between natural capital and man-made capital. There is also a moral imperative to leave open as many doors of opportunity to future generations as is feasible (Taylor, 1992). References was also made to the Brundtland Report (Brundtland, 1987).

The following conclusions emerged from these discussions:

The nuclear option is probably not sustainable. Some climatologists (and ecologists) cannot support the use of nuclear energy.

Much more research on sustainable energy systems is required. Scientists and politicians have to consider sustainability as **the** major long term development criterion.

Different research from different viewpoints by different scientists (which may often produce different results) is necessary as a driving force for science and research.

Balancing effects do exist. We should not judge single effects to be either positive or negative but should take a broad view. We should also not try to "engineer" the effects.

Technology transfer from north-south and west-east is a necessity, but has to be oriented towards the users' needs with sustainability as one (but not the only) important criterion.

Developing the economy should not be a contradiction to natural systems but a part of a sustainable Earth system. We have to use natural resources, being aware that we are part of nature and do not govern nature. We need nature, nature does not need us.

An environmental tax on excessive use of nature has to be considered seriously as a means to steer the economy towards a sustainable system.

Scientists have to play an active role in politics (as experts, consultants, politicians etc.). Their international orientation and cooperation can be further used as a vehicle to drive international politics in the desired direction. One-voice reports (like the IPCC report) are very valid and important prerequisite for political decisions.

In summary, a number of points can be emphasised. Sustainability is a glittering but very valid criterion for judging the developing of energy systems and economics. The "Business as usual" scenario is not sustainable and therefore not acceptable. Renewable energies therefore seem to be the only presently available sustainable alternative. Technically, and from the resources point of view, renewable sources can provide sufficient energy for an acceptable quality of life. However, economical constraints have to be reconsidered according to the sustainability law. The consequence of this would be a substantial increase in energy prices. A transition period is inevitable and consequently non-sustainable solutions are temporarily permitted in particular for LDCs. Energy saving is the most efficient energy "source" and should have highest priority in MDCs. In addition reduction of emissions is a essential. Scientific effort should emphasise sustainable energy systems.

4 Working Group 3 - An Environmental Tax

The group commenced by being presented with several thought-provoking statements from Professor Kondratyev. The first of these was that energy consumption and development are the main causes of the destruction of the biosphere and consequently of environmental pollution. Secondly, humans should consume no more than 1% of the Earth's primary production. Lastly, the suggestion was made that countries which consume more than their share of per capita primary production should be taxed.

Assuming that, as a consequence of all economic activity, environmental pollution necessarily leads to biosphere destruction, the group then considered the need to find a method of discouraging activities detrimental to the environment that are in excess of those needed for economic development and maintenance and also to assist developing countries to grow in an environmentally sound manner.

As a broad approach to the problem, it was suggested that monetary penalties or fines should be imposed on countries that exceed agreed environmental standards. Two national examples were considered, the pollution of ground water in China and SO_2 emission in the USA. It is necessary to consider who should pay, who should monitor, and hence police the situation, and also on what to use any money so obtained. The details of such an approach would require both objective scientific information and political support.

The problems first need to be identified. What are the specific environmental issues we want to address? Within these issues, what are the specific problems? Is the IPCC scenario a suitable solution, or should there be WMO/UN scientific panels? We need international agreement before we can start, and dedicated institutions to monitor the situation. What about Agenda 21?

Another approach could be to establish standards rather than to impose a tax on energy consumption or emissions. Medium-term (5-10 year) improvement-based goals could be established. Emission reductions should be reviewed on a case-by-case basis. Appropriate use of non-renewable resources should be considered, and habitat preservation should be taken seriously.

Developed countries should be asked to assist developing countries with transfer of state-of-the-art technology and expertise. International cooperation and order must be promoted and scientific establishments should be helped in developing countries by joint programmes to help to achieve such goals.

Monetary penalties need to be assessed. Developing countries cannot afford to pay a tax on activities but need investment to increase industrial and environmental efficiency. Time needs to be allowed for improvement goals to be met, but penalties must be imposed if goals are not met.

In order to monitor the situation, such bodies as the IPCC, WMO and UN should be used at international level, but we would have to depend on internal agencies to deal with local problems. We should aim to improve the universal standard of living while protecting the global environment.

5 Working Group 4 - North-South Relations in the Management of Climate Change

A number of key problems were identified in the north-south relationships in connection with the allocation of responsibilities for managing the environment:

5.1 Economy

The lack of finance in the south creates a situation of a non-sustainable use of natural resources to finance their development and, in the worst cases, to finance their debts. Usually the exploitation of their natural resources is for the benefit of the consumers in the north. However, in relation to global trade, the economic contribution from the south is only marginal and so the laws of the free market cannot be used to regulate environmental degradation.

Due to development in eastern Europe, much needed resources for environmental management, especially in Africa, are being diverted. The attention of western Europe is very much focused on the problems of eastern Europe at present.

5.2 Consumption

There is a global inequality of consumption between the north and south. The over-consumption of the north needs to be reduced which entails a change of lifestyle and attitude. The developing countries of the south need to remodel their economic plans to a more sustainable use of resources so that they do not follow the same track of development as has been taken by the developed countries of the north.

5.3 Technology Transfer

Technology transfer from the north to the south should cover in particular the fields of industry and agriculture. More energy-efficient technology should be made available to the south to curb growing energy consumption as it develops. It should be noted, however, that the technology transferred should be suited both physically and culturally to the particular country in the south. This transfer should not only consider the technical aspects but also the cultural, social and political aspects. Therefore, in tackling environmental issues, there is a need for greater cooperation amongst scientists, engineers and social scientists.

5.4 Deforestation

One particular case study of north-south relationship is the issue of deforestation.

5.4.1 General Perspectives

The issue of deforestation and its contribution to global climate change is still controversial but tropical rainforests are known to be an important terrestrial sink of carbon dioxide and other trace gases. Deforestation by itself creates local environmental impacts that need immediate attention. Since the geographical location of these forests is in the south, while the consumers of the timber are usually in the north, we have chosen this issue to illustrate some of the key problems mentioned above in north-south relationship in climate change.

Deforestation of temperate forests of the north started with early agriculture and became extensive during the industrial revolution (0.5 Gt C per year at maximum). This accounts for the lower percentage of forest cover and the lack of primary natural forests in huge areas of the north. The forest here is also threatened by acid rain. Current deforestation in the north tends to be well managed and the wood is mainly used for the manufacture of paper and furniture. Forest resources in the north, however, is not a major contributor to the economy of the country in comparison to industry. Therefore there is no great economic pressure to use this particular natural resource. In fact, due to the greater extent of urbanisation, forests now have an important recreational value.

Deforestation of tropical forests has only begun extensively in the last few decades. An estimate of the carbon flux from this activity is about 0.6 - 2.5 Gt C for the year 1980. However, in comparison with fuel combustion, which is estimated to be 5.4 Gt C per year, its contribution to global carbon dioxide increase is only about 20%.

5.4.2 Economics

The tropical forests are considered by the south to be one of its natural resources that could be exploited to fund its imports and development costs. Most of the timber and timber products are being exported to the north. Areas that could be used for agriculture are usually planted with cash crops such as cocoa and cotton thereby producing pressure to deforest more areas for local food production. However, in terms of northern perspectives, this exploitation and trade in tropical forest products plays only a marginal role in their economy.

5.4.3 Consumption

The logical answer is for the north to decrease its consumption of tropical forest products. However, this action will deprive the developing south of much-needed foreign exchange. Therefore, in order to be successful, the south must be given the chance to develop its economy such that it is less dependent on tropical forests.

5.4.4 Technology Transfer

Local Low Level Technology Development: – The local use of forest resources is an important component in the issue of deforestation, for example as a source of fuelwood. Therefore, sustainable local management or replacement by other fuels such as biogas needs to be encouraged. This technology should be simple and suitable for the local community on self-help basis.

High Technology: – High technology forest management using remote sensing needs to be developed in the south. Research in the north on the effect of deforestation in climate change needs to be intensified. There is a necessity to build up local expertise and adequate funding to avoid unsuitable application, adapt the technology transferred (e.g. SEAWATCH THAILAND, see chapter 22) and to prevent brain-drain to the north. The development of local expertise on forest management should not only cover purely technical aspects but should consider other factors such as land pressure, peasant poverty and other social problems that can influence forest use in the south. There is a need for cooperation between natural, physical and social scientists to address this issue.

5.4.5 Further comments

Sometimes deforestation is an imposition on the local community based on politics and macro-economic planning with no consideration of scientific advice and local community aspirations.

6 Conclusion

It would be foolish to suppose that in these modest discussions the participants at this course had either obtained all the answers to the human problems associated with climate change or that we had the power suddenly to wave a magic wand and solve all these problems. However, it was heartening to the organisers to see that a group of young scientists from many different countries, while obviously deeply involved in their own scientific work, could spend some time and energy on discussing the wider social and political issues associated with their scientific work.

References

Brundtland, G., 1987, *Our Common Futures.* Report the United Nations (NY world Commission on the Environment and Development).

Taylor, A., 1992, *Choosing our Future, a Practical Politics of the Environment* (London: Routledge).

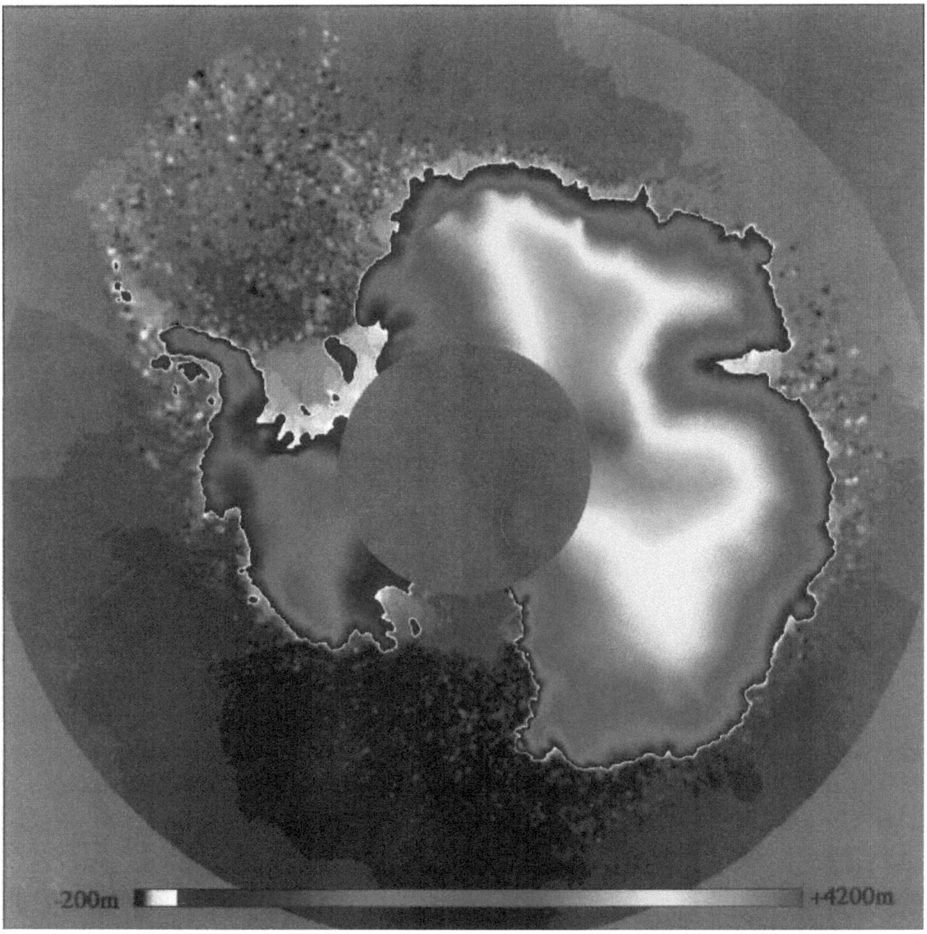

Plate 1 (D. R. Mantripp, P 341) Surface elevation contour map of Antarctica from the ERS-1. 35 day repeat.

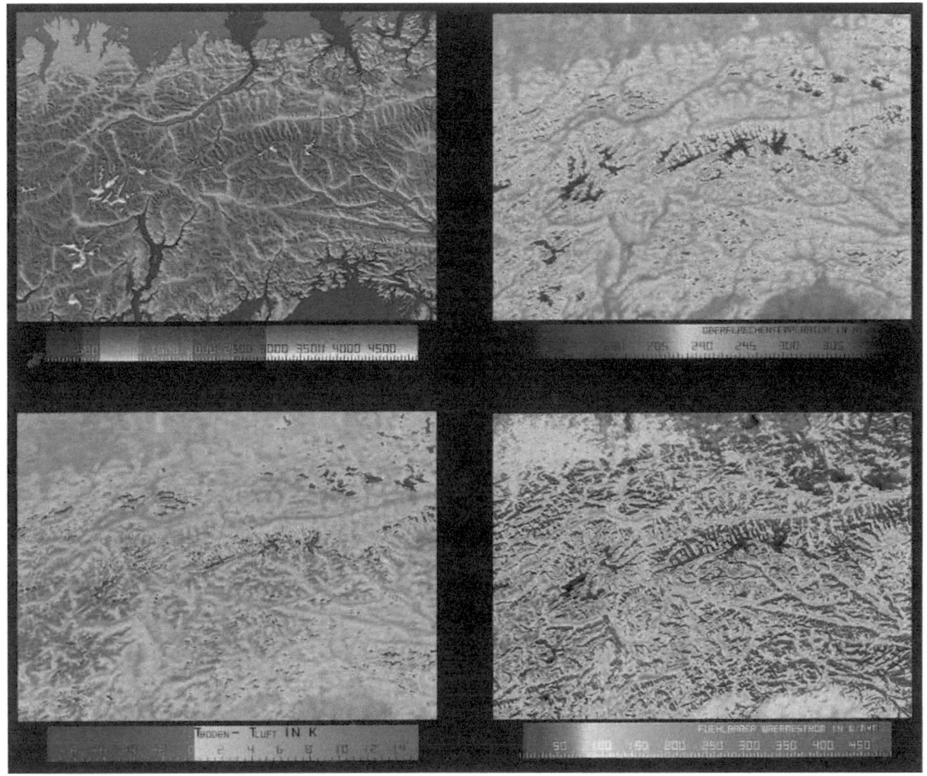

Plate 2 (H. Mannstein, P 367) Sensible heat flux in Wm^{-2} (bottom right) derived from AVHRR Ch.4 surface temperatures (top right), their difference from ambient air temperature (bottom left) using a digital terrain model (top left) in the Alps (46 - 48°N. 10 - 14°N) 12:00 UTC on September 1985.

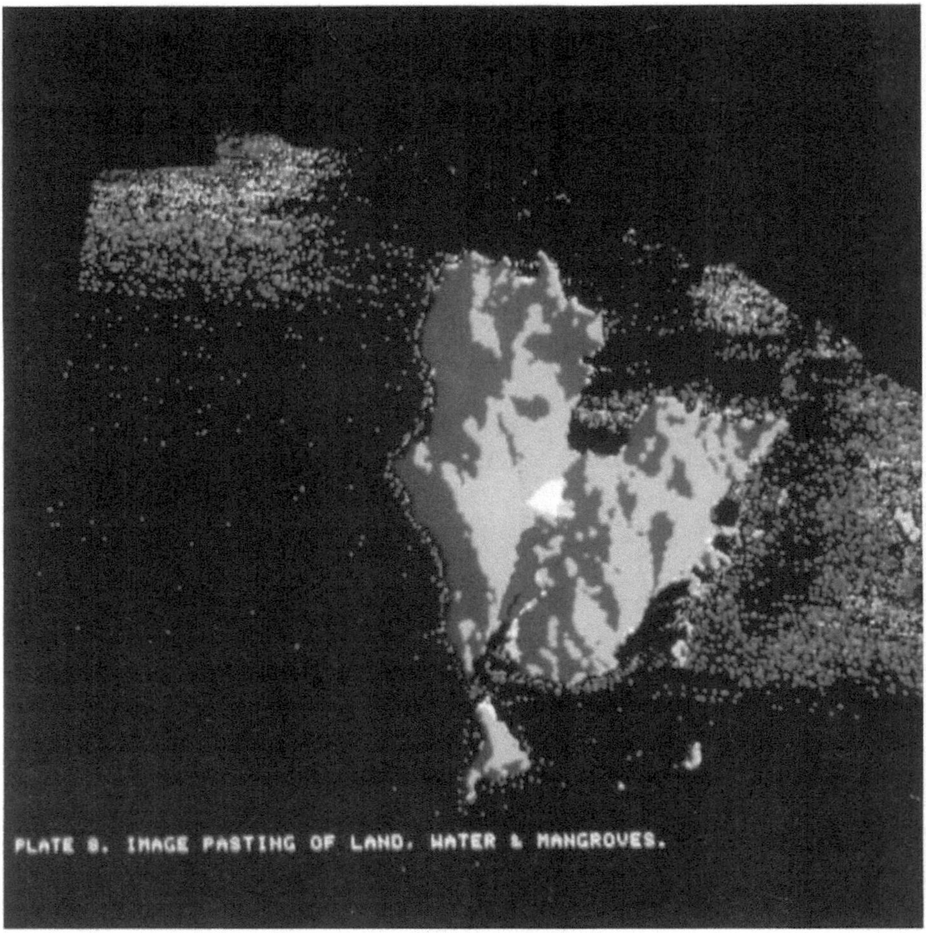

Legend:

- Forested Area
- Mangrove Area
- Cleared Land Areas
- Shallow Water
- Deep Water

Plate 3 (Mohd. Mohamed, P 429) Classified composite image of Pulau Redang 7 September 1990.

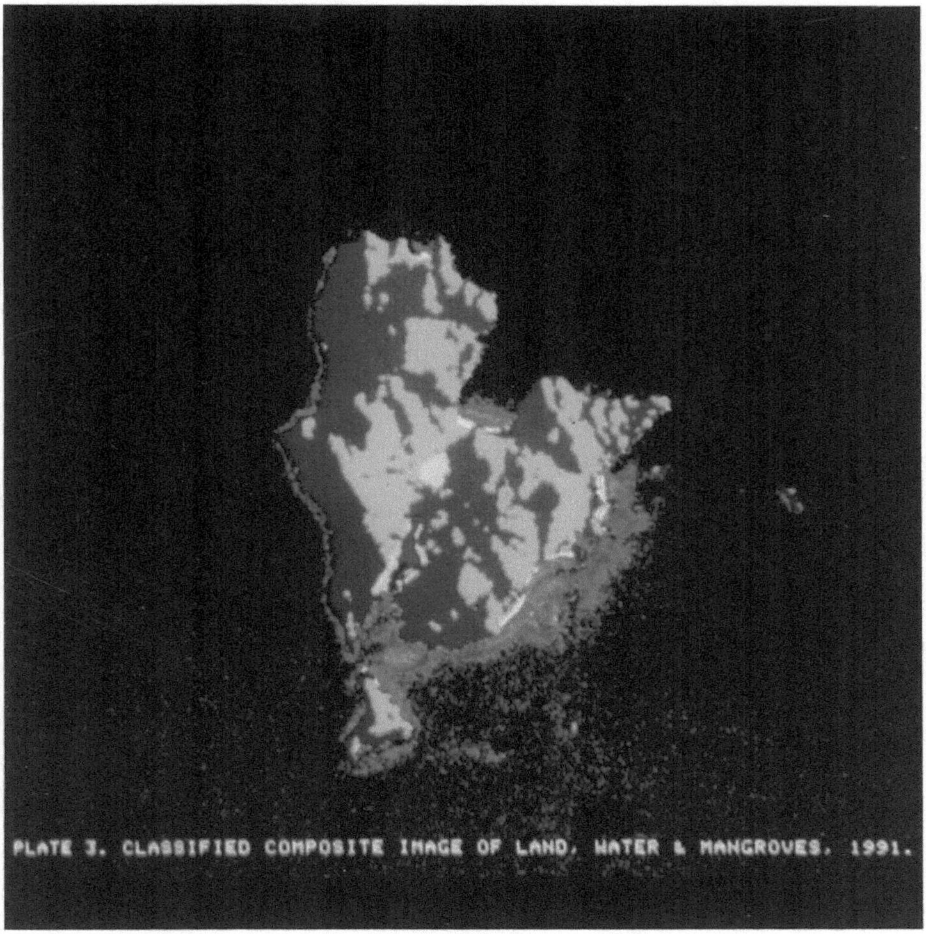

Legend:

- Forested Area
- Mangrove Area
- Cleared Land Areas
- Shallow Water
- Deep Water

Plate 4 (Mohd. Mohamed, P 429) Classified composite image of Pulau Redang 9 August 1991.

Participants

Mr. P. Ashcroft
Dept. of Engineering and
Public Policy
Carnegie Mellon University
Pittsburgh
Pennsylvania 15213, U.S.A.

Miss J. J. Bauman
Marine Sciences Research Center
State University of New York
and Stonybrook
New York 11794-500, U.S.A.

Mr. S. G. Beaven
University of Kansas Radar Systems and
Remote Sensing Lab
2291 Irving Hill Road # 308
Lawrence, Kansas 66045, U.S.A.

Mr. E. S. Benson
Federal Dept. of Forestry and Agricultural
Land Resources
Nat. forest Resources Sur. Project
PMB 5011
Jericho, Ibadan
Nigeria.

Dr. R. Benz
Dep RSX-222
Dornier GmbH
Postfach 142 Germany
7990 Friedrichshafen
Germany.

Miss R. Biagio
Imperial College
Centre for Environmental Tech.
58 Prince's Gardens
London SW7 2PE, U.K.

Miss M. Bijanzadeh
Meteorological Organization (IRIMO)
PO Box 13185 - 461
Tehran
Mehrabad
Iran.

Mr. F. Caiffa
Institute Fisbat - CNR
Via de Castagnoli 1
I-40126 Bologna
Itay.

Mr. D. M. Casey
Room A/11 Research Institute
University of Alabama-Huntsville
Huntsville AL 35899, U.S.A.

Dr. N. W. T. Chisholm
Institute of Earth Studies
University of Wales
Aberystwyth
Dyfed SY23 3DB, U.K.

Dr. F. Colao
ENEA
Dept INN SVIL
PO Box 65
00044 Franscati
Italy.

Prof. A. P. Cracknell
Dept. of APEME
University of Dundee
Dundee DD1 4HN
Scotland, U.K.

Mr. I. Csiszar
Satellite Research Lab of the
Hungarian Meteorological Service
PO Box 32
H-1675 Budapest
Hungary.

Mr. T. B. Davis
3-110 Tweedsmuir Street
Flin Flon
Manitoba R8A 0R8
Canada.

Mr. J. R. De Franca
Laboratoire D' Aerologie
Universite Paul Sabatier
118 Route de Narbone
31062Toulouse
Cedex
France.

Prof. T. De Jong
Laboratory for Space Research
PO Box 8070 Gronirgun
Netherlands.

Dr. L. V. De Santis
Institute Fisbat-CNR
Via de' Castagnoli 1
I 40126 Bologna
Italy.

Mr. V. R. Deshpande
Forecasting Research Division
Indian Inst. of Tropical Meteorol.
Pashan
Pune 411008
India.

Miss M. O. Domingues
INPE/DCT
Av Astronautas 1758
CP 515 - 12201
Sao Jose Dos Campos SP

Mr. D. Dong
Inst. of Atmospheric Physics
Academia Sinica
PO Box 2718
Beijing 10008
China.

Dr. R. J. Engelen
Institute for Marine and Atmospheric Research Utrecht
PO Box 80005
3508 TA Utrecht
Netherlands.

Mr. H. Erdun
Department of Meteorology
Istanbul Technical University
Faculty of Aeronautics and Astronautics
80626 Maslak
Turkey.

Dr. M. Fea
Earthnet
Via Galileo Galilei
00044 Frascati
Italy.

Miss P. Fernandez
Servicio de Teledeteccion
Instituto Nacional de Meteorologia
AP 285
28071 Madrid
Spain.

Mr. J. C. Fortea
Remote Sensing Unit
University of Valencia
Faculty of Physics
Calle Doctor Moliner 50
46100 Valencia
Spain.

Mr. A. N. Fox
Dept of Geological Sciences
Cornell University
Ithaca
New York 14853-1504, U.S.A.

Mr. R. Fraile
Lab Fisica de la Atmosfera
(VETER)
Universidad de Leon
24071 Leon
Spain.

Mr. G. B. Franca
Dept. of APEME
University of Dundee
Dundee DD1 4HN
Scotland, U.K.

Dr. K. Furmanczyk
Institute of Marine Sciences
University of Szczecin
71-412 Szczecin
Felczaka str 3A
Poland.

Miss A. L. Grossinho
Imperial College
Centre for Remote Sensing
The Biackett Laboratory
Prince Consort Road
London SW7 2BT, U.K.

Mr. P. Hampl
Czech Hydrometeorological Institute
Na Sabatce 17
14306 Praha 4
Czechoslovakia.

Dr. M. A. Haque
Sparrso
Agargaon
Sher-E-Bangla Nagar
Dhaka 1207
Bangladesh.

Mr. T. Heinemann
Institut fur Welltraumwissen- schaften
Freie Universitat Berlin
Fabeckstr 69
W-1000 Berlin 33
Germany.

Miss I. D. Ivanova
Institut of Electronics
72 Trakia blvd
1784 Sofia
Bulgaria.

Mr. B. P. Jallow
Dept. of Water Resources
7 Marine Parode
Bainjul
The Gambia

Dr. S. G. Jennings
Dept. of Physics
University College
Galway
Ireland

Mr. M. Kampel
INPE/DCT
Av Astronautas 1758
CP 515 - 12201
Sao Jose Dos Campos
Brazil.

Mr. E. Kaperanios
Inst. of Applied Informatics
Nuclear Research Center
Karlsruhe Gmbh
Postfach 3640
W-7500
Germany.

Mr. Khirrudin Abdullah
Dept. of APEME
University of Dundee
Dundee DD1 4HN
Scotland, U.K.

Academician K. Ya. Kondratyev
Counselor of Directorate
Institute for Lake Research
Sevastyanov street 9
196199 St Petersburg
Russia

Cap. V. Kostopoulos
Hellenic National Met. Service
PO Box 73502
GR 16603
Hellinkon
Athens
Greece.

Dr. T. V. Kumaran
Department of Geography
University of Madras
Madras 600 005
India.

Mrs. B. Lapeta
Institute of Meteorology and Water Management
ul P Borowego 14
30-215 Krakow
Poland.

Dr. B. M. Lesht
Environmental Research Div Bldg 203
Argonne National Laboratory
9700 S Cass Ave
Argonne IL 60439, U.S.A.

Mr. U. W. Lourens
Dept. of Agrometeorology
University of the Orange Free State
PO Box 339
Bloemfontein 9300
South Africa.

Mr. D. G. Low
Glenarder, Mossie Road
Grantown-on-Spey
Morayshire, U.K.

Dr. Kamlesh P Lulla
Space Shuttle Earth Observation Office
Code SN5
NASA-JSC
Houston
Texas 77058 U.S.A.

Dr. T McNally
ECMWF
Shinfield Park
Reading RG2 9AX U.K.

Dr. McGuffie
Dpte. of Applied Physics
University of Technology
Sydney
New South Wales 2007
Australia

Dr. H. Mannstein
Institut fur Fhysik der Atmosphaere
DLR Cberpfaffenhoffen
8031 Wessling
Germany

Mr. S. Mansor
Dept. of APEME
University of Dundee
Dundee DD1 4HN
Scotland, U.K.

Dr. D. Mantripp
MSSL
University College
Holmbury
St Mary
Dorking Surrey RH5 6NT
U.K.

Mr. J. L. G. Merayo
Serv de Investi. Y Analisis
del Clima
Subd De Climatologia Y Aplicacionei
Instit Nac De Meteorologia
Spain.

Mr. C. R. Milner
Dept of Plant Sciences
Downing Street
Cambridge CB2 3EA, U.K.

Dr. J. M. F. Morais
(Tropical Research Institute)
Centro de Pre-Hostoria e
Arquelogia - IICT
TV Conde da Ribeira 9
Portugal
P-1300 Lisboa
Portugal.

Mr. N. I. Nickolaev
Institute of Electronics
Bulgarian Academy of Sciences
Boul. Tradia 72
Sofia 1784
Bulgaria.

Dr. A. R. Palmer
Grasslands Research Centre
PO Box 101
Grahamstown 6140
South Africa.

Mr. J. Perez Marrero
Centro de Tecnologia Pesquera
PO Box 56
Telde (35200)
Gran Canaria
Spain.

Dr. J. Noll
ESA/ESTEC
PO Box 299
NL 2200
AG Noordwijk
Netherlands

Miss P. Post
Department of Geophysics
Tartu University
Tahe 4
Tartu
Estonia.

Dr. A. A. Samah
Department of Geography
University of Malaya
Kuala Lumpur 59100
Malaysia.

Dr. M. Saraber
Department of Meteorology
Wageningen Agricultural University
Duivendaal 2
6701 AP Wageningen
Netherlands.

Mr. L. S. Schuller
Institute for Meteorologie
Free University of Berlin
Dietrich-Schafer-Weg 6-10
D-1000 Berlin 41
Germany.

Mr. V. Schumacher
dept. of APEME
University of Dundee
Dundee DD1 4HN
Scotland, U.K.

Mr. C. G. Selles
Servei C T DE Gestio i EVO
lucio del Paisatge
Universitat de Barcelona
c/Baldiri i Reixac S/N
Barcelona 08028
Spain.

Mr. G. A. Serritella
Via Boranico 196
60029 Varano
Ancona
Italy.

Mr. J. Silva
Dept. of APEME
University of Dundee
Dundee DD1 4HN
Scotland, U.K.

Dr. A. Slingo
Hadley Centre
Meteorological Office
London Road
Bracknell RG12 2SZ
U.K.

Dr. P. Singhaseneh
Nation Research Council of
Thailand (NRCT)
196 Paholyotin Road
Chatuchak
Bangkok 1090
Thailand.

Mr. S. K. Soo
Malaysian Meteorological Service
Jalan Sultan
46667 Pataling Jaya
Malaysia.

Mr. N. A. Soulakellis
Dept of Geology and Physical
Geography
School of Geology
Aristotle Univ. of Thessaloniki
54006 Greece
Greece.

Prof. M. Ph. Stoll
ENSPS
University of Strasbourg
7 rue de l'Université
F67000 Strasbourg
France

Miss H. Sui
PO Box 775
Beijing 100101
China.

Mr. X. Sun
Dept. of APEME
University of Dundee
Dundee DD1 4HN
Scotland, U.K.

Dr. C. Tsolakidis
27 Gregori Lambraki St
85100 Rhodes
Greece.

Mr. H. T. Uitermark
Postbus 196
7400 AD Deventer
Netherlands.

Dr. A. C. A. Van Lammeren
Royal Netherlands Meteorological Institute
PO Box 201
3730 AE De Bilt
Netherlands.

Miss A. J. C. Van Riet
Boomstraat 116
2880 Bornem
Belgium.

Mr. M. Van Weele
IMAU
Princetonplein 5
3584 CC Utrecht
Netherlands.

Miss M. Varinou
Dept of Applied Physics
Laboratory of Meteorology
University of Athens
33 Ippokratous Str
10680 Athens
Greece.

Prof. C. Varotsos
Laboratory of Meteorology
Dept. of Applied Physics
University of Athens
33 Ippokratous Street
10680 Athens
Greece

Dr. R. A. Vaughan
Dept. of APEME
University of Dundee
Dundee DD1 4HN
Scotland, U.K.

Dr. S. Victorov
State Oceanographic Institute
Laboratory for Satellite Oceanography
23 Linia, 2A
St Petersburg 199026
Russia

Dr. P. Wadhams
Scott Polar Research Institute
University of Cambridge
Lensfield Road
Cambridge CB2 1ER
U.K.

Miss J. Wang
Argometeorological Institute of
Chinese Academy of Agri. Science
30 Baishiqiao Road
Beijing 100081
China.

Mr. J. Wang
Agrometeorological Institute of
Chinese Academy of Agri. Science
30 Baishiqiao Road
Beijing 100081
China.

Miss S. Wewetzer
Dept. of Geology
University of St. Andrews
St. Andrews
Fife, U.K.

Mr. I. H. Woodhouse
Department of Physics
Heriot-Watt University
Riccarton
Edinburgh EH14 4AS, U.K.

Mr. A. Yurtaikin
1/14-3-32 Novocheremushkinskaya St
Moscow 117449
Russia.

Mr. S. Zientek
Free University of Berlin
Dietrich Schafer Weg 6-10
1000 Berlin 41
Germany.

Index

The ASI Series Books Published as a Result of Activities of the Special Programme on Global Environmental Change

---

This book contains the proceedings of a NATO Advanced Research Workshop held within the activities of the NATO Special Programme on Global Environmental Change, which started in 1991 under the auspices of the NATO Science Committee.

The volumes published as a result of the activities of the Special Programme are:

Vol. 1: **Global Environmental Change.** Edited by R. W. Corell and P. A. Anderson. 1991.
Vol. 2: **The Last Deglaciation: Absolute and Radiocarbon Chronologies.** Edited by E. Bard and W. S. Broecker. 1992.
Vol. 3: **Start of a Glacial.** Edited by G. J. Kukla and E. Went. 1992.
Vol. 4: **Interactions of C, N, P and S Biogeochemical Cycles and Global Change.** Edited by R. Wollast, F. T. Mackenzie and L. Chou. 1993.
Vol. 5: **Energy and Water Cycles in the Climate System.** Edited by E. Raschke and D. Jacob. 1993.
Vol. 6: **Prediction of Interannual Climate Variations.** Edited by J. Shukla. 1993.
Vol. 7: **The Tropospheric Chemistry of Ozone in the Polar Regions.** Edited by H. Niki and K. H. Becker. 1993.
Vol. 8: **The Role of the Stratosphere in Global Change.** Edited by M.-L. Chanin. 1993.
Vol. 9: **High Spectral Resolution Infrared Remote Sensing for Earth's Weather and Climate Studies.** Edited by A. Chedin, M.T. Chahine and N.A. Scott. 1993.
Vol. 10: **Towards a Model of Ocean Biogeochemical Processes.** Edited by G. T. Evans and M.J.R. Fasham. 1993.
Vol. 11: **Modelling Oceanic Climate Interactions.** Edited by J. Willebrand and D.L.T. Anderson. 1993.
Vol. 12: **Ice in the Climate System.** Edited by W. Richard Peltier. 1993.
Vol. 13: **Atmospheric Methane: Sources, Sinks, and Role in Global Change.** Edited by M. A. K. Khalil. 1993.
Vol. 14: **The Role of Regional Organizations in the Context of Climate Change.** Edited by M. H. Glantz. 1993.
Vol. 15: **The Global Carbon Cycle.** Edited by M. Heimann. 1993.
Vol. 16: **Interacting Stresses on Plants in a Changing Climate.** Edited by M. B. Jackson and C. R. Black. 1993.
Vol. 17: **Carbon Cycling in the Glacial Ocean: Constraints on the Ocean's Role in Global Change.** Edited by R. Zahn, T. F. Pedersen, M. A. Kaminski and L. Labeyrie. 1994.

Vol. 18: **Stratospheric Ozone Depletion/UV-B Radiation in the Biosphere.** Edited by R. H. Biggs and M. E. B. Joyner. 1994.
Vol. 19: **Data Assimilation: Tools for Modelling the Ocean in a Global Change Perspective.** Edited by P. O. Brasseur and J. Nihoul. 1994.
Vol. 20: **Biodiversity, Temperate Ecosystems, and Global Change.** Edited by T. J. B. Boyle and C. E. B. Boyle. 1994.
Vol. 21: **Low-Temperature Chemistry of the Atmosphere.** Edited by G. K. Moortgat, A. J. Barnes, G. Le Bras and J. R. Sodeau. 1994.
Vol. 22: **Long-Term Climatic Variations – Data and Modelling.** Edited by J.-C. Duplessy and M.-T. Spyridakis. 1994.
Vol. 23: **Soil Responses to Climate Change.** Edited by M. D. A. Rounsevell and P. J. Loveland. 1994.
Vol. 24: **Remote Sensing and Global Climate Change.** Edited by R. A. Vaughan and A. P. Cracknell. 1994.
Vol. 25: **The Solar Engine and Its Influence on Terrestrial Atmosphere and Climate.** Edited by E. Nesme-Ribes. 1994.
Vol. 26: **Global Precipitations and Climate Change.** Edited by M. Desbois and F. Désalmand. 1994.
Vol. 27: **Cenozoic Plants and Climates of the Arctic.** Edited by M. C. Boulter and H. C. Fisher. 1994.

Springer-Verlag and the Environment

We at Springer-Verlag firmly believe that an international science publisher has a special obligation to the environment, and our corporate policies consistently reflect this conviction.

We also expect our business partners – paper mills, printers, packaging manufacturers, etc. – to commit themselves to using environmentally friendly materials and production processes.

The paper in this book is made from low- or no-chlorine pulp and is acid free, in conformance with international standards for paper permanency.